Handbook of Organic Compounds
NIR, IR, Raman, and UV-Vis Spectra Featuring
Polymers and Surfactants (a 3-volume set)

Volume 1
Methods and Interpretations

Handbook of Organic Compounds
NIR, IR, Raman, and UV-Vis Spectra Featuring
Polymers and Surfactants (a 3-volume set)

Volume 1
Methods and Interpretations

Jerry Workman, Jr.
Kimberly-Clark Corporation
Neenah, WI

ACADEMIC PRESS

A Harcourt Science and
Technology Company

San Diego   San Francisco   New York   Boston
London   Sydney   Tokyo

This book is printed on acid-free paper. ∞

COPYRIGHT © 2001 BY ACADEMIC PRESS
All rights reserved.
No part of this publication may be reproduced or transmitted in any form or by any means,
electronic or mechanical, including photocopy, recording, or any information storage
and retrieval system, without permission in writing from the publisher.

Requests for permission to make copies of any part of the work should be mailed to the
following address: Permissions Department, Harcourt, Inc., 6277 Sea Harbor Drive, Orlando,
Florida, 32887-6777.

ACADEMIC PRESS
A Harcourt Science and Technology Company
525 B Street, Suite 1900, San Diego, CA 92101-4495, USA
http://www.academicpress.com

ACADEMIC PRESS
Harcourt Place, 32 Jamestown Road, London, NW1 7BY, UK

Library of Congress Catalog Card Number: 00-105503
International Standard Book Number: 0-12-763561-0

PRINTED IN THE UNITED STATES OF AMERICA
00　01　02　03　04　IP　9　8　7　6　5　4　3　2　1

I am forever grateful to:

Norman Colthup, Clara Craver, Bill Fateley, Connie Paralez, Jeanette Grasselli, Robert Hannah, Dana Mayo, and Foil Miller, who have tirelessly and successfully taught the complexities of infrared interpretation to so many scientists, both young and old, at the Fisk and Bowdoin infrared interpretive courses

Professor Larry Dieterman, who passed along his truly contagious love for chemistry to so many students

Bertie, the best proofreader, and friend, on the planet

My parents, for their loving care and for engendering my intellectual drive

CONTENTS

Preface	ix
Acknowledgments	xi
Measurement Conditions for Spectral Charts (Vols. 2 and 3)	xiii
Permissions and Credits	xvii

I. Optical Spectroscopy
	1. Optical Spectrometers	3
	2. Sources, Detectors, Window Materials, and Sample Preparation for UV-Vis, NIR, IR, and Raman Spectroscopy	21

II. Sample Identification and Specialized Measurement Techniques
	3. Infrared Microspectroscopy	29
	4. Dichroic Measurements of Polymer Films Using Infrared Spectrometry	35
	5. Chemical Identification and Classification Tests	39

III. UV-Vis Spectroscopy
	6. Comparing Ultraviolet-Visible and Near-Infrared Spectroscopy	49
	7. UV-Vis Spectroscopy Charts	63
	8. Practical Guide for Evaluating Ultraviolet Spectra	65
	9. Functional Groupings and Band Locations for UV-Vis Spectroscopy in Nanometers (nm)	69
	10. UV-Vis Spectral Correlation Charts	71

IV. Introduction to NIR, Infrared, and Raman Spectra
	11. Comparing NIR, Infrared, and Raman Spectra	77
	12. Review of Near-infrared and Infrared Spectroscopy	79

V. NIR Spectroscopy
	13. Short-Wave Near-Infrared Spectroscopy	133
	14. SW-NIR for Organic Composition Analysis	137
	15. Interpretive Spectroscopy for Near-Infrared	143
	16. Functional Groupings and Calculated Locations in Nanometers (nm) for NIR Spectroscopy	183

CONTENTS

17.	NIR Spectral Correlation Charts	193
18.	NIR Band Assignments for Organic Compounds, Polymers, and Rubbers	197

VI. Infrared and Raman Spectroscopy

19.	Review of Interpretive Spectroscopy for Raman and Infrared	211
20.	Functional Groupings and Calculated Locations in Wavenumbers (cm^{-1}) for IR Spectroscopy	229
21.	Infrared Spectral Correlation Charts	237
22.	Functional Groupings and Calculated Locations in Wavenumbers (cm^{-1}) for Raman Spectroscopy	243
23.	Raman Spectral Correlation Charts	251
24.	Surfactant and Polymer Spectra	255

VII. Dielectric Spectroscopy

25.	Measurement of Dielectric Constant, Dielectric Loss, and Loss Tangent, and Related Calculations for Liquids and Film Samples	269
26.	Dielectric Constants of Materials	275

VIII. Chemometrics and Data Processing

27.	Practices of Data Preprocessing for Optical Spectrophotometry	295
28.	Review of Chemometrics Applied to Spectroscopy: Quantitative and Qualitative Analysis	301
29.	Review of Chemometrics Applied to Spectroscopy: Data Preprocessing	327
30.	Review of Chemometrics Applied to Spectroscopy: Multiway Analysis	339

Numerical Index to Spectra 1–2130	353
Alphabetical Index to Spectra 1–2130	379
Abbreviations Used in Surfactant and Polymer Spectra Names	419
Index	421

SPECTRAL ATLAS

Spectra Numbers 1–560

UV-Vis (200–900 nm) and SW-NIR (650–850 nm): Organic Compounds and Polymers

Spectra Numbers 561–592

SW-NIR (800–1100 nm): Organic Compounds

Spectra Numbers 593–1006

LW-NIR (1000–2600 nm): Organic Compounds and Polymers

Spectra Numbers 1007–2000

Infrared (4000–500 cm^{-1}): Organic Compounds, Polymers, Surfactants, and HATR

Spectra Numbers 2001–2130

Raman (4000–500 cm^{-1}): Organic Compounds and Polymers

PREFACE

This *Handbook of Organic Compounds: NIR, IR, Raman, and UV-Vis Spectra Featuring Polymers and Surfactants* is a compendium of practical spectroscopic methodology, comprehensive reviews, and basic information for organic materials, surfactants, and polymer spectra covering the ultraviolet, visible, near-infrared, infrared, Raman, and dielectric measurement techniques. It represents the first comprehensive multivolume handbook to provide basic coverage for UV-Vis, 4th-overtone NIR, 3rd-overtone NIR, NIR, infrared, and Raman spectra and dielectric data for organic compounds, polymers, surfactants, contaminants and inorganic materials commonly encountered in the laboratory. The text includes a description and reviews of interpretive and chemometric techniques used for spectral data analysis. The spectra found within the atlas are useful for identification purposes as well as for instruction in the various interpretive and data-processing methods discussed. This work is designed to be of help to students and vibrational spectroscopists in their daily efforts at spectral interpretation and data processing of organic spectra, polymers, and surfactants. All spectra are presented in terms of wavenumber and transmittance; ultraviolet, visible, 4th-overtone NIR, 3rd-overtone NIR, and NIR spectra are also presented in terms of nanometers and absorbance space. In addition, horizontal ATR spectra are presented in terms of wavenumber and absorbance space. All spectra are shown with essential peaks labeled in their respective units. Several individuals contributed to the material in this handbook, and comments were received from a variety of workers in the field of molecular spectroscopy. This handbook can provide a valuable reference for the daily activities of students and professionals working in modern molecular spectroscopy laboratories.

ACKNOWLEDGMENTS

The editor is grateful to the following individuals from the Analytical Science & Technology Group at Kimberly-Clark Corporation for their valuable assistance in compiling spectral data included in this work:

Tom Last, Mike King, Maurice Pheil, Mike Faley (student coop from the University of Wisconsin—Oshkosh), **Maria Delgado** (student summer intern from the University of Wisconsin—Madison), **Rick Beal, Jerry Kochanny, Tom Eby,** and **Wade Thompson.**

The editor also acknowledges **Paul R. Mobley** and **Bruce R. Kowalski** of the Laboratory for Chemometrics, Dept. of Chemistry, The University of Washington, Seattle, WA 98105, and **Rasmus Bro** of the Food Technology Chemometrics Group, Dept. of Dairy and Food Science, Rolighedsvej 30, 1958 Frederiksberg, Copenhagen, Denmark, who worked with me to write a three-part review of chemometrics in spectroscopy for *Applied Spectroscopy Reviews*.

MEASUREMENT CONDITIONS
FOR SPECTRAL CHARTS (VOLS. 2 AND 3)

VOLUME 2

Ultraviolet-Visible Region

Liquids

Spectral Region: 174 nm to 900 nm
1456 data points
Source: deuterium (to 350 nm), quartz-tungsten-halogen (to 900 nm)
Detector: R-928 photomultiplier (red sensitive)
Scan rate: 200 nm/min.
Integration time: 0.30
Slit height: 1/3 full
2 nm resolution (slit bandwidth)
Varian Cary 5G in transmittance mode with 1 cm pathlength cells for liquids. The measurements for liquids were made using a dual channel optical geometry with dry air as the initial and second channel background reference.

Solids

Spectral Region: 174 nm to 900 nm
1456 data points
Source: deuterium (to 350 nm), quartz-tungsten-halogen (to 900 nm)
Detector: R-928 photomultiplier (red sensitive)
Scan rate: 200 nm/min.
Integration time: 0.30
Slit height: 1/3 full
2 nm resolution (slit bandwidth)
Varian Cary 5G in reflectance mode for solids using Labsphere DRA CA-50, 150 mm (inner diameter) integrating sphere with photomultiplier and PbS detectors (useful range 250 – 2500 nm). The measurements for solids were made using a Spectralon® coated sphere with a background reference of Spectralon® SRS-99 (99% reflectance).

MEASUREMENT CONDITIONS FOR SPECTRAL CHARTS (VOLS. 2 AND 3)

Short wave-Near Infrared Region

Liquids only

Spectral Region: 800 nm to 1080 nm
799 data points
Source: tungsten-halogen (Vis-NIR)
Detector: 1024-element silicon DA
3.3 nm resolution
Default measurement values
Integration time per spectrum: 128 ms, 20 second data collection time.
Perkin-Elmer PIONIR 1024 Diode-Array transmittance with 10 cm pathlength for liquids; 1024 element silicon linear diode array detector. Self-referencing dual-path liquid cell was used. Dry air was used as initial background reference.

Long wave-Near Infrared Region

Liquids

Spectral Range A: 12000 cm^{-1} to 3500.17 cm^{-1}
6377 data points
Source: tungsten-halogen
Detector: NIR-PE
Beamsplitter: KBr
Phase resolution: 128
Phase correction: Power spectrum
Apodization: Blackman-Harris 4-term
Zero filling factor: 4
8 cm^{-1} resolution
1.0 mm aperture
1 cm pathlength quartz cell for liquids
Bruker Model FTS-66 FT-NIR, 3 minute data collection (215 co-added scans per measurement). Dry air was used as background reference.

Solids

Spectral Range B: 12000 cm^{-1} to 3498 cm^{-1}
3498 data points
Source: tungsten-halogen
Detector: NIR-PE
Beamsplitter: KBr
Phase resolution: 128
Phase correction: Power spectrum
Apodization: Blackman-Harris 4-term
Zero filling factor: 4
16 cm^{-1} resolution
Bruker Model FTS-66 FT-NIR
Specular Reflectance Accessory (30° incidence and reflectance angle)
3 minute data collection (215 co-added scans per measurement)
Polymer pellets and powders were measured "as received" using a cylindrical sample cell with silica windows. A gold-coated reflectance mirror was used for the background reference.

VOLUME 3

Mid-Infrared Spectral Region

Organic Compounds and Polymers

Spectral Range: 4000 cm^{-1} to 500 cm^{-1}
1816 data points
Source: Globar
Detector: DTGS KBr
Beamsplitter: KBr
Autogain velocity: 1.5825
Apodization: Happ-Genzel
Zero filling factor: 1 level
Aperture: 25
4 cm^{-1} resolution
Nicolet Model 510 FT-IR Spectrometer
32 and 64 co-added scans per measurement
KBr pellet – typical sampling for transmittance mode
Blank KBr was used as the typical background reference material

Surfactants

Spectral Range: 3750 cm^{-1} to 650 cm^{-1}
1551 data points
Source: Globar
Detector: DTGS KBr
Beamsplitter: KBr
Autogain velocity: 1.5825
Apodization: Happ-Genzel
Zero filling factor: 1 level
Aperture: 25
2 cm^{-1} resolution
Nicolet Model 710 and Model SX FT-IR Spectrometers
64 co-added scans per measurement
Capillary films, cast films, or KBr disks for transmittance mode using blank transmittance plate materials for the background reference

HATR (Horizontal Attenuated Total Reflectance) Measurements

Liquids

Spectral Range: 4000 cm^{-1} to 650 cm^{-1}
1738 data points
Source: Globar (Everglo™ Mid-IR source)
Detector: DTGS KBr
Beamsplitter: KBr
Autogain velocity: 0.6329
Apodization: Happ-Genzel
Zero filling factor: None
Phase correction: Mertz
4 cm^{-1} resolution
Nicolet Avatar Model 360 FT-IR Spectrometer
32 co-added scans per measurement

MEASUREMENT CONDITIONS FOR SPECTRAL CHARTS (VOLS. 2 AND 3)

zinc selenide (ZnSe) 45° - 10 bounce horizontal ATR crystal
No sample present for background reference measurements

Raman Spectral Measurements

Liquids and Solids

Spectral Range: 3800 cm^{-1} to 200 cm^{-1}
1801 data points
Source: Nd:YAG laser at 1064 nm (0-300 mW power)
Detector: Raman - germanium (Ge)
Beamsplitter: calcium fluoride (CaF2)
Apodization: Blackman-Harris 4 term
Zero filling factor: 4
Aperture: 6 mm
Phase correction: Power spectrum
2 cm^{-1} resolution
Bruker FTS-66 FT-NIR Spectrometer
Neodymium: Yttrium Aluminum Garnet (Nd:YAG) excitation laser at 1064 nm
1000 co-added scans per measurement, focused beam diameter approximately 2 mm.
Samples for measurement were contained in silica NMR tubes with 11 mm (outer diameter) by 9 mm (inner diameter); approximately 30 mm height.

PERMISSIONS AND CREDITS

Permission to reprint copyrighted material has been granted by the individual copyright owners.

The following articles have been adapted with permission from Marcel Dekker, Inc., New York:

> J. Workman. Interpretive spectroscopy for near-infrared. *Applied Spectroscopy Reviews* 31(3), 1996.
>
> J. Workman, P. Mobley, B. Kowalski, R. Bro. Review of chemometrics applied to spectroscopy, part 1. *Applied Spectroscopy Reviews* 31 (1 & 2), 73–124, 1996.
>
> P. Mobley, B. Kowalski, J. Workman, R. Bro. Review of chemometrics applied to spectroscopy, part 2. *Applied Spectroscopy Reviews* 31(4), 1996.
>
> Rasmus Bro, J. Workman, Paul Mobley, Bruce Kowalski. Review of chemometrics applied to spectroscopy: 1985–1995, part 3. *Applied Spectroscopy Reviews* 32(3), 1997.
>
> J. Workman. Review of process and non-invasive near-infrared and infrared spectroscopy: 1993–1999. *Applied Spectroscopy Reviews* 34 (1 & 2): 1–91, 1999.

The following was adapted with Permission from Academic Press, Boston:

> J. Workman, A. Springsteen, eds. *Applied Spectroscopy: A Compact Reference for Practitioners.* Boston: Academic Press, 1998.
>
> Chapter 1. Optical Spectrometers (J. Workman)
>
> Chapter 2. Ultraviolet, Visible, and Near-Infrared Spectrometry (J. Workman)
>
> Appendix A: Sources, Detectors, and Window Materials for UV-Vis, NIR, and IR Spectroscopy (J. Workman)
>
> Appendix B: Practices of Data Preprocessing for Optical Spectrophotometry (J. Workman)
>
> Appendix C: Infrared Microspectroscopy (J. Workman)

The practical guide for evaluating ultraviolet spectra is reprinted with permission from D. Pavia, G. Lampman, and G. Kriz, *Introduction to Spectroscopy: A Guide for Students of Organic Chemistry* (Philadelphia: Saunders College, 1979).

I.
OPTICAL SPECTROSCOPY

1.

OPTICAL SPECTROMETERS

A common assumption in spectroscopic measurements is that Beer's law relationship applies between a change in spectrometer response and the concentration of analyte material present in a sample specimen. The *Bouguer, Lambert, and Beer relationship* assumes that the transmission of a sample within an incident beam is equivalent to 10 exponent the negative product of the molar extinction coefficient (in mol^{-1} cm^{-1}), times the concentration of a molecule in solution (in mol^{-1}) times the pathlength (in cm) of the sample in solution. There are some obvious (and not-so-obvious) problems with this assumption. The main difficulties in the assumed relationship are that the molecules often interact, and the extinction coefficient may vary due to changes in the molecular configuration of the sample. The obvious temperature, pressure, and interference issues also create a less-than-ideal situation for the analyst. However, for many (if not most) analytical problems the relationship holds well enough.

Properties of the Bouguer, Lambert, and Beer (Beer's law) relationship:

$$T = \frac{I}{I_0} = 10^{-\varepsilon c l}$$

where T = transmittance, I_0 = intensity of incident energy, I = intensity of transmitted light, ε = molar extinction coefficient (in L · mol^{-1} cm^{-1}), c = concentration (in mol · L^{-1}), and l = pathlength (in cm). To simplify this equation into its more standard form, showing absorbance as a logarithmic term and, used to linearize the relationship between spectrophotometer response and concentration, gives the following expression as the relationship between absorbance and concentration.

$$\text{Abs.} = A = -\log\left(\frac{I}{I_0}\right) = -\log(T) = \varepsilon c l$$

The following statements hold true for what is most often termed Beer's law: (1) The relationship between transmittance and concentration is nonlinear, (2) yet the relationship between absorbance and concentration is linear. Beer's law is the common basis for *quantitative analysis.* Knowledge of Beer's law allows us to calculate the maximum theoretical dynamic range for an instrument using a few simple mathematical relationships.

The goal in the design of an optical spectrometer is to maximize the energy (or radiant power) from a light source through the spectrometer to the detector. The optical throughput for a spectrometer is dependent upon multiple factors, such as the light source area, the apertures present within the light path, lens transmittance and mirror reflectance losses, the exit aperture, and the detector efficiency. A simplified model to describe such a system is given below. From *paraxial optical theory* it is known that:

$$A_s \Omega_{\text{Ent}} = A_d \Omega_{\text{Exit}}$$

This relationship applies when A_s = the illuminated area of the light source, A_d = the illuminated detector area, Ω_{Ent} = the solid angle subtended by the source at the entrance aperture of the spectrometer, and Ω_{Exit} = the solid angle subtended by the detector by the exit aperture. This relationship is essential in determining the distances of the source and detector from the main spectrometer optical system, as well as the size of the entrance and exit apertures (or slits).

Several terms are useful in any discussion of spectrometry: *selectivity* = specific sensor response to the *analyte of interest; sensitivity* = the *quantifiable level of response* from a sensor with respect to the concentration of a specified analyte; and *detection limit* = the smallest concentration difference that can be detected above the background noise level of a sensor. This is often estimated using 2–3 times the background RMS noise as a signal and estimating the concentration for this signal level using a calibration curve. Additional details on the physics of spectrometers can be found in Blaker (1970), Bracey (1960), Braun (1987), Ditchburn (1965), Fogiel (1981), and Wist (1986).

TYPES OF SPECTROMETERS

The reader is referred to Bracey (1960), Braun (1987), and Wist (1986) for details on spectrometer design.

Discrete Photometers

Discrete photometers consist of an irradiance source, discrete interference filters, a sample compartment, and detector, along with appropriate electronics for signal amplification and stabilization of detector signals. The optical configuration for a generic photometer is shown in Figure 1.1.

Single-Beam Spectrometers

Single-beam spectrometers have a single optical channel that is configured to measure either the sample or reference channel, but not both simultaneously. The resultant spectrum is the ratio of the transmission spectra from sample and reference measurements. In practice the response of the detector measured with zero photon energy flux is measured as the *dark current*. The final transmission (in *T* units) spectrum from this device is given as:

$$\text{Reflectance or Transmittance Spectrum} = \frac{\text{Sample}_T - \text{Dark}}{\text{Reference}_T - \text{Dark}}$$

If the spectrometer has any instabilities (either optical, mechanical, or electronic), the time constant between sample and reference measurements becomes critical. Single-beam instruments must either be more inherently stable or alternate between sample and reference measurements at

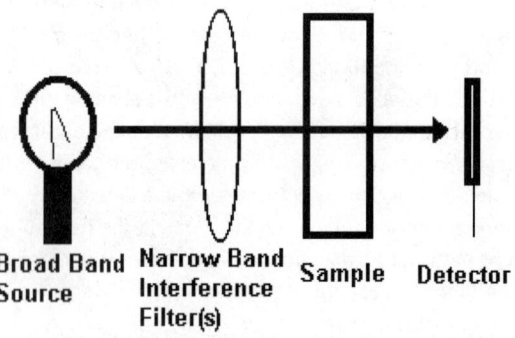

Fig. 1.1   Interference filter photometer optical configuration.

TYPES OF SPECTROMETERS

a frequency that makes negligible the rate of change of the spectrometer. Designs for single and double monochromator spectrophotometers, as well as a diode array design, are shown in Figures 1.2a–1.2c. These designs can be configured as single- or double-beam spectrometers.

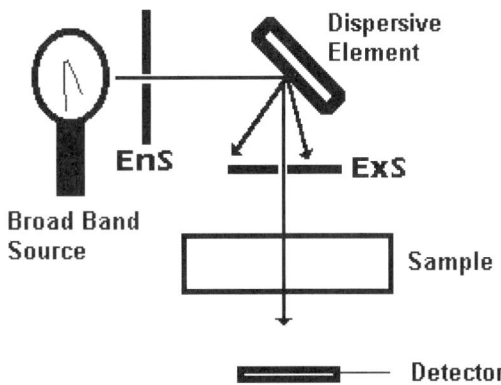

Fig. 1.2a Single monochromator system (dispersive) optical configuration. EnS and ExS designate entrance slit and exit slit, respectively.

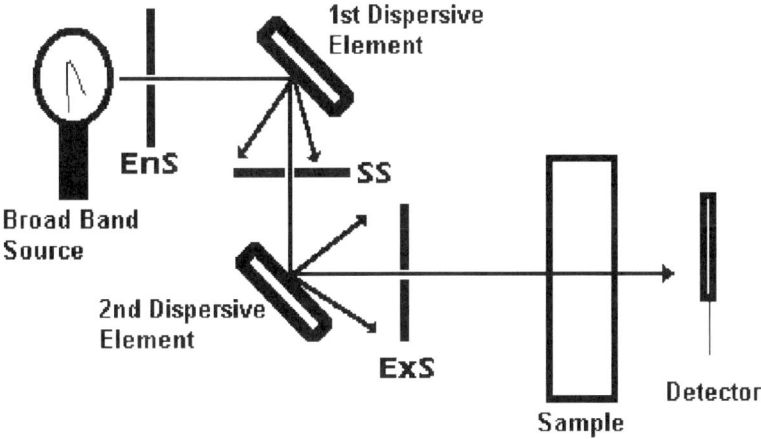

Fig. 1.2b Double monochromator system (dispersive) optical configuration. EnS, SS, and ExS designate entrance slit, second slit, and exit slit, respectively.

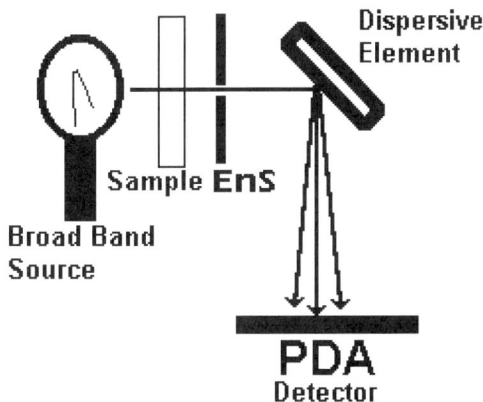

Fig. 1.2c Diode array spectrophotometer (dispersive) optical configuration. EnS designates the entrance slit.

Double-Beam Spectrometers

A double-beam spectrometer consists of both sample and reference channels and measures both channels simultaneously. The separation of the light beam from the source is accomplished using a beam splitter such that roughly 50% of the source's emitted energy is divided to both sample and reference channels. The use of the dual-beam concept compensates for instrument instabilities inherent in all spectrometers. The resultant spectrum is the ratio of the sample and reference channels in transmission. As in the case of single-beam instruments, in practice the final spectrum from this device is given as:

$$\text{Reflectance or Transmittance Spectrum} = \frac{\text{Sample}_T - \text{Dark}}{\text{Reference}_T - \text{Dark}}$$

Double-Beam/Dual-Wavelength Spectrometers

Double-beam/dual-wavelength spectrometers are capable of making measurements at two nominal frequencies simultaneously. They possess the capability of illuminating a specimen with two wavelengths (λ_1 and λ_2) while measuring the spectrum at both wavelengths. This is accomplished by using two dispersive elements (e.g., diffraction gratings) to disperse the incident energy from the source onto the sample specimen at the two different wavelengths. A shutter interrupts one of the beams while the other is incident to the detector. The signals from λ_1 and λ_2 are processed in such a manner that the displayed signal is a differential absorbance of $D\lambda_1 - \lambda_2$; this differential absorbance is proportional to concentration. Dual-wavelength scanning is used to cancel the effects of background when measuring turbid samples, for quantitative determination of a single component in multicomponent mixtures, or for quantitative determination of high-speed reactions.

Interferometer-Based Spectrometers

Interferometry is often used to measure wavelengths; this type of measurement is conducive to spectroscopic measurements. In spectroscopy, the accuracy of wavelength measurements can be critical to from 6 to 10 significant figures. Wavelength measurements from interferometer-type devices are much more accurate than those generally made using dispersive-type instruments. The most common type of interferometer is the Michelson type shown in Figure 1.3. A parallel beam of light from the spectrometer source (SS) is directed into a bean splitter (BS) at an angle of 45° from normal. The beam splitter consists of a 50% reflective surface. The transmitted portion of light from the beam splitter is directed to a highly reflective movable mirror (MM), whereas the light reflected from the beam splitter strikes a fixed mirror (FM). From the perspective of the detector (D) the following phenomena are "observed." If the pathlengths to FM and MM are the same, an observer at the D position will see a bright fringe. If the pathlengths are not the same, the phase difference will determine whether the fringe observed at the detector is light or dark. The fixed mirror is adjusted only for tilt, to keep it normal to the incident beam. As MM is moved, the path becomes different and the phase changes. A movement of MM (d) is related to the number of fringes in the interference pattern ($\#f$) observed at D by the relationship:

$$\frac{d}{\#f} = \frac{\lambda}{4}$$

when λ = the wavelength of parallel incident light entering the interferometer and d is expressed in units identical to wavelength units. The optical design configuration for an interferometer-based instrument is shown in Figure 1.3.

Open-Path and Emission Spectrometers

An open-path spectrometer is designed according to the features of dispersive and interferometer-based instruments, with the exception that the sample is located remotely from the instrument

DETAILS OF SPECTROMETER COMPONENTS

and the light source is either sunlight or laser power. The basic configuration for sampling in the open path design is shown in Figure 1.4.

DETAILS OF SPECTROMETER COMPONENTS

Light Sources

All materials emit electromagnetic radiation, the nature of which depends upon the material and its temperature. Gases or vapors of atoms at less than atmospheric pressure emit radiation at discrete wavelengths when an electric current is passed through them. Excited gases containing molecules emit spectra consisting of multiple lines very close together, which comprise emission bands. Solid materials emit or radiate continuous spectra across all frequencies. The following relationships define the physics of light sources.

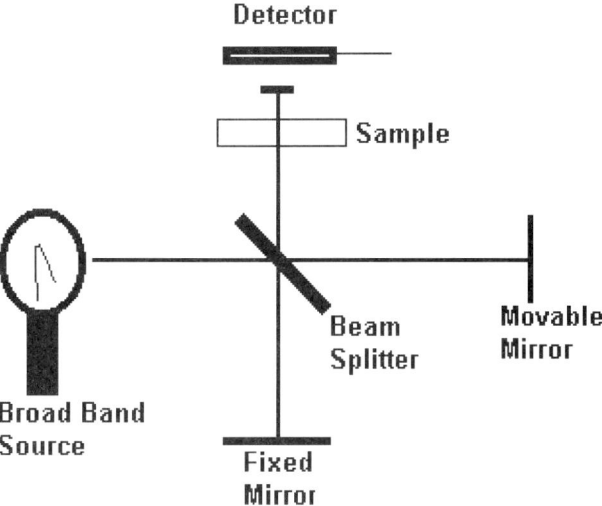

Fig. 1.3   Fourier transform spectrophotometer (interferometer) optical configuration.

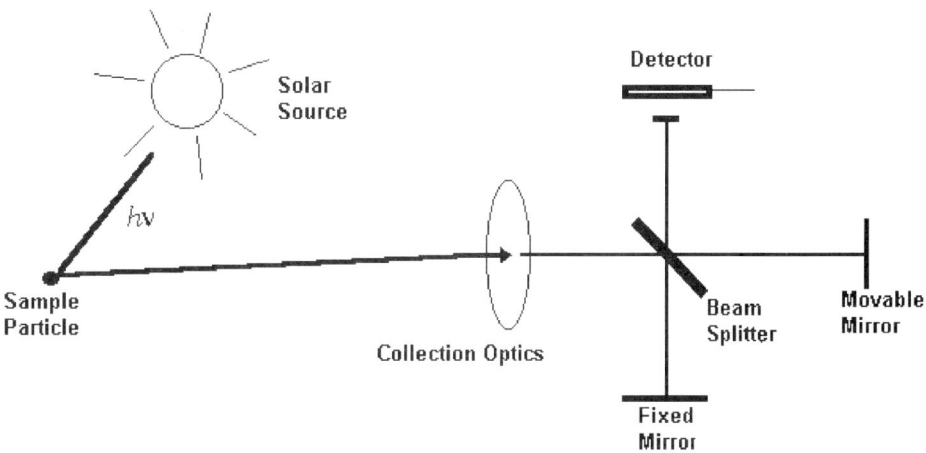

Fig. 1.4   Open-path sampling configuration using an interferometer.

Kirchoff's law states that for a thermal radiator at constant temperature and in thermal equilibrium, the emissivity (*e*) at any given frequency is equal to the absorptance (*a*) for radiation from the same direction; so $e = a$, where *absorptance* is defined as the ratio of energy absorbed by the surface to that of incident energy striking the surface.

An ideal surface absorbing all the energy that strikes its surface could be termed a *black body* (no radiation is reflected or emitted, thus it appears perfectly black at all frequencies). It would also stand from Kirchoff's law that this black body would also be the ideal emitter of radiation.

The *emissivity* (ε) for any black body source is defined as the ratio of the emitted radiance (ρ) by the source to the radiance of a black body (BB) at the same temperature and frequency (wavelength). The following equation illustrates this:

$$\varepsilon = \frac{\rho}{\rho_{BB}}$$

The spectral radiance (*P*) is defined as the radiant power (or photon flux density) emitted per unit source area per unit solid angle (the conventional units for expressing this are watts/m²/steradians).

Max Planck derived the principles of quantum mechanics and expressed this using the concept of light quanta as discrete energy "packets" that are transferred with an energy (*E*) proportional to the frequency of the electromagnetic radiation involved. Thus *Planck's law* has become one of the most widely known concepts in the physics of electromagnetic radiation. The expression is given as:

$$E = h\nu$$

where E = the energy content of each light quantum, and Planck's constant (h) = 6.63 . . . × 10^{-34} joule-sec.

From Planck's work, the *Planck's radiation formula* is used to calculate the radiance emitted by from a black body at a given wavelength (λ), emissivity (*e*), and temperature (*T*) as:

$$\rho_{BB} = 3.745 \times 10^4 \lambda^{-5} \left(\frac{1.4388 \times 10^4}{e^{\lambda T} - 1} \right)$$

where the units for ρ_{BB} are watts/cm²/μm.

From Planck's radiation formula other relationships for a black body spectrum can be derived. The *Stefan Boltzmann law* gives the total radiance emitted (ρ_{BB}) from a black body as a function of the black body temperature:

$$\rho_{BB} = \sigma \times T^4$$

where σ = Stefan's constant = 5.67×10^{-8} watts/m² °K⁴. *Wien's displacement law* demonstrates the shift (to shorter wavelengths) of the maximum peak position for a black body spectrum as a function of temperature:

$$\lambda_{max} = \frac{2898\mu \times °K}{T}$$

where λ_{max} is the peak maximum position from the black body spectrum in microns (μ) and *T* is in °K. Note: $T(°K) = T(°C) + 273$.

Detectors

There are two basic types of photon detectors: photoemissive and solid state. The photoemissive type is generally represented by the photomultiplier tube detectors; the solid-state type is represented by photodiode detectors, pyroelectric detectors, and infrared detectors.

In defining the physics of detector devices, several terms deserve explanation. First, the term *specific detectivity* (or *D**) is essential. This *D-star* value is defined as the detectivity of a radiation detector as a function of the square root of the product of the active detector element area (*A*) and the bandwidth (*ω*, in cycles per second), divided by the noise equivalent power (NEP) of the detector element. D* is reported in cm. Hz$^{1/2}$ watts^{-1} units.

$$D^* = \frac{\sqrt{A \times \omega}}{\text{NEP}}$$

Other concepts important to the discussion of detectors include: *spectral sensitivity, radiant sensitivity, detectivity, dark current,* and *quantum efficiency*.

Filters

Two basic types of interference filters exist: *bandpass filters* and *edge filters*. Bandpass filters transmit light for only a defined spectral band. The transmitted spectral bands may be from less than 1 nm FWHM (**f**ull band**w**idth at **h**alf **m**aximum transmission band height) to 50 nm or more FWHM. Edge filters transmit light either above or below a certain wavelength region; these are referred to as "cut-on" and "cut-off" types, respectively. These filters transmit efficiently throughout a broad region until the transmission limit of the filter substrate material is reached.

Interference filters consist of a solid Fabry–Perot cavity. This is a device made of a sandwich of two partially reflective metallic layers separated by a transparent dielectric spacer layer. The partially reflective layers are of higher refractive index than the dielectric spacer layer and are $\lambda/4$ in thickness, where λ is the peak wavelength (wavelength of maximum transmission) for the filter. The lower-refractive-index spacer layer is made to $\lambda/2$ thickness. The thickness of the dielectric spacer layer determines the actual peak transmission wavelength for the filter. Only the $\lambda/2$ light transmits with high efficiency; the other wavelengths experience constructive interference between the multiple-order reflections from the two partially reflective layers.

The wavelength position of the transmittance peak (λ_t) through either a Fabry–Perot interferometer or a bandpass interference filter is given as:

$$\lambda_t = \frac{2 \times n_\varepsilon t_\sigma \cos\alpha}{o}$$

where λ_t = the wavelength of maximum transmittance for the filter, n_ε = the refractive index of the surrounding medium (air = 1.0003), t_σ = the thickness of the dielectric spacer in microns, sometimes referred to as the *effective refractive index*; α = the angle of incidence of the light impinging onto the dielectric spacer; o = the order number for the interference (a nonzero integer as 0, 1, 2 . . .).

The assumption for all interference filters is that incident energy striking the filter is collimated and at a normal incidence. The wavelength of peak transmittance for the filter can be moved by varying the angle of incidence (α) of light impinging upon the surface of the filter. The relationship defining the peak position of the maximum transmission is given by:

$$\lambda_{\text{new}} = \lambda_{\max} \sqrt{1 - \left(\frac{n_\varepsilon}{n_\sigma}\right) \sin^2\alpha}$$

where λ_{New} = the wavelength of the new peak transmission position at incident angle α (when the incident angle, α, is nonnormal ≥ 0°), λ_{Max} = the wavelength of the current peak maximum for the interference filter with impinging light at an incident angle α = 0°, n_ε = the refractive index of the surrounding medium (air = 1.0003), n_σ = the refractive index of the dielectric spacer, sometimes referred to as the *effective refractive index*, and α = the angle of incidence of the light impinging onto the filter surface.

Gratings

When incident light strikes a diffraction grating, the light is separated into its component wavelengths, with each wavelength scattered at a different angle. To calculate the particular angle

(θ_d) at which each wavelength of light (λ_a) is scattered from a diffraction grating, the following expression is used:

$$\theta_d = \sin^{-1}\left(\frac{o \times \lambda_a}{s}\right)$$

where θ_d = the angle of diffracted light from the normal angle, o = the order number (integers as 1, 2, 3, etc.), s = the spacing of the lines on the grating (in the same units as wavelength), λ_a = the wavelength of the incident light in air (for white light it represents the wavelength of interest for calculation of the diffraction angle).

The dispersion of a grating refers to how broadly the monochromator disperses (or spreads) the light spectrum at the sample specimen position. Dispersion is generally expressed in units of nm per mm. Dispersion depends upon the groove density (number of grooves per mm) of the grating.

The intensity distribution of light (I) from the surface of a diffraction grating is given by:

$$I = \left[\left(\frac{I_0}{S^2}\right)\left(\frac{\sin b}{b}\right)\left(\frac{\sin S \times a}{\sin a}\right)\right]$$

where I = the intensity from the surface, I_0 = the incident energy intensity, S = the number of slits passed by the light following diffraction, $b = \frac{\pi \times s}{\lambda_a}$, s = the spacing of the lines on the grating (in the same units as wavelength), λ_a = the wavelength of the incident light in air (for white light it represents the wavelength of interest for calculation of the diffraction angle).

$$a = \frac{\pi \times \sigma}{\lambda_a}$$

where σ is the spacing of the slits (in the same units as wavelength) and λ_a = the wavelength of the incident light in air (for white light it represents the wavelength of interest for calculation of the diffraction angle).

The resolution (P) of a diffraction grating is given by the following expression:

$$P = oN = \frac{\lambda}{\Delta\lambda}$$

where o = the order of the interference pattern, N = the total number of lines on the diffraction grating surface, λ = the wavelength at which the resolution is determined, $\Delta\lambda$ = the distance in wavelength between the lines or optical phenomena to be resolved.

Beam splitters

Beam splitters are optical devices used to divide and recombine an optical beam (or beams) of light. They can be produced using half-silvered mirrors, which reflect approximately 50% of the energy incident to them; the remaining 50% is transmitted through the beam splitter.

Prisms

Prisms can be used either to disperse light into its spectral components or as right-angle prisms with reflective coatings on the hypotenuse side to bend light at a 90° angle. The dispersive properties of prisms have been known since the late 17th century. The deflection or dispersion of a prism is described using *Snell's law* at each optical surface of the prism, taking into account the refractive index of the prism at each wavelength. Snell's law describes the change in the path of light crossing an interface between two different materials (with different refractive indices) when the incident angle of the first surface is other than 90°. The wavefronts or wave propagation angle through the interface must move toward the normal angle as shown in Figure 1.5.

The numerical explanation of Snell's law is given by:

DETAILS OF SPECTROMETER COMPONENTS

$$n_1 \sin \alpha_1 = n_2 \sin \alpha_2$$

where n_1 = the refractive index of the first medium at the interface of two materials of varying refractive indices, n_2 = the refractive index of the second medium at the interface of two materials of varying refractive indices, α_1 = the angle of incidence and reflection of the impinging light onto the surface of the second medium, and α_2 = the angle of refraction of the light passing through the interface of two materials of varying refractive indices.

For an equilateral (dispersive) prism, the *wave propagation angle* (δ) (shown in Figure 1.6) through the prism is given by:

$$\delta = \alpha_1 + \sin^{-1}[(\sin \theta)(n^2 - \sin^2 \alpha_1)^{\frac{1}{2}} - (\sin \alpha_1 \cos \theta] - \theta$$

where α_1 = the angle incident to the surface of the prism, θ = the wedge angle of the prism (60° for an equilateral triangle), n = the refractive index of the prism at the frequency (wavelength) of incident energy, and δ = the wave propagation angle (angle of refraction) through the prism.

Interferometer Assemblies

Multiple interferometer types exist in modern spectrometers, the reader is referred to Candler (1951), Jamieson (1963), Steel (1983), and Strobel (1989) for a more exhaustive description of

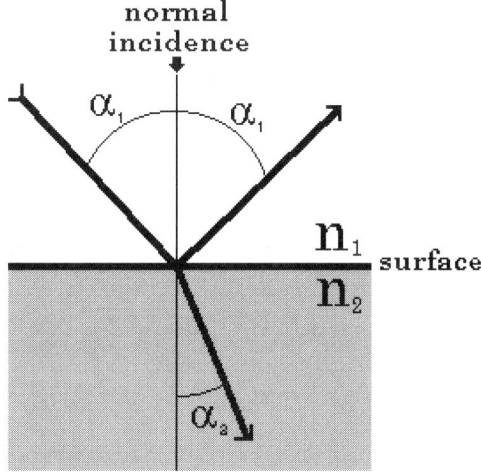

Fig. 1.5 Demonstration of Snell's law showing the light wave propagation angles of reflection and refraction at the interface between materials of differing refractive indices n_1 and n_2.

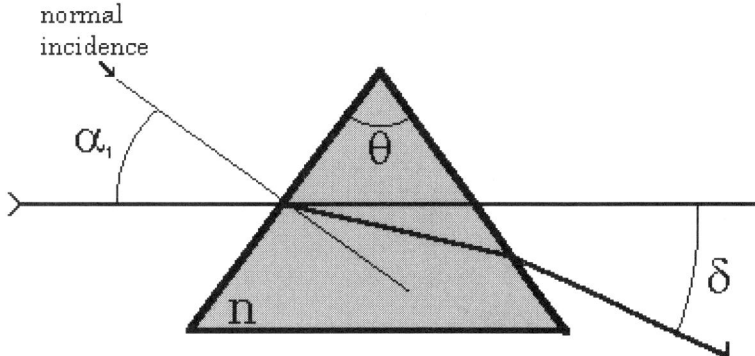

Fig. 1.6 Wave propagation angles through an equilateral prism for light dispersion.

the optical configurations for these devices. The classical interferometer design is represented by the Michelson interferometer, as shown in Figure 1.3. A movable mirror (MM) is displaced linearly minute distances to yield an interference pattern (or interferogram) as a series of sine waves when the interference pattern is observed from a specific field of view Figure 1.7a.

The interferogram is plotted as the intensity of light (y-axis) versus the mirror position (x-axis); thus the signal from the interferometer is a function of time (because the mirror is in motion at a constant velocity (Figure 1.7b). The raw interferogram, as it is sometimes termed, is converted to a spectrum using the Fourier transform, and a spectrum is determined by ratioing a spectrum determined with a sample in the beam (as the sample spectrum) to a spectrum determined with no sample in the beam (as the background spectrum).

A laser beam of known frequency is used to signal a separate sensor to provide near-perfect sampling of the mirror position. The resolution of the interferometer depends upon the distance of motion in the movable mirror. The precise distance traversed by the movable mirror can be determined using the number of fringes observed to pass a given field of view in time (t), given the wavelength of light from the laser. The distance traversed by the mirror is given as

$$d = \frac{f\lambda}{2}$$

where d = the distance traveled, f = the number of fringes in an interference pattern passing a specified field of view, and λ = the wavelength of laser light used.

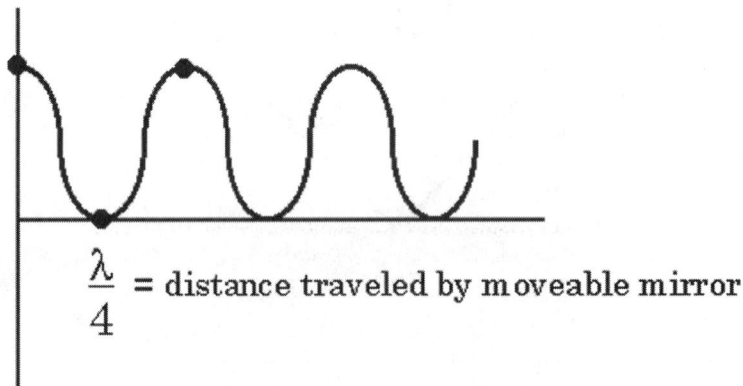

Fig. 1.7a   Interferometric output as a sine wave function.

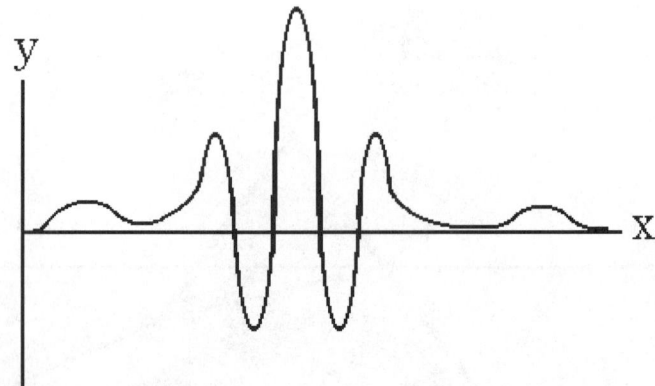

Fig. 1.7b   The interferogram is plotted as a function of the light intensity (y-axis) versus the mirror position (x-axis); thus the signal is a function of time (because the mirror is moved at a constant rate). The raw interferogram is subjected to the multiple-step Fourier transformation, and a spectrum results.

High resolution involves moving the mirror at integral cycles over greater distances per unit time as compared to low-resolution measurements. A complete explanation of the physics of interferometry is beyond the scope of this chapter. The reader is referred to Griffiths (1986).

Polarizers

A diversity of polarizing elements exists for the purpose of rotating or selecting light of a specific electronic vector orientation. When the electronic direction vector of light incident to a surface is parallel to the electronic field vector of the surface, increased interaction of the incident light (absorption) occurs. This principle is important in characterizing the surface chemistry for optical components, thin films, metal surfaces, and semiconductor interfaces. The use of a quartz plate can act as a polarizer that will rotate the plane of linearly polarized light. This rotation (P) can be described using the following relationship:

$$P = \left[\pi t \left(|n_\Lambda - n_P|\right)\right] \times \frac{1}{\lambda_0}$$

where t = the thickness of a quartz plate cut perpendicularly to the optical axis, n_Λ = the refractive index for right circularly polarized light, n_P = the index of refraction for left circularly polarized light, and λ_0 = the vacuum wavelength of light entering the quartz plate. *Note:* for quartz, $n_\Lambda = 1.5582$ and $n_P = 1.5581$.

Electronic Components Used in Spectrometry

For references describing the electronic components of optical spectrometers, see Braun (1987), Jamieson (1963), and Strobel (1989). Two special categories of electronic devices deserve representation when discussing spectrometers. The first category is detectors and detector electronics that produce a current of voltage signal proportional to the photon flux striking the detector. Detector stability is provided by correct electronic circuitry that allows the detector signal to be selectively amplified with the minimum introduction of noise; thus electronic circuitry enhances the signal-to-noise ratio of the detector signal.

Digital microcomputers comprise the second essential electronic element for modern spectrometers. With the addition of appropriate software, sophisticated instrument control and data processing can enhance the usefulness and user friendliness of spectrometers. These issues are described in detail in Jamieson (1963) and Strobel (1989).

PROPERTIES OF SPECTROMETERS

Aperture Diameter

To calculate the aperture (a) required by an optical system to resolve two objects with known linear separation, the Rayleigh criterion for resolution is used:

$$a = \frac{1.22 \times \lambda}{\alpha_R}$$

where α_R = the angle of separation from the measuring device exit aperture to the objects to be resolved (which is calculated as $\tan \alpha = \frac{\text{opp.}}{\text{adj.}}$, where α_R is expressed in radians), λ = the wavelength of light observed from the objects, and a = the aperture of the optic system.

Entrance and Exit Pupils

The *entrance pupil* refers to the size and location of the entrance aperture between the light source and the remainder of an optical system. The *exit pupil* refers to the size and location of the exit aperture within an optical train just prior to the detector.

Bandpass and Resolution

The terms *bandpass* and *resolution* are used to express the capability of a spectrometer to separate spectral bands or lines that are separated by some finite distance. For an instrument that disperses energy over a prespecified spectral region of the electromagnetic spectrum, the bandpass of a spectrometer is used to describe which portion of the spectrum can actually be isolated by the spectrometer in a "pure" wavelength form. The spectrometer bandpass is dependent upon the dispersion of the grating (see the earlier section on gratings) and the entrance and exit slit widths. An illustration is often given to elucidate the problem associated with measuring monochromatic light using conventional spectrometers. If the ideal spectrometer were used to measure a bright-line emission spectrum at a single wavelength (λ_1), the spectrum would appear as a single line (Figure 1.8). What really occurs when such a spectrum is measured using a conventional spectrometer is a broad band spectrum, as shown in Figure 1.9. The spectrum assumes a Gaussian-like (or bell-shaped) curve. This characteristic broadening of a line spectrum through the spectrometer is an illustration of the spectrometer bandpass. The actual bandpass for any instruments assigned a value by determining the full width at half maximum (FWHM) height of the bell-shaped spectrum. Thus for the band in Figure 1.9 the FWHM could be empirically determined by finding the wavelength where maximum intensity occurs and measuring the peak height at this position. This height measurement is divided in half and the bandwidth measured at this height on the band, as illustrated in Figure 1.10.

The actual shape of a band is the result of several instrumental characteristics, including the overall quality of the optics and detector systems as well as the width and positions of the entrance and exit slits. Every dispersive spectrometer consists of a dispersive element (e.g., diffraction grating) in combination with an entrance and an exit slit. The image of the entrance slit and exit slit determines the spectrometer bandpass, which is sometimes referred to as the *slit function*. Actually the slit function is the result of the convolution (combination) of the images of these two slits. The bandshape of a dispersive spectrometer is shown in Figures 1.9 and 1.10. Other factors associated with optical and electronic quality cause a rounded overall shape. The bandpass of a spectrometer is equal to the FWHM. Often texts dealing with instrumentation will state that the bandpass of a spectrometer is approximated by the product of the linear dispersion of the monochromator and the entrance or exit slit width (whichever is larger).

The resolution of a spectrometer can be defined as the minimum distance between two peaks that can be detected by the spectrometer under designated operational performance settings. Resolution is calculated by multiplying the slit width (generally expressed in mm) by the dispersion of the monochromator (in nm per mm). Due to practical issues and nonideal optics, the actual resolution of a spectrometer must be slightly greater (poorer) than the theoretical value.

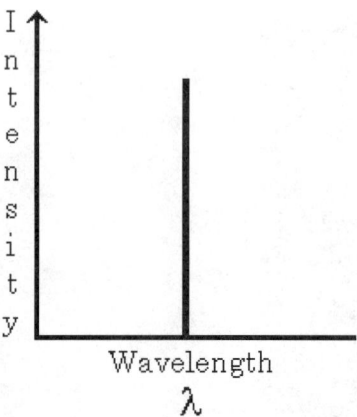

Fig. 1.8 Bright-line emission spectrum at a single wavelength as it would appear in an ideal spectrophotometer.

PROPERTIES OF SPECTROMETERS

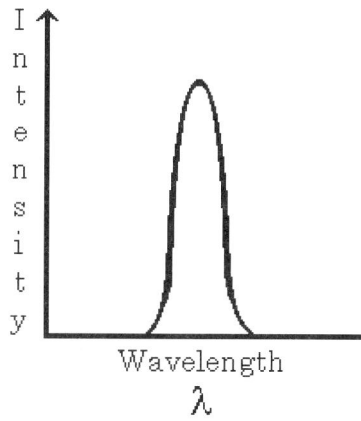

Fig. 1.9   Spectrum of a bright-line emission source (e.g., deuterium lamp). The characteristic broadening is an illustration of the bandpass of a spectrophotometer.

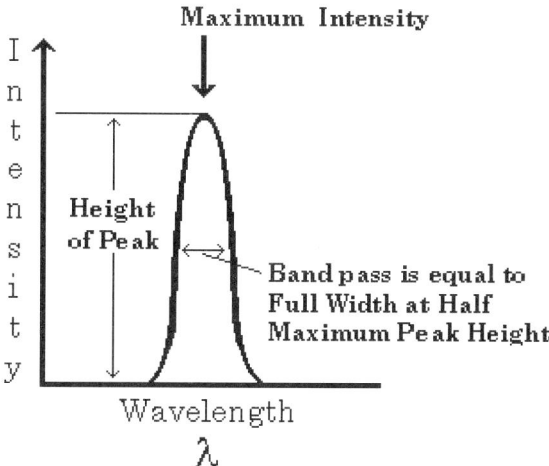

Fig. 1.10   Illustration of the determination of bandpass using the bell-shaped peak obtained by using a bright-line source projected through a monochromator optical system.

To summarize, bandpass and resolution are identical in practice. Only the resolution specification of a spectrometer is the expression of bandpass under the specified measuring conditions of an instrument dependent upon the slit width settings.

The empirical resolution of a spectrometer is determined by measuring the FWHM in mm for two narrow bands that are completely resolved (to the baseline) using the spectrometer. The spatial difference between the maximum absorbance (lambda max.) is determined between the bands (in mm), simultaneously noting the difference between the lambda max. points in nm. The various measurements required for this calculation are shown in Figure 1.11 and illustrated by the following relationship:

$$\text{Bandpass} = \text{Resolution} = \frac{\text{Band difference in nm}}{\text{Band difference in mm}} \times \text{FWHM in mm}$$

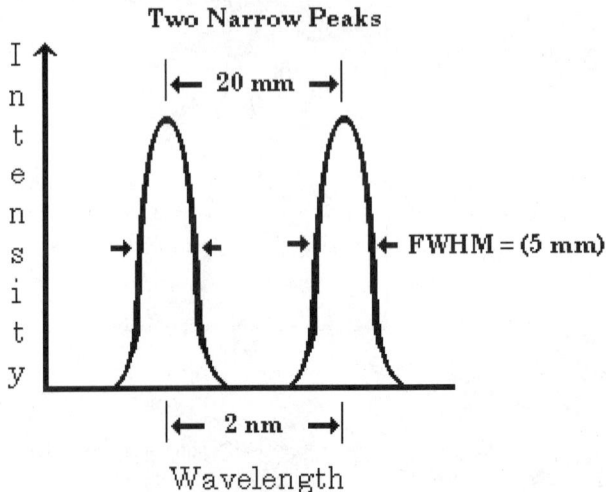

Fig. 1.11 Method to determine maximum resolution of a spectrometer under specific measurement conditions. In the example shown, Bandpass = Resolution = FWHM × Dispersion = (2 nm) / (20 mm) × 5 mm = 0.50 nm resolution.

Numerical Aperture

The numerical aperture (NA) is a measure of how much light can be collected by any optical system. The NA is expressed as the product of the refractive index of the incident material (n_i) and the sine of the ray angle maximum (θ_{Max}) from normal incidence; the function is given by

$$NA = n_i \times \sin\theta_{Max}$$

Attenuation (Light Transmittance Losses in Optical Systems)

Losses in transmitted light through spectrometers are due to absorption, reflection, scattering, and optical misalignment; the losses can vary with temperature and wavelength. The quantity of optical loss is expressed as an attenuation rate in decibels (dB) of optical power per unit distance (cm). Typical losses result from launch optics, temperature variations, optical couplings within the optical path, aging of mirrored surfaces, and soiled optical surfaces. The losses in energy transmitted through a spectrometer can be calculated by using Beer's law. Beer's law states that the irradiance of energy through an absorbing medium falls exponentially with the distance of transmission according to the following relationship:

$$I_d = I_0 \times 10^{(\alpha d/10)}$$

where I_d = the irradiance at distance (d) from the source, I_0 = the source irradiance at $d = 0$, α = the attenuation (absorption) coefficient in units of dB/cm, and d = distance in cm.

Attenuation losses are wavelength dependent; thus the value for α is a function of the incident wavelength (λ).

Etendue

The etendue (or relative throughput advantage) for an optical system is the product of the potential illuminated surface area (A) and the solid angle of the optic. Traditionally, this is represented by the two following equations, where ε' represents the etendue, and Ω_s represents the solid angle. Thus the etendue is represented as

PROPERTIES OF SPECTROMETERS

$$\varepsilon' = A \times \Omega_S$$

and the solid angle is given by

$$\Omega_S = 2\pi\left(1 - \sqrt{1-(\mathrm{NA})^2}\right)$$

Therefore these two equations allow us to calculate the relative improvement for an optical spectrometer optical system. As can be seen, the NA and aperture diameter are preeminent factors for throughput in optical systems, as shown in Table 1.1.

Throughput

The relative throughput (T) represents the overall effectiveness of an optical system to transmit light relative to the amount of light introduced into the system (I_0) from the light source. It is defined as the ratio of light energy passing into an optical system to the light energy passing out of the optical system. For dispersive spectrometers, this relationship is defined by

$$T = \frac{\pi D w_s (R_1 \times R_2 \times \cdots \times R_k)(\Sigma_g)}{4f^2}$$

where D = the dispersion constant in mm/nm, w_s = the slit width in mm (or exit slit), f = the f/number of the optical system (f/number = $\frac{1}{2(\mathrm{NA})}$), R_k = the reflectivity of mirrors or other optical surfaces, Σ_g = the spectral efficiency of the grating (approximately 0.80 at the blaze wavelength).

Signal-to-Noise Ratio

The theoretical total signal (S) from an optical system can be given by:

$$S = R_S B_\lambda \varepsilon' \tau q$$

where R_s = the light source spectral radiance, B_λ = the spectral bandwidth, ε' = the etendue of the spectrometer optical system, τ = transmission losses (and emissivity), and q = quantum efficiency.

The measured signal-to-noise ratio (s/n) from an optical system can be calculated as the full transmitted signal divided by the RMS noise (in transmittance units). Thus for a 100% line with RMS noise as 0.001%T, s/n = 100/0.01 = 10,000 : 1. This applies when RMS noise is calculated as

$$\mathrm{RMS} = \sqrt{\frac{1}{n}\sum_{i=1}^{n}(T_i - \overline{T})^2}$$

where T_i = the individual transmission value at each data channel i and $\overline{T}$ = the mean transmission value for all data channels.

Table 1.1   Etendue and Relative Throughput as a Function of Numerical Aperture (NA)

NA (Numerical Aperture)	Relative Etendue (ε') for 1-mm-diameter aperture	Relative Throughput
0.20	0.10	1
0.40	0.40	4
0.60	1.00	10

Dynamic Range

The range of a specified analyte concentration over which a sensor response is directly proportional to a change in concentration is the dynamic range of a spectrometer. Dynamic range is limited by stray light and noise. Knowledge of Beer's law (previously discussed) allows us to calculate the maximum theoretical dynamic range for an instrument using a few simple mathematical relationships and a calculation of the relationship to stray light and maximum observable absorbance value, given by the following. (*Note:* The error in a measurement due to stray light can be computed using this expression.)

$$A_i = \log\left(\frac{1+\frac{S_l}{I_0}}{\frac{I}{I_0}+\frac{S_l}{I_0}}\right) = \log\left(\frac{100+I_s}{T+I_s}\right)$$

where I_0 = incident light intensity, I = transmitted light intensity, S_l = stray light intensity, I_s = stray light as a percentage of I_0, and T = percent transmittance of the measurement under test.

Error in a measurement is also attributable to the combined noise of the measurement system, given as:

$$\text{Noise (as \%)} = \frac{\text{RMS} \times 100}{A}$$

The relative dynamic range of a spectrometer is written as:

$$A_{DR} = \log\left(\frac{100+I_s}{I_s}\right) - k(\text{RMS}_A)$$

where k = multiplier for desired confidence level and RMS_A = root mean square noise measurement (in A). *Note:* To simplify calculations, the $-k(\text{RMS})$ term can be dropped, yielding an estimated value for dynamic range.

Stray Radiant Energy

The relationship between absorbance and stray light is given by:

$$A_i = \log\left(\frac{1+\frac{S_l}{I_0}}{\frac{I}{I_0}+\frac{S_l}{I_0}}\right) = \log\left(\frac{100+I_s}{T+I_s}\right)$$

This calculation relationship in absorbance units applies where I_0 = incident light intensity, I = transmitted light intensity, S_l = stray light intensity (as a fraction of I_0), I_s = stray light (as fraction of I_0), and T = percentage of transmittance. The relationship between transmittance (T) and Absorbance (A) is given in Table 1.2.

Table 1.2   Relationships between %T, T, and A

% Transmittance	Transmittance	Absorbance
100.0	1.0	0.0
10.0	0.1	1.0
1.0	0.01	2.0
0.1	0.001	3.0
0.01	0.0001	4.0
0.001	0.00001	5.0
0.0001	0.000001	6.0

The calculation of percentage error in a measurement due to stray radiant energy is given by

$$E\,(\%) = 100\left[1 - \left(\frac{\log\left(\frac{100 + I_S}{T + I_S}\right)}{A_t}\right)\right]$$

where I_s = stray light as a percentage of I_0, T = percentage transmittance of measurement, and A_t = true absorbance level of sample specimen measured.

Wavelength Accuracy

The accuracy in wavelength measurements is determined by taking a standard reference material (or emission line spectrum) of known wavelength position (λ_k) and making measurements of these known positions using the spectrometer. The difference between the known position(s) and the positions as measured using the spectrometer (λ_s) is reported as the wavelength accuracy of the spectrometer, expressed as ($\lambda_k - \lambda_s$).

Wavelength Repeatability

Wavelength repeatability (λ_r) is the precision with which a spectrometer can make repeated measurements at the same nominal wavelength over temporal and environmental changes. This specification is calculated as

$$\lambda_r = \sqrt{\frac{1}{n}\sum_{i=1}^{n}\left(\lambda_i - \bar{\lambda}\right)^2}$$

where λ_i = wavelength determined at each of multiple measurements taken n times and $\bar{\lambda}$ = mean wavelength determined using each of the multiple measurements taken n times.

Photometric Accuracy

The accuracy in photometric measurements is determined by taking a standard reference material of known transmission (T_k) values and making measurements at specific wavelengths of these known photometric values using the spectrometer. The difference between the known transmission and the transmission as measured using the spectrometer (T_s) is reported as the photometric accuracy of the spectrometer, expressed as ($T_k - T_s$).

Photometric Repeatability

Photometric repeatability (T_r) is the precision with which a spectrometer can make repeated measurements at the same nominal transmission value over temporal and environmental changes. This specification is calculated as

$$T_r = \sqrt{\frac{1}{n}\sum_{i=1}^{n}\left(T_i - \bar{T}\right)^2}$$

where T_i = transmission determined at each of multiple measurements taken n times and $\bar{T}$ = mean transmission determined using each of the multiple measurements taken n times.

REFERENCES

Blaker, J. W. (1970). *Optics II—Physical and Quantum Optics.* New York: Barnes and Noble, 91 pp.
Bracey, R. J. (1960). *The Techniques of Optical Instrument Design.* London: English Universities Press, 316 pp.
Braun, R. D. (1987). *Introduction to Instrumental Analysis.* New York: McGraw-Hill, 1004 pp.

Candler, C. (1951), *Modern Interferometers.* Glasgow: Hilger and Watts, 502 pp.
Ditchburn, R. W. (1965). *Light,* 2nd ed. London: Blackie, 833 pp.
Fogiel, M. (ed.) (1981). *The Optics Problem Solver.* New York: Research & Education Association, 817 pp.
Griffiths, P.R., de Haseth, J.A. (1986). *Fourier Transform Infrared Spectrometry.* New York: Wiley, pp. 1–55.
Jamieson, J. A., McFee, R. H., Plass, G. N., Grube, R. H., Richards, R. G. (1963). *Infrared Physics and Engineering.* New York: McGraw-Hill, 673 pp.
Steel, W. H. (1983). *Interferometry.* 2nd ed., Cambridge: Cambridge University Press, 308 pp.
Strobel, H. A., Heineman, W. R. (1989). *Chemical Instrumentation: A Systematic Approach,* 3rd ed. New York: Wiley, 1210 pp.
Wist, A. O., Meiksin, Z. H. (1986). *Electronic Design of Microprocessor-Based Instruments and Control Systems.* New York: Prentice-Hall, 287 pp.

FURTHER READING

Ball, C. J. (1971). *An Introduction to the Theory of Diffraction.* Oxford: Permagon Press, 134 pp.
Clark, G. L. (ed.) (1960). *The Encyclopedia of Spectroscopy.* New York: Reinhold, 787 pp.
Cohen, J. B. (1966). *Diffraction Methods in Material Science.* New York: Macmillan, 357 pp.
Guild, J. (1960). *Diffraction Gratings as Measurement Scales.* London: Oxford University Press, 211 pp.
Jenkins, F. A., White, H. E. (1957). *Fundamentals of Optics.* New York: McGraw-Hill, 637 pp.
Johnson, Jr., C. S., Pedersen, L. G. (1986). *Problems and Solutions in Quantum Chemistry and Physics.* New York: Dover, 429 pp.
Knowles, A., Burgess, C. (eds.) (1984). *Practical Absorption Spectrometry.* New York: Chapman and Hall, 234 pp.
Mach, E. (1926). *The Principals of Physical Optics.* New York: Dover, 324 pp.
Smith, R. J. (1980). *Electronics: Circuits and Devices.* 2nd ed. New York: Wiley, 494 pp.

2.

SOURCES, DETECTORS, WINDOW MATERIALS, AND SAMPLE PREPARATION FOR UV-VIS, NIR, IR, AND RAMAN SPECTROSCOPY

Table 2.1 UV-Vis, NIR, IR Emission Sources.

Source type	Useful emission range (microns)	Useful emission range (cm^{-1})
Quartz tungsten-halogen monofilament lamp	0.22–2.7	45,455–3,704
DC deuterium lamp for UV	0.185–3.75	54,054–2,667
Pulsed xenon arc lamp	0.18–2.5	55,556–4,000
DC arc lamp	0.20–2.5	50,000–4,000
Globar	1.0–100	10,000–100
Nernst glower	0.30–35	33,333–286
Carbon arc	0.50–100	20,000–100
Mercury lamp	0.30–100	33,333–100
Visible and NIR lasers	Helium-neon (He:Ne) @ 632.8 nm and visible lasers at 768 nm	15,802.8 13,020.8
	Neodymium:yttrium aluminum garnet (Nd:YAG) @ 1064 nm or 532 nm, generally 0–3 W output power	9,398.5 18,797.0

Useful wavelength and frequency working ranges are given.

Table 2.2 UV-Vis, NIR, and IR Detectors.

Detector type	Useful detection range (microns)	Useful detection range (cm^{-1})
Silicon	0.30–1.1	33,333–9,091
PbS (Lead sulfide)	1.1–3.0	9,091–3,333
InAs (Indium arsenide)	1.7–5.7	5,882–1,754
InGaAs (Indium gallium arsenide)	0.90–1.7	11,111–5,882
Ge:X (Germanium)	2–40	5,000–250
Ge:Au (Germanium gold)	2–9	5,000–1,111
Ge:Cd (Germanium cadmium)	2–24	5,000–417
PbSe (Lead selenide)	1.7–5.5	5,882–1,818
Ge:Zn (Germanium zinc)	2–40	6,667–250
InSb (Indium antimonide)	1.8–6.8	5,556–1,471
PbTe (Lead telluride)	1.5–4.5	6,667–2,222
DTGS/KBr (Deuterated triglycine sulfate)	0.83–25	12,050–400
DTGS/PE (Deuterated triglycine sulfate)	10–120	1,000–83
MCT (Mercury cadmium telluride) or HgCdTe (photovoltaic)	1–17	10,000–588
TGS (Triglycine sulfate)	10–120	1,000–83
PLT (Pyroelectric lithium tantalate) (LiTaO$_3$)	1.5–30	6,667–333

Useful wavelength and frequency working ranges are given.

Table 2.3 UV-Vis, NIR, and IR Window Materials.

Window material	Useful transmittance range (microns)	Useful transmittance range (cm^{-1})
CsI (Cesium iodide)	0.3–50	33,333–200
PbS (Lead sulfide)	1.1–3.0	9,091–3,333
KBr (Potassium bromide)	0.25–26	40,000–385
KCl (Potassium chloride)	0.25–20	40,000–500
NaCl (Sodium chloride)	0.25–16	40,000–625
KRS-5 (Thallium Bromide-Iodide)	0.6–40	16,667–250
Ge (Germanium)	1.1–30	9,091–333
As$_2$S$_3$ (Arsenic sulfide)	0.6–15	16,667–667
MgF$_2$ (Magnesium fluoride) (IRTRAN-1)	0.6–9.5	16,667–1,053
ZnSe (Zinc selenide) (IRTRAN-4)	0.6–26	16,667–385
BaF$_2$ (Barium fluoride)	0.2–13	50,000–769
ZnS (Zinc sulfide) (Cleartran or IRTRAN-2)	0.6–15	16,667–667
CaF$_2$ (Calcium fluoride) (IRTRAN-3)	0.2–9	50,000–1,111
Al$_2$O$_3$, Aluminum oxide (Sapphire)	0.2–7	50,000–1,429
SiO$_2$ (Fused silica or quartz)	0.2–4	50,000–2,500
AgBr (Silver bromide)	0.5–35	20,000–286
Polyethylene (high density)	16–333	625–30

Useful wavelength and frequency working ranges are given.

Table 2.4 Fiber-Optic Materials

Fiber-optic material	Useful transmittance range (microns)	Useful transmittance range (cm^{-1})
SiO$_2$ (Fused silica or quartz)	0.2–1.25	50,000–8,000
SiO$_2$ (Anhydrous quartz)	0.4–2.64	25,000–3,788
ZrF (Zirconium fluoride)	0.9–4.76	11, 111–2,100
Chalcogenide	2.22–11.1	4,505–901

Useful wavelength and frequency working ranges are given.

SELECTION OF USEFUL MEASUREMENT PATHLENGTHS (BASED ON PURE HYDROCARBONS)

Fourth-overtone NIR, 650–800 nm, 10 cm

SW-NIR, 800–1100 nm, 5–10 cm

LW-NIR, 1050–3000 nm, 0.1–2 cm

MIR, 2500–20K nm, 0.1–4 mm

Raman, 2500–20K nm, N/A

MATRIX/MEASUREMENT TECHNIQUES OF CHOICE

Gases: long-path MIR (0.5–20 m)

Solids: diffuse reflectance, diffuse transmittance, HATR

Liquids: all techniques

Organics: all techniques

Wastewater: sparging MIR

Caustics: MIR-ATR

Optically dense materials/opaque materials: HATR

SOLVENTS FOR SPECTROSCOPIC MEASUREMENTS

UV-Vis Solvent Cut-off Wavelengths (in Nanometers)

Chloroform (235–240 nm)

Hexanes (195–202 nm)

Methanol and ethanol (205 nm)

Water (190 nm)

1,4-Dioxane (205 nm)

Acetonitrile (190 nm)

Solvents for Near-Infrared Spectroscopy

For liquid measurement in the NIR, solvents most nearly transparent from 1.0 to 3.0 microns include: carbon tetrachloride, tetrachloroethylene, carbon disulfide, and chlorofluorocarbons.

None of these materials can be considered safe from an environmental or health perspective, and thus they are not recommended. The only signal for any of the solvents above occurs at 2.21 microns for carbon disulfide (which should not be used at pathlengths above 1 cm). All of the preceding solvents are transparent, even to 10-cm pathlengths, over the NIR spectral region, with the only exception so noted. Solvents such as chloroform, methylene chloride, dioxane, di (*n*-butyl) ether, triethylene glycol (dimethyl ether), heptane, benzene, acetonitrile, dimethylformamide, and dimethyl sulfoxide have all been used as solvents for a portion of the NIR region, generally from 1.0 to 1.6 microns and from 1.8 to 2.2 microns. Silicone lubricants and Nujol (liquid paraffin) have also been suggested as solvents, meeting the requirements of nontoxicity, low cost, and availability. These materials, however, are far from ideal spectroscopically.

For solid samples, oven-dried (105°C for 2 hours) sodium chloride, potassium bromide, or potassium chloride can be used to dilute samples across the entire 1.0–3.0 micron range. *Note:* Diffuse reflectance is often used for near-infrared measurements of solid samples. The instruments can be easily configured for such measurements, and the requirements for optimum spectra include small sample size (preferably less than 50-micron-diameter mean particle size) and an infinite pathlength (generally 5–10 mm).

Solvents for Infrared Spectroscopy

Common solvents used for infrared analysis include chloroform, acetone, methanol, and hexane. Cast films are often made by dissolving or extracting a soluble solid or liquid sample in one of the above solvents, filtering the extract, evaporating most of the solvent in a suitable hood, and casting a thin film (by evaporation) onto an infrared-transparent window material. Chloroform has IR bands at 3020 (weak), 1215, and 755 cm^{-1} (strong). Acetone has several sharp IR bands near 3000 (weak), 1715 (strong), 1420 (weak), 1365 (strong), 1220 (strong), 1090, and 535 wavenumbers. Methanol has bands at 3650–2750 (three broad and strong bands), 1450 (medium broad), 1120, 1030 (strong), and 640 cm^{-1} (broad). Hexanes have bands at 2750–3000 (strong and sharp), 1445 (strong), 1250 (weak), 900 (weak), and 850 cm^{-1} (weak).

For solid samples, oven-dried (105°C for 2 hours) potassium chloride or potassium bromide can be used to dilute samples across the entire range from 40,000 to 385 cm^{-1}.

ANALYTICAL METHOD AND SAMPLE PREPARATION FOR INFRARED SPECTROSCOPY

Infrared spectra must be measured in such a manner as to reduce to negligible, or preferably to eliminate altogether, the carbon dioxide doublet at 2349 cm^{-1} and the multiple bands due to water vapor, indicated by multiple sharp bands centered near 3600 cm^{-1} and 1640 cm^{-1}. This can be accomplished either by means of extended sample compartment purge times using dry air or dry nitrogen as a purge gas or by carefully referencing out the background under identical reference and measurement ambient air conditions. *Note:* A background reference using air can serve to accommodate for small amounts of water vapor and carbon dioxide, but the resultant sample spectrum should not exhibit these bands as noticeable.

The spectrum of the unknown and the reference spectrum should in no case obtain a minimum percentage transmittance (%T) of less than 2% so as to provide structural detail of the most prominent peaks. If spectra exceed this limit (by exhibiting a %T of less than 2%), then a thinner film should be recast in the case of thin-film measurements, or the sample must be further diluted in the case of KBr pellet sample preparation. If these procedures are followed closely, an accurate comparison of the test sample spectrum and reference spectrum can be made.

Powders

Each sample is prepared by finely grinding a small quantity of each powder (~0.01 g) using a mortar and pestle and then adding several drops of mineral oil (Nujol) and continuing to grind the sample until it becomes a white paste, something like cold cream. The paste is then evenly distributed between a pair of infrared windows (either KBr or NaCl) to form a thin translucent film. If the film is totally opaque, then the particle size of the powder is still too large and the material must be subjected to further grinding. Once the thin translucent film is formed between the windows, the transmission spectrum is measured. The maximum absorbance of the infrared spectrum for each sample must not exceed 1.2 Au. If this occurs, the windows must be further compressed (taking care not to break the windows) to decrease the pathlength between the windows until the absorbance is within range.

Pastes and Creams

Each sample is prepared by placing a small droplet of material onto an infrared window (either KBr or NaCl). The windows are gently pressed together to form a thin transparent film. Once the thin transparent film is formed between the windows, the transmission spectrum is measured. The maximum absorbance of the infrared spectrum for each sample must not exceed 1.2 Au. If this occurs, the windows must be further compressed (taking care not to break the windows) to decrease the pathlength between the windows until the absorbance is within range.

Solvent-Soluble Materials

Each sample is prepared by weighing approximately 0.05 g of sample into a 50-mL beaker and dissolving it with 20 mL chloroform on a hot plate (~60°C). Once dissolved, indicated by a clear solution with no solid present (~5–10 minutes), a few drops of the solution are added to an infrared window (KRS-5, KBr or NaCl) and a slight film of the material is cast onto the window. Once the thin translucent film is formed on the window, the transmission spectrum is measured. The maximum absorbance of the infrared spectrum for each sample must not exceed 1.2 Au. If this occurs, a drop of chloroform is added to the window to disperse the film further until the absorbance is within range.

Liquids

A few tiny dropletss of the solution are added to an infrared window (KRS-5, KBr or NaCl) and a slight film is cast onto the window. Once the thin translucent film of the material is formed on the window, the transmission spectrum is measured. The maximum absorbance of the infrared spectrum for each sample must not exceed 1.2 Au. If this occurs, a drop of chloroform is added to the window to disperse the wax further until the absorbance is within range.

SOLVENTS FOR RAMAN SPECTROSCOPY

Most organic solvents have some spectral signature using Raman spectroscopy. Water and alcohols are low-scattering materials for Raman emission spectroscopy and thus are useful as solvents. Glass and silica are also useful as containers for measuring solids or neat liquids using Raman spectroscopy, for the Si–O–H is a poor scattering material and thus has little to no Raman signal intensity. Small glass NMR sample vials are often used for Raman measurements and are especially serviceable because they are considered disposable.

II.

SAMPLE IDENTIFICATION AND SPECIALIZED MEASUREMENT TECHNIQUES

3.

INFRARED MICROSPECTROSCOPY

The IR microscope can be used for many samples typically measured using the IR macro bench. It can be faster and easier than using the benchtop system for routine samples due to the lesser requirements for sample preparation. The only restrictions for microscope use is the size and hardness of the sample. Most samples are compatible for measurement using the microscope. A variety of objectives are available for IR measurements using transmittance, diffuse and specular reflectance, reflectance-absorption, ATR (attenuated total reflectance), and grazing angle. Visible objectives are used for finding and aligning the sample into the proper position for measurement. A description of microscope terms and procedures is outlined next. Microscopic measurements are more optimized by purging the measurement area using dry nitrogen to eliminate absorption bands due to carbon dioxide and water vapor. A plastic ring is often employed between the microscope objective and sample stage for purging. Purge gases can also blow the sample out of the field of view.

ALIGNMENT

Alignment is the procedure used to maximize energy throughput and thus to increase the signal-to-noise ratio of the microscope. The noise is measured and recorded over time to ensure optimal microscope performance.

Alignment Procedure for Microscope Optics for Use in Reflectance Measurements

A gold-coated mirror is generally used for microscope alignment in the reflectance mode. Use the upper source (reflectance configuration on single-illumination systems) to focus a bright image on the gold surface. Perform standard noise-performance measurement as outlined next.

Alignment Procedure for Microscope Optics for Use in Transmission Measurements

The alignment procedure for IR microscopes follows three main steps and uses the optical survey mode of the microscope.

1. Focus the objective onto the sample (generally a 100-micron pinhole) by moving the stage in the *x*-, *y*-, and *z*-directions to adjust the sample to the focal spot of the objective. This step is performed with upper illumination (reflectance configuration on single-illumination systems)

and with the upper aperture removed. With proper alignment the pinhole should be exactly on center and appear as a dark spot.
2. Focus the condenser to the focal point of the sample. This step is performed using the lower source (transmittance configuration on single-illumination systems) and with the lower aperture present. No sample or pinhole should be present for this step. Move the condenser focus knob in the z-direction until the brightest possible spot is observed. The condenser thumbscrews are then used to complete the x- and y-adjustments to exactly center the bright spot.
3. Position both the upper and lower apertures in place and align the condenser image on the center focal point of the objective. This step is performed using the 100-micron pinhole in exact center position with both the upper and lower light sources (transmittance configuration on single-illumination systems). Once completed, this configuration should provide the maximum energy throughput for the microscope optics and thus provide the highest signal-to-noise ratio.

APERTURE

The aperture is the opening within the microscope optics that is responsible for producing an optical image and for specifying the area of the sample to be measured using transmittance or reflectance. The aperture is generally made of aluminum or steel coated with carbon black to absorb extraneous IR light. Coated-glass apertures are available to allow visual observation of a sample without allowing IR energy to pass the area around the opening. The size of the aperture should be compatible with the image size generated by the specific objective and condenser optics. Table 3.1 presents the objective or condenser magnification and the corresponding compatible aperture.

Table 3.1 Objective/Condenser Magnification and Compatible Aperture.

Objective/condenser magnification	Compatible aperture
10 times	1000 microns (1.0 mm)
15 times	1500 microns (1.5 mm)
20 times	2000 microns (2.0 mm)
25 times (some ATR)	2500 microns (2.5 mm)
30 times (some grazing angle)	3000 microns (3.0 mm)
32 times	3200 microns (3.2 mm)

COMPENSATION RING

The compensation ring is a focusing mechanism for both the objective and condenser optics that is used to adjust for the refraction of light caused by a specific window thickness over or under the sample. If a window is used over the sample, the compensation ring of the objective is used to adjust for the refraction of light. If a window is placed under the sample, the compensation ring of the condenser is used to adjust for the refraction of light. To calculate the compensation setting required for each window thickness and material, use the following relationship:

$$\text{Compensation setting} = \text{Thickness of window (in mm)} \times \frac{\text{Refractive index of window}}{1.5}$$

COMPRESSION CELL

The compression cell allows two sample windows with flat surfaces to be squeezed together for the flattening of samples to make them transparent or to flatten them for easier transmittance measurement. KBr and diamond are common window materials.

DICHROISM

Dichroism is the use of infrared polarized light (as p- and s-polarized light) to measure the molecular orientation of crystalline polymers and other highly oriented molecules. For microscope work, the requirements are (1) the capability to generate p- and s-polarized infrared energy, and (2) the capability of minimizing the stray radiant energy (stray light) by using redundant apertures (see earlier discussion of this term). The difference between a spectrum of a material taken with p-polarized light ($A_{\parallel}$) and the spectrum of a material illuminated with s-polarized light ($A_{\perp}$) can reveal molecular orientation sensitivities of the measured molecules. When the electronic field vector of the sample molecules is perpendicular to the field direction vector of the infrared energy there is very little interaction and slight or no absorption of the light. When the electronic field vector of the sample molecules is parallel to the field direction vector of the infrared energy there is a large interaction and significant absorption of the light. One can measure the effect of stretching on molecular orientation by making dichroic measurements before and after stretching the sample. The dichroic ratio in (arbitrary units) can be measured from +1.0 to –1.0 by using the following ratio:

$$\text{Dichroic ratio} = \frac{A_{\parallel} - A_{\perp}}{A_{\parallel} + A_{\perp}}$$

GAIN SETTING OF DETECTOR

The gain is the voltage amplification of a detector (above a nominal voltage setting) that has the effect of increasing the signal strength of the detector in a linear fashion over the linear range of a detector. For example, if the signal-to-noise ratio of a measurement A is 50% of a previous measurement B, moving the gain setting to 2 for this more difficult measurement A will yield the same voltage at the detector as measurement B. This gain enhancement will provide increased detector sensitivity for measurements over a limited linear detection range.

LIGHT SOURCES

Microscopes are generally provided with an upper light source and a lower light source. The upper source is used to view the objective image for alignment or for the reflectance measurement mode. The lower source is used to view the condenser image for alignment or for the transmittance mode measurement. Some microscopes have a single visible light source. For aligning this microscope configuration, the reflectance mode is used for objective focusing and the transmittance mode is used for condenser focusing.

MOTORIZED STAGE

The motorized stage allows the microscope to "memorize" the point position (or positions) on a sample and to find a position reproducibly.

NOISE PERFORMANCE MEASUREMENT

The optical performance of the microscope is evaluated by completing the alignment procedure to optimize energy throughput and the signal-to-noise ratio. Once optimized, the noise of the microscope system is recorded for later performance comparisons. The noise and spectrum are recorded over time to verify continuous optical performance of the system. The measurement parameters should be measured in exactly the same manner for each measurement. The number of scans, resolution, apodization function, data format, and frequency range should always be adjusted to the same parameter setup each time a noise measurement is made. The noise should be recorded as the peak-to-peak (maximum percentage transmittance minus minimum percentage transmittance) over a prespecified frequency region of the spectrum. Generally, peak-to-peak noise should be less than 0.1 %T for most IR microscopes. *Note:* Peak-to-peak noise should approximately double when moving from benchtop IR transmittance measurements to microscope IR transmittance measurements. The peak-to-peak noise factor should increase approximately four times when moving from the benchtop IR transmittance measurements to microscope IR reflectance measurements.

OBJECTIVE

The optical device taking the light from the microscope and focusing it onto the sample is called the objective. The higher the magnification, the lower the throughput. A 15× objective is a general-use objective. Higher magnification allows more specific sample position measurements, but reduces the energy throughput and signal-to-noise ratio. Types of objectives include standard, specular reflectance, ATR, and grazing-angle.

ATR Objective

ATR is used for surface analysis using physical contact with a sample surface. The objective utilizes a crystal of material for the actual physical contact with the sample. Typical crystal materials include diamond (Di), germanium (Ge), silicon (Si), and zinc selenide (ZnSe). ATR measurements can yield excellent-quality spectra provided that the contact pressures of the ATR crystal and the sample are held constant; reproducible data requires reproducible contact pressure. Commercial contact gauges are available from suppliers of microscope accessories. One percent reproducibility is typically achieved when keeping the contact pressure constant. ATR crystals are generally ZnSe with a refractive index of 2.42 and an angle of incidence of 45°. Sample penetration is around 40 microns using this ATR crystal. An air background is taken for ATR measurements. The depth of penetration is calculated for ATR crystals using the following relationship:

$$\text{Depth of penetration for ATR} = \frac{\text{Wavelength of incident light}}{2\pi \times n_1 (\sin^2 \alpha - n_2^2)^{1/2}}$$

Where n_1 = refractive index of the ATR crystal; refractive indices for typical ATR crystals are: ZnSe (2.42), Di (2.2), Si (3.6), Ge (4.0)
n_2 = refractive index of the sample
α = angle of incidence (and reflection) of the ATR crystal, typically 45°

Grazing-Angle Objective

A grazing-angle objective involves a measurement configuration where light from the objective strikes the sample at a high angle of incidence for measuring the coating or surface characteristics of a sample. This measurement technique has a sample penetration of typically less than 1 micron. This technique can be used to measure very thin surface characteristics of reflective samples. The technique is qualitative and is not particularly reproducible and thus not useful for quantitative analysis. The grazing-angle objective typically is configured for a 65–80° angle of incidence. Apertures are not generally used when using grazing-angle measurements; there is a fivefold pathlength increase when using grazing-angle as compared to standard transmittance measurement geometry.

Visible Objective

A visible microscope objective is for use in a visual survey of the sample for the purpose of inspection, sample positioning, and alignment. Crosshair features are often available for measuring sample distances, sample size, and sample alignment.

REDUNDANT APERTURES

A pair of apertures can reduce the diffraction of light caused when the light passing through an aperture is of a wavelength close to the size of the aperture. A second aperture matched to the first aperture can reduce or eliminate the spurious energy due to diffraction. Use of a second aperture (termed the *redundant aperture*) allows a single sample layer less than 50 microns to be measured on a routine basis.

REFRACTION OF LIGHT

Refraction is the change in direction of light due to a change in the velocity of the light while passing through two or more materials of differing refractive indices. In microscope systems this phenomenon causes the offset of the light based on the refractive index and thickness of the sample and sample accessory window material(s). The refracted light must be adjusted for by using a compensation ring, which is a feature of both the objective and condenser optics.

SPURIOUS ENERGY

Spurious energy in a microscope system is unfocused light caused by the diffraction of IR energy passing through an aperture that is nearly the same size as the wavelength of the IR energy. If the wavelength of light is longer than the size of an aperture, the incident light will be reflected back. The closer the wavelength is to the size of the aperture, the greater is the percentage of light diffracted. Typically for microscopes, apertures of less than 50–60 microns are too small and create diffraction problems. The use of multiple apertures can reduce the diffraction of light passing through the microscope optics.

STRAY LIGHT

Stray light is dealt with in other sections of this text; however, the main source of stray light (also termed *stray radiant energy*) within the IR microscope is the spurious energy caused by diffraction of infrared energy through small apertures.

WINDOW MATERIALS FOR MICROSCOPY

Typical window materials for microscopy include: barium fluoride (BaF_2) for use with polar solvents (including water); potassium bromide (KBr) for solids and nonpolar solvent use; zinc selenide (ZnSe) with its high refractive index for use in diamond cell background measurement; and diamond for compression cell work, where higher pressures are required.

4.

DICHROIC MEASUREMENTS OF POLYMER FILMS USING INFRARED SPECTROMETRY

INTRODUCTION

Infrared dichroism is the use of infrared polarized light as p- (90°) and s- (0°) polarized light to measure the molecular orientation of polymer films and other highly oriented molecules. The capability to generate p- and s-polarized infrared energy is required to complete these measurements. The difference between a spectrum of a polymer film taken with s-polarized light ($A_{\parallel}$) and the spectrum of a film illuminated with p-polarized light ($A_{\perp}$) will reveal the molecular orientation of the polymer backbone and attached side groups. When the electric field vector of the infrared-active molecule (dipole) is perpendicular to the field direction vector of the infrared energy there is no interaction of the incident infrared energy with the dipole and thus no infrared absorption. When the electronic field vector of the infrared-active molecule (dipole) is parallel to the field direction vector of the infrared energy there is a large interaction and significant infrared absorption. The molecular orientation of the various molecules in a polymer film can be measured by making dichroic (s- and p-polarized light) measurements before and after stretching of the film sample.

THEORY

The intensity of an infrared absorption band is proportional to the square of the transition moment (or infrared-active dipole moment). The absolute intensity of an infrared band also depends upon the direction of the transition moment (dipole electric field vector) and the field direction vector (electric field vector) of the incident infrared radiation. The proportion of the transition moment (TM_p) in the direction of the infrared electric field direction vector (**E**) is given as

$$TM_p = tm \times \cos \beta$$

noting that $(tm \times \cos \beta)^2$ is proportional to the infrared absorbance, that tm is the transition moment of the molecule (dipole) of interest, and that β is the included angle. An illustration of the dipole electric field vector and the electric field vector of the s-polarized infrared radiation (with respect to the angle of polymer stretching) is illustrated in Figures 4.1 and 4.2.

Polymer films may be stretched, resulting in uniaxial elongation (orientation along the stretching axis). During the process of stretching, the polymer backbone will align in a parallel manner along the direction of stretching. However, under the stretching conditions, the attached groups

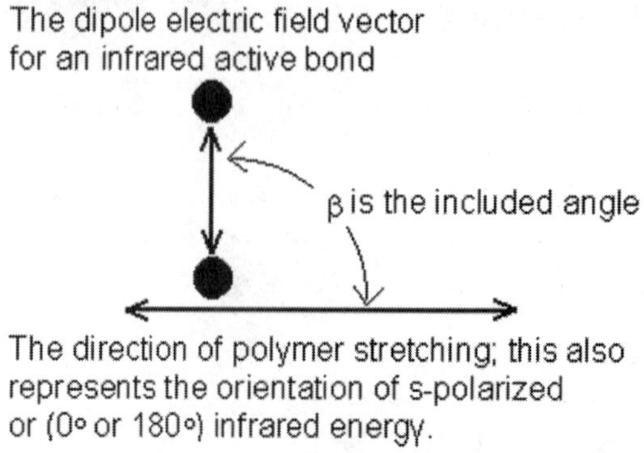

Fig. 4.1 Illustration of the bond angle between the dipole electric field vector and the direction of polymer stretching.

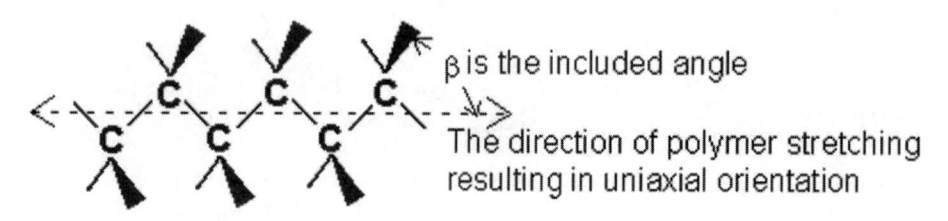

Fig. 4.2 Illustration of Fig. 4.1 for a molecular example.

may assume no preferred orientation with respect to the direction of stretching. For polymer films the direction of stretch is generally chosen as the horizontal (↔) direction. When this stretching direction is chosen, the s-polarized or (0° or 180°) infrared energy is designated as the infrared polarizer orientation for measuring $A_{\parallel}$ (absorbance spectrum representing parallel polarized light). Likewise, the p-polarized or (90° or 270°) IR energy is the orientation for measuring $A_{\perp}$ (absorbance spectrum representing perpendicularly polarized light) [1].

METHODS OF CALCULATING DICHROIC PARAMETERS

The dichroic calculations (in arbitrary units) can be reported using the following relationships for polymer films:

$$\text{Dichroic ratio } (R) = \frac{A_{\parallel}}{A_{\perp}} \qquad (1)$$

$$\text{Dichroic difference } (\Delta A) = A_{\parallel} - A_{\perp} \qquad (2)$$

$$\text{Dichroic difference ratio } (\Delta A_r) = \frac{A_{\parallel} - A_{\perp}}{A_{\parallel} + A_{\perp}} \text{ (from } +1.0 \text{ to } -1.0) \qquad (3)$$

$$\text{Polymer orientation ratio } (P_r) = K_1 \frac{R-1}{R+2} \qquad (4)$$

Where $K_1 = 2/(3\cos^2 \beta^{-1})$ and β = the bond moment angle (in degrees) between the absorption band of interest and the polymer backbone. *Note:* if β is unknown, replace constant K_1 with the value 1.0 for calculation of P_r. *Note:* R is the Dichroic Ratio.

Calculation of Bond Moment Angles (β) [1]

$$\text{Dichroic ratio } (R) = \frac{A_\parallel}{A_\perp} = 2\left(\frac{\cos^2 b}{\sin^2 b}\right) = 2\cot^2 b \tag{5}$$

Thus by using Eq. (5), the values for R at any angle β can be calculated, where β is the included angle or the infrared dipole angle relative to the direction of stretch (i.e., s-polarized orientation). The angle β and the corresponding R values are shown in Table 2.2.

Table 2.2 Dichroic Ratio Versus Infrared Dipole Angle Relative to Direction of Stretch.

β (infrared dipole angle relative to direction of stretch)	R (dichroic ratio)
90°	0.000
80°	0.062
70°	0.265
60°	0.667
54°44'	1.000
50°	1.407
40°	2.843
30°	3.000
20°	15.094
10°	64.667
$\beta \to 0°$	$R \to \infty$

GRAPHICAL REPRESENTATION

The polymer stretching ratio (α) is plotted on the abscissa versus the (P_r) polymer orientation ratio as the ordinate. The polymer stretching ratio (α) is given by $\alpha = L_s/L_o$ = Stretched length/Initial length. The initial length should be set to approximately 1.0 cm by making two marks 1.0 cm apart directly on the polymer area to be stretched. The value for the polymer stretching ratio (α) for no stretching should be 1.0.

METHOD

Infrared Polarizers

Two basic types of infrared polarizers are commercially produced. These include the Brewster angle cross-plate polarizers and the wire grid polarizers. The Brewster angle cross-plate polarizers use a high-refractive-index material (such as germanium) and have a useful transmission range of 5000–500 cm^{-1}. The wire grid polarizers consist of 0.2-micron-wide aluminum strips separated by a space of approximately 0.4 microns. The matrix for placing the aluminum strips is often KRS-5 with a useful working range of 5000–285 cm^{-1} [2].

All polarizers are placed within the sample compartment directly in the infrared beam. They are mounted for complete 360° rotation. At 0° or 180° rotational settings the transmitted electric field vector (**E**) of the infrared energy is horizontal (s-polarized). This represents $A_\parallel$, or the absorbance spectrum for parallel polarized light. Likewise, at 90° or 270° rotational settings the transmitted electric field vector (**E**) of the infrared energy is vertical (p-polarized). This represents $A_\perp$, or the absorbance spectrum for perpendicularly polarized light.

Polymer Stretching

If an automated polymer stretcher is available, this is used per manufacturer's instructions. If no special apparatus is available, the following manual technique is used. A manual device is constructed for stretching the polymer film over a small area for measurement. The polymer films are cut in strips approximately 4 cm (width) by 6 cm (length). Each side of the strip is mounted, using standard adhesive tape, to two metal pieces (approximately 4 cm by 7 cm). The polymer is measured at initial length (without stretching) using both 0° ($A_{||}$) and 90° ($A_{\perp}$) polarizer settings. Then, by gently stretching the polymer film in approximately 2–3-mm increments, the sample is repeatedly measured for each length change using both 0° ($A_{||}$) and 90° ($A_{\perp}$) polarizer settings. The stretch length change is determined by marking the film at initial length with two marks approximately 1 cm apart. These marks are best made using a fine-point permanent marker and are manually measured as each 2–3 mm stretch is completed. The spectral measurements are best made using a standard infrared film holder during measurement.

Infrared Measurements and Reporting

Measurements are made as 0° ($A_{||}$) and 90° ($A_{\perp}$) starting at $\alpha = 1.0$ (initial) and at 5 or more increasing stretch lengths (L_s). Select the functional group bands of interest on the infrared spectrum (*Note:* The absorbance of the band should be less than or equal to 0.2A and the bandwidth at half-height should be less than or equal to 30 cm^{-1}. This ensures the highest linearity and photometric accuracy of the infrared instrument). The absolute absorbance values are measured for these selected infrared bands for both the 0° ($A_{||}$) and 90° ($A_{\perp}$) spectra. When these results are tabulated, calculate and plot P_r (ordinate) versus α (abscissa). Report final dichroic results in tabular form as R, ΔA, ΔA_r, and P_r for all six measurements for two bands per sample following the equations given in the previous equations.

REFERENCES

1. N. B. Colthup, L. H. Daly, and S. E. Wiberley. *Introduction to Infrared and Raman Spectroscopy.* San Diego, CA: Academic Press, 1990, p. 99.
2. J. Harric. Harrick Scientific Corp., Ossining, NY, 1995.

5.

CHEMICAL IDENTIFICATION AND CLASSIFICATION TESTS

INTRODUCTION

Identification tests are used to supplement spectroscopic information to make positive identification of materials as belonging to specific classes of compounds. Confirmatory tests are useful to ensure accuracy when identifying materials.

The alkanes, alkenes, alkynes, and benzenoid aromatic compounds, aldehydes, ketones, amines, carbohydrates, alcohols, and halogen organics can be identified using standard chemical tests. For each of these tests, special emphasis is placed on the reactivity or nonreactivity of each of these different hydrocarbon classes. As a result of these classification tests and the determination of certain physical constants (i.e., melting point, boiling point, density, and refractive index) the chemist is able to identify unknown compounds with reasonable certainty even without additional sophisticated analytical instrumentation.

REAGENTS

The following reagents are required for the testing of each compound class mentioned.

Tests for aromaticity
1. Concentrated sulfuric acid
2. Fuming sulfuric acid
3. Chloroform and aluminum chloride

Tests for unsaturation
4. Bromine solution (in water or chloroform)
5. 2% Potassium permanganate in water

Test for terminal C≡C triple bond
6. Copper ammonium chloride

Tests for aldehydes and ketones
7. 2, 4-Dinitrophenylhydrazine
8. Tollen's reagent
9. Sodium bisulfite reagent

Tests for amines
10. Dilute (0.1 N) Hydrochloric acid
11. Hinsberg reaction (benzenesulfonyl chloride)

12. Nitrous acid
13. Bromine (5%) in water

Tests for carbohydrates
14. Fehling's reagent

Tests for alcohols
15. Lucas' test reagent

GENERALIZED TEST PROCEDURE

A few milliliters (1–2) of the reagent is placed in a large-mouth test tube (i.e., 7–10 cm) and the test material is added dropwise (1–5 drops). Careful observation of the results is made using safety glasses and appropriate safety protection. *Note:* Refer to each test for specific procedures.

SAFETY ISSUES

Individuals performing these tests must be adequately trained in the use, storage, and disposal of reagent chemicals. DANGER! Some reagents are poisonous and react violently with water. Refer to an appropriately trained chemical safety specialist for safety guidelines. The chemist performing these tests must be cautioned to use appropriate safety protection and to comply with all safe practices in the use of chemicals. This test document does not purport to address all of the safety concerns, if any, associated with its use. It is the responsibility of the user of this method to establish appropriate safety and health practices and to determine the applicability of regulatory limitations prior to use. Appropriate professional training is essential for the proper use of these procedures. Refer to local, state, and federal laws affecting the proper use and disposal of chemicals.

TESTS FOR AROMATICITY

1. Concentrated Sulfuric acid

Procedure

To 40–60 drops (2–3 mL) of concentrated sulfuric acid add 3–4 drops of test solution.

> *Insoluble or no reaction* (i.e., no heat liberated) indicates an alkane, benzene, monoalkyl-aromatic, or possibly a dialkyl-aromatic.
>
> *Soluble or reaction* (i.e., heat liberated) indicates polyalkylaromatic, alkene, cyclopropane, or alkynes.

Examples
a. Toluene yields no reaction.
b. 2-Pentene forms a clear yellow liquid and evolves heat.
c. *p*-Xylene yields no reaction.
d. Styrene forms a clear yellow liquid and evolves heat.
e. Mesitylene forms a clear yellow liquid and evolves heat.
f. Napthalene yields no reaction.

2. Fuming Sulfuric acid

Procedure

To 40 drops (2 mL) of fuming sulfuric acid add 10 drops (0.5 mL) of test solution.

Soluble indicates aromatic compound.

Insoluble or reaction (i.e., heat liberated) indicates aliphatic compound.

Examples

a. Aalkanes (C21–C40) yield no reaction.
b. Cyclohexane decomposes into a yellow precipitate in solution.
c. Benzene forms a clear brown solution.
d. Napthalenes form a cloudy white solution.

3. Chloroform and Aluminum chloride

Procedure

To 2 drops (0.1 mL) of test solution add 40 drops (2 mL) chloroform. Following vigorous agitation, add 0.5 g of dry, anhydrous aluminum chloride. Add the aluminum chloride to the test tube so that some of the solid goes into the chloroform and some remains along the top of the liquid line in the test tube.

Reaction indicates an aromatic system that does not have any strong deactivating groups (for example, $-NO_2$, $-CXO$, $-CN$)

Examples

a. Benzene yields an orange or red solution.
b. Naphthalene yields a blue solution.
c. Biphenyl yields a purple solution.
d. Anthracene yields a green solution.
e. Phenanthrene yields a purple solution.
f. Cyclohexane yields no color reaction.
g. Cyclohexene yields no color reaction but gives a brown precipitate.
h. Toluene yields an orange solution.
i. Styrene yields an orange solution.

TESTS FOR UNSATURATION

4. Bromine (5%) Solution (in Water or Chloroform)

Procedure

To 4 drops (0.2 mL) of the test solution add 40 drops (2 mL) of chloroform. Add 5% bromine in chloroform, drop by drop, carefully shaking the test tube between each drop. If two or more drops are decolorized in less than 1 minute, the test is positive.

Reaction indicates unsaturated carbons (i.e., alkenes or alkynes); aromatics *do not* react. The brown bromine solution changes to colored or colorless quite dramatically.

No reaction indicates saturated carbons (i.e., alkanes).

Examples

a. Cyclohexene yields a colorless solution.
b. Styrene yields a yellow solution.
c. Toluene yields no reaction.

CHEMICAL IDENTIFICATION AND CLASSIFICATION TESTS

5. Potassium Permangenate (2%) in Water

Procedure

To 2 drops (0.1 mL) of test solution add 40 drops (2mL) ethanol. Add 2% potassium permangenate solution in water, drop by drop, carefully shaking the test tube between drops. A positive test is indicated if the purple permangenate color disappears rapidly after the addition of 2–3 drops of the 2% permangenate reagent.

> *Reaction* indicates unsaturated carbons (i.e., alkenes or alkynes); aromatics *do not* react. A positive reaction is also indicated if the purple solution changes to a brown precipitate quite dramatically.
>
> *No reaction* indicates saturated carbons (i.e., alkanes).

Examples

a. Benzene yields no reaction.
b. Cyclohexene yields a brown precipitate.
c. Toluene yields no reaction.
d. Terpentine yields a brown precipitate.
e. Naphthalene yields no reaction.

TEST FOR TERMINAL C≡C TRIPLE BOND

6. Copper Ammonium Chloride

Reagent Recipe

Dissolve 0.3 g sodium hydrogen sulfite in 60 drops (3 mL) of 5% aqueous sodium hydroxide. In a separate vessel, dissolve 1.2 g $CuSO_4 \cdot 5H_2O$ and 0.3 g NaCl in 4 mL of warm deionized water. With careful stirring, slowly add the bisulfate solution to the copper sulfate solution. Cool to room temperature, decant the supernatant, and wash the copper chloride precipitate with 2-mL portions of 5% aqueous sodium hydroxide solution. Dissolve the precipitate in a mixture of 3 mL water and 4 mL concentrated ammonium hydroxide. Keep this solution tightly stoppered so as to prevent air oxidation.

Procedure

For the test liquid add 2 drops (0.1 mL) of unknown into 1 mL of ethanol and 1 mL of ammoniacal cuprous chloride reagent. Formation of a green precipitate indicates a positive test.

> *Reaction* forming a green precipitate indicates a terminal **C≡C**.
>
> *No reaction* indicates the absence of such a bond.

Examples

a. 1-Hexyne yields a green precipitate.
b. Cyclohexene yields no reaction.
c. Phenylacetylene yields a green precipitate.
d. Styrene yields no reaction.

When the results of these tests are combined with boiling point and/or refractive index, many organic liquids can be positively identified.

TESTS FOR ALDEHYDES AND KETONES

7. 2, 4-Dinitrophenylhydrazine

Reagent Recipe

Add 1 mL of concentrated sulfuric acid to 0.2 g of 2, 4-dinitrophenylhydrazine. Water (2 mL) is added dropwise with mixing. To the warm solution, add 5 mL of ethanol (95%).

Procedure

Add 3 drops of the unknown liquid to 3 mL of 95% ethanol. Add this combined solution to the reagent mixture. Agitate, and observe for the following.

> *Reaction* forming a bright red or orange precipitate indicates that either functional group is present.
>
> *No reaction* indicates that no aldehyde or ketone groups are present.

Examples
a. Diethylketone yields a red-orange precipitate.
b. Benzaldehyde yields a red-orange precipitate.
c. Toluene yields no reaction.
d. Acetone yields a yellow precipitate

8. Tollen's Reagent

Reagent Recipe

Add 4 mL of a 5% silver nitrate solution to 2 drops of a 10% sodium hydroxide solution. To this mixture add a 2% solution of ammonia dropwise, with constant agitation, until the silver oxide precipitate dissolves. An excess of ammonia is to be avoided to retain the sensitivity of the test. The reagent should be prepared just prior to testing.

Procedure

In a very clean test tube add the Tollen's reagent and 3–4 drops of the test liquid, and warm the mixture gently. A mirrored surface indicates an aldehyde. If the inside of the tube is not clean, a precipitate will result, also indicating the presence of the aldehyde.

> *Reaction* forming a white silver precipitate indicates the presence of an aldehyde but not ketone functional group(s).
>
> *No reaction* indicates that no aldehyde groups are present.

Examples
a. Benzaldehyde yields a white precipitate.
b. Diethylketone yields no reaction.
c. Formaldehyde yields a mirrored surface or precipitate.

9. Sodium Bisulfite Reagent

Reagent Recipe

Add 5 mL of ethanol (95%) to 10 mL of a 40% aqueous sodium bisulfite solution. Decant the liquid from a small quantity of precipitate that forms, and use the liquid for the procedure.

Procedure

Add 5 drops of the test liquid to 1 mL of the sodium bisulfite reagent. Stopper the test tube (container) and shake vigorously.

Reaction liberating heat indicates that either functional group (aldehyde or ketone) is present. *Note:* Some methyl ketones and cyclic ketones yield a positive test. Aryl ketones and sterically hindered ketones do not react.

No reaction indicates that no aldehydes, methyl ketones, or cyclic ketones are present

Examples
a. Diethylketone yields no reaction.
b. Benzaldehyde yields heat.
c. Toluene yields no reaction.
d. Methylpropylketone yields heat.

TESTS FOR AMINES

10. Dilute (0.1 N) Hydrochloric Acid

Reaction forming amine hydrochloride salt occurs—sample is soluble.

No reaction indicates that no amine groups are present—sample remains insoluble.

Examples
a. Methylamine yields methylamine hydrochloride salt.
b. Aniline yields aniline hydrochloride salt.
c. Ethylamine yields ethylamine hydrochloride salt.

11. Hinsberg Reaction (Benzenesulfonyl chloride)

Reagent Recipe

To 0.6 mL of aniline add 5 mL of 10% sodium hydroxide solution and 0.8 mL of benzenesulfonyl chloride. This mixture is added to a test tube, stoppered, and shaken vigorously. The mixture should be maintained below 20°C to prevent the formation of a purple dye. Make certain the solution is basic and that any residue or solid material is filtered from the liquid and the liquid is retained.

Reaction of primary or secondary amines form N-substituted benzenesulfonamide precipitate.

No reaction indicates that no primary or secondary amine groups are present. *Note:* Tertiary amines do not react to form precipitate.

Examples
a. Methylamine (primary amine) yields N-substituted benzenesulfonamides precipitate.
b. Aniline (primary amine) yields N-substituted benzenesulfonamides precipitate.
c. Ethylamine (primary amine) yields N-substituted benzenesulfonamides precipitate.
d. Secondary amines yield N-substituted benzenesulfonamides precipitate.
e. Tertiary amines yield no reaction.

12. Nitrous Acid

Perform the test reaction at 0°C.

> *Reaction* of aromatic primary amines form brilliant orange aniline dyes; primary aliphatic amines liberate nitrogen gas bubbles. Secondary aliphatic and aromatic amines yield yellow N-nitroso derivatives.
>
> *No reaction* indicates that no primary amine groups are present. *Note:* Secondary and tertiary amines do not react to form highly colored dyes.

Examples
a. Primary aromatic amines form brilliant orange aniline dyes.
b. Primary aliphatic amines liberate nitrogen gas bubbles.
c. Secondary aliphatic and aromatic amines yield yellow N-nitroso derivatives.
d. Tertiary amines do not react.

13. Bromine (5%) in Water

> *Reaction* of aromatic amines removes the color from bromine water, yielding a yellow precipitate. Aromatic amines are oxidized to benzonitriles.
>
> *No reaction* indicates that no amine groups are present. *Note:* Phenols can also react with bromine water, and this test should not be used to distinguish phenols from amines

Examples
a. Aromatic amines remove the color from bromine water, yielding a yellow precipitate.
b. Benzylamine removes the color from bromine water, yielding a yellow precipitate.
c. Phenol removes the color from bromine water, yielding a yellow precipitate.

TESTS FOR CARBOHYDRATES

14. Fehling's Reagent

Recipe

Fehling's reagent is made by mixing equal volumes of each of the following solutions:

a. Copper sulfate solution: 6.9 g of copper sulfate (hydrated) in 100 mL of water
b. 34.6 g of sodium potassium tartrate plus 14 g of sodium hydroxide pellets in 100 mL water

Procedure

Add 0.1 g of test material to 3 mL water and 3 mL Fehling's reagent. Heat the mixture to boiling. Observe the results as the liquid cools.

> *Reaction* of all monosaccharides and most disaccharides with Fehling's reagent causes a red precipitate of cuprous oxide to form.
>
> *No reaction* indicates that no monosaccharide or disaccharide groups are present.

Examples
a. D-Fructose causes a red precipitate of cuprous oxide to form.
b. D-Glucose causes a red precipitate of cuprous oxide to form.
c. Starch yields no reaction.

TESTS FOR ALCOHOLS

15. Lucas' Test Reagent

Reagent Recipe

To 30 g of concentrated hydrochloric acid, dissolve 37.5 g of anhydrous (fused) zinc chloride. Stir gently in a water bath at near 0°C to avoid loss of HCl. This volume of material (approximately 39 mL) is sufficient for approximately 10–12 tests.

Procedure

To 0.5 mL of the test alcohol add 3 mL of the Lucas' reagent at 25–28°C. Add the mixture to a test tube, stopper, and shake vigorously for several seconds, and then allow it to stand. Observe immediately, in 5 minutes, and in 1 hour.

> *Reaction* of tertiary alcohols is at once, yielding a milky white precipitate of alkyl chloride. Secondary alcohols react within 5 minutes to give the same result, with a second layer separating in the test tube after several minutes. Primary alcohols do not react at near room temperature and the solution remains clear. Separation of a secondary layer of a milky white precipitate of alkyl chloride occurs only in the presence of secondary or tertiary alcohols.
>
> *No reaction* indicates that no secondary or tertiary alcohol groups are present.

Examples
a. Ethanol yields no reaction.
b. Methanol yields no reaction.
c. Isobutyl alcohol does react.
d. 2-methyl-1-propanol does react.

REFERENCES

1. James S. Swinehart. *Organic Chemistry and Experimental Approach.* Englewood Cliffs, NJ: Prentice-Hall, 1969.
2. Ervin Jungreis. *Spot Test Analysis: Clinical, Environmental, Forensic, and Geochemical Applications.* New York: Wiley, 1985.
3. Ralph L. Shriner, Reynold C. Fuson, David Y. Curtin. *The Systematic Identification of Organic Compounds: A Laboratory Manual.* New York: Wiley, 1948.
4. Emile M. Chamot, Clyde W. Mason. *Handbook of Chemical Microscopy.* New York: Wiley, 1931. *Note:* This reference is most useful for inorganic cation and anion identification.
5. Roger Adams, John R. Johnson, Charles F. Wilcox, Jr. *Laboratory Experiments in Organic Chemistry.* New York: Macmillan, 1967.

III.

UV-VIS SPECTROSCOPY

6.

COMPARING ULTRAVIOLET-VISIBLE AND NEAR-INFRARED SPECTROMETRY

INTRODUCTION

Ultraviolet (190–380 nm), visible (380–750 nm), short-wave near-infrared (750–1100 nm), long-wave near-infrared (1100–2500 nm), infrared (4000–400 cm^{-1}), and Raman spectroscopy comprise the bulk of the electronic and vibrational mode measurement techniques. The measurement modes for UV-Vis-NIR spectroscopy are given in Table 6.1.

Basic spectroscopic measurements involve the instrumental concepts of bandpass and resolution, signal-to-noise ratio, dynamic range, stray light, wavelength accuracy and precision, and photometric accuracy and precision. These concepts are described in Chapter 1.

INSTRUMENTATION

Optical configurations

The basic optical configuration for UV-Vis-NIR instrumentation is shown in the schematic diagrams in Figures 6.1 through 6.5. Double-monochromator instruments (dispersive, Figure 6.1) provide a traditional means for reducing the stray light of an instrument at the point of measurement. Double-monochromator instruments are generally used for applications where high degrees of photometric accuracy and repeatability are required. Optical components with strict specifications are often tested using double monochromator systems. These systems demonstrate extremely low-stray-light specifications on the order of 0.0001% *T* or better.

Table 6.1   Basic UV-Vis-NIR Measurement Modes.

Instrument measurement mode	Description of measurement
Scan	Absorbance vs. wavelength
Time drive	Absorbance vs. time at a specific wavelength
Individual Wavelength(s)	Individual absorbance(s) at selected wavelength(s)
Chemometrics and Quantitative methods	Concentration of analyte vs. absorbance
Kinetics	Kinetic rates of reaction

COMPARING ULTRAVIOLET-VISIBLE AND NEAR-INFRARED SPECTROMETRY

Single monochromators (dispersive, Figure 6.2) are lower cost than double-monochromator systems and are generally used as workhorses within the laboratory. These systems typically meet the basic requirements for routine quantitative and qualitative work for relatively undemanding applications. The dynamic range of these systems is stray light limited.

Diode array detection (dispersive, Figure 6.3) offers the advantage of the absence of moving parts extending the longevity and reliability of such systems over more traditional spectropho-

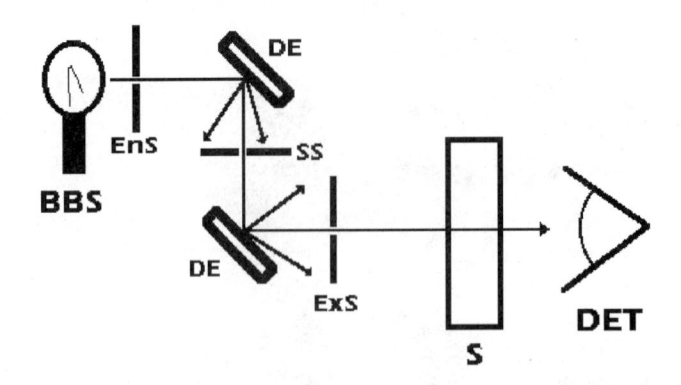

Fig. 6.1 Double-monochromator System (dispersive) optical configuration.

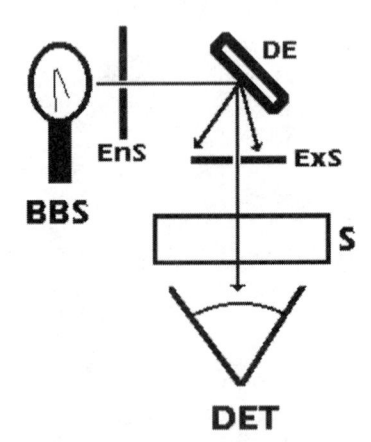

Fig. 6.2 Single-monochromator system (dispersive) optical configuration.

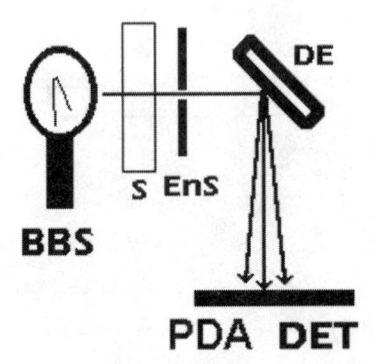

Fig. 6.3 Diode array spectrophotometer (dispersive) optical configuration.

tometers. The main advantage of these instruments is the rapidity with which data can be collected (e.g., in milliseconds) versus the traditional scanning instrument, which makes spectral measurements in seconds to minutes.

Interferometers (Figure 6.4) provide the main optical element for Fourier transform infrared spectrophotometry. Interferometer-based Fourier transform spectrophotometers are extremely accurate in the frequency scale, but have sometimes lacked precision in the photometric domain when compared to their older counterparts, the dispersive-based instruments. However, interferometry is an extremely accurate means for measuring spectra with respect to frequency-dependent measurements.

Interference filter photometers (dispersive, Figure 6.5) provide a low-cost, rugged alternative to grating- or interferometer-based instrumentation. These instruments typically contain from 5 to 40 interference filters that select the proper wavelengths for quantitative analysis based on previous work with scanning instruments, or based on theoretical positions for absorption.

Light-emitting diodes (LEDs) provide emission of prespecified wavelengths of light. Interference filters are sometimes used to further reduce the bandwidth of the light used for analysis.

Basic instrument configurations are shown in Figures 6.1 to 6.5. The key for abbreviations used in the figures are: BBS (broad band source), EnS (entrance slit), DE (dispersive-element,

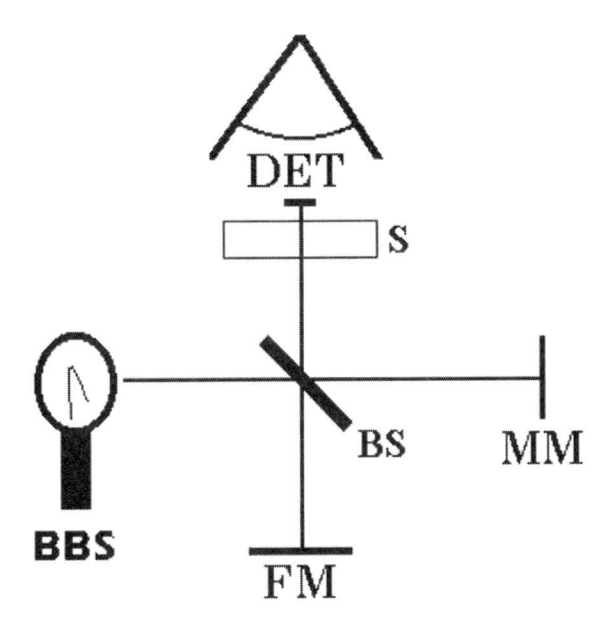

Fig. 6.4 Fourier transform spectrophotometer (interferometer) optical configuration.

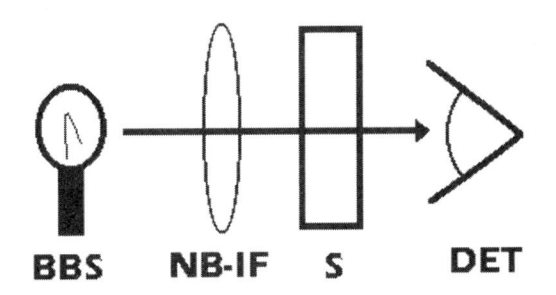

Fig. 6.5 Interference filter photometer optical configuration.

grating, or prism type), SS (second slit), ExS (exit slit), S (sample), DET (detector), PDA DET (photodiode array detector), bs (beam splitter), fm (fixed mirror), mm (movable mirror), NB-IF (narrow bandpass interference filter).

Typical Lamp Sources (Useful Working Ranges)

Most instrumentation used to measure the region from 190 to 2500 nm utilizes a combination of the quartz tungsten-halogen lamp for the visible and near-infrared regions (approximately 350 to 2500 nm), and the DC deuterium lamp for the ultraviolet region (from 190 to 350 nm). The useful working ranges for the most common sources are as follows:

Quartz tungsten-halogen filament Lamp (220–2700 nm)

DC deuterium lamp for UV (185–375 nm)

Pulsed xenon arc lamp (180–2500 nm)

DC arc lamp (200–2500 nm)

Unusual Sources

Unusual sources can be found when making measurements within the ultraviolet or near-infrared region. These can include lasers, such as the nitrogen laser (337.1 nm), and a variety of dye lasers (350–750 nm).

Calibration Lamps (with Emission Line Locations in Nanometers)

Calibration lamps are used to check the wavelength accuracy for ultraviolet and visible spectrophotometers. The main lamps used include the following two:

Mercury (argon) lamp (253.7, 302.2, 312.6, 334.0, 365.0, 404.7, 435.8, 546.1, 577.0, 579.0)

Mercury (neon) lamp (339.3, 585.2, 793.7, 812.9, 826.7, 837.8, 864.7, 877.2, 914.9, 932.7, 953.4)

Detectors (Useful Working Ranges)

A variety of detectors are available for UV-Vis-NIR measurements. High-performance instruments utilize photomultiplier tube technology from the ultraviolet into the visible region. Lead sulfide is the detector of choice for near-infrared measurements. The more common detectors follow, with their useful operating ranges indicated. More information on detector performance is given in Chapter 1 on Optical Spectrometers.

Silicon photodiode (350–1100 nm)

Photomultiplier tubes (PMTs) (160–1100 nm)*

PbS (lead sulfide) (1000–3000 nm)

CCDs (charge-coupled devices) (180–1100 nm)

Photodiode arrays (silicon-based PDAs) (180–1100 nm)

InGaAs (Indium gallium arsenide) (800–1700 nm)

Window and Cuvet Materials

A variety of window materials are available for sample cells and optical elements within spectrophotometers. These materials are selected for their optical clarity for use in specific wave-

---

*Total detection range using PMT technology.

length regions as well as their strength and cost characteristics. Table 6.2 lists data for the most common materials used for sample cells and optical components in UV-Vis-NIR instrumentation.

Fiber Optic Materials (Useful Working Range)

New core and cladding materials are becoming available for use in fiber-optic construction. The most common materials in current use include those listed here. The useful working range for typical samples of these materials is given in parentheses.

SiO_2 (pure fused silica or quartz) (0.2–1.25 microns)

Anhydrous quartz (0.4–2.64 microns)

ZrF (zirconium fluoride) (0.9–4.76 microns)

Chalcogenide (2.22–11.1 microns) for use in NIR-IR measurements

Methods for Testing UV/Vis Instrumentation

Linearity checks are performed by using three neutral-density glass filters available from NIST as SRM 930D. These glasses have nominal percentage transmittance at 10, 20, and 30%. Solutions of nickel and cobalt in nitric and perchloric acids are available as SRM 931. Metal on quartz transmittance standards are available as SRM 2031 with nominal percentage transmittance at 10, 30, and 90%.

Photometric accuracy is measured for UV using SRM 935 consisting of a solution of potassium dichromate in perchloric acid. An additional material consisting of potassium acid phthalate in perchloric acid is available as SRM 84.

Wavelength accuracy is measured using ASTM Practice E 275–83, Practice for Describing and Measuring the Performance of Ultraviolet, Visible, and Near-Infrared Spectrophotometers.

Table 6.2 Characteristics of Window/Cuvet Materials.

Optical material	Transmittance range (in nm)	Refractive index at 600 nm	Relative rupture strength (sapphire = 100)
Methacrylate	250–1100	—	—
UV-Grade fused silica	200–2500	1.4580	10.9
Synthetic fused silica	230–2500	1.4580	10.9
Crystalline quartz (SiO_2)	240–2500	1.5437	2.3
Quartz, extremely low O-H	190–2500	1.5437	2.3
Flint glass (SF 10)	380–2350	1.7268	3.8
Flint glass (SF 8)	355–2350	1.6878	3.8
BK 7 glass	315–2350	1.5165	3.7
Optical crown glass	320–2300	1.5226	3.7
Borosilicate crown glass	360–2350	1.4736	3.7
Pyrex®	360–2350	1.4736	3.8
Tempax®	360–2350	1.4736	3.8
Sapphire (Al_2O_3)	150–5000	1.7677	100
Sodium chloride	250 nm–16 μm	1.5400	0.5
Suprasil 300®	190–3600	1.54	3.8
Diamond	220–4000	2.40	83.7

A second method used is E 958–83, Practice for Measuring Practical Spectral Bandwidth of Ultraviolet-Visible Spectrophotometers. Potassium dichromate in perchloric acid at pH 2.9 exhibits known maxima at 257 and 350 nm, with minima at 235 and 313 nm. Samarium perchlorate can be used for the wavelength region 225–520 nm with excellent results. Holmium oxide glass filters exhibit bands at 279.3, 287.6, 333.8, 360.8, 385.8, 418.5, 446.0, 453.4, 536.4, and 637.5 nm. In addition, the holmium glass exhibits bands within the 750–1200-nm region. Didymium oxide glass filters are available for use from 250 to 2000 nm.

Stray light measurements are made using a sharp cut-off filter. Examples of such filter materials include saturated solutions of such compounds as potassium ferromanganate and lithium carbonate. Other solutions exhibiting abrupt cut-off wavelengths include: KBr, KCl, NaI, $NaNO_3$ solutions, and acetone. Refer to ASTM E 169-87, Practice for General Techniques of Ultraviolet-Visible Quantitative Analysis.

SAMPLING CONSIDERATIONS

Sample Presentation Geometry

A variety of sample presentation methods are available to the analyst. These include: transmission (straight and diffuse), reflectance (specular and diffuse), transflectance (reflectance and transmission), and interactance (a combination of reflectance and transmission).

Cuvet Cleaning

Light cleaning: Detergent wash, followed by multiple pure-water rinses.

Heavy cleaning: Repeat the light cleaning followed by cleaning with a chromic-sulfuric acid solution wash and multiple pure-water rinses.

Table 6.3 Typical Sampling Accessories.

Sample cells	Outer dimensions (in mm)	Pathlength (in mm)	Capacity (in mL)
Standard transmittance	45 (H) × 12.5 (W) × 3.5 (L)	1.0	0.3
	45 (H) × 12.5 (W) × 7.5 (L)	5	1.5
	45 (H) × 12.5 (W) × 12.5 (L)	10	3.0
Semimicro	45 (H) × 12.5 (W) × 12.5 (L)	10	1.0 or 1.5
Micro cell	25 (H) × 12.5 (W) × 12.5 (L)	10	0.5
Cylindrical cells	10 (L) × 22 (Diameter)	10	3.1
	20 (L) × 22 (D)	20	6.3
	50 (L) × 22 (D)	50	16
	100 (L) × 22 (D)	100	31
Microflow cell	50 (H) × 12.5 (W) × 12.5 (L)	10	0.4 or 0.6
Round	75 (H) × 12 (D)	~10	5.9
	105 (H) × 19 (D)	~17	23.8
	150 (H) × 19 (D)	~17	34.0

Selection of Pathlength

UV, 190–350 nm, 1 mm–10 cm
SW-NIR, 800–1100 nm, 5–10 cm
LW-NIR, 1050–3000 nm, 0.1–2 cm

Matrix/Measurement Techniques

Clear solids (optical materials): transmittance

Translucent or opaque solids: diffuse reflectance or diffuse transmittance (for turbid samples)

Reflecting optical surfaces: specular reflectance

Clear liquids: transmittance

Translucent or opaque liquids: reflectance or diffuse transmittance

High optical density (highly absorbing): tiny pathlengths in transmittance

Sample Optical Properties

Clear samples are measured using transmission spectroscopy.

Colored samples are generally measured using transmission spectroscopy, unless the optical density exceeds the linear range of the measuring instrument. At this point either dilution or narrowing of the pathlength is the technique of choice.

Fine scattering particulates within a sample are measured by diffuse transmission or diffuse reflectance methods. The scattering produces a pseudo-pathlength effect that must be compensated for by using scatter correction data-processing methods when making quantitative measurements on scattering materials.

Large scattering particulates present a challenge for measurements, for the particles intercept the optical path at random intervals. Signal averaging can be employed to compensate for random signal fluctuations as well as careful monitoring and tracking of signal changes. Reflectance spectroscopy can be used to measure the size, velocity, and concentration of scattering particulates within a flowing stream.

High-absorptivity (optically dense) materials with absorbances above 4–6 OD are difficult to measure accurately without the use of double-monochromator instruments with stray light specifications below 0.0001% *T*. Measurements can be made with extremely slow scanning speeds and by opening the slits during measurement. These measurements should be avoided by the novice unless high-performance instrumentation and technical support are available.

APPLICATIONS

The grouping of atoms producing a characteristic absorption is called a chromophore (*chromo* = color + *phore* = producer). A specific grouping of atoms produces a characteristic absorption band at specific wavelengths. The intensity and location of these absorption bands will change with structural changes in the group of atoms and with solvent changes. Typical UV solvents are given in Table 6.4. The location of bands associated with UV chromophores is shown in Table 6.5.

Table 6.4 Typical UV Solvents and Approximate UV Cut-Off Wavelengths.

Solvent	UV cut-off (in nm)
Acetonitrile	190
Water	190
Cyclohexane	195
Isooctane	195
n-Hexane	201
Ethanol (95 vol. %)	205
Methanol	205
Trimethyl phosphate	210
Acetone	220
Chloroform	240
Xylene	280

Table 6.5 Absorptions of UV Chromophores (160–210 nm).

Chromophore	Absorption Band (in nm)
Nitriles (R-C≡N)	160*
Acetylenes (—C≡C—)	170*
Alkenes (>C=C<)	175–185*
Alcohols (R—OH)	180 (175–200)*
Ethers (R—O—R)	180*
Ketones (R—C=O—R′)	180*, 280
Amines, primary (R—NH$_2$)	190* (200–220)
Aldehydes (R—C=O—H)	190*, 290
Carboxylic acids (R—C=O—OH)	205
Esters (R—C=O—OR′)	205
Amides, primary (R—C=O—NH$_2$)	210
Thiols (R—SH)	210
Sulfides (R$_2$S)	210–215
Unsaturated aldehydes (C=C—C=O)	210–250
Nitrites (R—NO$_2$)	270–275
Carbonyl (R$_2$>C=O)	270–310
Azo-group (R—N=N—R)	340–350

*Absorptions below the cut-off for common solvents would not be observed in solvent solution measurements.

APPLICATIONS

Use of UV-Vis for Life Sciences

Enzymatic Methods

Table 6.6 gives a summary of the various enzymatic methods of analysis using spectroscopic technique.

Table 6.6 Spectroscopic Assay Measurements for Enzymes.

Enzyme name	Reaction type and assay wavelength (in nm)
Lactate dehydrogenase	Direct absorbance at 350
Alkaline phosphatase	Direct absorbance at 550
NADH and NADPH	Direct absorbance at 340
Alcohol dehydrogenase	Increased Abs.—NADH at 340
Aldolase	Increased Abs. at 240; decreased Abs. at 340
D-Amino acid oxidase	Decreased Abs.—NADH at 340
L-Amino acid oxidase	Change in Abs. at 436
Alpha-Amylase	Color reaction at 540
L-Arginase	Color reaction at 490
Arylsulfatase	Hydrolysis reaction at 405
Catalase	Liberation of H_2O_2—Abs. at 240
Cholinesterases	Color reaction at 340
Alpha-Chymotrypsin	Hydrolysis products at 280
Creatine kinase	Increased Abs.—NADH at 340
Deoxyribonuclease I	Depolymerization product at 260
Diamine oxidase	Decreased Abs.—NADH at 340
Beta-Galactosidase	Hydrolysis Product at 405, or 436
Glucose oxidase	Color reaction at 436
Glucose-6-phosphate dehydrogenase	Increased Abs.—NADH at 340
Glucose phosphate isomerase	Increased Abs.—NADPH at 340
Beta-Glucosidase	Increased Abs.—NADPH at 340
Beta-Glucuronidase	Color reaction at 540
Glutamate-oxaloacetate transaminase	Decreased Abs.—NADH at 340
Glutamate pyruvate transaminase	Decreased Abs.—NADH at 340
Gamma-Glutamyl transferase	*p*-Nitroaniline at 400
Hexokinase	Increase in Abs.—NADH at 340
Alpha-Hydroxybutyrate dehydrogenase	Decreased Abs.—NADH at 340
Isocitrate dehydrogenase	Increase in Abs.—NADH at 340
Lactate dehydrogenase	Decrease in Abs.—NADH at 340
Leucine amino peptidase	Presence of *p*-nitroalinine at 405
Lipase	Color reaction at 540
Monoamine oxidase	Color reaction at 456
Pepsin	Color reactions at 578, 691, and 750
Peroxidase	Color reaction at 460
Acid phosphatase	Increased Abs. at 300
Alkaline Phosphatase	Color reaction at 530
Pyruvate kinase	Decreased Abs. at 340
Sorbitol dehydrogenase	Decrease in Abs.—NADH at 340
Trypsin	Increased Abs. at 280
Urease	Color reaction at 580, 630
Xanthine oxidase	Color reaction at 530–580

Use of SW-NIR for Organic Composition Analysis

Table 6.7 demonstrates the basic functional group measurements that have useful signal in the short-wavelength near-infrared (SW-NIR) region (800–1100 nm) of the electromagnetic spectrum. As can be seen from the table, the SW-NIR region is used to measure molecular vibrations as combination bands for C—H groups, for second overtones of O—H and N—H groups, and for third overtone C—H group measurements. All NIR spectroscopy is used to measure these basic organic functional groups resulting from molecular vibrations (Table 6.8). The advantages of SW-NIR include high signal-to-noise ratio from readily available technologies, typically 25,000 : 1; as well as high throughput using fiber-optic cabling. An additional advantage of SW-NIR over other IR regions includes the use of flow-cell pathlengths sufficiently large for industrial use (most often 5–10 cm). This range of pathlengths is useful in obtaining representative sample size measurements and in preventing fouling of internal cell optics. Representative SW/NIR spectra are shown in Fig. 6.6 on p. 60.

Approximate LW-NIR Band Location for Common Organic Compounds

Traditional LW-NIR spectroscopy included the use of filter instruments to measure the major constituents of nutritional interest in grain and forage materials. Selected wavelengths com-

Table 6.7 C—H, N—H, and O—H Stretch Absorption Bands for Specific Short-Wavelength NIR (800–1100 nm) functional groups (2nd or 3rd Overtones).

Structure	Bond vibration	Approx. band locations (in nm)
ArCH . (Aromatics)	C–H str. 3rd overtone	857–890
. CH—CH (Methylene)	C–H str. 3rd overtone	930–935
. CH_3 (Methyl)	C–H str. 3rd overtone	912–915
"	C–H combination	1010–1025
R—OH (Alcohols)	O–H str. 2nd overtone	940–970
ArOH (Phenols)	O–H str. 2nd overtone	947–980
HOH (Water)	O–H str. 2nd overtone	960–990
Starch	O–H str. 2nd overtone	967
Urea	Sym. N–H str. 2nd overtone	973
. $CONH_2$ (Primary amides)	N–H str. 2nd overtone	975–989
.CONHR (Secondary amides)	N–H str. 2nd overtone	981
Cellulose	O–H str. 2nd overtone	993
Urea	Sym. N–H str. 2nd overtone	993
$ArNH_2$ (Aromatic amines)	N–H str. 2nd overtone	995
. NH (Amines, general)	N–H str. 2nd overtone	1000
Protein	N–H str. 2nd overtone	1007
Urea	N–H str. 2nd overtone	1013
RNH_2 (Primary amines)	N–H str. combination	1020
Starch	O–H str. combination	1027
CONH (Primary amides)	N–H str. combination	1047
$=CH_2$ (Methylene)	C–H str. combination	1080

APPLICATIONS

Table 6.8 Typical UV/VIS/NIR Methods.

Analyte or analytical method	Wavelength(s) used for measurement (in nm)
Aromatics, amines, aldehydes, naphthalenes, phenols, ketones (in ppm levels)	254 or 313
Violet*	400
ASTM D1209**	430
International color (blue)	440
Blue*	450
ASTM D1209**	455
ASTM D1209**	480
Green*	500
ASTM D1209**	510
International color (green)	520
Yellow-green	550
Orange*	600
International color (red)	620
Red*	650
Dark red*	700
Cu(II) ion	820
Reference wavelength	830
Ar C–H 3rd overtone	875
Carbonyl C–H 3rd overtone	895
Turbidity	900
Methoxy C–H str. 3rd overtone	905–909
Methyl C–H str. 3rd overtone	915
Methylene C–H str. 3rd overtone	930
Alcohol O–H 2nd overtone	960–970
Water O–H str. 2nd overtone	958–960

*Used for tristimulus, chromaticity coordinates, color distance, and CIE coordinates.

**Used for ASTM E-346–78 (color in methanol), D2108–71 (color in halogenated solvents), and E 202–67 (color in glycols).

monly chosen for filter instrument measurements of these components included the following wavelengths (in nanometers) with the corresponding nutritional parameters:

2270 nm   lignin
2310 nm   oil
2230 nm   reference region
2336 nm   cellulose
2180 nm   protein

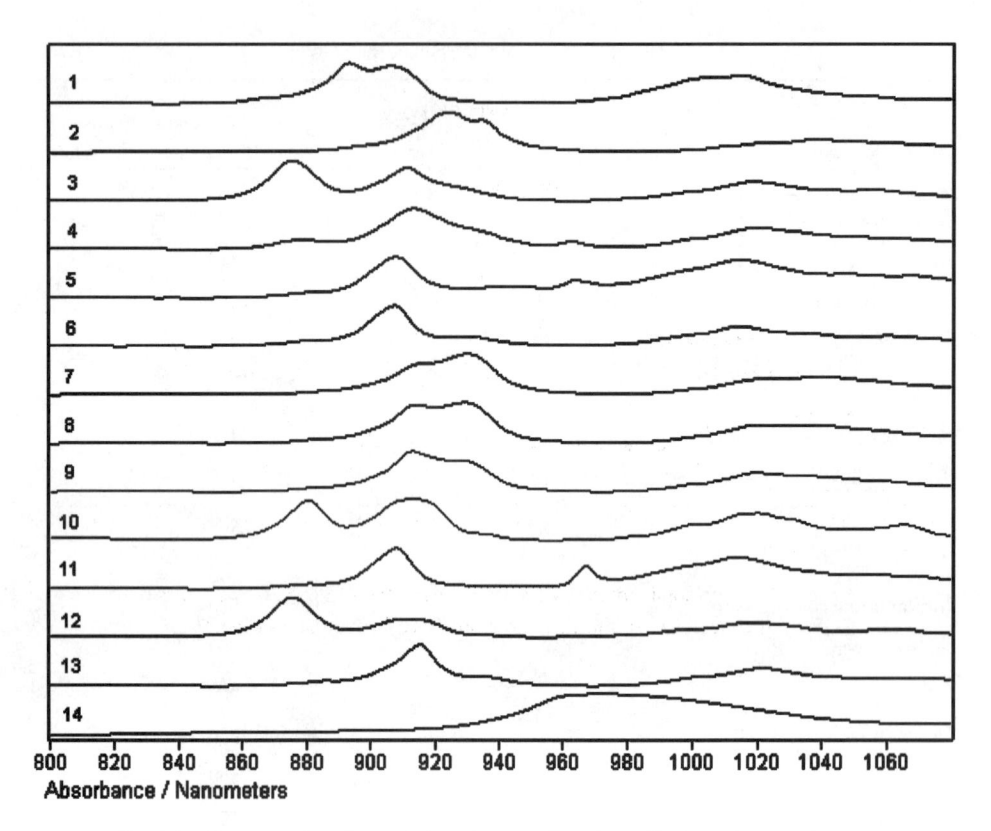

Fig. 6.6 Short-wavelength near-infrared spectra for selected hydrocarbons and water.

Key 1 acetone; 2 cyclohexane; 3 ethylbenzene; 4 gasoline with added ethanol; 5 isopropanol; 6 *tert*-butyl methyl ether; 7 *n*-decane; 8 *n*-heptane; 9 pentane; 10 *p*-xylene; 11 tert-butanol; 12 toluene; 13 trimethyl pentane; 14 water.

2100 nm	carbohydrate
1940 nm	moisture
1680 nm	reference region

Dominant near-infrared spectral features include the measurement of the overtones and combination absorption bands from the following functional groups. Figure 6.7 and Table 6.9 further define the locations of these groups in typical NIR spectra.

- Methyl C–H
- Methylene C–H
- Methoxy C–H
- Carbonyl C–H
- Aromatic C–H
- Hydroxyl O–H
- N–H from primary amides, secondary amides; both alkyl and aryl group associations
- N–H from primary, secondary, and tertiary amines
- N–H from amine salts

APPLICATIONS

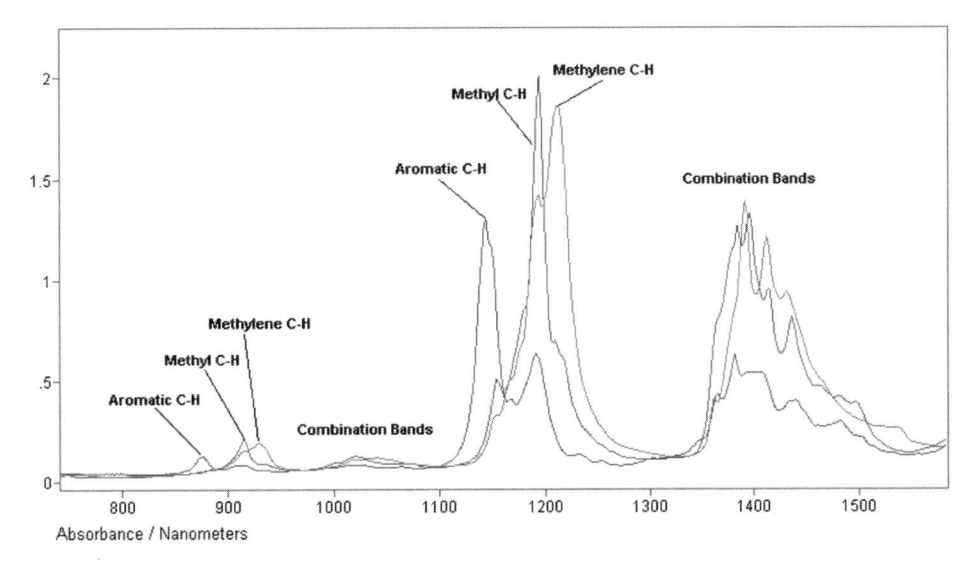

Fig. 6.7 Predominant long-wavelength near-infrared (LW-NIR) hydrocarbon spectral features (showing absorbance as a function of wavelength).

Table 6.9 C—H, N—H, and O—H Stretch Absorption Bands for Specific Long-Wavelength NIR (1100–2500 nm) Functional Groups (1st and/or 2nd Overtones).

Structure	Bond vibration	Location (in nm) of 2nd overtone	Location (in nm) of 1st overtone
ArCH . (Aromatics)	C–H str.	1143–1187	1714–1780
. CH—CH (Methylene)	C–H str.	1240–1247	1860–1870
. CH$_3$ (Methyl)	C–H str.	1216–1220	1824–1830
"	C–H combination	1347–1367	2020–2050
R—OH (Alcohols)	O–H str.	—	1410–1455
ArOH (Phenols)	O–H str.	—	1421–1470
HOH (Water)	O–H str.	—	1440–1485
Starch	O–H str.	—	1451
Urea	Sym. N–H str.	—	1460
. CONH$_2$ (Primary amides)	N–H str.	—	1463–1484
.CONHR (Secondary amides)	N–H str.	—	1472
Cellulose	O–H str.	—	1490
Urea	Sym. N–H	—	1490
ArNH$_2$ (Aromatic amines)	N–H str.	—	1493
. NH (Amines, general)	N–H str.	—	1500
Protein	N–H str.	—	1511
Urea	N–H str.	—	1520

FURTHER READING

Classics
Sawyer, R.A. (1963). *Experimental Spectroscopy.* New York: Dover Publications.
Edisbury, J.R. (1967). *Practical Hints on Absorption Spectrometry (Ultra-violet and visible).* New York: Plenum Press.

General
Vanasse, G.A. (ed.) (1981). *Spectrometric Techniques—Vol. II.* Boston: Academic Press.
Knowles, A., Burgess, C. (eds.) (1984). *Practical Absorption Spectrometry* (Techniques in Visible and Ultraviolet Spectrometry; Vol. 3). London: Chapman and Hall.

Color
Committee on Colorimetry: Optical Society of America. (1973). *The Science of Color,* Washington, D.C.: The Society.

Life Sciences
Guilbault, G.G. (1976). *Handbook of Enzymatic Methods of Analysis.* New York: Marcel Dekker.

Interpretive Spectroscopy
Bonanno, A., Griffiths, P. (1993). *Near Infrared Spectrosc.* 1: 13–23.
Hirschfeld T., Zeev Hed, A. (1981). *The Atlas of Near Infrared Spectra.* Philadelphia: Sadtler Research Laboratories.
Kelly, J. J., Barlow, C. H., Jinguji, T. M., Callis, J. B. (1983) *Anal. Chem.* 61: 313.
Kelly, J. J., Callis, J. B. (1990) *Anal. Chem.* 62: 1444.
Mayes, D. M., Callis, J. B. (1989) *Appl. Spectrosc.* 43: 27.

7.

UV-VIS SPECTROSCOPY CHARTS

The majority of organic compounds are somewhat transparent in the ultraviolet and visible regions. In some cases, specific types of organic materials absorb ultraviolet (UV) and visible (Vis) radiation, giving useful information toward quantitative analysis or identification of compounds. When combined with physical data such as melting point, solubility, and boiling point, identification can be enhanced using ultraviolet and visible spectroscopy. When UV-Vis spectra are combined with infrared or NMR spectral data, important structural features can be assessed.

UV spectra at lowered temperatures can yield significant increases in structure due to lowered collisional and rotational energy. Large molecules at room temperature can exhibit fine structure in the ultraviolet due to their high rotational energy levels.

Solvent interaction is another important characteristic of ultraviolet and visible spectra. The solvent selected should be "invisible" in the spectral region of interest as much as is possible. Solvents can have an effect on the position, intensity, and bandwidth of the various absorption bands for any UV-Vis absorbing solute. The polarity of a solvent can have a dramatic effect on the spectral structure of the solute. A nonpolar solvent (e.g., alkanes) will not cause hydrogen bonding with the solute material, and thus fine structure can generally be observed. Polar solvents (e.g., methanol, ethanol, water) will cause hydrogen bonding with the solute in most cases and result in slurred structure or "rounded" absorbance bands. The ultraviolet-visible spectra given in this volume are measured neat and thus demonstrate their own UV characteristics, including absorption bands, cut-off wavelengths, visible spectra, and 4th overtone NIR vibrational spectra (i.e., from 680 to 850 nm).

A variety of substituents added to a basic chromophore (i.e., UV-absorbing functional group) will cause the bands to shift in wavelength or frequency *position.* Typically these shifts are termed *red shift* (i.e., a band shift to a lower frequency or longer wavelength, also termed *bathochromic shift*); and a *blue shift* (i.e., a band shift to a higher frequency or shorter wavelength, also known as a *hypsochromic shift*). Changes in intensity can also occur as a decrease in absorbance (i.e., *hypochromic*) and as an increase in absorbance (i.e., *hyperchromic*).

The following practical guide for evaluating ultraviolet spectra is reprinted with permission from D. Pavia, G. Lampman, and G. Kriz, *Introduction to Spectroscopy: A Guide for Students of Organic Chemistry.* Philadelphia: Saunders College, 1979.

8.

PRACTICAL GUIDE FOR EVALUATING ULTRAVIOLET SPECTRA*

When an ultraviolet spectrum is used by itself, it is often difficult to extract a great deal of information from it. However, several generalizations can be formulated which are a good guide to the use of ultraviolet data. These generalizations will, of course, be a good deal more meaningful when combined with infrared data, which can, for instance, definitely identify carbonyl groups, double bonds, aromatic systems, nitro groups, nitriles, enones, and other important chromophores. In the absence of infrared data, the following observations should be regarded only as guidelines. Note that ultraviolet spectra are often displayed with the wavelength (in nanometers), or frequency (in wavenumbers) as the abscissa, and the molar absorptivity (ε) along the ordinate. The absorptivity (ε) is given by:

$$\varepsilon = \frac{A}{bc}$$

where ε is the molar absorptivity (i.e., the molecular extinction coefficient), b is the pathlength of the sample (in centimeters), c is the concentration of analyte in the sample (in moles per liter), and A is the absorbance of the given test solution in AU. *Note:* The final units for ε are

as $\frac{A \times \text{liter}}{\text{cm} \times \text{moles}}$ or, as often shown,

$$L \cdot mol^{-1} \cdot cm^{-1}$$

1. A single band of low to medium intensity (ε = 100–10,000) at wavelengths less than 220 nm usually indicates an $n \rightarrow \sigma^*$ transition. Amines, alcohols, ethers, and thiols are possibilities, provided that the nonbonded electrons are not included in a conjugated system. An exception to this generalization is that the $n \rightarrow \sigma^*$ transition of cyano groups (–C≡N:) appears in this region. However, this is a weak transition (ε less than 100), and, of course, the cyano group is easily identified in the infrared. Do not neglect to look for N-H, O-H, C-O, or S-H bands in the infrared spectrum.

2. A single band of low intensity (ε = 10–100) in the region 250–360 nm, with no major absorption at shorter wavelengths (200–250 nm), usually indicates an $n \rightarrow \sigma^*$ transition. Since the absorption does not occur at long wavelength, a simple, or unconjugated, chromophore is indicated, generally one which contains an O, N, or S atom. Examples of this might include C=O, C=N, N=N, –NO$_2$, –COOR, –COOH, or –CONH$_2$. Once again, the infrared spectrum should help with identification.

*This chapter is reprinted with permission from the copywrite owner for D. Pavia, G. Lampman, and G. Kriz, *Introduction to Spectroscopy: A Guide for Students of Organic Chemistry,* Philadelphia: Saunders College, 1979.

3. Two bands of medium intensity (ε = 1,000–10,000), both with λ_{max} above 200 nm, generally indicate the presence of an aromatic system. If an aromatic system is present, there may be a good deal of fine structure in the longer wavelength band (in nonpolar solvents only). Substitution on the aromatic rings will increase the molar absorptivity above 10,000, particularly if the substituent increases the length of the conjugated system.

    In polynuclear aromatic substances, a third band will appear near 200 nm, a band which in simpler aromatics occurred below 200 nm, where it could not be observed. Most polynuclear aromatics (and heterocyclic compounds) have very characteristic intensity and band shape (fine structure) patterns, and they may often be identified by comparison to spectra which are available in the literature.

4. Bands of high intensity (ε = 10,000 to 20,000) which appear above 210 nm generally represent either an α,β-unsaturated ketone (check the infrared spectrum) or a diene or polyene. The longer the length of the conjugated system, the longer the observed wavelength will be. For dienes, the λ_{max} may be calculated using the Woodward–Fieser rules or, if there are more than four double bonds ($n > 4$), the Fieser–Kuhn rules. Enones are discussed below.

5. Simple ketones, acids, esters, amides, and other compounds containing both π systems and unshared electron pairs will show two absorptions: an $n \to \pi^*$ transition at longer wavelength ($\lambda > 300$ nm, low intensity) and a $\pi \to \pi^*$ transition at shorter wavelengths ($\lambda < 250$ nm, high intensity). With conjugation (enones), the λ_{max} of the $\pi \to \pi^*$ band moves to longer wavelengths and can be predicted by Woodward's rules. The ε value usually rises above 10,000 with conjugation, and, as it is very intense, it may obscure or bury the weaker $n \to \pi^*$ transition.

    For α,β-unsaturated esters and acids, Nielsen's rules may be used to predict the position of λ_{max} with increasing conjugation and substitution.

6. Compounds which are highly colored (have absorption in the visible region) are likely to contain a long-chain conjugated system or a polycyclic aromatic chromophore. Benzenoid compounds may be colored if they have enough conjugating substituents. For nonaromatic systems, usually a minimum of 4 to 5 conjugated chromophores are required to produce absorption in the visible region. However, some simple nitro, azo, nitroso, α-diketo, polybromo, and polyiodo compounds may also exhibit color, as may many compounds with a quinoid structure:

WOODWOOD–FIESER RULES

Conjugated dienes exhibit an intense band (ε = 20,000 to 26,000) in the spectral region from 217 nm to 245 nm, due to a $\pi \to \pi^*$ transition. The position of this band appears to be quite insensitive to the nature and type of solvent. Basically the rules indicate the positions of λ_{max} based on the structure and attached groups forming the overall diene molecule. A cis (cisoid) structure exhibits bands at a higher wavelength than a trans (transoid) structure. Table 8.1 illustrates the basic positional changes for λ_{max} based on molecular configuration and attached groups.

FIESER–KUHN RULES

The Woodward–Fieser rules work well for polyenes with from one to four conjugated double bonds. Fieser and Kuhn have developed a simple set of empirical rules that work well for polyene systems such as those found in carotenoid pigments like β-carotene and lycopene. Table 8.2

demonstrates the use of these rules for estimating λ_{max} based on the structure and attached groups forming the overall polyene molecule.

NIELSEN'S RULES FOR α,β-UNSATURATED ACIDS AND ESTERS

A simple set of empirical rules have been developed for λ_{max} positions for unsaturated acids and esters. This set of rules is described in Table 8.3.

Table 8.1 Woodward–Fieser Empirical Rules for Dienes (Additive for λ_{max})

Molecular type	Cis (λ_{max})	Trans (λ_{max})
Parent compound	253 nm	214 nm
Double-bond-extending conjugation	+30 nm	+30 nm
Alkyl substituent or ring residue	+5 nm	+5 nm
Exocyclic double bond	+5 nm	+5 nm
Polar groupings:		
$-OCOCH_3$	0	0
$-OR$	+6 nm	+6 nm
$-Cl_2$, $-Br$	+5 nm	+5 nm
$-NR_2$	+60 nm	+60 nm

Table 8.2 Fieser–Kuhn Empirical Rules for Polyenes

λ_{max} (in hexane) = $114 + 5M + n(48.0 - 1.7n) - 16.5R_{endo} - 10R_{exo}$

ε_{max} (in hexane) = $(1.74 \times 10^4)\,n$

where:
M = total number of alkyl-substituted groups, n = total number of conjugated double bonds, R_{endo} = number of rings with endocyclic double bonds, R_{exo} = number of rings with exocyclic double bonds.

Table 8.3 Nielsen's Empirical Rules for α,β-Unsaturated Acids and Esters (for λ_{max})

Base values for:

$$\begin{array}{c}\beta\alpha\\ C=C\\ \betaCOOR\end{array}\qquad\begin{array}{c}\beta\alpha\\ C=C\\ \betaCOOH\end{array}$$

With α or β alkyl group	208 nm
With α,β or β,β alkyl group	217 nm
With α,β,β alkyl group	225 nm
For an exocyclic α,β double bond	+5 nm
For an endocyclic α,β double bond in a 5- or 7-membered ring	+5 nm

9.

FUNCTIONAL GROUPINGS AND BAND LOCATIONS FOR UV-VIS SPECTROSCOPY IN NANOMETERS (NM)

Group	Molecular structure	Log ε	Relative band intensity (ε) 0–1500	Primary band	Secondary band	Comments
I. Functionality						
Alcohols	R–OH	2.5	101	180	—	
Ethers	R–O–R	3.6	1000	180	—	
Thiols	R–SH	2.9	317	210	—	
Alkenes	R'–C=C–R''	3.1	317	175	—	
Acetylenes	R–C≡C–R	3.0	317	170	—	
Nitriles	R–C≡N	0.8	3	160	—	
Azo compounds	R–N=N–R	0.9	4	340	—	
Nitrates	R–NO$_2$	0.9	4	271	—	
Aldehydes	R–C=O–H	2.0	32 / 3	190	290	In hexane
Ketones	R–C=O–R'	3.0	317 / 10	180	280	In hexane
Carboxylic acids	R–C=O–OH	1.6	10	205	—	In water
Esters	R–C=O–O–R'	1.5	10	205	—	In ethanol
Amides	R–C=O–N–	1.5	10	210	—	In water
Amines, Primary	R–NH$_2$	3.5	997	190	—	
Conjugated Alkenes	RC=C–C=C–R'	4.1–5.1	1417	175–465		

Group	Molecular structure (group)	Band intensity (Actual ε)	Primary band	Secondary band	Comments
II. Applications					
Ethylene	Alkene	15,000	175	—	
1,3-Butadiene	Alkene	21,000	217	—	
1,3,5-Hexatriene	Alkene	35,000	258	—	
Beta-carotene	Alkene	125,000	465	—	
Acetone	Ketone	900 / 12	189	280	
3-Buten-2-one	Ketone	7,100 / 27	213	320	
Formaldehyde (in hexane as solvent)	Aldehyde	—	293	—	In hexane
Acetone (in hexane as solvent)	Ketone	—	279	—	In hexane
Amides (in water as solvent)	Amides	—	214	—	In water
Esters (in water as solvent)	Esters	—	204	—	In water
Carboxylic acids (in ethanol as solvent)	Carboxylic acids	—	205	—	In ethanol

10.

UV-VIS SPECTRA CORRELATION CHARTS

Ultraviolet Primary Band Locations by Functional Group

(See color plate 1.)

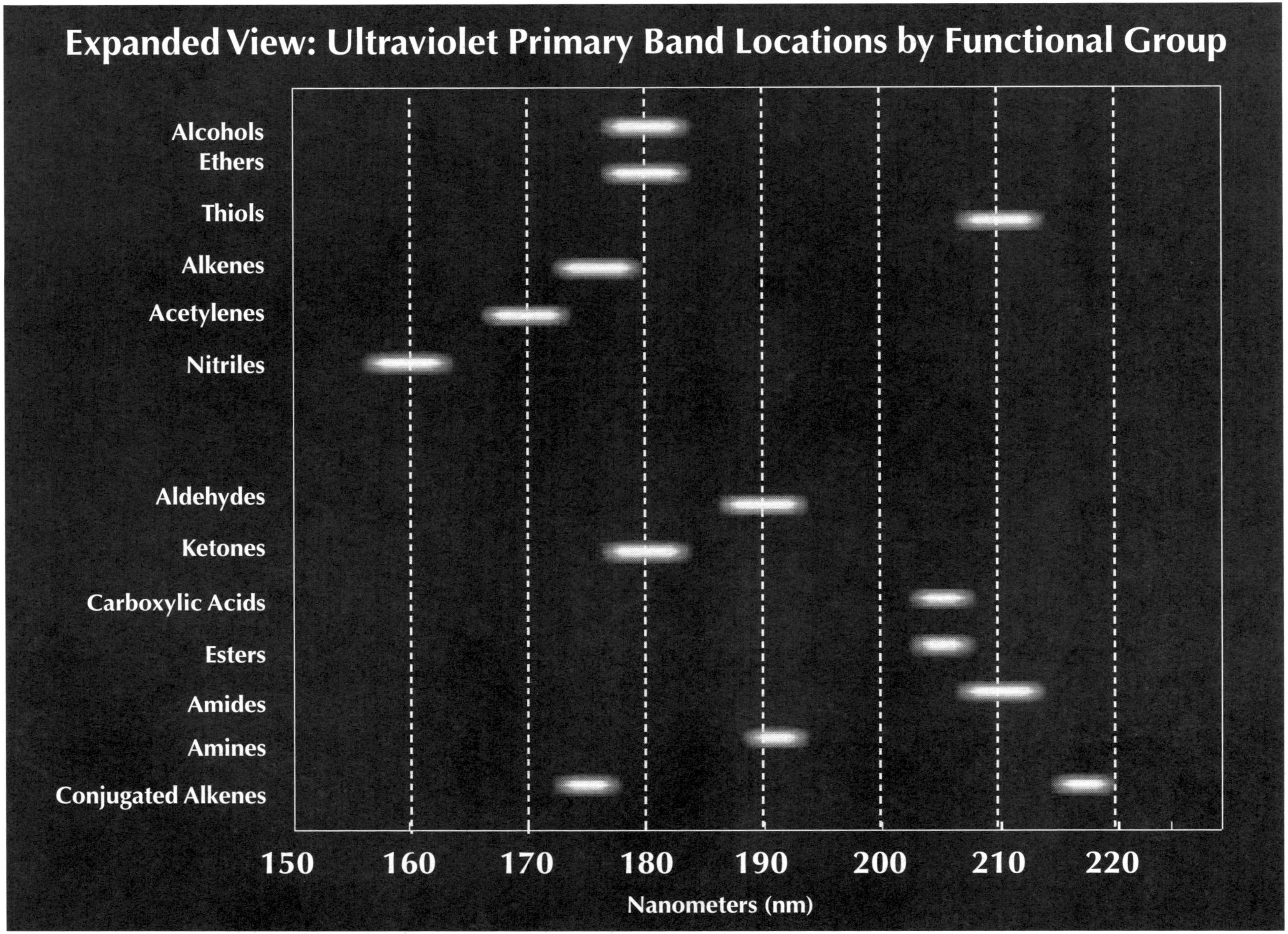

Expanded View: Ultraviolet Primary Band Locations by Functional Group

(See color plate 2.)

UV-Vis Chromophores
RELATIVE INTENSITIES

Chromophore	Relative Intensity
RC=C–C=C–R' Conjugated Alkenes	1417
R–NH2 Amines, Primary	997
R–C=O–N– Amides	10
R–C=O–O–R' Esters	10
R–C=O–OH Carboxylic acids	10
R–C=O–R' Ketones	317
R–C=O–H Aldehydes	32
R–NO2 Nitrates	3
R–N=N–R Azo Compounds	3
R–C≡N Nitriles	3
R–C≡C–R' Acetylenes	317
R'–C=C–R" Alkenes	317
R–SH Thiols	317
R–O–R Ethers	1000
R–OH Alcohols	101

(See color plate 3.)

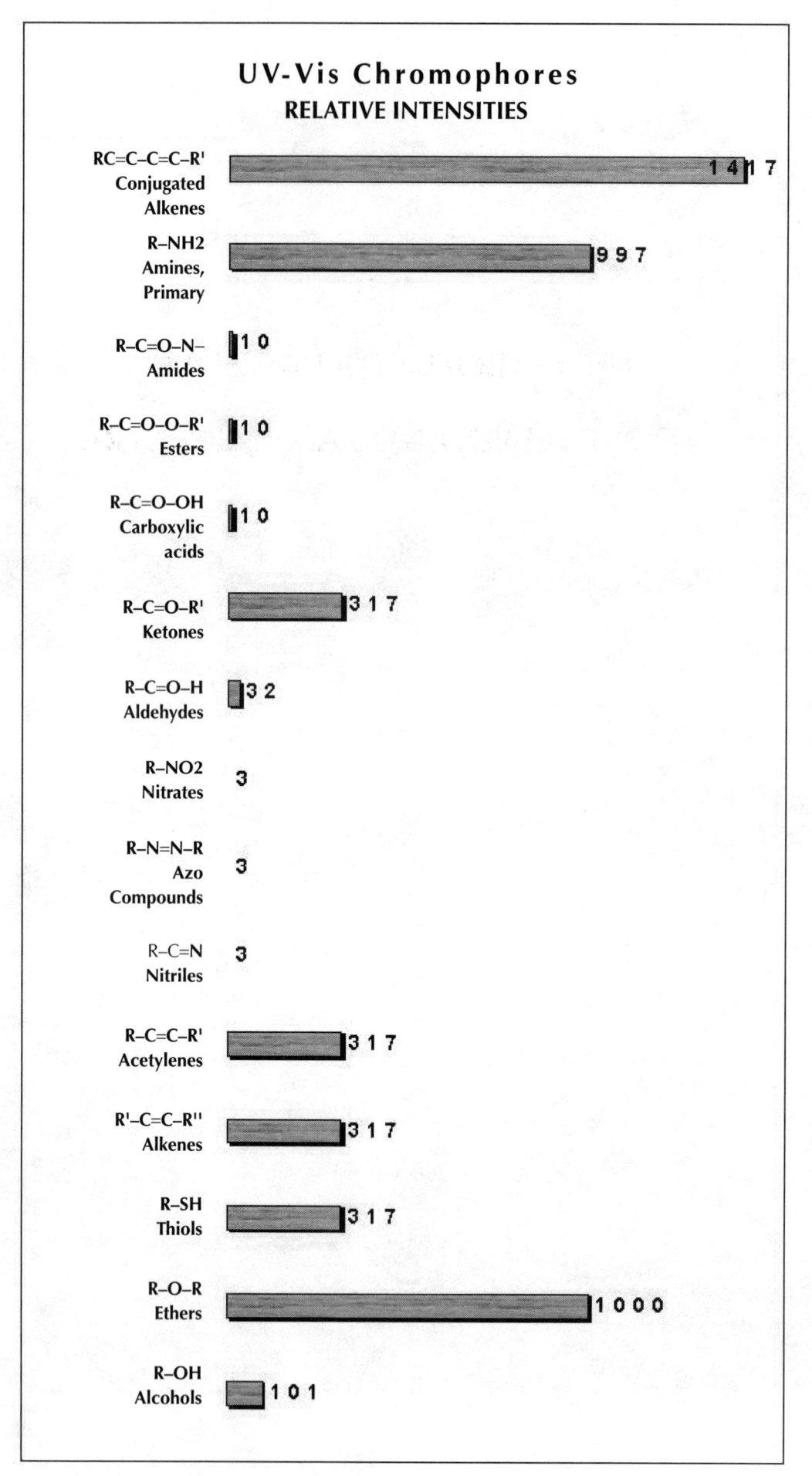

IV.
INTRODUCTION TO NIR, INFRARED, AND RAMAN SPECTRA

11.

COMPARING NIR, INFRARED, AND RAMAN SPECTRA

NEAR-INFRARED SPECTROSCOPY

Near-infrared spectroscopy is used where multicomponent molecular vibrational analysis is required in the presence of interfering substances. The near-infrared spectra consist of overtones and combination bands of the fundamental molecular absorptions found in the mid-infrared region. Near-infrared spectra consist of generally overlapping vibrational bands that are nonspecific and poorly resolved. Chemometric mathematical data processing can be used to calibrate for qualitative or quantitative analysis despite these apparent spectroscopic limitations. Traditional near-infrared spectroscopy was used in agricultural product analysis for lignin polymers (2270 nm), paraffins and long-alkane-chain polymers (2310 nm), glucose-based polymers such as cellulose (2336 nm), amino acid polymers such as proteins (2180 nm), carbohydrates (2100 nm), and moisture (1440 and 1940 nm).

The dominant near-infrared spectral features include the methyl C–H stretching vibrations, methylene C–H stretching vibrations, aromatic C–H stretching vibrations, O–H stretching vibrations, methoxy C–H stretching, and carbonyl-associated C–H stretching. In addition, N–H from primary amides, secondary amides (both alkyl and aryl group associations), N–H from primary, secondary, and tertiary amines, and N–H from amine salts predominate in near-infrared spectral features of polymers and organic compounds.

INFRARED SPECTROSCOPY

Infrared spectroscopy provides a measurement technique for intense, isolated, and reliable absorption bands of fundamental molecular vibrations from polymers and other organic compounds. The spectrometric methodology allows for univariate calibration with the higher signal strength (absorptivities) required for solid-, liquid-, or gas-phase measurements. Relatively small pathlengths of 0.1–1.0 mm are required for hydrocarbon liquids and solids. The technique is generally incompatible with the use of fiber optics, but specialized fiber materials exist. Instrumentation is of higher cost than near-infrared spectrophotometers for the most part.

Dominant mid-infrared spectra include the C–H (methyl, methylene, aromatic, methoxy, and carbonyl) fundamental stretching and bending molecular vibrations, O–H (hydroxyl) stretch fundamental vibrations; N–H (amine) stretching, C–F (fluorocarbon) stretching, —C=N (nitrile) stretching, —C=O (carbonyl) stretch from esters, acetates, and amides; C–Cl stretch from chlorinated hydrocarbons, and —NO$_2$ from nitro-containing compounds.

RAMAN SPECTROSCOPY

Raman spectroscopy can be used for a variety of measurements on samples that are aqueous in nature or where glass sample holders are present. Carbon dioxide, water, and glass (silica) are weak scatterers, and thus there is generally no problem in analyzing samples having these properties. There is typically no sample preparation involved with samples measured using Raman. Raman spectroscopy is complementary to mid-infrared spectroscopy in the measurement of fundamental molecular vibrations. Raman measurements are compatible with fiber optics. Raman measurements exhibit high signal-to-noise ratios and involve a reasonable cost for instrumentation. The dominant Raman spectral features are acetylenic C≡C stretching, olefinic C=C stretch at 1680–1630 cm^{-1}; N=N (azo-) stretching, S–H (thio-) stretching, C=S stretching, C–S stretching, and S–S stretching bands. Raman spectra also contain such molecular vibrational information as CH_2 twist and wagging, carbonyl C=O stretch associated with esters, acetates, and amides, C–Cl (halogenated hydrocarbons) stretching, and —NO_2 (nitro-/nitrite) stretching. In addition, Raman yields information content on phenyl-containing compounds at 1000 cm^{-1}.

Raman spectroscopy is most useful for the identification of symmetrical dipole molecular vibrations. This is in contrast to infrared spectroscopy, where the molecular vibrations exhibiting the highest absorption coefficients are the asymmetrical ones. Note that, as in infrared molecular vibrations, the Raman frequency increases as the mass of the atoms within a molecular bond and as the bonds become more fixed or rigid. As Hooke's law illustrates, for molecular vibrational spectra, combination bands are most often representative of two stretch vibrational bands in *combination* with a single bending vibration.

The Raman effect is illustrated as elastic and inelastic collisions of photons with the atoms of a vibrating molecule. *Elastic scattering,* also termed *Rayleigh scattering,* occurs as the majority of photon interactions with the molecular vibration. In addition to elastic, or Rayleigh, scattering, a broad background of photons can exist from fluorescence emission. Inelastic collisions also occur, some with an increase in energy (i.e., blue-shifted anti-Stokes shift) and others resulting in a decrease in photon energy (i.e., red-shifted, or Stokes shift). There is most often more Stokes than anti-Stokes photon shifts in Raman scattered light, and thus the Stokes-shifted energy is used for Raman measurements.

Typical standards proposed for Raman spectroscopy include Kopp 2412, also known as NIST c540a. For infrared spectroscopy, NIST SRM 1921 (polystyrene film) is used for wavelength calibration; carbon dioxide, nitrous oxide (NO), and NO_2 in the gaseous phase are used for infrared spectrophotometer frequency calibration. For near-infrared spectrometer calibration, NIST SRM 1920 is proposed as an NIR reflectance standard, and SRM 2035 is used as a transmission wavelength standard.

Raman group frequencies are often complementary to infrared group frequencies. When Raman bands are intense, infrared bands often can be weak. The converse relationship often holds true as well. When there is almost no coincidence between the band locations and intensities of a Raman spectrum when compared to an infrared spectrum, it is indicative of a molecular structure with a center of symmetry (i.e., a symmetrical molecule). Asymmetrical molecular structures exhibit more similarity between the Raman and the infrared spectrum.

12.

REVIEW OF NEAR-INFRARED AND INFRARED SPECTROSCOPY

INTRODUCTION

A proliferation of work involving near-infrared, infrared, and Raman spectroscopy in process and image analysis has occurred recently. More papers are being written about the application of the near-infrared spectral region to all types of analyses than ever before. This review includes the aspects of near-infrared and infrared spectroscopic measurements to the analysis of materials categorized into distinct applications areas, including (in alphabetical order): agriculture, animal sciences, biotechnology, cosmetics, earth sciences/atmospheric science/mineralogy, environmental monitoring, fine chemicals/chemical production, food and beverages, forensic science, gas phase analysis, instrument development physics, medical/clinical chemistry, military related research, moisture only, petroleum/natural gas/fuel research, pharmaceutical production, plant sciences and pulp and paper, polymer science, semiconductors and materials science, surface analysis, and textiles.

In this present review an expanded definition of the *eras* for the implementation of process analyzers has been included. The earlier definition for these eras was encompassed by the terms *off-line, at-line, on-line, in-line,* and *noninvasive* [1]. A new, more expanded descriptive list of applicable analyzer/era terminology, as extracted from the various titles of recent research papers, includes (in alphabetical order): *airborne, at-line, automatic/automated, hyperspectral imaging, imaging/image analysis, in situ, in-line, near-line, noninvasive/nondestructive, on-line, open-path, portable/hand-held, process monitoring/production control, quality control/quality assurance, quality monitoring, rapid, real-time,* and *remote.*

The basic interest in and growth of two-dimensional imaging spectroscopy, including airborne imaging, hyperspectral imaging, remote monitoring, microspectroscopic imaging, and basic applications of standard spectroscopic techniques modified to produce a spectral image, deserve special recognition and description. Thus the early portion of this review will be devoted to an imaging section. The second portion will describe and review the current literature from an applications perspective, but will also address the descriptive terms regarding instrument and process eras. This review was intended to be more comprehensive than descriptive, so the limited space requires shortened, sentence-long descriptions for many of the papers cited. Papers of unique interest are attributed in longer descriptions.

The search criteria for this review are listed in Table 12.1. Note the number of hits for each set of search and selection criteria. The *Analytical Abstracts* were searched for the Royal Society of Chemistry (U.K.) and the American Chemical Society. The search criteria with number of hits listed under "Items" is given in the "Description" text.

Table 12.1 Search Criteria

Set	Items	Description
S1	30999	IR OR INFRARED OR INFRA()RED OR NIR OR FTNIR
S2	13411	AUTOMATIC OR IN()SITU OR INSITU OR PORTABLE
S3	11825	RAPID OR REMOTE OR HAND()HELD OR HANDHELD
S4	4426	((AT OR IN OR OFF OR NEAR)()LINE) OR INLINE OR ONLINE OR OFFLINE
S5	377312	PROCESS? OR NON()INVASIVE OR NONINVASIVE
S6	2243	QUALITY()(CONTROL? OR MEASUREMENT)
S7	240	ON()LINE
S8	5859	S1 AND S2:S7
S9	8222	S1(5N)(SPECTROMET? OR SPECTROPHOTOMET? OR SPECTROSCOP?)
S10	1697	S9 AND S2:S7
S11	1473	10/ENG
S12	1433	RD (unique items)
S13	69933	PROCESS OR NON()INVASIVE OR NONINVASIVE
S14	921	S9 AND (S2:S4 OR S6:S7 OR S13)
S15	787	14/ENG
S16	754	RD (unique items)

Measurement Techniques Reviewed

From the foregoing search criteria it can be seen that this review covers the use of near-infrared, infrared, far-infrared, and NIR-Raman measurement techniques. A particular emphasis is placed on these measurement techniques as associated with a process analytical term such as those previously described. The work related to in situ surface measurements was included for comprehensiveness. The large quantity of work occurring in this area can provide information useful for the purpose of monitoring or controlling chemical reactions at surfaces for a variety of practical applications.

General Imaging

This review includes a brief look at the recent scientific literature directly associated with imaging spectroscopy using the near-infrared and infrared spectral regions. Key words used for this literature search included *near-infrared (NIR)*, *Fourier transform near-infrared (FT-NIR)*, *infrared*, and *Fourier transform infrared (FT-IR)*. These measurement technique search words were used in combination with image(s), imaging, and mapping terms. Imaging is particularly useful because the human mind interacts well with spatial images demonstrating physical or chemical phenomena. It is also precisely that imaging that defines the spatial relationship between the physical and the compositional information for a solid sample, yielding information on the structure and function of the material measured. Found later is a sample of the work being conducted in this technology area. These papers represent some of the most recent published work on the subject; notice the incredible diversity of each analytical technique and the broad potential exhibited by its applications.

A descriptive article surveying the field of fast FT-IR imaging, including theory and applications, is presented by Koenig and Snively [2]. This basic review describes instrumental aspects,

INTRODUCTION

imaging methods, data processing, and applications, with 20 references. An *Applied Spectroscopy* focal-point article describing infrared spectroscopic imaging in terms of focal-plane arrays, remote sensing, and astronomy, terrestrial, and medical applications is described in detail, with 35 references, by Colarusso et al. in Ref. 3.

W. van den Broek and coworkers [4], from Catholic University, Nijmegen, Netherlands, have completed a work entitled, "Plastic identification by remote sensing spectroscopic NIR imaging using kernel partial least squares (KPLS)." The authors describe an identification system for sorting plastics from nonplastics in waste recycling. Partial least squares (PLS) is used for data reduction in the classification of NIR spectral images. Given multidimensional NIR images, PLS projects the high-dimensional space into a low-dimensional latent space using the coded class information of the sample objects. PLS can be considered a supervised latent variable analysis, and data reduction by PLS increases the speed of on-line classification useful for process control. In order to apply this method in near real time, a rapid PLS version, kernel PLS (KPLS), was investigated. The method was successfully demonstrated for discrimination between plastics, nonplastics, and image backgrounds.

Wienke et al. [5], of the same institution, have presented and published another paper, entitled "Near infrared imaging spectroscopy (NIRIS) and image rank analysis for remote identification of plastics in mixed waste." This paper describes a method for the discrimination of plastics from nonplastics in household waste on the basis of a set of images taken at six wavelength ranges between 1100 and 2500 nm. The authors explain the use of multivariate image rank analysis (MIRA) and report that the use of this technique provided correct classification in 80%, or better of the test cases. The NIRIS experiments were performed using a traditional NIR source, with an InSb focal-plane diode array camera. The wavelength selection was performed with a filter wheel equipped with narrow-bandpass filters at 1600 nm, 1700 nm, 1562 nm, and 2200–2300 nm) as well as broad-bandpass filters at 1700–2150 nm and 2115–2550 nm. Samples of five different plastics and cotton, glass, paper, wood, metal, and ceramics were examined at distances from up to 2 meters. The technique of MIRA is reported for 51 samples in this study.

In a third publication the group from Catholic University, Nijmegen, Netherlands [6], published "Identification of plastics among non-plastics in mixed waste by remote-sensing near-infrared imaging spectroscopy. 2. Multivariate image rank analysis for rapid classification." Household waste was characterized by a sequence of images taken in four wavelength regions using NIR imaging spectrometry. Each sample was represented by a 3-dimensional stack of NIR images. A rapid data-compression method, followed by an abstract factor rotation of the stack into an intermediate four-element vector, was accomplished using multivariate image rank analysis (MIRA). This data reduction provided a single number used as a decision limit for the classification of the plastics. The method is insensitive to sample size and relative position within the camera cone of focus. The method was useful even with slight sample movement.

In a fourth publication, the preceding workers [7] published a paper entitled "Identification of plastics among non-plastics in mixed waste by remote-sensing near-infrared imaging spectroscopy. 1. Image improvement and analysis by singular value decomposition." The paper describes an improved data-processing method for reduction of experimental artifacts in a multivariate stack of remotely sensed NIR images. The images provide data for real-time plastic identification and classification. Reference and dark current spectral images were used to correct the raw spectral images for normal fluctuations in background, illumination geometry, lamp source intensity, and optical transmission. Shadow artifacts and specular reflection were separated from sample structures in the images by using singular value decomposition (i.e., Eigen analysis). The improved images were applied to a classification algorithm called multivariate image rank analysis (MIRA).

Ning et al. [8] published "Five novel applications of imaging visible and short near-infrared spectrophotometry and fluorimetry in the plant sciences. Part II. Non-invasive in vivo applications." A charge-coupled device (CCD) instrument for imaging spectrophotometry and fluorimetry is discussed, together with two plant science applications, in Part I [13]. These three potential applications include: (1) in vivo effects of a fungal pathogen; (2) following changes in water in vivo; and (3) estimation of quantum yield of photosynthetic fluorescence in variegated leaves. They are discussed together with possible limitations of the technique.

Sowa et al. [9] completed work entitled "Non-invasive assessment of regional and temporal variations in tissue oxygenation by near-infrared spectroscopy and imaging." The work shows how the techniques can be used to map regional variations in the tissue oxygen saturation in the forearm under conditions of interrupted or restricted blood flow. The data processing involves both multivariate analysis of temporal imaging and spectral data and fuzzy C-means clustering of temporal imaging data.

Villringer and Chance [10] published "Noninvasive optical spectroscopy and imaging of human brain function." Crowley and Zimbelman, of the U.S. Geological Survey, Reston, VA [11], completed work entitled "Mapping hydrothermally altered rocks on Mount Rainier, Washington, with airborne visible/infrared imaging spectrometer (AVIRIS) data." Makipaa, Tanttu, and Virtanen [12], of Tampere University of Technology, Pori, Finland, published "IR-based method for copper electrolysis short-circuit detection." Ning et al. [13], of Oregon State University, Corvallis, OR, published "Five novel applications of imaging visible and short near-infrared spectrophotometry and fluorometry in the plant sciences. Part I: Photographic and histological applications." Rowlands and Neville [14], of CAL Corporation, Ottawa, ON, published "Calcite and dolomite discrimination using airborne SWIR imaging spectrometer data." Otten et al. [15], of Kestrel Corporation, Albuquerque, NM, published "Measured performance of an airborne Fourier transform hyperspectral imager."

Valdez et al. [16], of the Electrical and Computer Engineering University of New Mexico, Albuquerque, NM, published "Selection of spectral bands for interpretation of hyperspectral remotely sensed images." Silk and Schildkraut [17], of Block Engineering, Marlborough, MA, completed work entitled "Imaging Fourier transform spectroscopy for remote chemical sensing." Rowan et al. [18] of the U.S. Geological Survey, Reston, VA, completed "Analysis of airborne visible-infrared imaging spectrometer (AVIRIS) data of the Iron Hill, Colorado, carbonatite-alkalic igneous complex."

Mapping

Mapping technologies provide a detail of data analysis in which individual spatial images are combined across a broad 2- or 3-dimensional landscape to produce an actual map of the spectrally resolved data across a spatially resolved surface. Images are a part of the map, but images alone do not comprise a map until they are directly related to the spatial dimension. Heekeren et al. [19], from the Department of Neurology, Charite, Humboldt-University, Berlin, Germany, completed work entitled "Towards noninvasive optical human brain mapping—Improvements of the spectral, temporal and spatial resolution of near-infrared spectroscopy." Kruse [20], of the Analytical Imaging and Geophysics LLC, Boulder, CO, published "Geologic mapping using combined analysis of airborne visible/infrared imaging spectrometer (AVIRIS) and SIR-C/X-SAR data." Watanabe et al. [21], Department of Neurosurgery, Tokyo Metropolitan Police Hospital, Fujimi, Chiyoda-ku, Tokyo, Japan, published "Noninvasive functional mapping with multichannel near-infrared spectroscopic topography in humans."

APPLICATIONS

Agricultural Sciences

Forage samples were dried and ground fine using standard methods, and NIR spectra were obtained from 1100 to 2500 nm. For calibration purposes the ash content was determined by combustion in a muffle furnace at 450°C for 12 h. A modified stepwise multiple linear regression technique was used for calibration. The paper by Vazquez de Aldana et al. [22] reported positive results for using NIR spectrometry for rapid, accurate estimation of ash content in forage material. A general article by Johnsen [23] addresses the use of on-line near-infrared spectroscopy in the feed and forage industries. A general article by Marvik [24] describes the uses of near-infrared spectroscopy for on-line feed analysis. Enzymatic degradation of wheat starch is monitored using near-infrared spectroscopy in a paper by Sinnaeve et al. [25]. Near-infrared reflectance spectroscopy is used to measure protein, oil, and moisture in whole-grain soybeans in work by Takahashi et al. [26].

Sernevi [27] reported on the use of a near-infrared transmittance network and on-line systems in grain segregation and process control. Takacs and coworkers reported on a portable near-infrared wheat analyzer in [28]. Rapid classification schemes are described in tobacco mixtures by using near-infrared spectroscopy and neural network classification techniques by Lo [29]. Rutherford and Van Staden [30] reported on the prediction of plant resistance to insect infestation based on analysis of the stalk surface wax using near-infrared spectroscopy.

Animal Sciences

Photoacoustic infrared spectroscopy has been demonstrated to be useful in the analysis of surface-treated wool by Carter and colleagues [31].

Biotechnology

The use of near-infrared measurement for process analysis in fermentation processes is discussed in work by Brookes et al. [32]. Near-infrared spectroscopy has shown promise in identification of various strains of yeast in Halasz et al. [33]. Zakim and Diem [34] have shown that spectral cytometry has value for the rapid analysis of cells. Macaloney et al. [35] have reported that near-infrared spectroscopy has shown itself to be valuable for the simultaneous measurement of several constituents essential for production control of a high-cell-density recombinant *Escherichia coli* process. Near-infrared spectroscopy is proposed as a method for estimating acetate, biomass, glycerol, and ammonium in an unmodified whole broth by Hall et al. [36]. The method is reported to measure and compute the concentrations of these components in 1 min (i.e., nearly in real time). The method could potentially be used on-line for process monitoring and control. The estimations for acetate and glycerol were calculated using a multiple linear least squares regression calibration algorithm. The estimations for ammonium and biomass were determined using a partial least squares calibration technique. The standard error of prediction (1σ) for each of the acetate, biomass, glycerol, and ammonium components were 0.7, 1.4, and 0.7 g/L and 7 mM, respectively.

DRIFTS and artificial neural networks have been reportedly combined for the rapid identification of bacterial species by Goodacre et al. [37]. NIR spectroscopy has been shown by Macaloney et al. to be valuable in control and fault analysis during high-cell-density *E. coli* fermentation [38]. A general discussion by Dosi et al. of the use of on-line near-infrared analysis for control of fermentation processes is given in Ref. 39. FT-IR is used by Szalontai et al. [40] to investigate fatty acyl chains in model and biological membrane structures. The air–water interface has been studied using polarization modulated Fourier transform infrared spectroscopy by Cornut et al. [41] in order to understand the basic structure and orientation of lipids and amphipathic peptides. This work is undertaken to gain a more fundamental understanding of the structure of biological membrane. Raman spectroscopy has been used by Berger et al. to determine concentration of biological materials in aqueous solutions [42]. FT-IR spectroscopy has been used by Henderson et al. [43] to identify *Candida* at the species level using various spectroscopic discrimination techniques.

Polarization modulation has been described by Fournier et al. for in situ measurements of monolayer films in the study of biological membranes at the air–water interface [44]. Lipid monolayers have been studied by Axelsen et al. in situ using internal reflectance FT-IR [45]. NIR has been used by Carlsson et al. for quality-control testing of hyaluronan as an alternative to biological testing [46]. FT-IR spectroscopy has been used to study biological membranes on solid surfaces as well as at the air–water interface by Pezolet and coworkers [47].

Cosmetics

None

Earth Sciences and Mineralogy

A review by Aldstadt and Martin containing 64 references describes the use of a cone penetrometer for in situ analysis of contaminants in soil and groundwater. The review contains details of samplers for soil gas, groundwater, and soil and methods for determination of organic contaminants. Methods covered for organic analyses include laser-induced fluorescence, Raman and infrared spectrometry, and sensors. Inorganic contaminants are analyzed using laser-induced breakdown spectrometry, XRF, and passive gamma-ray spectrometry [48]. Infrared analysis of coals by Matuszewska is described in Ref. 49. Speciation of water types in haplogranitic glasses and melts was determined by Nowak and Behrens using in situ near-infrared spectroscopy [50]. Airborne infrared spectroscopy of wildfires in the western United States is described by Worden and coworkers [51].

An FT-IR open-path air monitoring system was used to monitor volcanic fumarole gases by Chaffin et al. [52]. The spectrometer described included a 1-mm^2 Hg-Cd-telluride detector cooled with liquid N$_2$ and an infrared field source. The spectrometer and notebook computer were operated from a 12-V DC power pack, allowing portability. The applications of active and passive monitoring techniques are discussed. Of the potential analytes in volcanic plume gases, SO$_2$ and HCl were the only gases detected the best. The use of the system is discussed. Diode-laser-based near-infrared (NIR) spectrometry is described as an ideal and inexpensive method for gas monitoring by Martin and Feher [53]. NIR diode lasers for use in a transmission spectrometer are described for the analysis of atmospheric or waste gases. These gases can be measured in situ or by using extractive measurement techniques. Examples of continuous monitoring of NH$_3$ in stack gas are presented, as are issues of spectrometer sensitivity and design details.

FT-IR spectrometers were used by Watson and coworkers [54] for environmental analyses. Instrumentation was carried on tethered and free-flying manned hot-air balloons. These instruments were configured to include: mid- and far-IR cameras, multispectral imaging spectrometers, radiometers, forward-looking IR cameras, as well as a variety of other sensors. The instrumentation was powered from a portable generator suspended from the gondola or by deep-cycle batteries placed in the gondola. The experiments described include the detection of chemical effluents by FT-IR spectrometry and atomic emission spectrometry (AES). Other experiments are described, such as the detection of Hg contamination, stack testing, multisite air quality monitoring, and the in situ vertical profiling of the atmospheric concentration of various pollutants.

Atmospheric volatile organic compounds have been monitored by Hammaker et al. [55] using remote FT-NIR. The use of far-infrared spectroscopy for studies of terrestrial and planetary or extraterrestrial atmospheres is described by Griffin [56]. Forestry studies conducted in North Carolina by Zwicker report using the remote sensing capabilities of open-path FT-IR spectrometry to survey field data [57]. NIR has been used for rapid assessment and quantitative measurements of suspended C, N, and P from Precambrian shield lakes in Malley et al. [58]. Near-infrared spectroscopy and visible spectroscopy have been used to characterize asteroid composition via surface mineralogy, as described in a dissertation by Howell [59].

FT-IR spectroscopy has been used to characterize interfacial water species in situ at hydrophobic and hydrophilic surfaces in mineralogy by Yalamanchili et al. [60]. Remote FT-IR has been mounted on balloon-borne platforms for remote sensing of various analytes in the atmosphere, as described by Camy-Peyret et al. [61]. Atmospheric free radicals have been studied in situ by Blake using a mid-infrared magnetic rotation spectrometer [62]. Short-wave infrared spectrometry has been reported for use in airborne prospecting for gas and oil resources by Yang [63]. Infrared microspectroscopy has been shown by Gao et al. to be useful in the rapid identification of gemstones [64]. Multispectral instrumentation has been used to record the burning of biomass in West Africa by De A. Franca et al. [65]. Visible spectroscopy and near-infrared spectroscopy have been used to study the exposure levels and iron concentration of lunar soils by Fischer and Pieters [66].

Environmental Sciences

Open-path FT-IR spectrometry is reviewed for field use in remote sensing of airborne gas and vapor contaminants by Levine and Russwurm [67]. The review includes 27 references covering open-path FT-IR applications and describing where additional work is required for further development of the technique. The advantages of open-path FT-IR spectroscopy are described by Marshall et al. [68]. These benefits include (1) its versatility, (2) remote, long-path measurement, (3) in situ applications, and (4) near-real-time measurement. The technology is described in terms of the physical limitations of the instrumentation and the limitations in current data-processing techniques. FT-IR is combined with column liquid chromatography for the identification of herbicide contamination in samples of 50–100 ml river water by Somsen et al. [69]. Enough analyte was obtained for good-quality spectra down to water concentrations of 1–2 $\mu g/L$. Of six library search programs tested for solute identification, the best results (39 out of 45 correctly identified) were given by the peak-search-intensity-ignored algorithm. FT-IR emission spectroscopy has been employed by Wang et al. for the study of Freon-12 in an alcohol-plus-air flame [70].

Infrared spectroscopy has been reported by Haschberger for use as a remote method for monitoring trace gases from aircraft emissions [71]. Environmental monitoring applications using fiber-optic sensing and FT-IR spectroscopy is described by Druy et al. [72]. Ground-based FT-IR spectroscopy was used for remote sensing of pollution emission sources for application in environmental compliance monitoring by Schaefer [73]. Marshall et al. describe the effects of resolution on the performance of classical least squares (CLS) spectral interpretation when applied to volatile organic compounds (VOCs) of interest in remote sensing using open-air long-path Fourier transform infrared (FT-IR) spectrometry [74]. A 1995 patent [75] describes the method and apparatus for remote infrared sensing and analysis of motor vehicle exhaust gases.

A real-time background correction algorithm is described for passive FT-IR spectroscopy of chemical plumes for environmental monitoring applications by Polak et al. [76]. Organic wastewater contaminants were measured using acousto-optical tunable filter (AOTF)–based NIR spectroscopy by Eilert et al. [77]. Various trace gases are measured in situ using airborne infrared diode laser spectrometry by Toci and coworkers [78]. Soil contamination was rapidly determined using reflectance Fourier transform infrared spectroscopy by Adams and Bennett [79]. Theriault and Bissonnette report on the use of ground-based FT-IR for monitoring weather cloud parameters [80]. Remote FT-IR was used to remotely monitor motor vehicle exhaust gases by Davies et al. [81].

Fine Chemicals and Chemical Production

The potential difference measured in situ using FT-IR spectroscopy is presented by Bae et al. [82] for the reduction of nitrite on a solid polycrystalline Pt electrode in aqueous 0.1M $HClO_4$. The intensity of the N_2O asymmetric stretch band at 2231 cm^{-1} is monitored during a voltammetric sweep. The FT-IR results correlated with dynamic differential electrochemical MS observations previously published by Nishimura et al. (*Electrochim. Acta,* 36: 877, 1991). An on-line NIR method is presented by Norris and Aldridge for determination of the steady-state endpoint of homogeneous and heterogeneous organic reactions for chemical and pharmaceutical production [83]. The method uses periodic NIR monitoring over the progress of the reactions. The steady-state point is detected when the NIR spectra do not change significantly over time. The endpoint is confirmed by applying multivariate calibration methods to the spectral data, such as dendrograms (hierarchical cluster analysis), and scores plots from principal component analysis (PCA).

Richmond et al. compare photoelastic modulation (PEM) to real-time polarization modulation (RTPM) for the acquisition of cyclic voltammograms [84]. The voltammograms were obtained using a crystalline Cu rod electrode versus Ag/AgCl in KCl (0.1M) in the presence of 25mM thiocyanate ion (I), glucose (II) or imidazole (III) with a 50-mV/s sweep rate. Infrared spectroscopy using a 60° incidence angle was focused through a calcium fluoride (CaF_2) cell window via a photoelastic modulator (PEM) at 37 kHz. Real-time polarization modulation (RTPM) waveforms were sampled using a liquid N_2-cooled multichannel plate detector. The center frequency of the RTPM was 1975 cm^{-1} and 1575 cm^{-1} for I, II, and III, respectively. The PEM and the sampling electronics were removed for the acquisition of static linear polarization waveforms. The modulation technique produced overall increases in sensitivity relative to the static procedure, with excellent rejection of signals associated with atmospheric gases. Both techniques gave equivalent information with respect to surface species but differed in the extent to which the electrolyte solution was sampled. It was thus possible to discriminate spectral features associated with adsorbed species from those due to soluble components.

May et al. [85] report the use of an FT-IR spectrophotometer equipped with an 8-m gas cell having solid nickel mirrors. The system is used for the determination of H_2O in anhydrous HCl samples. An on-line determination of impurities in gaseous Cl_2 using a temporary system is described. In addition, the on-line monitoring of impurities in anhydrous HCl as it passed through a carbon bed using a permanently installed system is described. Conventional methods for the analysis of chlorine and hydrogen chloride are also reviewed. Bae et al. [86] report the use of two different in situ FT-IR spectroscopic techniques to study the vibrational properties of the model redox-active species 2,5-dihydroxybenzyl mercaptan adsorbed on Au in aqueous 0.1M $HClO_4$. The different spectroscopic methods included (1) potential-difference ATR FT-IR using a 4-nm-thick gold layer sputtered onto a Au-patterned ZnSe internal reflection element as the electrode, and (2) potential-difference FT-IR external reflection absorption spectroscopy on a solid Au electrode. The spectra obtained using the two methods were nearly identical.

The combination of chemometrics with FT-IR spectrometry was used by Lindblom et al. [87] to monitor diphenylamine stabilizer concentrations in the presence of nitrocellulose, the main component of rocket propellants. The shelf life of these propellants is normally determined by HPLC analysis of the diphenylamine stabilizer and its derivatives. A more rapid method of analysis was desirable, so the feasibility of using a partial least squares calibration model of the IR spectra as calibrated using concentrations determined by HPLC was investigated. The combination of chemometrics with FT-IR spectrometry allowed the monitoring of stabilizer concentrations to 0.2% diphenylamine. However, this spectroscopic method was not as accurate as HPLC. Standard error using the FT-IR method as compared to the HPLC method was 10% relative, with a correlation of 0.94 for diphenylamine and its first three derivatives.

FT-IR was used by Bandekar and coworkers [88] to quantitatively measure sulfur oxide in white liquor. Liquid samples of approximately 1.5 mL were placed into Circle (cylindrical internal reflectance cell for liquid evaluation) cells comprising an open-boat arrangement with a zinc selenide crystal at 45°. The infrared beam was configured for 12 internal reflections. The method was suitable for quantitative analysis calibration. Ranges to 70, 45, and 42 g/L were obtained for sulfate, thiosulfate, and sulfite species, respectively. Corresponding minimum detection limits were 0.01, 0.02, and 0.02 g/L. The method should be useful for in situ analysis but is limited to samples that are water-soluble and have IR-active bands in the mid-IR region.

IR spectroelectrochemical measurements were performed by Faguy and Marinkovic [89] on single-crystal Pt electrodes for a variety of electrochemical systems. Use of a ZnSe hemispherical window in the dual roles as a lens and as an IR-transparent wall of the electrochemical cell allowed beam collimation, near-critical-angle reflection, and low first-surface-reflection losses. Derivation of the expected figure of merit for in situ reflection/absorption spectroscopy is presented. Examples are given of adsorption processes from H_2SO_4 solutions and for the oxidation of glucose on Pt(111) and Pt(100) electrode surfaces.

An FT-IR spectroscopic method was developed by Ge et al. [90] using an attenuated total reflectance (ATR) sample boat with a zinc selenide (ZnSe) or zinc sulfide (ZnS) crystal. Methanol/acetonitrile (1:1) and air were used as the background spectra for the two types of

ATR crystals, respectively. The method was used to monitor the epoxidation step in the synthesis of *cis*-1-amino-2-indanol when indene is oxidized to indene oxide. For the ZnSe ATR a portion of the reaction mixture was analyzed by FT-IR spectroscopy without sample preparation, and for the ZnS ATR the sample was centrifuged to separate the organic and aqueous layers, with FT-IR spectrometry being performed on the organic layer. Results were calibrated and cross-validated by HPLC, with details described. Spectral data was collected from 1500 to 900 cm^{-1} and processed using multiple linear regression (MLR) and partial least squares (PLS).

An FT-IR method is described, with resultant data analysis techniques presented, for the prediction of organic vapor concentrations by Ruyken et al. [91]. The method includes techniques for identifying interferents present within the vapor mixtures. Basic principles of multivariate calibration using principal component regression (PCR) are outlined, as is the effect of interfering species on the predicted concentration. The infrared spectra of organic vapors at 60°C were recorded at 4 cm^{-1} resolution. A simple two-component calibration model was constructed for hexane and ethanol, with additions of dichloroethane, acetone, methyl acetate, and toluene as interferents. The prediction errors depended on the concentration of the interferent and the overlap of each interferent's spectrum onto the analyte regression vector. A residual library search method was used to identify each interferent.

The FT-IR spectrum of a ternary aqueous solution of chlorate, perchlorate, and dichromate was measured from 900 to 1250 cm^{-1} by Kargosha et al. [92]. After subtraction of the water spectrum and standard baseline correction, the spectrum was resolved using multicomponent analysis (i.e., partial least squares regression and multiple linear regression). The infrared bands at 973, 1110, and 950 cm^{-1} were used to quantify chlorate, perchlorate, and dichromate, respectively. Calibrations were linear from 0.2 to 63.8, from 0 to 73.47, and from 0 to 149 g/L of chlorate, perchlorate, and dichromate, respectively. The corresponding detection limits for each of the analytes were 1.064, 0.64, and 0.95 g/L. The relative standard deviation as 1 σ (RSD) for eight samples were from 1.19 to 2.7% relative. The method was applied to electrolytic solutions of chlorate and perchlorate.

In situ FT-IR spectroscopy was used to study the structures of powdered humic acid salts by Woelki and Salzer [93]. Samples were prepared by placing the salt samples with KBr in a DRIFTS (diffuse reflectance) spectrometer accessory. The sample reaction chamber was flushed for 15 min with dry nitrogen (N_2), and the sample was heated under stopped- and continuous-flow N_2 at 10° and 65°C/min, respectively. The spectra were obtained at 4 cm^{-1} resolution by coadding 50 scans at slow heating and 200 scans at fast heating. The structural changes caused by heating were recorded, with results presented.

Remote high-resolution FT-IR was used by Wang et al. [94] for spectral characterization of infrared flare material combustion. The flare materials were pressed into a 30-g pellet (20 × 15 mm) and positioned 15 m from a remote sensing (open-path) FT-IR instrument with a combustion flame 50 × 50 cm. The emission spectra were recorded from 4700 to 740 cm^{-1} with a resolution of 0.24 cm^{-1}. Peaks corresponding to HCl, HF, H_2O, CO_2, CO, and SiF_4 were obtained, allowing qualitative and quantitative determination of these combustion products. The flame temperature was estimated and the spectral radiation distribution was reported.

Rapid IR-spectrophotometric determination of total nitrogen in silicon dioxide–silicon nitride (SiO_2–Si_3N_4) mixtures is presented by Wisniewski and coworkers [95]. Samples of 1–1.5 mg were ground with 200 mg KBr and pressed into a tablet. The IR spectra were recorded from 400 to 2000 cm^{-1}. The N content was quantitated from the transmittance at 575 and 1450 cm^{-1} (background) using a formula described. The calibrations were linear over the 20–35% nitrogen levels. The relative standard deviation (RSD) of 1 σ was 7.5–44.5% relative. These results were compared to those obtained using a volumetric method where samples are dissolved in acid and the ammonia (NH_3) is released using distillation. The analysis time was 0.5 h. The method was applicable to mixtures containing lithium and magnesium salts and is suitable for evaluating the yield of Si_3N_4 synthesis from SiO_2 and NH_3.

In situ temperature-programmed diffuse-reflectance infrared Fourier-transform spectroscopy (TPDRIFTS) is used by Centeno et al. [96] to monitor chemical changes in the surface structure of vanadium oxide/titanium oxide catalysts during reactions. A general article is presented by

Brimmer describing the development and applications of near-infrared spectroscopy in the chemical and pharmaceutical industries [97]. Fiber-optic-based Fourier transform infrared spectroscopy (FT-IR) was used by Salim et al. [98] for in situ measurements to monitor the concentration of various reactants in a vertical OMVPE reactor. In situ attenuated total reflection (ATR) infrared spectroscopy was used to evaluate and study the various aspects of laminar flow in a channel-type electrochemical cell by Tolmachev et al. [99].

In situ IR reflectance spectroscopy was used by Lamy et al. [100] to study and monitor the electro-oxidation of methanol at Pt-Ru electrodes. In situ diffuse-reflectance infrared spectroscopy (DRIFTS) was used by Benitez et al. [101] for the study of the reversibility of CdGeON sensors toward oxygen. In situ ATR FT-IR spectroscopy was described in a doctoral dissertation by Dunuwila [102] for the measurement of crystallization phenomena for research and development of batch crystallization processes. In situ infrared spectroscopy was used to study the solid-state formation reaction of polyglycolide by Epple et al. [103]. In situ near-infrared spectroscopic investigation of the kinetics and mechanisms of reactions between phenyl glycidyl ether (PGE) and multifunctional aromatic amines is described by Xu et al. in [104].

Remote mid-infrared spectroscopy is used by Mijovic and Andjelic to monitor reactive processing for polymers [105]. Developments of in situ molecular spectroscopy in catalysis studies is given by Xin and Gao [106]. Cross-dispersion infrared spectrometry (CDIRS) for remote chemical sensing is described by Stevens et al. [107]. In situ infrared spectroscopy is used to by Eder-Mirth to study the mechanisms of selective alkylation of aromatics over zeolites [108]. Rapid infrared sampling methods are described by Scameborn for the organic chemistry laboratory using a diffuse reflectance (DRIFTS) accessory [109]. In situ FT-IR and FT-Raman studies of reactions in the hydrothermal environment are described by Schoppelrei in [110]. The use of near-infrared monitoring of dimethyl terephthalate in a low-pressure methanolysis process is presented by Agyare [111]. In situ infrared studies on dehydroxylation of MCM-41 are described by Li et al. [112].

Infrared absorption spectroscopy is used as a method for rapid fault detection in borophosphosilicate glass thin films by Zhang et al. [113]. In situ FT-IR studies of the acidity and catalytic properties of Pt/zeolites is described by Chien et al. [114]. In situ FT-IR spectroscopy is related to kinetic studies of methanol synthesis from CO/H_2 over $ZnAl_2O_4$ and $Cu-ZnAl_2O_4$ catalysts by Le Peltier et al. [115].

Fourier transform near-infrared (FT-NIR) is used by Binette and Buijs for process monitoring of multiple inorganic ions in aqueous solution. A scheme is presented for a near-infrared multipoint monitoring system for a three-stream plant producing soda using the Solvay process [116]. Near-infrared spectroscopy is used by Yeboah et al. for quantitative analysis of resorcinol in aqueous solution for chemical production in the photographic industry [117].

Food and Beverages

A presentation of the uses of ATR FT-IR spectroscopy for process control in the sugar industry is given by Veronique and Gilles in [118]. A patent is presented for real-time on-line analysis of organic and nonorganic compounds for food, fertilizers, and pharmaceutical products [119]. Dispersive near-infrared spectroscopy was used by van den Berg et al. to analyze 30 alcoholic beverage samples for ethanol. Samples were contained using a 2-mm-thick quartz cuvet, with spectral measurements made over the wavelength range of 1100–2498 nm. The absorbance spectra were preprocessed using a second derivative to remove offset and sloping baseline. The reported results indicated the suitability of the method for ethanol determination in alcohol-containing beverages. Increasing the variability in the training set by increasing the number of samples improved model performance [120].

Neural networks combined with near-infrared spectroscopy are described by Borggaard and Thorup for on-line quality measurement of unhomogenized meat products [121]. FT-IR combined with a liquid ATR cell heated to 50°C was used by Mossoba et al. to determine the total trans content of neat hydrogenated edible oils. The sloping background of the infrared trans band at 966 cm^{-1} was eliminated by ratioing the single-beam spectrum of the sample to the corresponding spectrum of the unhydrogenated material; the trans band could then be integrated between 990 and 945 cm^{-1}. The limits of detection and determination were 0.2 and 1%, respectively [122].

Near-infrared diffuse reflectance spectroscopy was used by Lee et al. for the rapid analysis of curds during a cheddar cheese–making process. Blended curd samples were packed into rotating 5-g sample cells. The samples were measured using a tilting filter NIR scanner equipped with three tilting filters for scanning from 1900 to 2320 nm. Data analysis was performed using multivariate computer software. Estimated values for moisture, fat, and lactose were calculated. Multilinear regression (MLR) equations were used to compare the results by NIR spectroscopy to classical chemical analysis. NIR spectrometry was reported to be suitable for rapid monitoring of chemical changes occurring in curds during the making of cheddar cheese [123].

An FT-IR method is reported by Hewavitharana and van Brakel for the direct determination of casein in milk. FT-IR spectra were recorded for homogenized milk samples while maintaining a sample temperature of 40°C. Data-acquisition parameters, such as resolution and the types of background spectra, were studied to achieve optimum conditions. Two chemometric methods, viz., partial least squares (PLS) and principal components regression (PCR), were used for data processing. The best results were obtained using a resolution of 4 cm^{-1} with PLS calibration over the spectral regions 3000–2800, 1600–1500, and 1300–1000 cm^{-1}. For comparison, the results obtained agreed closely with those obtained by the International Dairy Federation reference method [124].

Near-infrared spectroscopy is described by Berding and Brotherton [125] as a means to measure heterogeneous sugar cane samples. The typical labor-intensive subsampling necessary to overcome the within-sample heterogeneity inherent in the batch sampling of agricultural products, specifically sugar cane, has prompted the development of an at-line device for the presentation of large (1.5–3-kg) samples to the measuring instrument. An NIR spectrophotometer fitted with a remote reflectance probe was used to demonstrate the analysis of 30 samples per hour.

Near-infrared transmission (NIT) spectroscopy was applied by Lovasz et al. for the determination of several quality parameters in apples. Apples were analyzed for firmness, RI, pH, titratable acid, dry matter, and alcohol-insoluble solids using NIR transmission spectrometry. Short-wave near-infrared spectra were acquired over the wavelength range of 800–1100 nm. Data were processed using a chemometrics software package that included partial least squares regression (PLSR), multiplicative scatter correction (MSC), automatic outlier diagnostics, and standard cross-validation. The NIT results for each parameter (except titratable acid) were similar to those obtained using the standard reference methods [126].

Near-infrared transmittance (NIT) spectroscopy was used by Wold et al. to determine the average fat content of Atlantic salmon. Salmon were collected, killed, bled, and kept on ice; samples were shaped into cylinders of intact tissue (23 mm thick, 20 mm diameter). NIT spectra were over the wavelength range of 850–1048 nm, with the bandwidth set at 6 nm. Samples with skin intact were oriented with the skin side facing the incident light; samples were kept at 4–6°C throughout the measurement cycle. A good correlation was obtained between average fat content of salmon and transmission spectra measured on intact fish muscle with skin and scales. The method is rapid and nondestructive [127].

NIR spectroscopy was performed by Downey on fresh salmon, through the skin, to determine its moisture and fat content. Six sample sites were analyzed on the ventral side and six on the dorsal side of each salmon. Spectra were recorded on a scanning spectrophotometer with a surface interactance fiber-optic probe and a transmittance detector module at wavelengths from 400 to 1100 nm at 2-nm intervals. The best calibration was produced by a partial least squares treatment of the first-derivative spectra. The method should be useful for optimizing the oil and moisture content of farmed salmon [128].

A combined visible and short-wave near-infrared transportable spectrophotometer system was developed for on-line classification of poultry carcass quality by Chen et al. [129]. The spectrophotometer is a bifurcated optical-fiber system for measuring the reflectance of carcasses in the wavelength region of 471–963.7 nm. A diagram of the probe that collects light energy from the skin and muscle tissue of the carcass is presented, and the remaining components comprise a tungsten halogen lamp, a grating spectrograph with a photodiode array detector, and a PC. A neural network for classification into normal, septicemic, and cadaver carcasses was developed. When this device was used to classify normal and abnormal (septicemic or cadaver) carcasses, an overall accuracy of 97.4% was obtained.

A review is presented by Lipp, containing 125 references, describing the developments of methods for the determination and quantification of triglycerides in milk. The referenced papers appear in the literature during the period 1990–1994. In particular, gas- (GC) and high-pressure liquid-chromatographic (HPLC) methods are highlighted. Analytical methods such as infrared (IR) spectroscopy, mass spectrometry (MS), 12C/13C isotopic analysis, differential scanning calorimetry (DSC), gravimetry, and titrimetry are also discussed. An evaluation of the different approaches to the detection of adulteration of milk fat was also described [130].

A rapid, automated Fourier-transform infrared spectroscopic method was developed by van de Voort et al. for the determination of cis and trans content of edible fats and oils. A sample handling system for use in routine quality-control measurement of fats and oils is described. Prior to analysis, samples were warmed to within 5°C of the operating temperature of the instrument, using a microwave oven. The heated samples were aspirated into IR measurement liquid cell. The spectra were measured at a resolution of 4 cm^{-1} for 128 scans per spectrum. The liquid cell was emptied and reloaded with the next sequential sample. Spectra were calibrated by partial least squares (PLS) analysis for the cis and trans content, with results given [131].

Two infrared spectroscopic methods for cheese analysis are compared by McQueen et al. The methods were NIR filter (i.e., named optothermal spectroscopy; cf., *Int. Lab.* 20: 16, 1990) and ATR FT-IR spectroscopy. In NIR filter spectroscopy, spectra were recorded using optical transmission filters of 1740, 1935, and 2180 nm. The ATR spectra were recorded using a conventional FT-IR spectrometer at a resolution of 8 cm^{-1}. Partial least squares (PLS) regression analysis was used to correlate the spectral data to the composition of fat, protein, and moisture determined in 24 cheeses using reference analytical methods. With the NIR filter spectroscopy, correlation coefficients (*r*) were 0.95–0.97; the corresponding values obtained using ATR FT-IR were 0.97–0.99. The NIR technique was more rapid and somewhat easier to use than FT-IR [132].

Carbon dioxide in water samples was analyzed using an analytical development nondispersive IR CO_2 analyzer, as reported by Abdullah and Eek, was coupled to a ChemLab segmented-flow Auto-Analyzer 2 system. The sample stream was introduced to the analyzer using a flow rate of 0.4 mL/min. Once introduced the sample was mixed with deionized H_2O (0.8 mL/min) and 0.5M HCl (0.1 mL/min). The stream was sent into a gas stripper, with N_2 as carrier gas flowing at 800 mL/min. The gas phase was passed over H_2SO_4 and then through a magnesium perchlorate column before IR analysis; calibrations were linear from 0–3mM total CO_2. This technique was used to analyze seawater, estuarine water, and fresh water and was reported superior to alternative methods based on measurements of total alkalinity [133].

FT-IR spectroscopy was investigated by Bellon-Maurel et al. for on-line monitoring of the concentrations of glucose, maltose, maltotriose, and maltodextrin in starch hydrolysis mixtures. Measurements were made using a conventional FT-IR spectrometer with zinc selenide (ZnSe) ATR flat crystal accessory. Calibrations were accomplished using a partial least squares (PLS) regression approach with a calibration set of 30 mixtures. Acceptable results were obtained for the determination of glucose and maltose: the standard errors of prediction for glucose, maltose, maltotriose, and maltodextrin were 4.1, 3.4, 4.9, and 4.4 g/kg, respectively, compared to the required precision of 8, 10, 5 and 5 g/kg, respectively. Both repeatability and reproducibility were satisfactory [134].

A near-infrared method for rapid yeast trehalose measurement has been described by Moonsamy et al. The yeast slurry samples were prepared for analysis by dilution in beer; dried yeast samples were prepared by a method described previously (cf. *J. Am. Soc. Brew. Chem.* 52:145, 1994). Dried yeast samples were ground to a powder before spectral measurement using a dry sample fiber-optic probe. The slurry sample probe was introduced directly to the diluted yeast slurries. Both trehalose and glycogen spectra were preprocessed using a second-derivative data analysis over a range of 1000–2222 nm (4500–9996 cm^{-1}) using 459 data points. The method was reported to produce acceptable performance and may be suitable for rapid in-line measurements of yeast intracellular components [135].

NIR is reported by King-Brink et al. for use in the rapid analysis of meat composition over the typical raw material ranges for fat (6.8–58.8%), water (31.4–72.2%), and protein (8.9–21.0%) [136]. NIR has been used for many years in the routine analysis and discrimination of wheat products for quality control or assessment of the end products as reported by Bertrand et al. [137]. NIR is presented by Li et al. as a means for rapid assessment of the potential malting quality of barley and malt. In this work, 10 different components are monitored in the barley and malt for direct measurements of malting quality [138]. NIR has been proposed as a rapid method for determination of free fatty acid in fish and its application in fish quality assessment by Zhang and Lee [139].

FT-IR is described as a method for the rapid determination of casein in raw milk by Hewavitharana and van Brakel [140]. NIR and a customized large cassette presentation module are used by Berding and Brotherton for analysis of heterogeneous samples from sugarcane evaluation trials [141]. A summary of the use of NIR for rapid nondestructive raw material identification in the cosmetic industry is described by Grunewald [142]. A general description of the uses for NIR in the food industry is given by Hoyer [143]. Song and Otto describe rapid determination of constituents in sausage products by near-infrared transmission spectroscopy [144]. Near-infrared transmission spectroscopy over the silicon detection region is presented by Tenhunen et al. for use in the malting industry [145]. NIR assessment of curds in cheese making is described by Lee et al. in [146].

NIR has also been demonstrated by Pascual et al. to be a useful measurement technique for automatic determination of protein fractions in Manchega ewe's milk [147]. Rapid detection of insect contamination in cereal grains is reported by Chambers and Ridgway in [148]. In-line NIR measurements are compared to density/sound velocity measurements for the determination of alcohol and extract in brewed alcoholic beverages by Stokes and Blazier in [149]. A variety of molecular spectroscopic techniques, such as UV, IR, and Raman spectroscopy, are compared by Chmielarz et al. as measurement techniques for studies of the double-bond positional isomerization process in linseed oil [150]. NIR technology applied for on-line assessment and monitoring during the manufacture of sugar syrups is described by Jones et al. [151].

A general discussion and review of NIR applications in the milling industry is given by Brotherton and Berding in [152]. Rapid fruit juice analysis using infrared spectroscopy is described by Meurens et al. [153]. In-line NIR measurement of important beer components is described by Petersen et al. [154]. Near-infrared transmittance (NIT) spectroscopy is described as a "powerful tool for fast process and product control in breweries and malthouses" by Zahn [155].

Forensic Science

Forensic science continues to refine the ability to identify original automobile topcoats using infrared spectroscopy with in situ measurement techniques. With ever-increasing databases and pigment search libraries, identification becomes easier, as reported by Suzuki and Marshall [156]. Infrared and other analytical methods are extensively described to aid in the rapid and conclusive identification of I homologues in illicit drug preparations. The hydrochloride salts of 3,4-methylenedioxycathinone (I) homologues were analyzed by IR spectroscopy using a KBr matrix. CHCl$_3$ solutions of the I homologues (as free bases) were analyzed by GC-MS-IR spectrometry to obtain vapor-phase spectra of the compounds. MS, NMR, and gas chromatography (GC) measurement parameters, and results are also presented. Melting points of the compounds, and presumptive tests, were also completed, with details described by Dal Cason [157].

Gas-Phase Analysis

A patent describing the method and apparatus for automatic identification of anesthetic gas mixtures in respiratory gas mixtures is described [158]. The system uses diode array detection, combined with variable-transmission filters at 4–5 μm and 7–10 μm. Infrared spectroscopy combined with principal components–based digital filtering has been described by Hayden and Noll as a means for identifying trace gases in multispectral and hyperspectral imagery [159]. Near-infrared spectroscopy has been utilized by Helland for on-line analysis of hydrocarbon gases. A method is described for maintaining calibration models over time [160]. A portable FT-IR analyzer system is described by Ahonen et al. for gas analysis in industrial hygiene situations [161]. FT-IR is described by workers at Argonne National Laboratory, Argonne, IL, as a method for continuous monitoring of gas emissions by Mao et al. [162].

Instrument Physics

Shaffer and Small, at the Center for Intelligent Chemical Instrumentation, Department of Chemistry, Clippinger Laboratories, Ohio University, Athens, have described the optimization of algorithms for discriminant analysis of remote FT-IR spectra. Five optimization methods—simplex optimization, simulated annealing, generalized simulated annealing, genetic algorithms, and simplex–genetic algorithm hybrid—were compared for their ability to position the discriminant. Simplex optimization consistently outperformed the other discriminant methods [163]. The method and practice for describing the performance of an airborne Fourier transform hyperspectral imager are described by Otten et al. [164]. Remote FT-IR spectroscopy is described by Mijovic et al. as a means of in situ real-time monitoring of chemical processes [165].

A patented and improved near-infrared quantitative analysis instrument includes a number of removable finger inserts, each dimensioned for a different finger size, which facilitates proper aligning and fitting of an individual user's finger into the optical system of the analysis instrument, taking into account the size of the individual's finger. The insert, according to the present invention, can also be designed to accommodate samples of various substances for quantitative analyte measurement in blood [166]. Rutter et al., at Lockheed Martin IR Imaging Systems, Lexington, MA, have derived a multispectral MCT detection system for remote infrared sensing of atmospheric constituents [167]. A basic description of the uses of near-infrared spectroscopy for process and production control is described by Hansen [168].

Uses of silica fiber optics and fiber-interface sampling devices for near-infrared spectroscopy is presented by Streamer and DeThomas [169]. Sample-conditioning needs for process streams for use in near-infrared and mid-infrared spectroscopy are described by Gallaher et al. [170]. Foulgoc et al. present work on the use of TeX-based fiber optics for remote infrared spectroscopy [171]. A discussion on the selection of spectral bands to interpret hyperspectral remotely sensed images taken using hyperspectral methods is described by Valdez et al. [172].

An infrared reflectance device is described by Vohra et al. that is intended for eventual incorporation in a conventional cone penetrometer. Using a nichrome-wire source, an off-axis paraboloidal mirror focused the incident beam to a circular spot size of 5 mm at the sample surface. The diffuse reflectance from the sample was collimated using a second off-axis parabolic mirror and then focused by a third onto the entrance face of a 12-m-long clad As/S-based chalcogenide fiber (core diameter 0.25 mm). The fiber conducted the reflected light to an FT-IR spectrometer with an MCT detector. The operating wavenumber range for the system is 6000–2225 cm^{-1}. A calibration graph based on the C–H stretching absorption at 3100–2850 cm^{-1} was reported to be linear up to 1000 ppm of trichloroethylene when measured using dry sand, with the detection limit reported as 250 ppm. The H–S combination band at 3200 cm^{-1} was used as a reference [173].

Genetic algorithms (GAs) were used to optimize piecewise linear discriminant analysis (PLDA), as employed by Shaffer and Small in the automated classification of FT-IR spectrometry remote sensing data [174]. Workman and Brown describe an ASTM (American Society for Testing and Materials) standard practice for multivariate calibration that was previously discussed (cf. *Spectroscopy* 11 (2):48, 1996). Part II describes a systematic approach to the calibration process, including major sources of calibration and analysis error and the evaluation of the calibration model with statistical tests. Other official methods and practices are briefly presented [175].

A transmission cell based on fiber optics and a stainless steel union cross was developed by Tilotta et al. for use in on-line SFE-FT-IR spectrometry. The sample was placed in the sample cell, pressurized with CO_2, and heated to the extraction temperature. After extraction (~30 min) the valve between the SFE cell and the IR cell was opened and the IR spectra were recorded. The IR cell was cleaned by flushing with liquid CO_2. Example spectra of fresh orange peel, (+)–limonene, gasoline-contaminated soil, and potato chips are presented. The method should be suitable for the determination of organic compounds in solid samples. A schematic diagram of the on-line SFE-FT-IR system is given [176].

Wilks describes small, portable filter–based IR analyzers that are dedicated to specific process analysis applications. They weigh less than 5 lb and consume less than 20 watts of power drawn from a 12-V DC source. Applications described include the determination of fat in hamburger meat, oil and grease in water and soil samples. These measurements are completed using the C–H stretch band of fats and oils found at 3.4 microns [177].

An overview is presented by Deeley et al. discussing ways in which manufacturers are facilitating the validation process of FT-IR spectrometry. Three areas are examined: the monitoring of instrument performance, establishment of the validity of data-manipulation procedures, and the generation of well-controlled operating procedures [178]. FT-IR-ATR spectroscopy using 50 internal reflections at a Ge/I ATR interface prior to detection from 4400 to 700 cm^{-1} is described by Kulesza et al. for study of the spectroelectrochemistry of Prussian blue in the solid state [179].

A two-fiber probe suitable for fluorescence, phosphorescence, Raman, mid-IR, and NIR (reflectance) spectroscopic measurements is described by Lin et al. The efficiency of this probe for fluorescence measurement across a window was demonstrated by polishing the fiber ends at predetermined angles and using different probe geometries. Best performance was obtained by maximizing the overlap of the light cones of the two fibers at the interface of the window and sample [180].

A method is described in detail by Treffman and Morrison for hand grinding and polishing of KBr lenses designed for coupling an IR beam from a FT-IR spectrometer into a hot-filament chemical vapor deposition reactor for in situ measurements [181]. Workman describes the current glossary of the ASTM (American Society for Testing and Materials) E 13.11 Chemometrics Task Group on Terminology. Comments and suggestions should kindly be passed to the author on the Internet at *jworkman@kcc.com* [182].

The performance of a commercially available zirconium fluoride fiber (ZrF$_4$), with smaller amounts of five other fluorides (Furuuchi), was compared with that of quartz fibers commonly used for remote NIR (1000–2500 nm) measurements by Maeda et al. The fibers exhibited good mechanical properties. Spectra were measured using a FT-NIR spectrophotometer. The instrument used a halogen lamp source and lead sulfide (PbS) detector. The new fiber has excellent transmittance in the 2000–2500-nm region, whereas the transmittance of quartz drops abruptly over this same region. Absorbance and second-derivative spectra of palmitic acid, potassium ferricyanide, silicone grease, and sucrose are illustrated, showing the useful bands available in the expanded NIR region [183].

Different aspects of the optimization of NIR FT-Raman spectroscopy (excitation at 1064 nm) are presented by Schrader. Methods for overcoming the fluorescence of impurities, increasing Raman intensity, and the efficiency of Raman scattering are summarized. Applications of the technique for routine analysis, process control, and quality control are discussed, and possible future developments of process Raman are considered [184].

A fuzzy logic program is described by Ehrentreich et al. for the complete extraction of knowledge from tabulated IR spectrum–structure correlations. Computer-based rule generation with use of the SpecInfo database is also used, and the identities of structural fragments are deduced from spectral features by using fuzzy logic. The CorTab rule-base is used with automatic rule generation and interpretation modules based on PROLOG and C computer languages [185].

An expert system knowledge base (ESKB) was designed by Meng and Ma that employs a set of 360 rules (details given) to express the laws of spectral interpretation, in order to imitate the reasoning process of the human brain in relation to the IR spectrometry group-frequency correlation (GFC) of pentavalent organophosphorus compounds (OPC). The OPC are divided into three groups: (1) compounds containing a P = O group, (2) compounds containing a P = S group, and (3) compounds containing a P = Se group. The ESKB consists of IR band positions, intensities, and widths for each OPC and has been built up using data obtained in-house from a large number of OPC. The ESKB uses logic-based expressions to deduce GFC based on the three IR parameters. It is semiautomatic and convenient to maintain [186].

A review is presented by McKelvy et al. on the aspects of IR spectroscopy relevant to chemical analysis for the period November 1993 to October 1995. On-line and in-process applications are discussed as well as in situ and real-time analysis. Applications such as environmental analysis, food and agriculture, surface techniques, polymer applications, and biochemical analysis are discussed; 1958 references [187].

Advice is presented by Gold on the consideration, adoption, and success criteria of NIR methods in industry [188]. A mid-IR probe was fabricated by MacLaurin et al. from two single-strand 1.5-m × 750-micron core chalcogenide fibers, a zinc selenide (ZnSe) internal reflection element, and an adjustable optical interface for the spectrometer. An integral internal water-cooling system maintained the fibers at 35°C when the probe was immersed in a sample at 200°C. Its performance was illustrated by the in situ monitoring of the acid-catalyzed esterification reaction of but-2-enoic acid and butan-2-ol in toluene at 110°C for 8 h. The results were comparable with those obtained using an overhead ATR flow cell arrangement with an extractive sampling loop. Quantitative calibration of the spectroscopic data was achieved using GC reference data. Relative errors of less than 5% were obtained [189].

Some of the main factors that must be considered in the development and implementation of process optical-fiber NIR spectrometry, such as the choice of the most appropriate instrumentation and materials, are described by Catalano. The occupations of laboratory and process personnel involved are presented [190]. Several sample preparation procedures for measuring infrared spectra obtained for water-containing or water-soluble materials and a soluble solid are compared by Dirksen and Gagnon. Traditional IR accessories are used with a disposable IR card (3M, St. Paul, MN) that comprises a cardboard sample holder (5 × 10 cm) that fits into and aligns with most spectrometer optical systems. The use and performance of this technique is described [191].

Although measurements by NIR spectrometry are often more precise than those obtained by the reference method, it is also possible under certain circumstances for the NIR method to achieve better accuracy given specific means of calibration. This paper by Fearn describes the mathematical details of such claims [192]. The choice of whether to use an FT-IR instrument or a filter photometer depends on the application and the end-user requirements; however, there are some instances in which FT-IR performs better and is more cost-effective than filter IR. This concept is presented by Coates [193].

Wilks advocates the use of infrared filter photometers where possible when two or three wavelengths are needed for process analysis. Filter photometers are claimed to offer such advantages as: (1) very high signal levels, (2) better signal-to-noise ratios than with FT-IR, (3) no moving parts, (4) easy temperature control, (5) insensitivity to vibration, and (6) long-term unattended operation [194]. The merits of using an interferometer-based instrument as compared to a grating-based instrument depends on the type of sample and the end-user requirements. If an opaque sample has to be penetrated, then NIR spectrometry should be used. However, if surface measurements or clear liquid analysis (short path length) or gas analysis is required, then mid-IR or FT-IR spectrometry should be used. The only way to ensure proper calibration transfer between instruments is to have them built to the same performance specifications. Proper calibration of instruments should be provided by the manufacturer, as described by Ciurczak [195].

Workman writes that the SEP (standard error of prediction) from a test set is the standard deviation of the residuals between reference-laboratory-derived and predicted values for samples other than those of the calibration set, whereas the estimated SEP is that required for correct process control. The use of the chi-square test (cf. H. Mark and J. Workman, *Statistics in Spectroscopy*. Boston: Academic Press, pp. 143–168) is described to compare actual SEP to estimated SEP [196]. A channel-flow cell for attenuated-total-reflection Fourier-transform infrared spectroelectrochemistry is described by Barbour et al [197]. An overview is presented by Benson. The principles and instrumentation used for on-line NIR spectroscopy are described. The scope of on-line measurements (e.g., in process control) is discussed with reference to moisture measurement and multicomponent analysis; some experimental data from such analyses is presented [198].

Non-invasive in vivo measurement of a blood spectrum were studied by Nahm and Gehring using time-resolved near-infrared spectroscopy. The in vivo measurements were taken using light from a stabilized tungsten-halogen lamp guided by a 3-mm-diameter multiple-fiber bundle to the surface of a fingertip; transmitted light from the other side of the finger was collected by a single 0.6-mm-fiber connected to a diode array spectrometer. Scanning was over the silicon region from 600 to 1012 nm. The diode array had a data resolution of 0.8 nm per pixel. Spectra readings were taken with an integration time of 40 ms and a repetition rate of 16 Hz; the total measurement time was 10 min. The spectral transmission of the fingertip was corrected using a reference spectrum through a 1-mm layer of a barium sulfate ($BaSO_4$) standard [199].

A simple and inexpensive NIR spectrometer using 1-cm-thick layers of common organic liquids (e.g., ethanol, ethylene glycol, DMSO, and cyclohexane) as spectral filters is described by Fong and Hieftje. The source for this instrument was a 55-W tungsten-halogen incandescent lamp powered by a 12-V stabilized DC power supply. The source radiation was modulated by an optical chopper, and a 1500-nm long-pass NIR interference filter was employed to avoid short-wavelength NIR stray radiation. The filter liquids were contained in a series of 1-cm glass cuvets on a sliding holder, and the sample cuvet was placed in a stationary holder in front of the filter cuvets. Applications demonstrating uses for this spectrometer include the determination of methanol in water and of water in propylene glycol and DMSO. Best-case limits of detection were 0.02% (m/m) methanol in water and 0.0006 and 0.004% water in DMSO and propylene glycol, respectively. The instrument could be useful as a low-cost industrial process analyzer [200].

Univariate and multivariate partial least squares (PLS) calibration algorithms were applied by Miller to the modeling of the normal temperature fluctuations in a polyol process. Once the calibration equations were generated, they were installed using an on-line NIR fiber-optic spectrometer. Incorporating samples scanned at different temperatures provides the robust calibration data across all process temperatures [201]. A simple near-infrared-spectrometric sorption-based vapor sensor is described by Fong and Hieftje. The device uses silica-gel TLC (200-micron) plates, and Analtech plates (HP-RP2F and HP-RP8F bonded-phases, 150 microns thick) were used as NIR sensors. The procedures for construction and utilization of this device are presented. The original silica-gel plate was responsive for vapors ranging from water to carbon tetrachloride and generally performed better than the Alltech plates. Detection limits were 41–210 ppm. It was noted that polar compounds exhibited lower detection limits than nonpolar compounds. Remote vapor analysis may be possible by using fiber optics, NIR light-emitting diodes, or NIR lasers [202].

Remote detection of aluminum hydroxide was achieved by combining IR-transmitting fiber optics with an FT-IR spectrophotometer. Plastic-coated chalcogenide fibers 1 to 4 meters in length and having 15–25 cm of their length decoated were used for measurement. Five absorption bands were detected for aluminum hydroxide in the spectral range of 3350–3650 cm^{-1}. The method may be suitable for measuring industrial corrosion in aluminum alloys, as reported by Namkung et al. [203].

Calculations for establishing the bias between results obtained using a process instrumental method and a reference method is presented by Workman. The method for establishing the significance of any bias found is also described. The use of the standard error of the difference between the process instrumental results and the reference results for monitoring the performance of the instrument is also discussed [204]. Some examples of the use of NIR spectroscopy for on-line analysis in process and pilot plant production are presented by Aalijoki et al. NIR spectrometers, equipped with optical fibers, circumvent many sample-handling problems. The use of neural networks for these measurements is also briefly described [205].

The use of partial least squares (PLS) for the automatic classification of infrared spectra was presented by Luinge et al. The method was applied to IR spectra of pesticides divided into a training set and a test set of 175 and 184 spectra, respectively. Spectra were measured over the region of 4000–625 cm^{-1} and a data point resolution of 1 cm^{-1}. The PLS results were compared with those obtained using an error-back-propagation neural network method, and a combined principal component analysis (PCA) neural network method described earlier (cf. *Anal. Chim. Acta* 283:508, 1993). The methods yield roughly equivalent performance, but the training times are reduced from hours for neural networks to minutes for PCA neural networks and PLS methods, respectively. The PLS method as employed was also useful for detecting spectral outliers [206].

The use of FT-IR-ATR spectrometry for on-line analysis is presented by Wilson et al. Method development is described and several possible applications are presented [207]. A special computational method for prefiltering infrared spectral database searches, using the flashcard algorithm (*J. Chem. Inf. Comput. Sci.* 34:984, 1994) is described by Klawun and Wilkins. A back-propagation neural network was trained to recognize 35 different functional groups from a set of 609 matrix isolation FT-IR spectra, each with 752 data points. The method was also modified for application to a larger training set of 2651 spectra, with use of 40 functional groups. This set required 38.5 min of training on a Cray C90 supercomputer [208].

The application of FT-IR spectroscopy and discriminant analysis for quality control of industrial materials and products is discussed by Hartshorn. A compositional analysis testing system (CATS) is used to form training sets for discriminant analysis in order to assess its use in quality control using infrared spectra, with a discussion of results [209]. The development of an application using acousto-optic tunable-filter (AOTF) near-infrared spectrometry is presented by Huehne et al. The system is demonstrated for the determination of water in acetone [210].

Workman presents a review of process near-infrared applications (matrices and constituents) tabulated from 1980 through 1994. Sections are presented on the various instrument designs (e.g., optical filters, scanning grating monochromators, photodiode arrays, interferometers, and acousto-optical tunable-filter systems). The eras or instrument configurations used for process NIR are also described (e.g., at -line, automated, in-line, in situ, near-line, noninvasive, on-line, portable, process, quality-control, quality-measurement, rapid, and remote systems). The review contains 156 references [211].

A simple fiber-optic interface was studied by Heglund et al. [212] for on-line supercritical-fluid extraction–Fourier–transform infrared spectrometry, as previously described (cf. *Am. Lab.* 28(6):36R–36T, 1996). The interface consisted of a stainless steel union cross through which two pieces of commercially available chalcogenide infrared-transmitting fiber-optic cables were inserted along one of the axes of the cross. One end of this tubing was connected to the SFE extraction cell while the other end was connected to a vent-off valve. The IR spectra were recorded via the fiber-optic cables. The system was demonstrated for the analysis of caffeine in coffee and petroleum hydrocarbons in soil [212].

Infrared spectrometry is presented by Driver et al. for use in on-line process monitoring. Spectra are presented for several sampling accessory designs including (1) a side-stream flow-through cell, (2) a variable-path immersible probe, and (3) a high-pressure-gas cell. The parameters affecting measurement accuracy and process sampling are discussed [213]. A device used to focus energy from an infrared spectrometer (FT-IR) source into a fiber-optic cable and to refocus the energy returning via a second optical fiber to the spectrometer detection system is described by Lowry et al. [214].

An in situ DRIFTS (diffuse-reflectance infrared Fourier-transform spectroscopy) cell was used by Dossi et al. to provide infrared analytical evidence of the oxygen-induced formation of iron carbonyl from carbon monoxide and the steel walls of the cell. Formation of Fe(CO)$_5$ can be detected by the use of zeolite as a blank sample. Intense analyte bands occur at 2034, 2014, and 1979 cm^{-1} [215].

The philosophy and implementation of liquid sample handling for design of process NIR spectroscopy is presented by Crandall. A generic diagram of a process NIR system is presented, including a sample-conditioning system, a temperature conditioner, and a remotely mounted liquid-flow cell. The liquid-transmission cell is connected to the spectrometer via fiber-optic cables. (For Part 1, see *Anal. Abstr.* 56:11A20, 1994 [216].)

A review and discussion is presented by Workman on the use of NIR spectrometry in process analysis. Aspects related to the hitherto-slow acceptance of chemometrics-dependent methods in process analysis is discussed. The article contains 11 references [217]. A set of algorithms and a computer program based on these algorithms for comparing IR spectra from different instruments and/or different data intervals are presented in Kavak and Esen. The various mathematical considerations for this application are described [218].

A spectroelectrochemical cell with radial liquid flow for in situ external-reflection IR measurements is presented by Sundholm and Talonen [219]. Infrared remote sensing using FT-IR spectroscopy and spectral emissivity is described in a Ph.D. dissertation by Korb in [220]. A basic description of on-line sensing using NIR spectroscopy is described by Hindle [221]. A discussion of the acceptance of NIR as an analytical technique is given by Luttge [222]. A general description of the uses of NIR for process control is given by Pedersen [223]. A discussion of successful noncontact NIR filter photometer applications is given by Hammond [224]. Methods and apparatus for applying in-line NIR measurement in processes is presented by Mindel [225].

A patent describes the method and apparatus for determination of the concentration of an analyte present in a sample using multispectral analysis in the near-infrared region of approximately 1100–3500 nm. Diffuse reflectance and chemometrics are used to determine analyte concentration(s). Information obtained from each nonoverlapping region of wavelengths can be cross-correlated to remove background interferences [226]. A Russian-based universal near-IR rapid analyzer based on a high-stability FT-IR spectrometer is described by Matyunin and Morozova for measurement over the wavelength range 750–2500 nm [227]. Egevskaya, from the Institute of Semiconductor Physics, Siberian Branch of the Russian Academy of Sciences, Novosibirsk, Russia, describes the construction of a portable "double cat's-eye" FT-IR spectrometer [228].

A patent describing an optical probe for in situ detection of hydrocarbon concentration is given in Ref. 229. Transient infrared spectroscopy is described as a method for on-line analysis of solids and viscous liquids by Weeks et al. [230]. A presentation of the reliability and safety of NIR-based library search systems is discussed by Molt [231]. A patent describing hand-held infrared spectrometers is given Ref. 232. NIR spectra are subjected to a "polar qualification system" (PQS) for qualifying different food products. The application is described, using several examples, by Kaffka and Gyarmati in Ref. 233. Vials made from borosilicate glass are recommended for NIR sampling as a possible substitute for fiber-optic probes. The glass vials are low cost and are more flexible for analytical requirements. They also do not compromise analytical precision and accuracy, as reported by Simard and Buijs [234].

Near-infrared spectroscopy is described by Pandey and Kumar as a tool for noncontact and nondestructive process monitoring. A general discussion of this technique and of successful applications is presented [235]. A new software package called NIRQA is introduced by Westerhaus for production operators in quality-control situations. The software is compatible with NIRS1, 2, and 3 programs [236]. An introduction of a process analyzer capable of measurements from the ultraviolet through the infrared region is discussed by Rinke et al. [237]. The application of NIR instrumentation for implementing the Taguchi concept in a chemical process situation is presented by Wortel and Hansen. A method for optimizing the Taguchi concept by selection of the appropriate calibration parameters is presented [238].

A patent [239] describes a portable, battery-powered multiple-filter infrared spectrometer consisting of an infrared source with a chopper, from which IR energy passes through compound parabolic concentrators (CPCs) and is directed by a beam splitter to the sample. The beam interacts with the sample by reflectance and is returned through the CPC and beam splitter to a stationary filter assembly and discrete detector array, where the energy is converted to an AC (alternating current) signal. This current energizes an LED bar graph display for spectral identification. This method and apparatus are used to determine IR spectra of solids, liquids on a mirror, and gases in a container with a mirror to reflect light.

A discussion of the use of NIR and IR spectra for on-line process monitoring is given. Nonlinear calibration methods and inferring concentrations from NIR spectra are presented by Amrhein et al. [240]. A simple transmission infrared cell made of Swagelok components for in situ infrared spectroscopic studies of a heterogeneous catalysis process is presented by Komiyama and Obi [241]. A simple method to fabricate lenses for in situ Fourier-transform infrared spectroscopy is described by Treffman and Morrison [242]. A patent describing the method of property determination using near-IR spectroscopy is described in Ref. 243. A general review of space-borne, remote-sensing FT-IR instrumentation is given by Persky [244]. Wienke et al. compare an adaptive resonance theory–based neural network (ART-2a) against other classifiers for rapid sorting of postconsumer plastics by remote near-infrared spectroscopy using an InGaAs diode array [245].

A general description of in situ real-time monitoring of reactive systems by remote fiberoptic near-infrared spectroscopy for chemical or polymer process monitoring is given by Mijovic and Andjelic [246]. The description of the uses of TeX fiber optics for infrared spectroscopy of biological and chemical compounds is described by Foulgoc et al. [247]. Wilks describes portable infrared analyzers for repetitive analysis [248]. A description of the uses of FT-IR for a long-term quality assurance program is presented by Childers et al. [249]. A field-portable Fourier-transform infrared spectrometer system is described by Engel et al. [250]. A description of the advancements in near-infrared fiber-optic spectrometry probes was presented by Hansen and Khettry [251]. A general presentation of the use of on-line NIR measurements for process control is given by Curtin [252].

The vibration spectroscopy of surfaces using a nonlinear optical process is described by Hirose, and an introduction to the use of IR-visible sum frequency generation spectroscopy (VSFG) is presented in [253]. An improved near-infrared quantitative analysis instrument is disclosed comprising a removable finger insert that facilitates the proper aligning and fitting of an individual user's finger into the optical system of the analysis instrument. The finger insert also prevents the analysis instrument's optical system from being damaged by foreign matter typically introduced by a user's finger. The finger insert can also be provided with filters that will enable the insert to be used as an optical standard for the analysis instrument [254].

A discussion of the development of quality assurance/quality control (QA/QC) performance standards for field use of open-path FT-IR spectrometers is given by Cone et al. [255]. An optimized diffuse-reflectance sensor system utilizing fiber optics for remote sensing is described by Driver et al. [256]. A general article discussing the use of NIR for quality control in a chemical, pharmaceutical, or biomedical manufacturing process is given by Gonzalez [257].

A description of remote characterization of materials by vibrational spectrometry through optical fibers is given by Griffiths et al. [258]. The spectral characterizations of low- and high-resolution remote FT-IR emission spectroscopy is presented by Wang et al. [259]. An echelle grating spectrometer (EGS) for mid-IR remote chemical detection is described by Stevens et al. [260]. Tellurium halide (TeX) fibers for remote infrared spectroscopy is given by Zhang et al. [261]. A general description of FT-IR analysis and its uses is given by Wilson et al. [262]. A general discussion of the use of NIR technology to optimize plant operations in the chemical or petroleum industry is presented by Espinosa et al. [263]. The application of spectroscopy, especially near-infrared, for process and pilot monitoring is described by Aaljoki et al. [264].

Medicine and Clinical Chemistry

Cigarette smoke samples were collected on a glass-fiber filter. The filters were measured by Di Luzio et al. using the NIR region from 1445 to 2345 nm, with one measurement per filter. Calibration results were linear from 0.08 to 1.49, 0.02 to 2.46 and 1.13 to 17.15 mg/cigarette for nicotine, moisture, and tar, respectively. This technique is fast, simple, nondestructive, and safe. It can be used for prescreening samples [265]. Near-infrared Fourier-transform Raman spectroscopy has been used with partial least squares (PLS) regression to analyze cholesterol/cholesterol linoleate/cholesterol oleate and cholesterol palmitate/cholesterol stearate mixtures and to determine single components within the ternary mixtures. The method is being developed by Le Cacheux et al. for eventual application to in situ measurements of arterial walls [266].

Fiber-optic evanescent-wave spectroscopy was combined with FT-IR by Simhi et al. to measure the total protein, cholesterol, urea, and uric acid in 2 mL of human serum. Multivariate calibration models derived from a partial least squares (PLS) algorithm were compared to a neural-network calibration. The method was reported as applicable for analysis of glucose and triglycerides in serum [267]. Lipids were determined directly in fecal homogenate by FT-IR-ATR and partial least squares regression modeling by Franck et al. A zinc selenide ATR accessory was used for optical sampling of the slurries. The infrared spectral regions 1763–1730, 1680–1612, 1595–1568, 1564–1524, 1491–1448, 1429–1400, and 1390–1367 cm^{-1} were used for quantitation. The infrared spectral results were within 2.5% of those obtained using the reference gravimetry method [268].

An overview is presented by Stevens and Vadgama of the use of IR spectroscopy in clinical chemistry applications. The use of near-IR spectroscopy to determine hemoglobin, glucose, urea, and bilirubin in blood is described. Additional uses for NIR for monitoring oxygen levels and measuring body fat are presented. A general discussion for the development of near-IR analyzers for use in noninvasive near-infrared patient testing is discussed [269]. Near-infrared Raman spectrometer systems are proposed by Brennan et al. for in vivo studies of diseased human tissue. Safe and rapid data acquisition are benefits of such technology. Two designs of such systems are described: (1) a high-quality-spectra instrument for laboratory use, and (2) a trolley-mounted optical-fiber instrument for clinical use [270].

A portable FT-IR gas analyzer was evaluated by Ahonen et al. for industrial hygiene measurements made in different factories using organic solvent mixtures. The detection limit was approximately 1 mg per cubic meter. The results obtained were compared with those obtained by the generally used adsorption tube method. Continuous monitoring with this system allows exposure peaks to be identified and allows preventative measures to be taken [271]. Fourier-transform infrared spectrometry was used by Cole and Martin to determine gas-phase sidestream cigarette smoke components (i.e., the smoke generated during smolder from the lit end of a cigarette). Five gas-phase components—NH_3, CO_2, CO, HCN, and NO—were determined in the sidestream cigarette smoke from five different cigarette brands. Gas concentrations were calculated from absorbance measurements made every 30 s. A 20-fold decrease in sample analysis time was achieved with the described method [272].

Raman, mid-IR DRIFTS, and NIR spectrometry were used by Lewis et al. to analyze fibrous (asbestos) and nonfibrous forms of serpentine and amphibale minerals. Raman spectra were obtained with excitation at 785, 632.8, and 1064 nm. Results indicated that NIR spectrometry was the most useful method due to: (1) relatively easy-to-interpret spectra in the 7400–6900-cm^{-1} range, (2) a higher single-to-noise ratio; (3) the use of silica optical-fiber cables, and (4) the short analysis time in comparison to Raman and FT-IR. The paper reports that the method compares well with other conventional methods of asbestos analysis, such as optical microscopy and transmission electron microscopy [273]. NIR spectroscopy, or "near-infrared hemoglobinometry," is described by Kuenstner and Norris for human blood samples. NIR spectra were measured from 400 to 2500 nm at 2-nm intervals in an open-faced cell with a vertical light path. The combination of (A1638–A1628)/(A1638–A1626) gave highly accurate results, with R_2 = 0.99 and standard errors of 0.32 g/dL. It was noted that no diodes are available in this region, making it difficult to produce a dedicated low-cost instrument at this time [274].

The application of noninvasive near-infrared hemoglobin spectroscopy for in vivo monitoring of tumor oxygenation and response to oxygen modifiers is described by Hull and Foster [275]. Near-infrared analysis of dried human serum for determining human serum urea concentration is described by Hall and Pollard. Sera was allowed to reach room temperature, and a 400 µL portion was spotted onto filter paper. The filter was then dried at 80°C for 3 min. After cooling to room temperature for 30 s, the sample was measured using a spectral range from 400 to 2500 nm. Urea-related spectral bands were observed at 1450, 1470, and 1980 nm. Unmodified serum samples were also analyzed using a standard method for comparison. Results are presented and discussed [276].

Near-infrared spectroscopy was used for whole blood and blood serum sample analysis by Domjan et al. Examples for qualitative detection of protein and lipid in human sera, as well as distinction of albumin and globulin dissolved in physiological saline solution, are discussed. Other constituents measured were protein, lipoprotein, oxygen saturation, and carbon dioxide pressure in whole blood. Results indicate that NIR spectrometry is a rapid, accurate, and cost-effective method for determining quality parameters in either whole blood or serum [277].

Kromoscopic analysis is suggested by Sodickson and Block as a possible alternative to standard spectroscopic analysis for noninvasive measurement of analytes in vivo. Kromoscopy utilizes broad band, spectrally overlapping NIR detectors. The analyte bands are integrated using two or more detectors with different relative weightings. The response of the detectors to glucose was tested by measurement of the transmission of a series of glucose solutions in a 4-cm-path cell with a range up to 0.6M [278].

The use of NIR spectroscopy for process control of antithrombin III production in biotechnology manufacturing is discussed by Harthun et al. [279]. Specific details of wavelength selection and correlation data are given by Domjan et al. for the rapid analysis of beta-lipoprotein in human blood serum using near-infrared spectroscopy [280]. A spectrograph for noninvasive glucose sensing is described in a patent [281]. A general discussion of the uses of noninvasive infrared spectroscopic analysis in clinical chemistry is described by Stevens and Vadgama [282]. In situ characterization of beta-amyloid in Alzheimer's diseased tissue by synchrotron Fourier-transform infrared microspectroscopy is described by Choo et al. [283].

Noninvasive blood glucose monitoring by means of near-infrared spectroscopy is described, and methods for improving the reliability of the calibration models presented, by Mueller et al. [284]. A paper discussing the means of enhancing calibration models for noninvasive near-infrared "spectroscopical" blood glucose determination is presented by Fischbacher et al. [285]. Passive-remote Fourier-transform infrared spectroscopy is used for detection of emission sources by Demirgian et al. [286]. Noninvasive near-infrared spectroscopy is described by Wolf et al. for monitoring regional cerebral blood oxygenation changes during physiological experiments in the rat [287].

A dissertation by Pastrana-Rios describes the in situ infrared spectroscopic test of the pulmonary surfactant system [288]. Laser diodes and a photoacoustic sensor head are presented by Spanner and Niessner for the noninvasive determination of blood constituents [289]. Near-infrared spectroscopy is used by Hamaoka et al. for noninvasive measurement of oxidative metabolism on working human muscles [290]. A general discussion is given by Heise on the current state of the art using near-infrared spectroscopy for noninvasive monitoring of metabolites [291]. Fourier-transform infrared microspectroscopy is used by LeVine and Wetzel for in situ chemical analyses of frozen tissue sections of the white matter in the rat brain [292, 293].

A novel approach to using NIR spectroscopy for noninvasive reflectance spectroscopy is described by Schlager and Ruchti [294]. The use of near-infrared spectroscopy and neural networks for noninvasive determination of blood/tissue glucose is described by Jagemann et al. [295]. Noninvasive detection of hemoglobin oxygenation changes during cortical spreading depression in the rat brain is described by Wolf [296]. Second-derivative near-infrared spectroscopy is described by Cooper et al. as a method for the noninvasive measurement of absolute cerebral deoxyhemoglobin concentration and mean optical path length in the neonatal brain [297]. Near-infrared spectroscopy is presented for the noninvasive measurement of blood glucose by Noda et al. [298].

Military, Explosives, and Propellants

Remote high-resolution FT-IR was used to study the spectral characteristics of infrared flare material by Wang et al. [299]. NIR spectroscopy was presented as a method for analysis of propellants and explosives by Rohe et al. [300]. Process-related problems can be solved using acousto-optic tunable-filter (AOTF) NIR spectroscopy, according to Medlin et al. [301].

Moisture Measurement

The use of NIR spectroscopy for in-line measurement of high-moisture products is presented by Wust et al. The application discussed is the production of cottage cheese and the monitoring of solids content during the production process. However, the method described is applicable to other processes where water content is an important production parameter [302]. Moisture determination in hard gelatin capsules using NIR spectroscopy is described by Berntsson et al. The paper discusses the importance of moisture monitoring for quality production of pharmaceutics. No sample pretreatment was required, and the analysis time was 1–2 min. The best calibration model was obtained using MLR and three selected wavelengths over the range of 1100–2500 nm at 2-nm intervals. The root mean square error of prediction was 0.1% absolute over a range of 5.6–18% moisture [303].

Petroleum, Natural Gas, and Fuels Research

On-line information obtained from a photodiode array NIR absorption spectrometer and a linear CCD fluorescence spectrometer were combined by Litani-Barzilai et al. to predict 10 petroleum properties, including research and motor octane numbers, vapor pressure, API gravity, and aromatic content. The proposed combined spectral technique produced a standard error of prediction (SEP) for octane number of 0.2. When NIR was used alone, the SEP for octane number was 0.4 [304]. Near-IR reflectance spectroscopy was proposed by Stallard et al. for the determination of motor oil contamination in sandy loam. The sieved and dried sandy loam containing virtually no organic matter was weighed, treated with a few drops of Pennzoil 10W30, weighed again, mixed with a spatula, and mounted in a commercial FT-NIR spectrometer fitted with a diffuse-reflectance accessory. The angle of incidence was 45°, and both the specularly and diffusely reflected beams were collected. The spectrum was recorded over the range 1600–1900 nm. For 0.13–0.26% (w/w) contamination (95% confidence), the 1-sigma precision was 0.17% (w/w). The design of a portable instrument for field use having seven spectral channels, each with a resolution of 10 nm, is suggested [305].

Automation of FT-IR spectrometry for liquid samples is described, including hardware and software, by Stanek. Turnkey applications have been demonstrated for lubricating oil analysis, fuel, and edible oils and fats [306]. Qualitative analysis of oil sand slurries using on-line NIR spectroscopy and an optical-fiber-bundle probe was demonstrated by Friesen. Diffuse-reflectance NIR spectra were measured from 1100 to 2300 nm. The condition of the stream was indicated using principal component score plots. Changes in the steady state of the sand were indicated by directional changes in the score plots [307]. FT-IR spectrometry and multivariate regression were reported by Andrarde et al. to be useful in the quality control of kerosene production. Results are presented of the progress from sequential univariate tests to multivariate approaches, leading to improved laboratory performance and reduced delay times for critical production information [308].

The details for on-line advanced control of a gasoline blender using NIR spectroscopy are presented by Lambert et al. The NIR analyzer monitors the product stream from a gasoline blender; this information is passed to the control software. The software adjusts the flow rate of the components to meet the blending target specifications. The final product was also analyzed off line for quality control using NIR spectrometry [309]. On-line NIR analysis has also been demonstrated by Lambert et al. as a means for optimization of steam-cracking operations. NIR analytical technology has been used since 1991 for in-house on-line process control for a steam-cracker. The NIR analyzer consists of a sampling conditioner for sample filtration, water removal, temperature control, and the NIR spectrometer. A multiplexer is utilized, with two measurement cells linked by fiber optics. The measurement cells can be used to analyze up to 13 naphtha properties and 16 raw gasoline properties, respectively, in less than 1 min [310].

Hydrocarbon and CO_2 measurements were performed using GC, FID, and FT-IR spectrometry and the results compared by Stephens et al. [311]. The results obtained using a combination of FT-NIR and GC are used by Kania to optimize gasoline blending. FT-NIR spectrometry is used for on-line analysis of gasoline for octane number, while GC analysis is used to monitor distillation products, aromatics, total olefins, and oxygenates [312]. Acousto-optic tunable filter (AOTF) transmission spectrophotometry has been demonstrated by Pruefer and Mamma as a measurement method for the on-line analysis of motor octane number in gasoline [313]. Applications of on-line NIR are described, including steam-cracker feedstock optimization and gasoline blending by Lambert et al. in [314].

An in situ infrared spectroscopic investigation of selective alkylation of toluene over basic zeolites by Palomares et al. is found in Ref. 315. A review, with 52 references, of the use of NIR in refinery and petroleum process monitoring is given by Workman [316]. In-situ FT-IR monitoring of a solar-flux-induced chemical process is presented by Markham et al. [317]. A general description of the use of NIR analysis for refineries is given by Buttner [318]. The method for control of a steam-cracking process by near-infrared spectroscopy is described in a patent [319]. A general description of the use of spectroscopic techniques used to study combustion-chamber chemistry is given by Masi [320].

A patent using NIR to determine cracking properties is described by Descales et al. [321]. A method using NIR to determine lubricant properties is described in the patent literature [322]. An in situ infrared absorption method for determination of high-pressure phase diagrams of methane-heavy hydrocarbon mixtures is given [323]. Gas chromatographic and IR spectroscopic data combined with PLS (partial least squares) regression analysis is presented for quality control of jet fuels by Seinsche et al. [324]. An on-line application of an acousto-optic NIR analyzer for product monitoring and plant control in refineries is described by Buettner et al. [325]. The use of on-line near-infrared optimization for refining and petrochemical processes is given by Lambert et al. [326]. Nondestructive quality control of pharmaceutical tablets by near-infrared reflectance spectroscopy is presented by Conzen et al. [327].

The on-line determination of total olefins in gasoline by process gas chromatography and Fourier-transform infrared spectroscopy is found in a paper by Bade et al. [328]. In situ examination of coal macerals oxidation by infrared microspectroscopy is found in a paper by Landais [329]. A method and apparatus for nonintrusive in situ chemical analysis of a lubricant film in running reciprocating machinery using near-infrared spectroscopy is described in a patent [330]. A NIR method for in situ measurement and control of a low-inventory alkylation unit is described in a patent [331]. A method for FT-IR process monitoring of metal powder temperature and size distribution is detailed by Rosenthal et al. [332].

Pharmaceutical Industry

Quantitation of the active compound and major excipients in a pharmaceutical formulation by near-infrared diffuse-reflectance spectroscopy with a fiber-optic probe is presented by Blanco et al. Spasmoctyl samples, containing otilonium bromide as active agent and microcrystalline cellulose, maize starch, sodium starch glycolate, and glyceryl palmitostearate as excipients, were analyzed by NIR spectrometry [333]. A novel approach to the transfer of multivariate calibrations based on finite impulse response (FIR) filtering of a set of spectra is described by Blank et al. The method uses a spectrum (frequently the mean of a calibration set) on the target (secondary) instrument as a means to calculate the transfer filter. The NIR method was compared with direct transfer and piecewise direct transfer (PDT) on NIR reflectance spectra of different sample types [334].

A clean method using flow-injection Fourier-transform infrared spectrometry for the simultaneous determination of propyphenazone and caffeine in powdered pharmaceutical tablets is given by Bouhsain et al. [335]. On-line near-infrared spectroscopy is used to determine the end-point for polymorph conversions in pharmaceutical processes as described by Norris et al. [336]. The incorporation of HPLC-FT-IR spectrometry into pharmaceutical research programs in such a way that it can compete with, e.g., NMR and MS, and can be automated is discussed by Pivonka and Kirkland [337].

A near-infrared pattern-recognition method is described by Aldridge et al. for the identification and determination of the quality of the solid drug substance, polymorph A. The method can discriminate between the desired polymorphic form of the drug and another polymorph, detect samples containing trace levels of the undesired form, and discriminate between the desired polymorph and other crystalline forms. The Mahalanobis distance pattern-recognition method was more suitable than the SIMCA residual variance test. It was not sensitive to variation in response between NIR instruments, so multivariate standardization was not required [338].

NIR spectroscopy was evaluated by Sekulic et al. as an on-line technique for monitoring the homogeneity of pharmaceutical mixtures during the blending process. A model mixture of 10% sodium benzoate, 39% microcrystalline cellulose, 50% lactose, and 1% magnesium stearate was used to evaluate the measurement technique. A V-blender was modified to accept a fiber-optic probe located directly at the axis of rotation. Sample components were charged to the blender in the same order throughout the experiments. Transmission spectra were collected from 1100 to 2500 nm (25 spectra per run) over a 25-min period. The method was able to determine blend homogeneity long before the typical blending period was complete [339].

An NIR instrument was interfaced to a blending vessel using a fiber-optic cable. Data was fed to a PC via a parallel digital input–output interface and processed by use of a customized software package. The system was tested by preparing a three-component mixture of 10% active ingredient, 45% lactose, and 45% maize starch. The system was acceptable for controlling a fully automated blending system [340]. NIR spectroscopy was coupled with a polar qualification system (PQS) for detecting slight differences in the physical and chemical characteristics of pharmaceutical materials by Plugge and Van der Vlies. Measurements were performed using a spectrometer equipped with a rotating sample holder for powders. Each spectrum was transformed into the second derivative and then to a polar spectrum in which the absorbance valves were the radii and the angle of the function was designated as the wavelength. The center of gravity of the polar spectrum was taken as a quality parameter indicator. The software to run the PQS was written in PASCAL 7.0. The method was used to study the differences in particle size distributions between various lactose samples and also the effectiveness of a blending process for producing homogenous mixtures [341].

Quantitative analysis of headache tablet components was described by Schmidt using NIR reflectance spectroscopy. Powdered samples were introduced to the FT-NIR spectrometer by an automatic system. The measurement technique utilized the diffuse reflectance of the NIR radiation guided onto the sample using optical fibers; the reflected light was also directed onto the detector by optical fibers. A partial least squares (PLS) method was used for calibration. Spectra were acquired from 6500 to 5500 cm^{-1} at 4 cm^{-1} resolution and a 20-s measurement time. Samples were also analyzed through the plastic packaging material at 2 cm^{-1} resolution. Results agreed with those obtained by HPLC and gravimetry [342].

Near-infrared reflectance spectroscopy was used by Han and Faulkner to quantitatively analyze tablets during production. Spectra were acquired over the spectral range of 1100–2500 nm and converted to second-derivative spectra. For moisture determination the scan range was 1400–1450 nm. Partial least squares regression was used for calibration. For coating thickness measurements, the primary analysis wavelength was 2162 nm. Identification of tablets inside blister packaging was also demonstrated. Results are compared to those obtained using HPLC [343]. Qualifying pharmaceutical substances using near-IR spectroscopy and a polar qualification system (PQS) is described by van der Viles et al. The polar quantification system is a discriminant method used in detecting small differences in chemical and physical properties for qualifying pharmaceutical substances by fingerprinting. Full details of the investigation have been published previously (cf. *Pharm. Technol. Europe.* 7: 43, 1995) [344].

Tablets (pink, pentagonal) or film-coated tablets (white, oblong) were analyzed by NIR (1100–2350 nm) reflectance spectrometry through the blister pack plastic using a fiber-optic bundle, as described by Dempster et al. [345]. Wavelength distance, Mahalanobis wavelength distance, and SIMCA residual variance distance were investigated as discriminant techniques. Library validation and test set data are presented, along with the advantages and disadvantages of each discriminant method. The use of FT-NIR spectrometry for the quality control of solid pharmaceutical formulations was investigated by Dreassi et al. The quantitative performance of the method was demonstrated by testing for three solid drug formulations: powders containing benzydamine hydrochloride and tricetol, pills containing ibuprofen, and tablets containing paracetamol [346].

The coupling of chiral methods such as optical rotatory dispersion and circular dichroism with more conventional methods such as UV-visible and IR spectrometry was performed to obtain optical purity and concentration information by Erskine et al. Mathematical approaches such as partial least squares and principal component regression methods were used to predict the enantiomeric purities of test mixtures [347]. In situ Fourier-transform infrared and calorimetric techniques were used by Landau et al. to study the preparation of a pharmaceutical intermediate. A reaction effluent passed through an ATR flow cell and was monitored by FT-IR spectrometry. Periodic samples of the reaction effluent and the off-gas were analyzed by HPLC. The data obtained was used to give a mechanistic insight into the process, and statistical tests on the data showed when the process deviated from the desired operating conditions [348].

The use of NIR reflectance spectrometry for the quality control of ranitidine hydrochloride (I) tablet production is described by Dreassi et al. Discriminant analysis was applied to the IR spectral data to construct calibration maps to identify ranitidine hydrochloride and mixtures of ranitidine hydrochloride for tablets, cores, and coated tablets. Multiple linear regression was used to construct calibration systems to determine it and its H_2O content during the various production stages up to the finished product. The method gave satisfactory results in all instances, thereby allowing qualitative and quantitative checks at all stages of the production cycle [349].

NIR spectrometry was used by Gonzalez et al. to analyze a pharmaceutical mixture of a polyalcohol, a cellulose, and a thickener. Qualification of the spectra was achieved using pattern-recognition techniques. A computer program was developed for comparison of the sample spectra with a standard spectrum. The program enabled samples with a "fail" qualification to be classified into four groups according to the shape of the plot and the wavelength that presented the greatest distance from the mean [350]. NIR (1100–2500 nm) reflectance spectroscopy was used by Dreassi et al. to characterize several pharmaceuticals on the basis of their physical properties. NIR spectra were recorded and linear discriminant analysis was used to process the raw absorbance data. The method was used to distinguish substances with different crystalline states (nitrofurantoin, gemfibrozil, chenodeoxycholic acid, levamisole-tetramisole) and densities (paracetamol, ibuprofen) [351].

FT-IR spectrometry was used by Severdia to detect traces of lubricating grease on the product contact surfaces of pharmaceutical tablet presses. Press surfaces were swabbed with Whatman No. 1 filter paper soaked in *n*-hexane (HPLC solvent grade). The swab was allowed to dry in air, and the lubricant residue was extracted by sonication of the swab in *n*-hexane. After removal of the filter paper, the extract was gently evaporated to dryness, the residue was dissolved in a known volume of $CDCl_3$, and the transmission IR spectrum of the solution was recorded (32 scans) at a spectral resolution of 4 cm^{-1} against a solvent reference [352].

An NIR spectrometric method is described by Blanco et al. for the quality control of the manufacturing process for a solid pharmaceutical preparation. The parameters measured include the identification and qualification of the product and the quantitation of the active component. The NIR spectrum was recorded from 1100–2500 nm using a spinning cuvet sample holder or a fiber-optic probe. The second derivative of the spectrum was compared to the reference spectra contained in a library compiled for 163 pharmaceutical materials. The sample was identified by determining the match that produced the highest correlation. Once identified, a comparison of the sample and reference spectra yielded information concerning all the parameters affecting the NIR spectra that allowed qualification of the sample. The concentration of the active component was determined by partial least squares calibration using two strategies to select samples for inclusion into the calibration set, namely, flat calibration and sample selection [353].

A near-infrared reflectance analysis method for the noninvasive identification of film-coated and non-film-coated blister-packed tablets is described by Dempster et al. [354]. Quantitative Fourier-transform near-infrared spectroscopy is used in the quality control of solid pharmaceutical formulations, as presented by Dreassi et al. [355]. In-situ Fourier-transform infrared and calorimetric studies of the preparation of a pharmaceutical intermediate are detailed by Landau et al. [356]. NIR spectroscopy is used as a tool for in-process control in pharmaceutical production as presented by Steffens and List [357].

Nondestructive quality control of pharmaceutical tablets by near-infrared reflectance spectroscopy is described by Weiler and Sarinas [358]. Internet sites for infrared and near-infrared spectrometry are given by Kraemer and Lodder, including on-line instruction and direct communication [359]. Process control and endpoint determination of a fluid-bed granulation by application of near-infrared spectroscopy is described by Frake et al. [360]. NIR spectroscopy is described by Wilson et al. for quality control in the pharmaceutical industry [361]. Determination of moisture in hard gelatin capsules using near-infrared spectroscopy for at-line process control of pharmaceutics is given by Berntsson et al. [362]. Near-infrared spectroscopy as a tool to improve quality is described by Plugge and van der Vlies [363].

Plant Sciences and Pulp and Paper

Ultrafast web thickness and coating measurements with true, two-dimensional control are described by Hindle and Smith [364]. Leaf nitrogen determination using a portable near-infrared spectrometer is presented by Blakeney et al. [365]. Multivariate calibration models using near-IR reflectance (NIR) spectroscopy on pulp and paper industrial applications are described by Antti et al. [366]. Rapid lignin measurement in hardwood pulp samples by near-infrared Fourier-transform Raman spectroscopy is presented by Sun et al. [367]. Vibrational spectroscopy—a rapid means of estimating plantation pulpwood quality as described by Michell—is found in Ref. 368. Rapid determination of effective alkali and dead-load concentrations in kraft liquors is presented by Leclerc and Hogikyan using attenuated total reflectance (ATR) infrared spectrometry [369]. Rapid characterization of dirt specks in kraft pulp using infrared spectrometry is described by Leclerc and Ouchi [370].

Polymer Science

A new correction algorithm for FT-IR on-line film-thickness measurement is presented by Wang et al. [371]. Process stream interfacing of in-line NIR analyzers is described by Mindel [372]. Near-IR FT Raman spectroscopy is used by Ozpozan et al. for on-line monitoring of the polymerization of vinyl acetate at 54°C in H_2O-containing sodium lauryl sulfate, NaH_2PO_4, and $K_2S_2O_8$ in a 150-mL reaction vessel. The 1064-nm line from a Nd:YAG laser passed through the side arm of the reaction vessel using backscattering geometry. Spectra were recorded at 8 min intervals, with scanning from 20 to 3500 cm^{-1} and a resolution of 4 cm^{-1}. The method is applicable to the monitoring of industrial production of the polymer [373].

An NIR instrument using an InGaAs detector (i.e., 900–1700 nm) was used by Feldhoff et al. in conjunction with a 30-cm-wide conveyor belt moving at 1 m/s carrying waste polymer vessels for identification and sorting. Use of the FuzzyARTMAP neural-network classifier (per Wienke et al. in *Chemom. Intell. Lab. Syst.* 32:165, 1996). The device gave an accuracy of >97% in the classification of objects made of poly(ethylene terephthalate), polystyrene, PVC and cardboard, but was slightly less reliable at distinguishing between polyethylene and polypropylene [374]. An array detector was used by van den Broek et al. over several incremental wavelength regions of 1546–1578, 1545–1655, 1655–1745, 1700–2150, 2207–2321 and 2115–2550 nm defined by a set of interference filters, over a wavelength range of 1.2–4.6 microns. The system was used to discriminate a variety of plastic materials automatically [375].

The structure of the conducting polymer poly(paraphenylene), synthesized under different electrochemical conditions, was studied in situ by Damlin et al. using external reflection FT-IR spectrometry [376]. An FT-IR microspectrometric system is described for the real-time on-polymer-bead monitoring of multistep combinatorial syntheses carried out in a flow cell, a diagram of which is presented by Pivonka et al. [377]. A description of the use of NIR absorption spectroscopy to study inter- and intramolecular hydrogen bonding and other interactions of polymers is given by Lachenal and Stevenson. This information is useful for the curing of epoxy resins and for process control in the production of polymers and their blends [378].

An FT-IR spectrophotometer was used by Kirsch et al. to analyze the monomer and comonomer in the reactor feed system of a polyolefin process. The system was used in closed-loop control of the comonomer feed flow to attempt to control changes in product density and type [379]. Formation of toxic gases during pyrolysis of polyacrylonitrile and nylons is analyzed by Nielsen et al. using an on-line optoacoustic FT-IR gas analyzer with a fixed-bed pyrolysis reactor that consisted of a vertical quartz tube (1 cm i.d.) heated by a Lindberg furnace. The major toxic products of polyacrylonitrile were HCN and NH$_3$; nylons produced NH$_3$ but only minor amounts of HCN. Both types of polymer produced CO [380]. A review by Siesler containing 22 references is presented on applications of optical-fiber remote near-infrared spectroscopy for polymer reaction and process control, including synthesis, processing, and recycling [381].

An NIR spectrometer is connected by optical fibers by Brush to a stainless steel immersion-sampling absorption cell (pathlength 6 mm). The manufacturing conditions of polyoxyethylene and polyoxypropylene glycols necessitate remote in situ analysis at temperatures over 260°C, with turbulence. Calibration equations for hydroxyl number and acid value using MLR (multi-linear least squares regression) gave SECs (standard errors of calibration) of 0.41 and 0.28, respectively [382]. NIR spectroscopy is described by Scott for the automatic sorting of post-consumer plastic waste using two-color fixed filters transmitting at 1660 nm and 1716 nm. The ratio of the 1716 nm signal to the 1660 nm signal allows the identification and separation of plastics made from poly(ethylene terephthalate) and from PVC [383].

Near-infrared external-reflection spectroscopy was reported by Xu et al. for on-line remote monitoring of thermoset resin cure in a polymer manufacturing application [384]. A general article describing the uses of near-infrared spectroscopy in monitoring polymer reactions by Dallin is found in Ref. 385. The challenges associated with using near-infrared spectroscopy for on-line monitoring in polyol production is described by Stromberg [386]. Infrared microspectroscopy is combined with a microtome for assessing potential toxins in polymer packaging by Jickells. This instrumental method allows much more rapid analysis than is possible with dissolution techniques and can be used to study the penetration and migration of food and toxin materials within plastics [387]. Specular-reflectance infrared spectroscopy was used by Gaarenstroom et al. for rapid identification of automotive plastics in dismantling operations [388].

External-beam mid-infrared spectroscopy is presented as a method for nondestructive rapid identification of plastics by Mucci [389]. Mid-infrared reflectance spectroscopy is proposed by Zachmann and Turner as a method for rapid characterization of black polymeric material [390]. In-situ Fourier-transform infrared analysis of poly(ester-urethanes) at low temperature using a newly constructed liquid nitrogen cooled sample stage is described by Rehman et al. [391]. Evanescent-wave FT-IR spectroscopy is used by Linossier et al. for in situ study of the degradation of polymer-substrate systems [392]. NIR spectroscopy is used by Fischer et al. for in-line process monitoring on polymer melts [393]. NIR spectroscopy is also proposed by Hansen and Vedula as a means for in-line measurement of copolymer composition and melt index [394].

Step-scan photoacoustic FT-IR spectroscopic studies were used by Niu and Urban to study surfactant exudation and film formation in 50%/59% Sty/n-BA latex films [395]. Infrared and resonance Raman spectroscopic studies on the photopolymerization process of the Langmuir–Blodgett films of a diacetylene monocarboxylic acid are described by Saito et al. [396]. A patent describing near-infrared measurement and control of polyamide manufacture is given in Ref. 397. Time-dependent polarization-modulation infrared spectroscopy was used for an in situ study of photoinduced orientation in azopolymers by Buffeteau and Pezolet [398]. An ultrafast near-infrared sensor is described by Chilling and Ritzmann for rapid on-line identification of plastics [399].

On-line Fourier-transform infrared spectroscopy is described by Kirsch et al. as a means of improved process control in polyolefin polymerization [400]. Near-IR spectroscopy is applied by Long et al. as a process monitor for the synthesis of polyester graft copolymers via the macromonomer technique [401].

Spectroscopic attenuated total reflectance and photoacoustic Fourier-transform IR approaches are described by Niu et al. for use in monitoring latex film formation at surfaces and interfaces [402]. Infrared spectrometric classification techniques are used by Van Every and Elder for validation in polyolefin formulation [403]. In situ external-reflection Fourier-transform infrared spectroscopy is presented by Damlin et al. as a method to study the structure of the conducting polymer poly(para-phenylene) [404].

Molecular spectroscopy is described by Zachmann and Turner as a means for identification of black plastics [405]. Near-IR spectroscopy is described by DeThomas and Brush in general terms as a method for monitoring the quality and process conditions during polyol production [406]. A patent describing a process for measuring and controlling the neutralization of inorganic acids in an aromatic polyamide solution based on near-infrared spectroscopy is given by Moessner [407]. A patent using near-infrared spectroscopy for chemical property determination in polymerization and organic reactions is presented in Ref. 408. The use of near-IR spectroscopy for fast on-line identification of plastics for use in recycling processes is given by Eisenreich et al. [409]. In situ, quantitative FT-IR studies of polymers irradiated with synchrotron radiation are described by Wollersheim and Hormes [410].

Remote, in-line monitoring of emulsion polymerization of styrene by short-wavelength near-infrared spectroscopy is described by Wu et al. for aspects relating to performance in the face of normal runs [411] and process upsets [412]. Remote in situ real-time fiber-optic near-infrared spectroscopy is described by Mijovic and Andjelic in their study of the mechanism and rate of a bismaleimide curing process [413]. Near-infrared spectroscopy is described by Sekulic et al. as a method for on-line monitoring of powder blend homogeneity for the mixing of pharmaceuticals [414]. In situ infrared spectroscopy is used by Yu et al. for studying the electric field–induced dipole reorientation in oriented nylon 11 [415].

A dissertation by Khettry describing the use of infrared fiber optics for the in-line monitoring of molten polymeric processes is presented in Ref. 416. In situ real-time monitoring of epoxy/amine kinetics by remote near infrared spectroscopy is described by Mijovic et al. [417]. Near-infrared spectroscopy is described by Kirsch and Drennen for the monitoring of the film-coating process in pharmaceuticals [418]. In situ FT-IR spectroscopy, synchrotron SAXS, and rheology are used to study the structure development during the reactive processing of model flexible polyurethane foam systems by Elwell et al. [419]. Near-IR is described by Hansen and Khettry as a method for in-line composition monitoring of molten poly(ethylene vinyl acetate) processes [420]. A patent presenting a process for measurement of the degree of cure and percent resin of glass fiber–reinforced epoxy resin prepreg is given in Ref. 421.

Fourier-transform infrared spectroscopy is described by Ge et al. as a means for quantitative monitoring of an epoxidation process [422]. Near-IR spectroscopy is used for real-time process monitoring of solution polymerization kinetics by Long et al. [423]. The general use of on-line NIR monitoring of polyols is presented by Miller and Curtin [424]. The application of FT-IR spectrum method in the photocuring process for polyester acrylate is described by Cao et al. [425]. Remote fiber-optic near-infrared spectroscopy is described for in situ real-time monitoring of reactive polymer systems by Mijovic and Andjelic [426]. Molecular spectroscopy is used by Graham et al. to provide rapid identification of plastic components recovered from scrap automobiles [427]. NIR spectroscopy is described by Hansen and Khettry for in-line monitoring of molten polymers using robust fiber-optic probes and rapid data analysis [428].

Recent developments in on-line analytical techniques applicable to the polymer industry are described by Patel et al. in [429]. The use of near-infrared spectroscopy to monitor the production of polyurethanes is described by Hall and DeThomas [430]. A Ph.D. dissertation by Lew describing the use of size exclusion chromatography and in-line near-infrared spectroscopy for polyolefin extrusion process monitoring is presented in Ref. 431. The use of molecular spectroscopy for the in-line monitoring of titanium dioxide content in a poly(ethylene terephthalate) extrusion process is described by Batra and Hansen [432]. A world patent outlining the process for measurement of the degree of cure and percent resin of fiberglass-reinforced epoxy resin prepreg using molecular spectroscopy is given in Ref. 433. This is also reported in this review as a U.S. patent [421]. Near-infrared spectroscopy is presented as a means for an in-line monitoring of molten polymers by Hansen and Khettry [434].

FT-IR external reflection is used by Lutz et al. for a low-cost unit for rapid analysis of carbon-filled rubbers [435]. Process FT-IR spectroscopy is used for real-time analysis of PC/ABS blend composition by Fidler [436]. IR-RAS is used for real-time in situ observation of vapor deposition polymerization of N-methylolacrylamide by Tamada et al. [437]. Near-infrared spectroscopy is described as a means for at-line [438] and in-line [439] process analysis of a variety of properties for poly(ethylene terephthalate) chips. Differential scanning calorimetry (DSC) and infrared spectroscopy are used by Ziegler et al. for in situ investigations of thermal processes in polymers [440]. A description of the value offered by molecular spectroscopy of polymers is offered by Siesler [441].

Semiconductors and Electronics

An FT-IR in-line sensor is used by Liu et al. for modeling, simulation, and control of a single-wafer process [442]. A doctoral dissertation describing the in situ FT-IR studies of semiconductor/electrolyte interface (cadmium telluride, silicon) is given by Fan [443]. In situ FT-IR-ATR spectroscopy is used by Ping and Nauer to investigate the redox behavior of poly(thienylpyrrole) thin-film electrodes in nonaqueous solutions [444]. In situ Fourier-transform infrared spectroscopy and electrochemical analysis are used for the characterization of urea adlayers at Pt(100) electrodes by Climent et al. [445]. Fourier-transform infrared spectroscopy is used by Farquharson et al. for real-time process control of molecular beam epitaxial growth of mercury cadmium telluride (MCT) films [446].

A quantitative in situ analysis method was developed by Sperline et al. to determine the adsorption and rinsing/removal of surfactants from Si surfaces. The method used was FT-IR-ATR spectroscopy at the silicon/aqueous solution interface in the 1550–1100 cm^{-1} spectral region. The method is useful for the wet cleaning of Si in semiconductor manufacturing. The surfactants tested were: dodecyl trimethyl ammonium bromide(I) and Triton X-100(II). A zinc selenide (ZnSe) internal-reflection element (IRE) was used after sputtering with a thin film of Si and Al_2O_3 as a buffer layer [447].

A Ph.D. dissertation by Papapanayiotou describes in situ infrared spectroscopic studies of electrodeposition additive adsorption on copper [448]. On-line molecular spectroscopy of various metal organics used in compound semiconductors is described by Lee et al. [449]. Fourier-transform infrared spectroscopy used for process monitoring and control of integrated circuit manufacturing is given by Liu et al. [450]. Fiber optics–based Fourier-transform infrared spectroscopy used for in situ concentration monitoring in OMCVD is described by Salim et al. [451].

Surface Analysis

In situ Fourier-transform infrared–attenuated total reflection (ATR) spectroscopy is used by Fan and Ng for monitoring of the polyaniline synthesis mechanism on a p-type silicon electrode surface [452]. Spatially resolved infrared spectroscopy is used by Qin and Wolf for the in situ study of spatial surface coverage during CO oxidation on supported catalysts [453]. Lei et al. report on the use of diffuse-reflectance Fourier-transform infrared spectrometry of BN films deposited on steels by remote microwave plasma CVD from borazine [454]. In situ FT-IR spectroscopy was used by Savary et al. to investigate the role of the nature of the acid sites in the oxydehydrogenation of propane on a VPO/TiO$_2$ catalyst [455].

A patent describing the method and apparatus for analyzing minute foreign substances and the process for manufacturing semiconductor or LCD elements using spectroscopy is described in Ref. 456. The application of Fourier-transform infrared spectroscopy to the optimization of chemically amplified photoresist processing is described in a Ph.D. dissertation by Howes [457]. The surface corrosion of natural aluminum is studied via the use of optical fiber–based FT-IR evanescent-wave absorption spectroscopy by Namkung et al. [458]. In situ Fourier-transform infrared–diffuse-reflection spectroscopy of direct methanol fuel cell anodes and cathodes is described by Pan et al. [459]. In situ high-resolution infrared spectroscopy of a photopolymerized C60 film is presented by Onoe and Takeuchi [460].

In situ FT-IR/IRS spectroscopy is used by Kellar et al. for studying the surface phase transitions of adsorbed collector molecules [461]. In situ UV and infrared spectroscopic studies are used for the study of the kinetics of adsorption of ethyl xanthate on pyrrhotite by Fornasiero et al. [462]. In situ transmission FT-IR spectroscopy is used by Mueller and Calzaferri to study the preparation of thin Mo(CO)6-Y-zeolite layers [463]. Hanaoka et al. describe in situ monitoring of selective copper deposition processes in a metal–organic chemical vapor deposition using Fourier-transform infrared reflection-absorption spectroscopy [464]. The use of in situ far-infrared spectroscopy of electrode surfaces with a synchrotron source is described by Melendres et al. [465].

Infrared (IR) spectroscopy and atomic force microscopy (AFM) are used by Stephens and Dluhy for in situ and ex situ structural analysis of phospholipid-supported planar bilayers [466]. In situ external-reflection absorption FT-IR spectroscopy is used by Trettenhahn et al. to study lead electrode surfaces in sulfuric acid [467]. In situ infrared spectroscopy is proposed by Pan et al. as a real-time diagnostic for plasma polymer film deposition [468]. Infrared reflection spectroscopy is described by Mielczarski et al. as a method for qualitative and quantitative evaluation of heterogeneous adsorbed monolayers on semiconductor electrode surfaces [469]. In situ IR-ATR spectroscopy is used by Ozanam et al. to study the hydrogenated silicon surface in organic electrolytes [470]. The optics and cell design for in situ far-infrared spectroscopy of electrode surfaces is described by Russell et al. [471].

A dissertation describing the in situ Fourier-transform infrared spectroscopy of chemistry and growth in chemical vapor deposition is given by Salim [472]. The in situ FT-IR-diffuse-reflection spectroscopy of the anode surface in a direct methanol/oxygen fuel cell is given by Fan et al. [473]. In situ infrared spectroscopic study on the role of surface hydrides and fluorides in the silicon chemical vapor deposition process is described by Chung et al. [474]. In situ UV-visible and infrared spectroscopic techniques are described by Beden for studying corrosion products and corrosion inhibitors on materials surfaces [475]. Internal and external reflection modes are compared by Bae et al. for in situ Fourier-transform infrared spectroscopy of molecular adsorbates at electrode–electrolyte interfaces [476].

An in situ infrared spectroscopic study of the mechanism of CO electrooxidation on Pt-Ru alloys is presented by Olligs et al. [477]. A technical report describing the uses of scanning tunneling microscopy (STM) and infrared spectroscopy as combined in situ probes of electrochemical adlayer structure as cyanide on Pt(111) is given by Stuhlmann et al. [478]. An in situ IR spectroscopic study of the reaction of dimethylaluminum hydride with photochemically deposited amorphous silicon is presented by Wadayama et al. [479].

Textiles

A method using NIR spectroscopy is described by Ghosh for at-line and on-line characterization of textile products [480]. A patent describing the use of a hand-held infrared spectrometer for identifying recyclable carpet materials is given in Ref. 481. Near-infrared light–fiber-optic spectroscopy is used for in situ analysis of sizing agents on fiber reinforcements by Kitagawa et al. [482]. The use of NIR diffuse-reflectance spectrometry and partial least squares (PLS) multivariate calibration for determining the finishing oil content in acrylic fibers is described by Blanco et al. [483].

INSTRUMENT AND MEASUREMENT NOMENCLATURE

The following references, listed by category, are included as a means of locating applications described by each of the following measurement eras or measurement configurations. The pertinent references are listed under each category.

Airborne: 11, 14, 15, 18, 20, 51, 63, 67, 78, 164

At-line: 38, 362, 438, 480

Automatic: 147, 158

Hyperspectral imaging: 3

Imaging technology and image analysis: 2–20, 164, 172

In situ: 41, 44, 45, 47, 50, 60, 62, 78, 86, 93, 96, 98–104, 108, 110, 112, 156, 189, 215, 219, 241, 242, 283, 288, 292, 293, 315, 323, 329, 330, 331, 398, 404, 410, 415, 417, 426, 443, 444, 448, 452, 453, 455, 459–464, 467, 468, 470–476, 479, 482

In-line: 149, 154, 225, 238, 302, 372, 393, 394, 411, 412, 416, 420, 428, 431, 432, 434, 439, 442

Near-line: None

Noninvasive or nondestructive: 22, 128, 142, 166, 235, 254, 282, 284, 285, 291, 295, 327, 358

On-line: 23, 24, 25, 27, 39, 53, 69, 83, 91, 121, 129, 141, 143, 151, 160, 176, 198, 207, 212, 213, 221, 224, 230, 240, 252, 304, 307, 309, 310, 313, 314, 325, 326, 328, 336, 339, 340, 359, 371, 374, 375, 379, 384, 386, 399, 400, 410, 414, 424, 429, 449, 480

Open-path: 52, 57, 68, 255

Portable or hand-held: 28, 161, 177, 228, 239, 248, 250, 271, 365

Process monitoring or production control: 57, 116, 213, 235, 249, 255, 256, 332, 382, 393, 423, 450, 458

Quality control or quality measurement: 46, 113, 137, 236, 257, 308, 324, 327, 346, 350, 355, 358, 361

Quality maintenance: None

Rapid: 6, 26, 29, 30, 34, 37, 43, 58, 64, 77, 79, 95, 109, 113, 122, 123, 124, 135, 136, 138, 139, 140, 142, 144, 146, 148, 153, 208, 233, 245, 265, 268, 277, 280, 347, 367–370, 388–390, 399, 427, 428, 434, 435

Real-time: 76, 84, 119, 165, 246, 417, 423, 426, 436, 437, 446

Remote: 4–7, 16, 17, 54, 55, 57, 61, 63, 65–67, 70–75, 81, 94, 105, 107, 159, 163, 165, 167, 171–173, 203, 220, 244–247, 256, 258–261, 273, 286, 299, 304, 311, 384, 411–413, 417, 426, 454, 458

ACKNOWLEDGMENT

The author gratefully acknowledges the assistance of Mary Sutliff of Kimberly-Clark Corp. West Research and Engineering Library, for her help with the literature searches for this review.

REFERENCES

1. J. B. Callis, D. L. Illman, B. R. Kowalski. Process analytical chemistry. *Anal. Chem.* 59:624A, 1987.
2. J. L. Koenig, C. M. Snively. Fast FT-IR imaging: Theory and applications. *Spectroscopy* 13(11):22–28, 1998.
3. P. Colarusso, L. H. Kidder, I. W. Levin, J. C. Fraser, J. F. Arens, E. N. Lewis. Infrared spectroscopic imaging: From planetary to cellular systems. *Appl. Spectrosc.* 52(3):106A–120A, 1998.
4. W. H. A. M. van den Broek, E. P. P. A. Derks, E. W. van de Ven, D. Wienke, P. Geladi, L. M. C. Buydens. Plastic identification by remote sensing spectroscopic NIR imaging using kernel partial least squares (KPLS). *Chemom. Intell. Lab. Syst.* 35(2):187–197, 1996.
5. D. Wienke, W. van den Broek, W. Melssen, L. Buydens, R. Feldhoff, T. Huth-Fehre, T. Kantimm, F. Winter, K. Cammann. Near infrared imaging spectroscopy (NIRIS) and image rank analysis for remote identification of plastics in mixed waste. Presented at ANAKON '95 held in Schliersee, Germany, 24–26 Apr. 1995; *Fresenius' J. Anal. Chem.* 354(7–8): 823–828, 1996.
6. D. Wienke, W. H. A. M. Van den Broek, L. Buydens. Identification of plastics among non-plastics in mixed waste by remote sensing near infrared imaging spectroscopy. 2. Multivariate image rank analysis for rapid classification. *Anal. Chem.* 67(20):3760–3766, 1995.
7. W. H. A. M. van den Broek, D. Wienke, W. J. Melssen, C. W. A. de From, L. Buydens. Identification of plastics among non-plastics in mixed waste by remote sensing near infrared imaging spectroscopy. 1. Image improvement and analysis by singular value decomposition. *Anal. Chem.* 67(20):3753–3759, 1995.
8. L. Ning, A. M. Chozinski, A. Azarenko, L. S. Daley, W. J. Bowyer, T. Buban, T. Edwards, J. B. Callis, G. A. Strobel. Five novel applications of imaging visible and short near-infrared spectrophotometry and fluorimetry in the plant sciences. Part II. Non-invasive in vivo applications. *Spectroscopy* 12(1):37–46, 1997.
9. M. G. Sowa, J. R. Mansfield, G. B. Scarth, H. H. Mantsch. Non-invasive assessment of regional and temporal variations in tissue oxygenation by near-infrared spectroscopy and imaging. *Appl. Spectrosc.* 51(2):143–152, 1997.
10. Arno Villringer, Britton Chance. Noninvasive optical spectroscopy and imaging of human brain function. *Trends Neurosci.* 20(10):435–442, 1997.
11. James K. Crowley, David R. Zimbelman. Mapping hydrothermally altered rocks on Mount Rainier, Washington, with airborne visible/infrared imaging spectrometer (AVIRIS) data. *Geology* 25(6):559–562, 1997.
12. Esa Makipaa, Juha T. Tanttu, Henri Virtanen. IR-based method for copper electrolysis short-circuit detection. *Proc. SPIE-Int. Soc. Opt. Eng.* Vol. 3056: Thermosense XIX: An International Conference on Thermal Sensing and Imaging Diagnostic Applications, 100–104, 1997.
13. Li Ning, Annie M. Chozinski, Anita Azarenko, Larry S. Daley, Walter J. Bowyer, Tamas Buban, Gerald E. Edwards, James B. Callis, Gary B. Strobel. Five novel applications of imaging visible and short near-infrared spectrophotometry and fluorimetry in the plant sciences. Part I: Photographic and histological applications. *Spectroscopy* 11(9):44, 1996.
14. Neil Rowlands, Robert Neville. Calcite and dolomite discrimination using airborne SWIR imaging spectrometer data. *Proc. SPIE-Int. Soc. Opt. Eng.* Vol. 2819: Imaging Spectrometry II, 36–44, 1996.
15. L. John Otten, III, Andrew D. Meigs, R. Glenn Sellar, J. Bruce Rafert. Measured performance of an airborne Fourier transform hyperspectral imager. *Proc. SPIE-Int. Soc. Opt. Eng.,* Vol. 2819: Imaging Spectrometry II, 291–299, 1996.
16. P. F. Valdez, G. W. Donohoe, M. R. Descour, S. Motomatsu. Selection of spectral bands for interpretation of hyperspectral remotely sensed images. *Proc. SPIE-Int. Soc. Opt. Eng.,* Vol. 2819: Imaging Spectrometry II, 195–203, 1996.
17. Kevin Silk, E. R. Schildkraut. Imaging Fourier transform spectroscopy for remote chemical sensing. *Proc. SPIE-Int. Soc. Opt. Eng.,* Vol. 2763: Electro-Optical Technology for Remote Chemical Detection and Identification, 169–177, 1996.
18. Lawrence C. Rowan, Timothy L Bowers, James K. Crowley, Carmen Anton-Pacheco, Pablo Gumiel, Marguerite J. Kingston. Analysis of airborne visible-infrared imaging spectrometer (AVIRIS) data of the Iron Hill, Colorado, carbonatite-alkalic igneous complex. *Econ. Geol.* 90(7):1966–1982, 1995.
19. H. R. Heekeren, R. Wenzel, H. Obrig, J. Ruben, J.-P. Ndayisaba, Q. Luo, A. Dale, S. Nioka, M. Kohl, U. Dirnagl, A. Villringer, B. Chance. Towards noninvasive optical human brain mapping—

Improvements of the spectral, temporal and spatial resolution of near-infrared spectroscopy. *Proc. SPIE-Int. Soc. Opt. Eng.,* Vol. 2979: Optical Tomography and Spectroscopy of Tissue, 847–857, 1997.

20. Fred A. Kruse. Geologic mapping using combined analysis of airborne visible/infrared imaging spectrometer (AVIRIS) and SIR-C/X-SAR data. *Proc. SPIE-Int. Soc. Opt. Eng.,* Vol. 2819: Imaging Spectrometry II, 24–35, 1996.
21. Eiju Watanabe, Yuichi Yamashita, Atsushi Maki, Yoshitoshi Ito, Hideaki Koizumi. Noninvasive functional mapping with multichannel near-infrared spectroscopic topography in humans. *Neurosci. Lett.* 205(1):41–44, 1996.
22. B. R. Vazquez de Aldana, B. Garcia-Criado, A. Garcia-Ciudad, M. E. Perez-Corona. Non-destructive method for determining ash content in pasture samples: Application of near-infrared reflectance spectroscopy. *Commun. Soil Sci. Plant Anal.* 27(3–4):795–802, 1996.
23. Erik Johnsen. How to use on-line NIR in the feed and food industries. *Process Control Qual.* 9(4):205–206, 1997.
24. O. Marvik. Reasons for buying NIR on-line and experiences with the implementation in a feed factory. *Process Control Qual.* 9(4):127–131, 1997.
25. Georges Sinnaeve, Pierre Dardenne, Beatrice Weirich, Richard Agneessens. On-line monitoring of enzymic degradation of wheat starch by near-infrared. *Near-infrared Spectrosc.: Future Waves, Proc. Int. Conf. Near-infrared Spectrosc.,* 7th; 290–294, 1996.
26. Masakazu Takahashi, Makita Hajika, Kazunori Igita, Tetsuo Sato. Rapid estimation of protein, oil and moisture contents in whole-grain soybean seeds by near-infrared reflectance spectroscopy. *Near-infrared Spectrosc.: Future Waves, Proc. Int. Conf. Near-infrared Spectrosc.,* 7th; 494–497, 1996.
27. Marianne Sernevi. Using near-infrared transmittance network and on-line systems in grain segregation and process control. *Near-infrared Spectrosc.: Future Waves, Proc. Int. Conf. Near Infrared Spectrosc.,* 7th. 295–299, 1996.
28. Gabor Takacs, Tibor Pokorny, Istvan Banyasz. Application of a portable near-infrared analyzer for wheat measurement. *Near-infrared Spectrosc.: Future Waves, Proc. Int. Conf. Near-infrared Spectrosc.,* 7th. 92–97, 1996.
29. Su-Chin Lo. Rapid classification and blends analysis of tobacco mixtures using near infrared and artificial neural networks. *Near-infrared Spectrosc.: Future Waves, Proc. Int. Conf. Near Infrared Spectrosc.,* 7th. 441–447, 1996.
30. R. S. Rutherford, J. Van Staden. Towards a rapid near-infrared technique for prediction of resistance to sugarcane borer, Eldana saccharina Walker (Lepidoptera: Pyralidae) using stalk surface wax. *J. Chem. Ecol.* 22(4):681–694, 1996.
31. E. A. Carter, P. M. Fredericks, J. S. Church. Fourier transform infrared photoacoustic spectroscopy of surface-treated wool. *Text. Res. J.* 66(12):787–794, 1996.
32. I. K. Brookes, B. N. Gedge, S. V. Hammond. Applications of near-infrared spectroscopy to fermentation process analysis. *Near-infrared Spectrosc.: Future Waves, Proc. Int. Conf. Near Infrared Spectrosc.,* 7th. 259–267, 1996.
33. Anna Halasz, Amal Hassan, Arpad Toth, Maria Varadi. NIR techniques in yeast identification. *Lebensm.-Unters. Forsch. A.* 204(1):72–74, 1997.
34. David S. Zakim, Max Diem. System and method for rapid analysis of cells using spectral cytometry. USA: Inphocyte, Inc.; WO 9730338 A1 Issue Date:08/21/1997; Filing Date:02/04/1997, 24 pp.
35. G. Macaloney, J. W. Hall, M. J. Rollins, I. Draper, K. B. Anderson, J. Preston, B. G. Thompson, B. McNeil. The utility and performance of near-infrared spectroscopy in simultaneous monitoring of multiple components in a high cell density recombinant *Escherichia coli* production process. *Bioprocess Eng.* 17(3):157–167, 1997.
36. J. W. Hall, B. McNeil, M. J. Rollins, I. Draper, B. G. Thompson, G. Macaloney. Near-infrared spectroscopic determination of acetate, ammonium, biomass, and glycerol in an industrial *Escherichia coli* fermentation. *Appl. Spectrosc.* 50(1):102–108, 1996.
37. Royston Goodacre, Eadaoin M. Timmins, Paul J. Rooney, Jem J. Rowland, Douglas B. Kell. Rapid identification of Streptococcus and Enterococcus species using diffuse reflectance-absorbance Fourier transform infrared spectroscopy and artificial neural networks. *FEMS Microbiol. Lett.* 140(2–3):233–239, 1996.
38. G. Macaloney, I. Draper, J. Preston, K. B. Anderson, M. J. Rollins, B. G. Thompson, J. W. Hall, B. Mcneil. At-line control and fault analysis in an industrial high cell density *Escherichia coli* fermentation, using NIR spectroscopy. *Food Bioprod. Process.* 74(C4):212–220, 1996.
39. E. Dosi, G. Vaccari, A. L. Campi, G. Mantovani, R. Gonzalez-Varay; A. Trilli. Control of fermentations by means of on-line near-infrared spectrometry. *Near-infrared Spectrosc.: Future Waves, Proc. Int. Conf. Near-infrared Spectrosc.,* 7th. 249–254, 1996.
40. B. Szalontai, F. Joo, L. Vigh. In situ deuteration of fatty acyl chains in model and biological membranes as a new approach to investigate membrane structures. A Fourier transform infrared spectroscopic study. *Spectrosc. Biol. Mol.,* Eur. Conf., 6th, 377–378, 1995.
41. Isabelle Cornut, Bernard Desbat, Jean Marie Turlet, Jean Dufourcq. In situ study by polarization modulated Fourier transform infrared spectroscopy of the structure and orientation of lipids and amphipathic peptides at the air–water interface. *Biophys. J.* 70(1):305–312, 1996.

REFERENCES

42. Andrew J. Berger, Yang Wang, Michael S. Feld. Rapid, noninvasive concentration measurements of aqueous biological analytes by near-infrared Raman spectroscopy. *Appl. Opt.* 35(1):209–212, 1996.
43. D. O. Henderson, R. Mu, M. Gunasekaran. A rapid method for the identification of Candida at the species level by Fourier transform infrared spectroscopy. *Biomed. Lett.* 51(204):223–228, 1995.
44. Patrice Fournier, Thierry Buffeteau, Anna M. Ritcey, Michel A. Pezolet. Polarization modulation FT-IR spectrophotometer for in situ measurements of monolayer films of model biological membranes at the air-water interface. *Spectrosc. Biol. Mol.,* Eur. Conf., 6th, 373–374, 1995.
45. Paul H. Axelsen, W. David Braddock, Howard L. Brockman, Craig M. Jones, Richard A. Dluhy, Brad K. Kaufman, Francisco J. Puga, II. Use of internal reflectance infrared spectroscopy for the in situ study of supported lipid monolayers. *Appl. Spectrosc.* 49(4):526–531, 1995.
46. Annika E. Carlsson, Kjell L.-R. Janne. Near-infrared spectroscopy as an alternative to biological testing for quality control of hyaluronan: comparison of data preprocessing methods for classification. *Appl. Spectrosc.* 49(7):1037–1040, 1995.
47. Michel Pezolet, Josee Labrecque, Muriel Subirade, Bernard Desbat. Infrared spectroscopy of monolayer films of model biological membranes on solid substrates and in situ at the air–water interface. *Spectrosc. Biol. Mol.,* Eur. Conf., 6th, 363–366, 1995.
48. J. H. Aldstadt, A. F. Martin. *Mikrochim. Acta* 127:1–18, 1997.
49. Aniela Matuszewska. The estimation of changes occurring in the coalification process on the ground of the results of infrared spectroscopy and ultimate analysis of the selected coals and macerals. *Pol. J. Appl. Chem.* 41(1–2):75–94, 1997.
50. Marcus Nowak, Harald Behrens. The speciation of water in haplogranitic glasses and melts determined by in situ near-infrared spectroscopy. *Geochim. Cosmochim. Acta.* 59(16):3445–3450, 1995.
51. Helen Worden, Reinhard Beer, Curtis P. Rinsland. Airborne infrared spectroscopy of 1994 western wildfires. *J. Geophys. Res.* 102(D1):1287–1299, 1997.
52. C. T. Chaffin, T. L. Marshall, W. G. Fateley, R. M. Hammaker. Infrared analysis of volcanic plumes: A case study in the application of open-path FT-IR monitoring techniques. *Spectrosc. Eur.* 7(3):18, 20–24, 1995.
53. P. A. Martin, M. Feher. Application of near-infrared diode lasers in atmospheric and on-line process emission monitoring. *NIR News.* 7(3):10–12, 1996.
54. S. M. Watson, R. T. Kroutil, C. A. Traynor, E. S. Edgerton, J. J. Bowser, R. B. Zweidinger, R. N. Olson, R. P. Dalley, W. J. Bone, R. Price. Remote-sensing and in situ atmospheric chemistry studies with the use of a manned hot-air balloon platform. *Field Anal. Chem. Technol.* 1(1):13–22, 1996.
55. R. M. Hammaker, J. M. Poholarz, T. L. Marshall, M. D. Tucker, M. R. Witkowski, W. G. Fateley. Remote sensing of volatile organic compounds in the atmosphere using a Fourier transform near-infrared spectrometer. *Leaping Ahead Near-infrared Spectrosc.* 510–513, 1995.
56. Matthew Griffin. Detectors for far-infrared spectroscopy of the Earth's atmosphere and beyond. *Spectrum.* 34(2):22–23, 1996.
57. Judith O. Zwicker. Intercomparison of data from open-path Fourier transform infrared spectrometers collected during the EPA/API cooperative remote sensing/dispersion field study at Duke Forest, North Carolina, January 9 to 27, 1995. *Proc. SPIE-Int. Soc. Opt. Eng.* Vol. 2883:550–567, 1996.
58. Diane F. Malley, Philip C. Williams, Michael P. Stainton. Rapid measurement of suspended C, N, and P from precambrian shield lakes using near-infrared reflectance spectroscopy. *Water Res.* 30(6):1325–1332, 1996.
59. Ellen Susanna Howell. Probing asteroid composition using visible and near-infrared spectroscopy (surface mineralogy). Doctoral dissertation, Univ. of Arizona, Tucson, AZ; 1995, 186 pp.
60. M. R. Yalamanchili, A. A. Atia, J. Drelich, J. D. Miller. Characterization of interfacial water at hydrophilic and hydrophobic surfaces by in situ FT-IR/internal reflection spectroscopy. *Process. Hydrophobic Miner. Fine Coal,* Proc. UBC-McGill Bi-Annu. Int. Symp. Fundam. Miner. Process. 3–16, 1995.
61. C. Camy-Peyret, P. Jeseck, T. Hawat, G. Durry, S. Payan, G. Berube, L. Rochette, D. Huguenin. The LPMA balloon-borne FT-IR spectrometer for remote sensing of atmospheric constituents. *Eur. Space Agency* (Spec. Publ.) ESA SP, ESA SP-370: 323–328, 1995.
62. Thomas A. Blake, Charles Chackerian, James R. Podolske, Prognosis for a mid-infrared magnetic rotation spectrometer for the in situ detection of atmospheric free radicals. *Appl. Opt.* 35(6): 973–985, 1996.
63. Bailin Rex Yang. Airborne shortwave infrared spectral remote sensing as a direct prospecting method for oil and gas resources. *Chin. J. Geochem.* 13(2):156–164, 1994.
64. Yan Gao, Jingzhi Li, Beili Zhang. The infrared microscope and rapid identification of gemstones. *J. Gemmol.* 24(6):411–414, 1995.
65. J. Ricardo, De A. Franca, J.-M. Brustet, J. Fontan. Multispectral remote sensing of biomass burning in West Africa. *J. Atmos. Chem.* 22(1 & 2):81–110, 1995.
66. Erich M. Fischer, Carle M. Pieters. Remote determination of exposure degree and iron concentration of lunar soils using VIS-NIR spectroscopic methods. *Icarus* 111(2):475–488, 1994.
67. S. P. Levine, G. M. Russwurm. Fourier-transform infrared optical remote sensing for monitoring airborne gas and vapor contaminants in the field. *Trends Anal. Chem.* 13(7):258–262, 1994.

68. T. L. Marshall, C. T. Chaffin, R. M. Hammaker, W. G. Fateley. An introduction to open-path FT-IR. Atmospheric monitoring. *Environ. Sci. Technol.* 28(5):224A–232A, 1994.
69. G. W. Somsen, I. Jagt, C. Gooijer, N. H. Velthorst, U. A. T. Brinkman, T. Visser. Identification of herbicides in river water using on-line trace enrichment combined with column liquid chromatography–Fourier-transform infrared spectrometry. *J. Chromatogr., A* 756:145–157, 1996.
70. Junde Wang, Jianxia Kang, Zuoru Chem, Mei Huang, Jupei Zhang, Tianshu Wang. The study of Freon-12 in alcohol/air flame with remote sensing Fourier transform infrared emission spectroscopy. *J. Environ. Sci. Health A: Environ. Sci. Eng. Toxic Hazard. Subst. Control.* A30(10):2111–2122, 1995.
71. P. Haschberger. Remote measurement of trace gases from aircraft emissions using infrared spectroscopy. *Mitt.—Dtsch. Forschungssanst. Luft-Raumfahrt.* (94–06, Impact of Emissions from Aircraft and Spacecraft Upon the Atmosphere):100–105, 1994.
72. Mark A. Druy, Paul J. Glatkowski, Roy Bolduc, W. A. Stevenson, Thomas C. Thomas. Applications of remote fiber optic spectroscopy using infrared fibers and Fourier transform infrared (FT-IR) spectroscopy to environmental monitoring. *Proc. SPIE-Int. Soc. Opt. Eng.* Vol. 2293 (Chemical, Biochemical, and Environmental Fiber Sensors VI):23–31, 1994.
73. Klaus Schaefer, Rainer Haus, Joerg Heland. Remote sensing of emission sources by ground based Fourier-transform infrared spectrometry. *Transp. Transform. Pollut. Troposphere.* Proc. EUROTRAC Symp., 3rd. 944–948, 1994.
74. T. L. Marshall, C. T. Chaffin, V. D. Makepeace, R. M. Hoffman, R. M. Hammaker, W. G. Fateley, P. Saarinen, J. Kauppinen. Investigation of the effects of resolution on the performance of classical least squares (CLS) spectral interpretation programs when applied to volatile organic compounds (VOCs) of interest in remote sensing using open-air long-path Fourier transform infrared (FT-IR) spectrometry. *J. Mol. Struct.* 324(1–2):19–28, 1994.
75. Michael D. Jack, David R. Nelson, Robert D. Stephens, Geoffrey A. Walter, Christopher B. Tacelli, Jose A. Santana. Optical IR sensing apparatus for remote analysis of vehicle exhaust gases. Santa Barbara Research Center; General Motors Corp. Canada Pat Appl, CA 2131865 AA, 11/3/95; CA 2131865 (9/12/94), *US 119788 (10/9/93).
76. Mark L. Polak, Jeffrey L. Hall, Kenneth C. Herr. Passive Fourier-transform infrared spectroscopy of chemical plumes: an algorithm for quantitative interpretation and real-time background removal. *Appl. Opt.* 34(24):5406–5412, 1995.
77. Arnold J. Eilert, William J. Danley, Xiaolu Wang. Rapid identification of organic contaminants in pretreated waste water using AOTF near-IR spectrometry. *Adv. Instrum. Control.* 50(Pt. 2):87–95, 1995.
78. Guido Toci, Piero Mazzinghi, Matteo Vannini. Airborne infrared diode laser spectrometer for in situ measurement of stratospheric trace gas concentration. *AIP Conf. Proc.* 386:437–438, 1997.
79. Jane W. Adams, Hollis H. Bennett. Using reflectance Fourier transform infrared spectroscopy (FT-IR) for rapid analysis of contaminated soils. *Contam. Soils.* 1:241–252, 1996.
80. J.-M. Theriault, L. Bissonnette, G. Roy. Monitoring cloud parameters using a ground-based FT-IR system. *Proc. SPIE-Int. Soc. Opt. Eng.,* Vol. 3106 (Spectroscopic Atmospheric Monitoring Techniques): 7–16, 1997.
81. Nicholas M. Davies, Moira Hilton, Alan H. Lettington. Vehicle exhaust gas monitoring by remote FT-IR spectroscopy. *Mikrochim. Acta, Suppl.* 14:551–553, 1997.
82. I. T. Bae, R. L. Barbour, D. A. Scherson. In situ Fourier-transform infrared spectroscopic studies of nitrite reduction on platinum electrodes in perchloric acid. *Anal. Chem.* 69:249–252, 1997.
83. T. Norris, P. K. Aldridge. Endpoint determination on-line and reaction coordinate modeling of homogeneous and heterogeneous reactions in principal component space using periodic near-infrared monitoring. *Analyst* 121(8):1003–1008, 1996.
84. W. N. Richmond, P. W. Faguy, R. S. Jackson, S. C. Weibel. Comparison between real-time polarization modulation and static linear polarization for in situ infrared spectroscopy at electrode surfaces. *Anal. Chem.* 68(4):621–628, 1996.
85. L. J. May, L. A. Lentz, B. A. Kirsch, R. O'Reilly. Long-path (8-meter) gas cell analysis of corrosive gases by Fourier-transform infrared spectroscopy. *Process Control Qual.* 7(2):101–111, 1995.
86. I. T. Bae, M. Sandifer, Y. W. Lee, D. A. Tryk, C. N. Sukenik, D. A. Scherson. In situ Fourier-transform infrared spectroscopy of molecular adsorbates at electrode/electrolyte interfaces: a comparison between internal and external reflection modes. *Anal. Chem.* 67(24):4508–4513, 1995.
87. T. Lindblom, A. A. Christy, F. O. Libnau. Quantitative determination of stabiliser in a single base propellant by chemometric analysis of Fourier transform infrared spectra. *Chemom. Intell. Lab. Syst.* 29(2):243–254, 1995.
88. J. Bandekar, R. Sethna, M. Kirschner. Quantitative determination of sulfur oxide species in white liquor by FT-IR. *Appl. Spectrosc.* 49(11):1577–1582, 1995.
89. P. W. Faguy, N. S. Marinkovic. Sensitivity and reproducibility in infrared spectroscopic measurements at single-crystal electrode surfaces. *Anal. Chem.* 67(17):2791–2799, 1995.
90. Z. Ge, R. Thompson, S. Cooper, D. Ellison, P. Tway. Quantitative monitoring of an epoxidation process by Fourier-transform infrared spectroscopy. *Process Control Qual.* 7(1):3–12, 1995.

REFERENCES

91. M. M. A. Ruyken, J. A. Visser, A. K. Smilde. On-line detection and identification of interferences in multivariate predictions of organic gases using FT-IR spectroscopy. *Anal. Chem.* 67(13):2170–2179, 1995.
92. K. Kargosha, S. H. Davarani, R. Moosavi. Quantitative analysis of binary aqueous sodium chlorate and sodium perchlorate solution in the presence of sodium dichromate using Fourier transform infrared spectrometry. *Analyst* 120(7):1945–1948, 1995.
93. G. Woelki, R. Salzer. Thermal investigations of the structure of two humic acid salts by in situ FT-IR spectroscopy. *Fresenius' J. Anal. Chem.* 352(5):529–531, 1995.
94. T. S. Wang, C. J. Zhu, J. D. Wang, F. M. Xu, Z. O. Chen, Y. H. Luo. Study on spectral characterization of infrared flare material combustion with remote high-resolution Fourier-transform infrared spectrometry. *Anal. Chim. Acta.* 306(2–3):249–258, 1995.
95. W. Wisniewski, E. Malinowska, S. Podsiadlo, M. Jarosz. Rapid IR-spectrophotometric determination of total nitrogen in silicon dioxide–silicon nitride mixtures. *Quim. Anal. (Barcelona)* 14(1):64–66, 1995.
96. M. A. Centeno, J. J. Benitez, P. Malet, I. Carrizosa, J. A. Odriozola. In situ temperature-programmed diffuse-reflectance infrared Fourier-transform spectroscopy (TPDRIFTS) of vanadium oxide/titanium oxide catalysts. *Appl. Spectrosc.* 51:416–422, 1997.
97. P. J. Brimmer. NIR method development in the chemical/polymer industry: meeting the needs of the analysis and the analyst. *Leaping Ahead Near-Infrared Spectrosc.* 6th. 360–363, 1995.
98. S. Salim, C. A. Wang, R. D. Driver, K. F. Jensen. In situ concentration monitoring in a vertical OMVPE reactor by fiber-optics-based Fourier transform infrared spectroscopy. *J. Cryst. Growth* 169(3):443–449, 1996.
99. Yuriy V. Tolmachev, Zhenghao Wang, Daniel A. Scherson. Theoretical aspects of laminar flow in a channel-type electrochemical cell as applied to in situ attenuated total reflection-infrared spectroscopy. *J. Electrochem. Soc.* 143(10):3160–3166, 1996.
100. Claude Lamy, Jean-Miichel Leger, Francoise Hahn, Bernard Beden. Electrooxidation of methanol at Pt-Ru electrodes: an in situ IR reflectance spectroscopic study. *Laboratoire Chimie I,* Universite Poitiers, 86022, Poitiers, Fr. Proc.—Electrochem. Soc., 96–98:356–370, 1996.
101. J. J. Benitez, M. A. Centeno, C. Louis dit Picard, O. Merdrignac, Y. Laurent, J. A. Odriozola. In situ diffuse reflectance infrared spectroscopy (DRIFTS) study of the reversibility of CdGeON sensors towards oxygen. *Sens. Actuators B* B31(3):197–202, 1996.
102. Dilum Devjke Dunuwila. An investigation of the feasibility of using in situ ATR ATIR spectroscopy in the measurement of crystallization phenomena for research and development of batch crystallization processes. Doctoral dissertation, Michigan State University, East Lansing, MI, 1996, 139 pp.
103. M. Epple, H. Kirschnick, J. M. Thomas. An in situ IR-spectroscopic study of the solid-state formation reaction of polyglycolide. Thermal elimination of NaCl from sodium chloroacetate. *Therm. Anal.* 47(2):331–338, 1996.
104. Lisheng Xu, J. H. Fu, John R. Schlup. In Situ Near-Infrared Spectroscopic Investigation of the Kinetics and Mechanisms of Reactions between Phenyl Glycidyl Ether (PGE) and Multifunctional Aromatic Amines. *Ind. Eng. Chem. Res.* 35(3):963–972, 1996.
105. Jovan Mijovic, Sasa Andjelic. Monitoring of reactive processing by remote mid-infrared spectroscopy. *Polymer* 37(8):1295–1303, 1996.
106. Qin Xin, Xingtao Gao. Developments of in-situ molecular spectroscopy in catalysis studies. *Prog. Nat. Sci.* 5(3):279–290, 1995.
107. C. G. Stevens, P. J. Kuzmenko, W. E. Conaway, F. Magnotta, N. L. Thomas, J. Galkowski, I. T. Lewis, T. W. Alger. Cross dispersion infrared spectrometry (CDIRS) for remote chemical sensing. *Proc. SPIE-Int. Soc. Opt. Eng.* Vol. 2552(Pt. 1):284–291, 1995.
108. Gabriele Eder-Mirth. Using in situ IR spectroscopy to understand selective alkylation of aromatics over zeolites. *Collect. Czech. Chem.* 60(3):421–427, 1995.
109. Richard Scamehorn. Rapid IR sampling methods for the organic chemistry laboratory using a diffuse reflectance accessory. *J. Chem. Educ.* 72(6):566, 1995.
110. Joseph William Schoppelrei. In situ FT-IR and FT-Raman studies of reactions in the hydrothermal environment. Doctoral dissertation, Univ. of Delaware, Newark, DE; 1997, 175 pp.
111. S. Agyare Yeboah. Near-infrared monitoring of dimethyl terephthalate in a low pressure methanolysis process. *Process Control Qual.* 9(4):189–195, 1997.
112. Qinghua Li, Yulan Zhu, Hongxian Han, Zhushi Li. In situ infrared studies on dehydroxylation of MCM-41. *Yanbian Daxue Xuebao (Bianjibu), Ziran Kexueban* 23(1):35–37, 1997.
113. S. Zhang, J. E. Franke, T. M. Niemczyk, D. M. Haaland, J. N. Cox, I. Banerjee. Testing of a rapid fault detection model for quality control: Borophosphosilicate glass thin films monitored by infrared absorption spectroscopy. *J. Vac. Sci. Technol. B* 15(4):955–960, 1997.
114. Shu-Hua Chien, Shaw-Yo Yang, Yung-Kuan Tseng, Hsiu-Wei Chen. In-situ FT-IR studies of the acidity and catalytic properties of Pt/zeolites. *Bull. Inst. Chem. Acad. Sin.* 44:35–44, 1997.
115. F. Le Peltier, P. Chaumette, J. Saussey, M. M. Bettahar, J. C. Lavalley. In-situ FT-IR spectroscopy and kinetic study of methanol synthesis from CO/H_2 over $ZnAl_2O_4$ and $Cu-ZnAl_2O_4$ catalysts. *J. Mol. Catal. A: Chem.* 122(2–3):131–139, 1997.

116. J. F. Binette, H. Buijs. Fourier transform near-infrared process monitoring of multiple inorganic ions in aqueous solution. *Near-infrared Spectrosc.: Future Waves, Proc. Int. Conf. Near Infrared Spectrosc.,* 7th. 287–289, 1996.
117. S. Agyare Yeboah, Regis Cinier, Jean Guilment. Quantitative analysis of resorcinol in aqueous solution by near-infrared spectroscopy-from the laboratory to the production floor. *Near-infrared Spectrosc.: Future Waves, Proc. Int. Conf. Near-infrared Spectrosc.,* 7th. 279–287, 1996.
118. Bellon-Maurel Veronique, Trystram Gilles. Process control in sugar industry by FT-IR spectrometer coupled with an ATR accessory. *Dev. Food Sci.* 36:11–18, 1994.
119. Sabrie Soloman. Real-time on-line analysis of organic and non-organic compounds for food, fertilizers and pharmaceutical product. WO Patent No. 9740361 A1; Issued: 10/30/97.
120. F. W. J. van den Berg, W. A. van Osenbruggen, A. K. Smilde. Process analytical chemistry in the distillation industry using near-infrared spectroscopy. *Process Control Qual.* 9:51–57, 1997.
121. Claus Borggaard, Jens Havn Thorup. On-line quality measurement of unhomogenized products with near-infrared spectroscopy and neural networks. *Near-infrared Spectrosc.: Future Waves, Proc. Int. Conf. Near-infrared Spectrosc.,* 7th. 81–85, 1996.
122. M. M. Mossoba, M. P. Yurawecz, R. E. McDonald. Rapid determination of the total trans-content of neat hydrogenated oils by attenuated total reflection spectroscopy. *J. Am. Oil Chem. Soc.* 73:1003–1009, 1996.
123. S. J. Lee, I. J. Jeon, L. H. Harbers. Near-infrared reflectance spectroscopy for rapid analysis of curds during cheddar cheese making. *J. Food Sci.* 62:53–56, 1997.
124. A. K. Hewavitharana, B. van Brakel. Fourier transform infrared spectrometric method for the rapid determination of casein in raw milk. *Analyst* 122:701–704, 1997.
125. N. Berding, G. A. Brotherton. Grabbing the BIG picture: a novel approach to beating within-sample material heterogeneity. *NIR News* 7(6):14–16, 1996.
126. T. Lovasz, P. Meresz, A. Salgo. Application of near-infrared transmission spectroscopy for the determination of some quality parameters of apples. *J. Near-infrared Spectrosc.* 2(4):213–221, 1994.
127. J. P. Wold, T. Jakobsen, L. Krane. Atlantic salmon average fat content estimated by near infrared transmittance spectroscopy. *J. Food Sci.* 61(1):74–77, 1996.
128. G. Downey. Non-invasive and non-destructive percutaneous analysis of farmed salmon flesh by near-infrared spectroscopy. *Food Chem.* 55(3):305–311, 1996.
129. Y.-R. Chen, R. W. Huffman, B. Park, M. Nguyen. Transportable spectrophotometer system for on-line classification of poultry carcasses. *Appl. Spectrosc.* 50:910–916, 1996.
130. M. Lipp. Review of methods for the analysis of triglycerides in milk fat: application for studies of milk quality and adulteration. *Food Chem.* 54(2):213–221, 1995.
131. F. R. van de Voort, A. A. Ismail, J. Sedman. A rapid, automated method for the determination of cis and trans content of fats and oils by Fourier-transform infrared spectroscopy. *J. Am. Oil Chem. Soc.* 72(8):873–880, 1995.
132. D. H. McQueen, R. Wilson, A. Kinnunen, E. Paaske Jensen. Comparison of two infrared spectroscopic methods for cheese analysis. *Talanta* 42(12):2007–2015, 1995.
133. M. I. Abdullah, E. Eek. Automatic method for the determination of total carbon dioxide in natural waters. *Water Res.* 29(5):1231–1234, 1995.
134. V. Bellon-Maurel, C. Vallat, D. Goffinet. Quantitative analysis of individual sugars during starch hydrolysis by FT-IR-ATR Spectrometry. Part I: Multivariate calibration study—repeatability and reproducibility. *Appl. Spectrosc.* 49(5):556–562, 1995.
135. N. Moonsamy, F. Mochaba, M. Majara, E. S. C. O'Connor-Cox, B. C. Axcell. Rapid yeast trehalose measurement using near-infrared reflectance spectrometry. *J. Inst. Brew.* 101(3):203–206, 1995.
136. Marcia King-Brink, Zoraida Defreitas, Joseph G. Sebranek. Use of near-infrared transmission for rapid analysis of meat composition. *Near-infrared Spectrosc.: Future Waves, Proc. Int. Conf. Near Infrared Spectrosc.,* 7th. 142–148, 1996.
137. Dominique Bertrand, Bruno Novales, Marie-Francoise Devaux, Paul Robert, Joel Abecassis. Discrimination of durum wheat products for quality control. *Near-infrared Spectrosc.: Future Waves, Proc. Int. Conf. Near-infrared Spectrosc.,* 7th. 430–435, 1996.
138. Yueshu Li, Gordon Laycock, Walter Fernets. Rapid assessment of potential malting quality of barley by near-infrared diffuse reflectance spectroscopy. *Near-infrared Spectrosc.: Future Waves, Proc. Int. Conf. Near-infrared Spectrosc.,* 7th. 475–478, 1996.
139. Hui-Zhen Zhang, Tung-Ching Lee. Rapid near-infrared spectroscopic method for the determination of free fatty acid in fish and its application in fish quality assessment. *J. Agric. Food Chem.* 45(9): 3515–3521, 1997.
140. Amitha K. Hewavitharana, Bram van Brakel. Fourier transform infrared spectrometric method for the rapid determination of casein in raw milk. *Analyst* 122(7):701–704, 1997.
141. N. Berding, G. A. Brotherton. Analysis of samples from sugarcane evaluation trials by near-IR spectroscopy using a new on-line, large cassette presentation module. *Sugarcane: Res. Effic. Sustainable Prod.* (Sugar 2000 Symp.), 57–58, 1996.
142. H. Grunewald. Rapid non-destructive raw material identification in the cosmetic industry with near-infrared (NIR) spectroscopy. *In-Cosmet.* Conf. Proc.:133–149, 1997.

REFERENCES

143. Hartmut Hoyer. NIR analysis in the food industry. *Process Control Qual.* 9(4):143–152, 1997.
144. Chunyu Song, Reinhard Otto. Rapid determination of the value-determining constituents in sausage products by near-infrared transmission spectroscopy. *Z. Lebensm.-Unters. Forsch.* 201(3):226–229, 1995.
145. J. Tenhunen, M. Svinhufvud, K. Sjoholm, K. Pietila. Use of NIT spectroscopy for monitoring the malting process. *Proc. Congr.—Eur. Brew. Conv.* 25:585–592, 1995.
146. S. J. Lee, I. J. Jeon, L. H. Harbers. Near-infrared reflectance spectroscopy for rapid analysis of curds during cheddar cheese making. *J. Food Sci.* 62(1):53–56, 1997.
147. Juan J. Pascual, Pilar Molina, Rosa Puchades. Automatic determination of protein fractions in Manchega ewe's milk by near-infrared reflectance spectroscopy. *Near-infrared Spectrosc.: Future Waves, Proc. Int. Conf. Near-infrared Spectrosc.,* 7th. 559–564, 1996.
148. John Chambers, Chris Ridgway. Rapid detection of contaminants in cereals. *Near-infrared Spectrosc.: Future Waves, Proc. Int. Conf. Near-infrared Spectrosc.,* 7th. 484–489, 1996.
149. C. H. Stokes, J. K. Blazier. In-line measurement of alcohol and extract comparing near-infrared versus density/sound velocity. *Monogr.—Eur. Brew. Conv.* 20:73–80, 1993.
150. B. Chmielarz, K. Bajdor, A. Labudzinska, Z. Klukowska-Majewska. Studies on the double bond positional isomerization process in linseed oil by UV, IR and Raman spectroscopy. *J. Mol. Struct.* 348:313–316, 1995.
151. Russell L. Jones, Geoff Parkin, Charles W. C. Harvey. On-line NIR for sugar-house syrups: A technological step too far? *Proc. Sugar Process. Res. Conf.* 224–243, 1994.
152. G. A. Brotherton, Nils Berding. Near-infrared spectroscopic applications for milling: prospects and implications. *Proc. Conf. Aust. Soc. Sugar Cane Technol.* (17th) 21–29, 1995.
153. Marc Meurens, Weijie Li, Michel Foulon, Victor Acha. Rapid analysis of fruit juice by infrared spectrometry. *Cerevisia* 20(3):33–36, 1995.
154. P. B. Petersen, J. K. Andersen, J. T. Johansen, E. H. Roege. In-line measurement of important beer variables with near-infrared spectroscopy. *Monogr.—Eur. Brew. Conv.* 20:56–72, 1993.
155. T. Zahn. Near-infrared-transmittance-spectroscopy. A powerful tool for fast process and product control in breweries and malthouses. *Monogr.—Eur. Brew. Conv.* 20:23–31, 1993.
156. Edward M. Suzuki, William P. Marshall. Infrared spectra of U. S. automobile original topcoats (1974–1989): III. In situ identification of some organic pigments used in yellow, orange, red, and brown nonmetallic and brown metallic finishes-benzimidazolones. *J. Forensic Sci.* 42(4):619–648, 1997.
157. T. A. Dal Cason. The characterization of some 3,4-methylenedioxycathinone (MDCATH) homologues. *Forensic Sci. Int.* 87:9–53, 1997.
158. Russell John Fischer, Patrick Francis Crane, Clement Lim Yu, Stephen Daniel Walker. Apparatus for automatic identification of gas samples. Ohmeda Inc., US Patent No. 5,731,581, Issued 03/24/98.
159. Andreas Hayden, Robert Noll. Remote trace gas quantification using thermal IR spectroscopy and digital filtering based on principal components of background scene clutter. *Proc. SPIE-Int. Soc. Opt. Eng.* Vol. 3071:158–168, 1997.
160. K. Helland. On-line analysis of hydrocarbon gases with emphasis on calibration model maintenance. *Proc. Int. Conf. Near Infrared Spectrosc.* 6th. 355–359, 1995.
161. Ilpo Ahonen, Hannu Riipinen, Aappo Roos. Portable Fourier transform infrared spectrometer for use as a gas analyzer in industrial hygiene. *Analyst* 121(9):1253–1255, 1996.
162. Zhuoxiong Mao, Jack Demirgian, Alex Mathew, Richard Hyre. Use of Fourier transform infrared spectrometry as a continuous emission monitor. *Waste Manage.* 15(8):567–577, 1995.
163. Ronald E. Shaffer, Gary W. Small. Comparison of optimization algorithms for piecewise linear discriminant analysis: application to Fourier transform infrared remote sensing measurements. *Anal. Chim. Acta* 331(3):157–175, 1996.
164. L. John Otten, III, Andrew D. Meigs, R. Glenn Sellar, J. Bruce Rafert. Measured performance of an airborne Fourier transform hyperspectral imager. *Proc. SPIE-Int. Soc. Opt. Eng.,* Vol. 2819:291–299, 1996.
165. Jovan Mijovic, Sasa Andelic, Srdan Pejanovic. In-situ real-time monitoring of processes by remote Fourier transform infrared spectroscopy. *J. Serb. Chem. Soc.* 61(12):1193–1202, 1996.
166. Reynaldo Quintana. Non-invasive near-infrared quantitative measurement instrument. Futrex, Inc., Reynaldo Quintana, US Patent No. 5,574,283, 11/12/96.
167. James Rutter, Dave Jungkman, James Stobie, Eric Krueger, James Garnett, Marion Reine, Brian Denley, Mark Jasmin, Anthony Sofia. A multispectral hybrid HgCdTe FPA/Dewar assembly for remote sensing in the atmospheric infrared Sounder (AIRS) instrument. *Proc. SPIE-Int. Soc. Opt. Eng.,* Vol. 2817:200–213, 1996.
168. W. G. Hansen. Production control by near-infrared spectroscopy. *Proc. Int. Conf. Near-infrared Spectrosc.* 6th. 376–381, 1995.
169. R. W. Streamer, F. A. DeThomas. Fiber optic interfaces for process measurements. *Proc. Int. Conf. Near-infrared Spectrosc.* 6th. 19–23, 1995.
170. Kenneth L. Gallaher, Fred Baudais, Rick Lester. The need for sample conditioning in process streams analyzed with NIR or MID-IR spectrometers. *Adv. Instrum. Control* 51(Pt. 1):139–148, 1996.
171. K. Le Foulgoc, L. Le Neindre, X. H. Zhang, J. Lucas. Tapered TeX glass optical fibers for remote IR spectroscopic analysis. *Proc. SPIE-Int. Soc. Opt. Eng.,* Vol. 2836:26–36, 1996.

172. P. F. Valdez, G. W. Donohoe, M. R. Descour, S. Motomatsu. Selection of spectral bands for interpretation of hyperspectral remotely sensed images. *Proc. SPIE-Int. Soc. Opt. Eng.,* Vol. 2819:195–203, 1996.
173. S. T. Vohra, F. Bucholtz, G. M. Nau, K. J. Ewing, I. D. Aggarwal. Remote detection of trichloroethylene in soil by a fiber-optic infrared reflectance probe. *Appl. Spectrosc.* 50:985–990, 1996.
174. R. E. Shaffer, G. W. Small. Genetic algorithms for the optimization of piecewise linear discriminants. *Chemom. Intell. Lab. Syst.* 35:87–104, 1996.
175. J. J. Workman, Jr., J. Brown. A new standard practice for multivariate quantitative infrared analysis—Part II. *Spectroscopy* 11(9):24, 26–30, 1996.
176. D. C. Tilotta, D. L. Heglund, S. B. Hawthorne. On-line SFE-FT-IR spectrometry with a fiber optic transmission cell. *Am. Lab.* 28(6):36R–36T, 1996.
177. P. Wilks. Portable infrared analysers for repetitive analysis. *Int. Lab.* 26:11A–11B, 1996.
178. C. M. Deeley, J. Sellors, R. A. Spragg. Systems validation for analytical FT-IR measurements. *Int. Labmate* 21:20–22, 1996.
179. P. J. Kulesza, M. A. Malik, A. Denca, J. Strojek. In situ FT-IR/ATR spectroelectrochemistry of Prussian blue in the solid state. *Anal. Chem.* 68:2442–2446, 1996.
180. J. Lin, S. J. Hart, J. E. Kenny. Improved two-fiber probe for in situ spectroscopic measurement. *Anal. Chem.* 68:3098–3103, 1996.
181. L. M. Treffman, P. W. Morrison, Jr. A simple method to fabricate lenses for in situ Fourier-transform infrared spectroscopy. *Rev. Sci. Instrum.* 67(4):1454–1457, 1996.
182. J. Workman, Jr. Process chemometrics/spectroscopy terminology. *NIR News* 7(4):15–16, 1996.
183. H. Maeda, Y. Ozaki, Y. Noda, Y. Mimura, T. Nakai, T. Tani. Usefulness of a newly developed fluoride-glass fiber in near-infrared light-fiber spectroscopy. *J. Near-infrared Spectrosc.* 3(1):43–52, 1995.
184. B. Schrader. Raman spectroscopy in the near-infrared: a most capable method of vibrational spectroscopy. *Fresenius' J. Anal. Chem.* 355(3–4):233–239, 1996.
185. F. Ehrentreich, U. Dietze, U. Meyer, H. Schulz, H.-M. Kloetzer, S. Abbas, M. Otto. IR-spectroscopy—Computer program for the interpretation of infrared spectra. *Fresenius' J. Anal. Chem.* 354(7–8):829–832, 1996.
186. Z. Y. Meng, Y. J. Ma. An expert system knowledge base for the analysis of infrared spectra of organophosphorus compounds. *Microchem. J.* 53(3):371–375, 1996.
187. M. L. McKelvy, T. R. Britt, B. L. Davis, J. K. Gillie, L. A. Lentz, A. Leugers, R. A. Nyquist, C. L. Putzig. Infrared spectroscopy. *Anal. Chem.* 68(12):93R–160R, 1996.
188. H. S. Gold. Process near-infrared spectroscopy in industry: Believers, sceptics, converts and preachers. *NIR News* 7(2):10–12, 1996.
189. P. MacLaurin, N. C. Crabb, I. Wells, P. J. Worsfold, D. Coombs. Quantitative in situ monitoring of an elevated-temperature reaction using a water-cooled mid-infrared fiber-optic probe. *Anal. Chem.* 68(7):1116–1123, 1996.
190. V. Catalano. In process near-IR spectroscopy people count as much as technology. *Analusis* 23(10):M8–M9, 1995.
191. T. A. Dirksen, J. E. Gagnon. A sample-preparation technique for infrared analysis of aqueous samples. *Spectroscopy* 11(2):58–62, 1996.
192. T. Fearn. How accurate can you get? *NIR News* 7(2):3, 1996.
193. J. P. Coates. Clarifying the issues. Further reflections on filter analyzers in process IR spectroscopy. *Spectroscopy* 10(9):24–26, 1995.
194. P. A. Wilks. Filter analysers vs. process FT-IR spectrometers, continued. *Spectroscopy* 10(9):23, 1995.
195. E. W. Ciurczak. Questions and answers on process analysis. Tips for selecting and calibrating IR instruments. *Spectroscopy* 10(9):19–20, 1995.
196. J. Workman, Jr. Comparing the SEP from a test set to an estimated SEP using the (chi-square) statistic. *NIR News* 7(1):14–15, 1996.
197. R. Barbour, Z. Wang, I. T. Bae, Y. V. Tolmachev, D. A. Scherson. Channel flow cell for attenuated-total-reflection Fourier-transform infrared spectroelectrochemistry. *Anal. Chem.* 67(21):4024–4027, 1995.
198. I. B. Benson. The characteristics and scope of continuous on-line near-infrared measurement. *Spectrosc. Eur.* 7(6):18, 20–24, 1995.
199. W. Nahm, H. Gehring. Non-invasive in vivo measurement of blood spectrum by time-resolved near-infrared spectroscopy. *Sens. Actuators B* B29(1–3):174–179, 1995.
200. A. Fong, G. M. Hieftje. Near-IR multiplex bandpass spectrometer utilizing liquid molecular filters. *Appl. Spectrosc.* 49(4):493–498, 1995.
201. K. L. Miller. Real world effects on univariate and multivariate calibrations. *Analusis* 23(4):M26–M27, 1995.
202. A. Fong, G. M. Hieftje. Simple near-infrared-spectrometric sorption-based vapor sensor. *Appl. Spectrosc.* 49(9):1261–1267, 1995.
203. J. S. Namkung, M. Hoke, R. S. Rogowski, S. Albin. FT-IR optical fiber remote detection of aluminium hydroxide by evanescent wave absorption spectroscopy. *Appl. Spectrosc.* 49(9):1305–1310, 1995.

REFERENCES

204. J. Workman, Jr. Statistical process control (SPC) charting: Using bias significance (based on average residuals) for instrument control monitoring. *NIR News* 6(5):8–9, 1995.
205. K. Aalijoki, H. Hukkanen, P. Jokinen. Application of spectroscopy, especially near infrared, for process and pilot monitoring. *Process Control Qual.* 6(2):125–131, 1994.
206. H. J. Luinge, J. H. van der Maas, T. Visser. Partial least squares regression as a multivariate tool for the interpretation of infrared spectra. *Chemom. Intell. Lab. Syst.* 28(1):129–138, 1995.
207. R. H. Wilson, J. Holland, J. Potter. Lining up for FT-IR analysis. FT-IR spectroscopy is no longer just a laboratory technique, it can now be used to provide "on-line" quality and process control in industrial processes. *Chem. Br.* 30(12):993–996, 1994.
208. C. Klawun, C. L. Wilkins. Neural network assisted rapid screening of large infrared spectral databases. *Anal. Chem.* 67(2):374–378, 1995.
209. J. H. Hartshorn. Infrared discriminant analysis techniques for determining product quality. *Spectroscopy* 10(3):32–35, 1995.
210. M. Huehne, U. Eschenauer, H. W. Siesler. Performance and selected applications of an acousto-optic tunable-filter near-infrared spectrometer. *Appl. Spectrosc.* 49(2):177–180, 1995.
211. J. Workman, Jr. A review of process near-infrared spectroscopy: 1980–1994. *J. Near-infrared Spectrosc.* 1(4):221–245, 1993.
212. D. L. Heglund, D. C. Tilotta, S. B. Hawthorne, D. J. Miller. Simple fiber-optic interface for on-line supercritical-fluid extraction-Fourier-transform infrared spectrometry. *Anal. Chem.* 66(20):3543–3551, 1994.
213. R. D. Driver, G. L. Dewey, D. A. Greenberg, J. D. Stark. Sample interface in on-line process monitoring. Liquid- and gas-phase sampling of optically homogeneous processes. *Spectroscopy* 9(4):36–40, 1994.
214. S. Lowry, T. May, A. Bornstein, Y. Weissman, R. Harman, I. Tugenthaft. New accessory for characterizing optical fibers with an FT-IR spectrometer. *Appl. Spectrosc.* 48(7):852–856, 1994.
215. C. Dossi, S. Recchia, A. Fusi. Infrared analytical evidence of oxygen-induced formation of iron carbonyl from carbon monoxide and steel walls of an in situ DRIFT cell. *Ann. Chim.* (Rome) 84(7–8):347–352, 1994.
216. J. Crandall. Sample systems in process near-infrared spectroscopy. Part 2. *NIR News* 5(5):6–7, 13, 1994.
217. J. Workman, Jr. Where is NIR process analysis? *NIR News* 5(6):13–14, 1994.
218. H. Kavak, R. Esen. Spectrum comparison of IR data taken from different spectrometers with various precision. *J. Chem. Inf. Comput. Sci.* 33(4):595–597, 1993.
219. G. Sundholm, P. Talonen. Modeling of a spectroelectrochemical cell with radial liquid flow for in situ external-reflection IR measurements. *J. Electroanal. Chem.* 377(1–2):91–99, 1994.
220. Andrew Robert Korb. Experimental methods in infrared remote sensing (spectral emissivity, Fourier transform spectrometer, thermal gradient. Doctoral dissertation, Johns Hopkins Univ., Baltimore, MD, 1997, 111 pp.
221. Peter H. Hindle. Towards 2000. The past, present and future of on-line NIR sensing. *Process Control Qual.* 9(4):105–115, 1997.
222. Jorn Luttge Jensen. Acceptance of NIR as an analytical method. *Process Control Qual.* 9(4):161–165, 1997.
223. Joan Gronkjaer Pedersen. Combining NIR data and production data for process control. *Process Control Qual.* 9(4):153–159, 1997.
224. Robert P. Hammond. Practical and successful on-line near-infrared applications. *Process Control Qual.* 9(4):117–121, 1997.
225. Brian D. Mindel. Process interfacing of in-line NIR spectrophotometers. *Process Control Qual.* 9(4):173–178, 1997.
226. Stephen F. Malin, Gamal Khalil. Method and apparatus for multi-spectral analysis in noninvasive infrared spectroscopy. Instrumentation Metrics, Inc., U. S. Patent No. 5,747,806, Issued 05/05/98.
227. Dmitry V. Matyunin, Marina V. Morozova. Universal near IR rapid-analyzer based on a high-stability FT-IR spectrometer for the wavelength range 750–2500 nm. *Mikrochim. Acta, Suppl.* 14:773–775, 1997.
228. Tatiana B. Egevskaya. Portable "double cat's eye" FT-IR spectrometer. *Mikrochim. Acta, Suppl.* 14:767–768, 1997.
229. Robert W. Dibble, Rajiv K. Mongia, Quang-viet Nguyen. Optical probe for in-situ detection of hydrocarbon concentration. Regents of the University of California; WO Patent No. 9725609, Issued: 07/17/97.
230. Stephan J. Weeks, John F. Mcclelland, Steven L. Wright, Roger W. Jones. Transient infrared spectroscopy for on-line analysis of solids and viscous liquids. *Mikrochim. Acta,* Suppl. 14:801–802, 1997.
231. Karl Molt. How safe are NIR-library search systems? Information-theoretical and practical aspects. *Fresenius' J. Anal. Chem.* 359(1):67–73, 1997.

232. Kenneth H. Levin, Samuel Kerem, Vladimir Madorsky. Handheld infrared spectrometers. Infrared Fiber Systems, Inc. WO Patent No. 9708537 A1, Issued: 03/06/97.
233. Karoly J. Kaffka, Laszlo S. Gyarmati. Reduction of spectral data for rapid quality evaluation. *Near-infrared Spectrosc.: Future Waves, Proc. Int. Conf. Near-infrared Spectrosc.*, 7th. 209–213, 1996.
234. Christine Simard, Henry Buijs. An alternative to handheld fiber optic probes. *Near-infrared Spectrosc.: Future Waves, Proc. Int. Conf. Near-infrared Spectrosc.*, 7th. 98–99, 1996.
235. G. C. Pandey, Ajay Kumar. Modern infrared spectroscopy: a tool for noncontact and non-destructive evaluation with special reference to process monitoring. *Trends NDE Sci. Technol.*, Proc. World Conf. NDT, 3:1541–1544, 1997.
236. Mark O. Westerhaus. Process control-a new quality control option from Infrasoft International. *Near-infrared Spectrosc.: Future Waves, Proc. Int. Conf. Near-infrared Spectrosc.*, 7th. 316–319, 1996.
237. Guenter Rinke, C. Hartig, U. Hoeppener-Kramar. New process analyzers measure from the UV to the IR. *Chem. Technol. Eur.* 3(1):24–27, 1996.
238. Vincent A. L. Wortel, Wei G. Hansen. A chemometric approach: evaluating important parameters in the near-infrared in-line calibration. *Near-infrared Spectrosc.: Future Waves, Proc. Int. Conf. Near-infrared Spectrosc.*, 7th. 306–315, 1996.
239. Christopher C. Alexay, William L. Truett, Christopher D. Prozzo, Barry O'Dwyer. Portable filter infrared spectrometer. Janos Technology Inc., U. S. Patent No. 5,519,219, Issued: 05/21/96.
240. M. Amrhein, B. Srinivasan, D. Bonvin, M. M. Schumacher. Inferring concentrations on-line from near-infrared spectra: non-linear calibration via mid-infrared measurements. *Comput. Chem. Eng.* 20(Suppl. B):S975–S980, 1996.
241. Masaharu Komiyama, Yoko Obi. Simple transmission infrared cell made of Swagelok for in situ infrared spectroscopic studies of heterogeneous catalysis. *Rev. Sci. Instrum.* 67(4):1590–1592, 1996.
242. Lia M. Treffman, Philip W. Morrison, Jr. A simple method to fabricate lenses for in situ Fourier transform infrared spectroscopy. *Rev. Sci. Instrum.* 67(4):1454–1457, 1996.
243. Sylvie Bages, Bernard Descales, Didier Lambert, Jean-Richard Llinas, Andre Martens, Sebastien Osta, Michel Sanchez. Property determination by near-IR spectroscopy. BP Chemicals Limited, BP Oil International Limited, WO Patent No. 9611399 A1, Issued: 04/18/96.
244. M. J. Persky. A review of spaceborne infrared Fourier transform spectrometers for remote sensing. *Rev. Sci. Instrum.* 66(10):4736–4797, 1995.
245. D. Wienke, W. van den Broek, W. Melssen, L. Buydens, R. Feldhoff, T. Kantimm, T. Huth-Fehre, L. Quick, F. Winter, et al. Comparison of an adaptive resonance theory based neural network (ART-2a) against other classifiers for rapid sorting of post consumer plastics by remote near-infrared spectroscopic sensing using an InGaAs diode array. *Anal. Chim. Acta* 317(1–3):1–16, 1995.
246. Jovan Mijovic, Sasa Andjelic. Novel research trends 2: In-situ real-time monitoring of reactive systems by remote fiber optic near-infrared spectroscopy. *Polym. News* 20(10):317–321, 1995.
247. K. Le Foulgoc, L. Le Neindre, Y. Guimond, H. L. Ma, X. H. Zhang, J. Lucas. Analysis of biological and chemical compounds by remote spectroscopy, using IR TeX glass fibers. *Proc. SPIE-Int. Soc. Opt. Eng.* Vol. 2508:399–410, 1995.
248. Paul Wilks. Portable infrared analyzers for repetitive analysis. *Am. Lab.* (Shelton, Conn.) 28(3):52–53, 1996.
249. Jeffrey W. Childers, George M. Russwurm, Edgar L. Thompson, Jr. Quality assurance considerations in a long-term FT-IR monitoring program. *Proc. SPIE-Int. Soc. Opt. Eng.* Vol. 2365:389–395, 1995.
250. James R. Engel, Rick K. Dorval, David L. Carlson, Thomas G. Quinn. Field portable Fourier transform infrared spectrometer system. *AT-ONSITE* 1(1):71–75, 1995.
251. Marion G. Hansen, Atul Khettry. Advancements in near-infrared fiber optic spectrometry probes. *Proc. Annu. ISA Anal. Div. Symp.* 28:107–113, 1995.
252. David L. Curtin. Utilization of on-line NIR measurements for process control. *AT-PROCESS* 1(2):90–94, 1995.
253. Chiaki Hirose. Vibration spectroscopy of surfaces using non-linear optical process: IR-visible sum frequency generation spectroscopy (VSFG). *Adv. Multi-Photon Processes Spectrosc.* 9:145–197, 1995.
254. Robert D. Rosenthal, John J. Mastrototaro, Joseph K. Frischmann, Reynaldo Quintana. Non-invasive near-infrared quantitative measurement instrument. Futrex Inc., U. S. Patent No. 5,436,455, Issued: 07/25/95.
255. A. Laurie Cone, Shahla K. Farhat, Lori A. Todd. Development of QA/QC performance standards for field use of open-path FT-IR spectrometers. *Proc. SPIE-Int. Soc. Opt. Eng.* Vol. 2365:334–338, 1995.
256. R. D. Driver, K. P. Grim, G. Dewey, M. L. Brubaker. Fiber-remote reflectance spectroscopy with an optimized diffuse reflectance sensor system. *Proc. SPIE-Int. Soc. Opt. Eng.* Vol. 2367:159–170, 1995.
257. F. Gonzalez, R. Pous. Quality control in manufacturing process by near-infrared spectroscopy. *J. Pharm. Biomed. Anal.* 13(4/5):423–429, 1995.

REFERENCES

258. Peter R. Griffiths, Ian R. Lewis, Nathan C. Chaffin, Nelson W. Daniel, Jr., John D. Jegla. Remote characterization of materials by vibrational spectrometry through optical fibers. *J. Mol. Struct.* 347:169–185, 1995.
259. Junde Wang, Jianxia Kag, Tianshu Wang. The spectral characterizations of the low- and high-resolution remote FT-IR emission spectroscopy. *Spectrosc. Lett.* 28(6):839–848, 1995.
260. Charles G. Stevens, Norman Thomas, Paul Kuzmenko, Terry Alger. An echelle grating spectrometer (EGS) for mid-IR remote chemical detection. *Proc. SPIE-Int. Soc. Opt. Eng.* Vol. 2266:2–12, 1994.
261. X. H. Zhang, H. L. Ma, C. Blanchetiere, Foulgoc, K. Le; J. Lucas, J. Heuze, P. Collardelle, P. Froissard, D. Picque, et al. Tellurium halide (TeX) IR fibers for remote spectroscopy. *Proc. SPIE-Int. Soc. Opt. Eng.* Vol. 2131:90–9, 1994.
262. Reginald H. Wilson, James K. Holland, Jason Potter. Lining up for FT-IR analysis. *Chem. Br.* 30(12):993–996, 1994.
263. A. Espinosa, D. Lambert, M. Valleur. Use of NIR technology to optimize plant operations. *Hydrocarbon Process. Int. Ed.* 74(2):86–89, 91–92, 1995.
264. Kari Aaljoki, Hannu Hukkanen, Petri Jokinen. Application of spectroscopy, especially near-infrared, for process and pilot monitoring. *Process Control Qual.* 6(2–3):125–131, 1994.
265. C. Di Luzio, S. Morzilli, E. Cardinale. Rapid near-infrared reflectance analysis (NIRA) of mainstream smoke collected on Cambridge filter pads. *Beitr. Tabakforsch. Int.* 16:171–184, 1995.
266. P. Le Cacheux, G. Menard, H. Nguyen Quang, P. Weinmann, M. Jouan, Nguyen Quy Dao. Quantitative analysis of cholesterol and cholesterol ester mixtures using near-infrared Fourier-transform Raman spectroscopy. *Appl. Spectrosc.* 50:1253–1257, 1996.
267. R. Simhi, Y. Gotshal, D. Bunimovich, B.-A. Sela, A. Katzir. Fiber-optic evanescent-wave spectroscopy for fast multicomponent analysis of human blood. *Sci. Total Environ.* 187:3421–3425, 1996.
268. P. Franck, J.-L. Sallerin, H. Schroeder, M.-A. Gelot, P. Nabet. Rapid determination of faecal fat by Fourier-transform infrared analysis (FT-IR) with partial least squares regression and an attenuated total reflectance accessory. *Clin. Chem.* 42:2015–2020, 1996.
269. J. F. Stevens, P. Vadgama. Infrared analysis in clinical chemistry: its use in the laboratory and in non-invasive near patient testing. *Ann. Clin. Biochem.* 34:215–221, 1997.
270. J. F. Brennan, III; Y. Wang, R. R. Dasari, M. S. Feld. Near-infrared Raman spectrometer systems for human tissue studies. *Appl. Spectrosc.* 51:201–208, 1997.
271. I. Ahonen, H. Riipinen, A. Roos. Portable Fourier-transform infrared spectrometer for use as a gas analyser in industrial hygiene. *Analyst* 121:1253–1255, 1996.
272. S. K. Cole, P. Martin. Determination of gas-phase sidestream cigarette smoke components using Fourier-transform infrared spectrometry. *Analyst* 121(4):495–500, 1996.
273. I. R. Lewis, N. C. Chaffin, M. E. Gunter, P. R. Griffiths. Vibrational spectroscopic studies of asbestos and comparison of suitability for remote analysis. *Spectrochim. Acta A* 52A(3):315–328, 1996.
274. J. T. Kuenstner, K. H. Norris. Near-infrared hemoglobinometry. *J. Near-infrared Spectrosc.* 3:11–18, 1995.
275. Edward L. Hull, Thomas H. Foster. Noninvasive near-infrared hemoglobin spectroscopy for in vivo monitoring of tumor oxygenation and response to oxygen modifiers. *Proc. SPIE-Int. Soc. Opt. Eng.* Vol. 2979:355–364, 1997.
276. J. W. Hall, A. Pollard. Near-infrared analysis of dried human serum for determining human serum urea concentration. *J. Near-infrared Spectrosc.* 1(3):127–132, 1993.
277. G. Domjan, K. J. Kaffka, J. M. Jako, I. T. Valyi-Nagy. Rapid analysis of whole blood and blood serum using near infrared spectroscopy. *J. Near-infrared Spectrosc.* 2(2):67–78, 1994.
278. L. A. Sodickson, M. J. Block. Kromoscopic analysis: a possible alternative to spectroscopic analysis for noninvasive measurement of analytes in vivo. *Clin. Chem.* 40(9):1838–1844, 1994.
279. Sven Harthun, Kathrin Matischak, Peter Friedl. Process control of antithrombin III production by near-infrared spectroscopy. *Anim. Cell Technol.: Vaccines Genet. Med.* (Proc. Meet. ESACT) 14th. 417–421, 1997.
280. Gyula Domjan, Janos Jako, Istvan Valyi-Nagy. Rapid analysis of beta.-lipoprotein in human blood serum using near-infrared spectroscopy. *Near-infrared Spectrosc.: Future Waves, Proc. Int. Conf. Near-infrared Spectrosc.*, 7th. 353–356, 1996.
281. Peter E. Raber, Jeff Santman. Methods and apparatus for noninvasive glucose sensing: Spectrograph. Diasense, Inc.; WO Patent No. 9725915 A1, Issued: 07/24/97.
282. John F. Stevens, Pankaj Vadgama. Infrared analysis in clinical chemistry: Its use in the laboratory and in non-invasive near patient testing. *Ann. Clin. Biochem.* 34(3):215–221, 1997.
283. Lin-P'ing Choo, David L. Wetzel, William C. Halliday, Michael Jackson, Steven M. LeVine, Henry H. Mantsch. In situ characterization of beta-amyloid in Alzheimer's diseased tissue by synchrotron Fourier transform infrared microspectroscopy. *Biophys. J.* 71(4):1672–1679, 1996.
284. U. A. Mueller, B. Mertes, C. Fischbacher, K. U. Jageman, K. Danzer. Non-invasive blood glucose monitoring by means of near-infrared spectroscopy: methods for improving the reliability of the calibration models. *Int. J. Artif. Organs* 20(5):285–290, 1997.

285. C. Fischbacher, K. U. Jagemann, K. Danzer, U. A. Muller, L. Papenkordt, J. Schuler. Enhancing calibration models for non-invasive near-infrared spectroscopical blood glucose determination. *Fresenius' J. Anal. Chem.* 359(1):78–82, 1997.
286. Jack C. Demirgian, Susan M. Macha, Shauna M. Darby, John Ditillo. Detection of emission sources using passive-remote Fourier transform infrared spectroscopy. *Field Screening Methods Hazard. Wastes Toxic Chem.*, Proc. Int. Symp. 1:626–639, 1995.
287. Tilo Wolf, Ute Lindauer, Uwe Reuter, Tobias Back, Arno Villringer, Karl Einhaupl, Ulrich Dirnagl. Noninvasive near-infrared spectroscopy monitoring of regional cerebral blood oxygenation changes during peri-infarct depolarizations in focal cerebral ischemia in the rat. *J. Cereb. Blood Flow Metab.* 17(9):950–954, 1997.
288. Belinda Pastrana-Rios. Lipid-lipid and lipid-protein interaction in the pulmonary surfactant system an in situ infrared spectroscopic test of the "squeeze-out" hypothesis (dipalmitoylphosphatidylcholine). Doctoral dissertation, Rutgers, State University, Newark, NJ, 1996, 153 pp.
289. G. Spanner, R. Niessner. Noninvasive determination of blood constituents using an array of modulated laser diodes and a photoacoustic sensor head. *Fresenius' J. Anal. Chem.* 355(3–4):327–328, 1996.
290. Takafumi Hamaoka, Hisao Iwane, Teruichi Shimomitsu, Toshihito Katsumura, Norio Murase, Shinya Nishio, Takuya Osada, Yuko Kurosawa, Britton Chance. Noninvasive measures of oxidative metabolism on working human muscles by near-infrared spectroscopy. *J. Appl. Physiol.* 81(3):1410–1417, 1996.
291. H. M. Heise. Non-invasive monitoring of metabolites using near-infrared spectroscopy: state of the art. *Horm. Metab. Res.* 28(10):527–534, 1996.
292. Steven M. LeVine, David L. Wetzel. In situ chemical analyses from frozen tissue sections by Fourier transform infrared microspectroscopy: Examination of white matter exposed to extravasated blood in the rat brain. *Am. J. Pathol.* 145(5):1041–1047, 1994.
293. Steven Mitchell LeVine, David Louis Wetzel. In situ chemical analysis of brain tissue by Fourier transform infrared microspectroscopy. *NeuroProtocols* 5(1):72–79, 1994.
294. Kenneth J. Schlager, Timothy L. Ruchti. TAMM—A reflective, noninvasive, near-infrared blood chemistry analyzer. *Proc. SPIE-Int. Soc. Opt. Eng.* Vol. 2386:174–184, 1995.
295. Kay-Uwe Jagemann, Christoph Fischbacher, Klaus Danzer, Ulrich A. Mueller, Bernardo Mertes. Application of near-infrared spectroscopy for non-invasive determination of blood/tissue glucose using neural networks. *Z. Phys. Chem.* (Munich) 191(2):179–190, 1995.
296. Tilo Wolf, G. Arnold, J. Dreier, Tobias Back, A. Villringer, U. Dirnagl. Noninvasive detection of hemoglobin oxygenation changes during cortical spreading depression in the rat brain. *Front. Headache Res.* 5:107–112, 1995.
297. C. E. Cooper, C. E. Elwell, J. H. Meek, S. J. Matcher, J. S. Wyatt, M. Cope, D. T. Delpy. The noninvasive measurement of absolute cerebral deoxyhemoglobin concentration and mean optical path length in the neonatal brain by second derivative near-infrared spectroscopy. *Pediatr. Res.* 39(1):32–38, 1996.
298. Mitsuhiko Noda, Mikio Kimura, Takeshi Ohta, Akihito Kinoshita, Fumiyoshi Kubo, Nobuaki Kuzuya, Yasunori Kanazawa. Completely noninvasive measurement of blood glucose using near-infrared waves. *Int. Congr. Ser.* 1100:1128–1132, 1995.
299. Tianshu Wang, Changjiang Zhu, Junde Wang, Fuming Xu, Zuoru Chen, Yunhua Luo. Study on spectral characterization of infrared flare material combustion with remote high resolution Fourier transform infrared spectrometry. *Anal. Chim. Acta* 306(2–3):249–258, 1995.
300. Thomas Rohe, Edna Gruenblatt, Norbert Eisenreich. Near-infrared-transmission spectroscopy on propellants and explosives. *Int. Annu. Conf.* 27:85.1–85.10, 1996.
301. Stephen Medlin, Charles Westgate, William Danley, Ursula Eschenauer. Solving process problems with AOTF-based NIR spectroscopy. *Proc. Army Res. Lab. Acousto-Opt. Tunable Filter Workshop*, 1st. 149–163, 1997.
302. E. Wust, A. Fehrmann, A. Hoffmann, L. Rudzik. In-line measurement of high moisture products. *Near-infrared Spectrosc.: Future Waves, Proc. Int. Conf. Near-infrared Spectrosc.*, 7th. 268–271, 1996.
303. O. Berntsson, G. Zackrisson, G. Oestling. Determination of moisture in hard gelatin capsules using near-infrared spectroscopy: applications to at-line process control of pharmaceutics. *J. Pharm. Biomed. Anal.* 15(7):895–900, 1997.
304. I. Litani-Barzilai, I. Sela, V. Bulatov, I. Zilberman, I. Schechter. On-line remote prediction of gasoline properties by combined optical methods. *Anal. Chim. Acta* 339:193–199, 1997.
305. B. R. Stallard, M. J. Garcia, S. Kaushik. Near-IR reflectance spectroscopy for the determination of motor oil contamination in sandy loam. *Appl. Spectrosc.* 50(3):334–338, 1996.
306. Z. Stanek. Automating laboratory FT-IR spectroscopy with a liquid analysis system. *Am. Lab.* 28(8):25–28, 1996.
307. W. I. Friesen. Qualitative analysis of oil sand slurries using on-line NIR spectroscopy. *Appl. Spectrosc.* 50:1535–1540, 1996.
308. J. M. Andrarde, S. Muniategui, P. Lopez-Maria, D. Prada. Use of multivariate techniques in quality control of kerosine production. *Fuel* 76(1):51–59, 1992.

REFERENCES

309. D. Lambert, B. Descales, A. Espinosa, M. Sanchez, S. Osta, J. Gil, A. Martens, M. Valleur. NIR on-line advanced control system for a gasoline blender. *Analusis* 23(4):M20–M25, 1995.
310. D. Lambert, B. Descales, S. Bages, S. Bellet, J. R. Llinas, M. Loublier, J. P. Maury, M. Loublier, J. P. Maury, A. Martens. Optimization of steam-cracking operations through on-line NIR analysis. *Analusis* 23(10):M10–M13, 1995.
311. R. D. Stephens, P. A. Mulawa, M. T. Giles, K. G. Kennedy, P. J. Groblicki, S. H. Cadle, K. T. Knapp. An experimental evaluation of remote sending-based hydrocarbon measurements: a comparison to FID measurements. *J. Chem. Inf. Comput. Sci.* 36(1):148–158, 1996.
312. A. L. Kania. Composition measurement in gasoline blending. *Analusis* 23(4):M18–M19, 1995.
313. H. Pruefer, D. Mamma. Near-infrared on-line analysis of motor octane number in gasoline with an acousto-optic tunable transmission spectrophotometer. *Analusis* 23(4):M14–M18, 1995.
314. D. Lambert, B. Descales, J. R. Llinas, A. Espinosa, S. Osta, M. Sanchez, A. Martens. On-line NIR monitoring and optimization for refining and petrochemical processes. *Analusis* 23(4):M9–M13, 1995.
315. A. E. Palomares, G. Eder-Mirth, J. A. Lercher. Selective alkylation of toluene over basic zeolites: an in situ infrared spectroscopic investigation. *J. Catal.* 168(2):442–449, 1997.
316. J. J. Workman. A brief review of near infrared in petroleum product analysis. *J. of Near Infrared Spectrosc.* 4(1–4), 69–74, 1996.
317. J. R. Markham, J. E. Cosgrove, C. M. Nelson, A. S. Bonanno, R. E. Schlief, M. A. Stoy, G. C. Glatzmaier, C. E. Bingham, A. A. Lewandowski. In-situ FT-IR monitoring of a solar flux induced chemical process. *J. Sol. Energy Eng.* 119(3):219–224, 1997.
318. G. Buttner. The use of NIR analysis for refineries. *Process Control Qual.* 9(4):197–203, 1997.
319. Didier Lambert, Jean Richard Llinas, Bernard Descales, Andre Martens, Claude Granzotto. Control of steam cracking process by near-infrared spectroscopy. BP Chemicals S. N.C., Naphtachimie S. A. EP patent No. 801298 A1, Issued: 10/15/97.
320. C. G. Masi. Spectroscopic techniques peer into combustion-chamber chemistry. *R&D* 38(13):14–16, 1996.
321. Bernard Descales, Andre Martens, Claude Granzotto, Didier Lambert, Jean-Richard Llinas. Cracking property determination. BP Chemicals S. N.C.; Naphtachimie S. A. EP Patent No. 706049 A1, Issued: 04/10/96.
322. Didier Lambert, Sylvie Bages, Bernard Descales, Jean-Richard Llinas, Andre Martens. Lubricant property determination. BP Chemicals S. N.C., EP Patent No. 706050A1, Issued: 04/10/96; U. S. Patent No. 5740073, Issued: 04/14/98.
323. Ph. Marteau, P. Tobaly, V. Ruffier-Meray, A. Barreau. In situ determination of high pressure phase diagrams of methane-heavy hydrocarbon mixtures using an infrared absorption method. *Fluid Phase Equilib.* 119(1–2):213–230, 1996.
324. Von K. Seinsche, F. Luigart, P. Bartl. Quality control of jet fuels. PLS (Partial least squares) regression analysis of gas chromatographic and IR spectroscopic data. *Erdoel, Erdgas, Kohle* 112(6):261–263, 1996.
325. G. Buettner, U. Grummisch, H. Pruefer. On-line application of an acousto-optic NIR analyzer for product monitoring and plant control in refineries. *Proc. Int. Conf. Near-infrared Spectrosc.*, 6th. 390–393, 1995.
326. Didier Lambert, Bernard Descales, Richard Llinas, Alain Espinosa, Sebastien Osta, Michel Sanchez, Andre Martens. On-line near-infrared optimization of refining and petrochemical processes. *Near-infrared Spectrosc.: Future Waves, Proc. Int. Conf. Near-infrared Spectrosc.*, 7th. 272–278, 1996.
327. J.-P. Conzen, A. Schmidt, J. Q. Wang. Non-destructive quality control of pharmaceutical tablets by near-infrared reflectance spectroscopy. *Near-infrared Spectrosc.: Future Waves, Proc. Int. Conf. Near-infrared Spectrosc.*, 7th. 378–385, 1996.
328. Robert K. Bade, Kenneth L. Gallaher, Stephen B. Hunt, Gerald L. Combs, Eugene L. Kesselhuth. The on-line determination of total olefins in gasoline by process gas chromatography and Fourier transform infrared spectroscopy. *Adv. Instrum. Control* 51(1):33–42, 1996.
329. P. Landais. In situ examination of coal macerals oxidation by infrared microspectroscopy. *Conf. Proc. Int. Conf. Coal Sci.*, 7th. 2:375–378, 1993.
330. Stephen Robert Nattrass. A method and apparatus for non-intrusive in situ chemical analysis of a lubricant film in running reciprocating machinery. Shell Internationale Research, Maatschappij B. V., EP patent No. 658757A1, Issued: 06/21/95.
331. Lawrence J. Altman, Rafi Jalkian. In situ measurement and control of low inventory alkylation unit. Mobil Oil Corp., U. S. Patent No. 5,407,830, Issued: 04/18/95.
332. Peter A. Rosenthal, Joe E. Cosgrove, John R. Haigis, James R. Markham, Peter R. Solomon, Stuart Farquharson, Philip W. Morrison, Jr.; Stephen D. Ridder, Francis S. Biancaniello. FT-IR process monitoring of metal powder temperature and size distribution. *Proc. SPIE-Int. Soc. Opt. Eng.* Vol. 2367:183–193, 1995.
333. M. Blanco, J. Coello, H. Iturriaga, S. Maspoch, C. de la Pezuela. Quantitation of the active compound and major excipients in a pharmaceutical formulation by near infrared diffuse-reflectance spectroscopy with fiber optical probe. *Anal. Chim. Acta* 333:147–156, 1996.

334. T. B. Blank, S. T. Sum, S. D. Brown, S. L. Monfre. Transfer of near-infrared multivariate calibrations without standards. *Anal. Chem.* 68:2987–2995, 1996.
335. Z. Bouhsain, S. Garrigues, M. de la Guardia. Clean method for the simultaneous determination of propyphenazone and caffeine in pharmaceuticals by flow-injection Fourier-transform infrared spectrometry. *Analyst* 122:441–445, 1997.
336. T. Norris, P. K. Aldridge, S. S. Sekulic. Determination of end-points for polymorph conversions of crystalline organic compounds using on-line near-infrared spectroscopy. *Analyst* 122:549–552, 1997.
337. D. E. Pivonka, K. M. Kirkland. Research strategy for the HPLC/FT-IR analysis of drug metabolites. *Appl. Spectrosc.* 51:866–873, 1997.
338. P. K. Aldridge, C. L. Evans, H. W. Ward, II, S. T. Colgan, T. Boyer, P. J. Gemperline. Near-infrared detection of polymorphism and process-related substances. *Anal. Chem.* 68(6):997–1002, 1996.
339. S. S. Sekulic, H. W. Ward, II, D. R. Brannegan, E. D. Stanley, C. L. Evans, S. T. Sciavolino, P. A. Hailey, P. K. Aldridge. On-line monitoring of powder blend homogeneity by near-infrared spectroscopy. *Anal. Chem.* 68(3):509–513, 1996.
340. P. A. Hailey, P. Doherty, P. Tapsell, T. Oliver, P. K. Aldridge. Automated system for the on-line monitoring of powder blending processes using near-infrared spectroscopy. Part I. System development and control. *J. Pharm. Biomed. Anal.* 14(5):551–559, 1996.
341. W. Plugge, C. Van der Vlies. Near infrared spectroscopy as a tool to improve quality. *J. Pharm. Biomed. Anal.* 14:891–898, 1996.
342. A. Schmidt. Quantitative analysis of the components in headache tablets by NIR reflection spectroscopy (NIRS). *Bruker Rep.* 142:12–14, 1996.
343. S. M. Han, P. G. Faulkner. Determination of SB 216469-S during tablet production using near-infrared reflectance spectroscopy. *J. Pharm. Biomed. Anal.* 14:1681–1689, 1996.
344. C. van der Viles, W. Plugge, K. J. Kaffka. Qualifying pharmaceutical substances by fingerprinting with near-IR spectroscopy and the polar qualification system. *Spectroscopy* 10(6):46–49, 1995.
345. M. A. Dempster, B. F. MacDonald, P. J. Gemperline, N. R. Boyer. A near-infrared reflectance analysis method for the non-invasive identification of film-coated and non-film-coated, blister-packed tablets. *Anal. Chim. Acta* 310(1):43–51, 1995.
346. E. Dreassi, G. Ceramelli, P. Corti, M. Massacesi, P. L. Perruccio. Quantitative Fourier transform near-infrared spectroscopy in the quality control of solid pharmaceutical formulations. *Analyst* 120(9):2361–2365, 1995.
347. S. R. Erskine, B. M. Quencer, K. R. Beebe. Rapid optical purity determination using chiral spectroscopy, achiral spectroscopy, and multivariate analysis. *Appl. Spectrosc.* 49(11):1682–1691, 1995.
348. R. N. Landau, P. F. McKenzie, A. L. Forman, R. R. Dauer, M. Futran, A. D. Epstein. In situ Fourier-transform infrared and calorimetric studies of the preparation of a pharmaceutical intermediate. *Process Control Qual.* 7(3–4):133–142, 1995.
349. E. Dreassi, G. Ceramelli, P. Corti, P. L. Perruccio, S. Lonardi. Application of near-infrared reflectance spectrometry to the analytical control of pharmaceuticals: ranitidine hydrochloride tablet production. *Analyst* 121(2):219–222, 1996.
350. F. Gonzalez, R. Pous. Quality control in manufacturing process by near-infrared spectroscopy. *J. Pharm. Biomed. Anal.* 13(4–5):419–423, 1995.
351. E. Dreassi, G. Ceramelli, P. Corti, S. Lonardi, P. L. Perruccio. Near-infrared reflectance spectrometry in the determination of the physical state of primary materials in pharmaceutical production. *Analyst* 120(4):1005–1008, 1995.
352. A. G. Severdia. Transmission FT-IR spectroscopy applied to the determination of hydrocarbon residues on pharmaceutical process equipment. *Appl. Spectrosc.* 49(4):540–541, 1995.
353. M. Blanco, J. Coello, H. Iturriaga, S. Maspoch, C. de la Pezuela, E. Russo. Control analysis of a pharmaceutical preparation by near-infrared reflectance spectroscopy. A comparative study of a spinning module and fiber optic probe. *Anal. Chim. Acta* 298(2):183–191, 1994.
354. Melissa A. Dempster, Brian F. MacDonald, Paul J. Gemperline, Nichole R. Boyer. A near-infrared reflectance analysis method for the noninvasive identification of film-coated and non-film-coated, blister-packed tablets. *Anal. Chim. Acta* 310(1):43–51, 1995.
355. Elena Dreassi, Giuseppe Ceramelli, Piero Corti, Maurizio Massacesi, Piero Luigi Perruccio. Quantitative Fourier transform near-infrared spectroscopy in the quality control of solid pharmaceutical formulations. *Analyst* (Cambridge, U. K.) 120(9):2361–2365, 1995.
356. Ralph N. Landau, Paul F. McKenzie, Andrew L. Forman, Richard R. Dauer, Mauricio Futran, Albert D. Epstein. In-situ Fourier-transform infrared and calorimetric studies of the preparation of a pharmaceutical intermediate. *Process Control Qual.* 7(3–4):133–142, 1995.
357. K.-J. Steffens, K. List. NIR spectroscopy as a tool for in-process control in pharmaceutical production. *World Meet. Pharm., Biopharm. Pharm. Technol.,* 1st, 155–156, 1995.
358. H. Weiler, S. Sarinas. Non-destructive quality control of pharmaceutical tablets by near infrared reflectance spectroscopy. *Proc. Int. Conf. Near Infrared Spectrosc.,* 6th. 412–416, 1995.
359. Elizabeth G. Kraemer, Robert A Lodder. Internet sites for infrared and near-infrared spectrometry. Part I: On-line instruction and direct communication. *Spectroscopy* 11(7):24, 26–29, 1996.

REFERENCES

360. P. Frake, D. Greenhalgh, S. M. Grierson, J. M. Hempenstall, D. R. Rudd. Process control and end-point determination of a fluid bed granulation by application of near-infrared spectroscopy. *Int. J. Pharm.* 151(1):75–80, 1997.
361. Heather Wilson, Thomas Byron, Jerry Sellors. NIR spectroscopy for quality control in the pharmaceutical industry. *Am. Lab* 29(20):17–20, 1997.
362. O. Berntsson, G. Zackrisson, G. Ostling. Determination of moisture in hard gelatin capsules using near-infrared spectroscopy: applications to at-line process control of pharmaceutics. *J. Pharm. Biomed. Anal.* 15(7):895–900, 1997.
363. W. Plugge, C. van der Vlies. Near-infrared spectroscopy as a tool to improve quality. *J. Pharm. Biomed. Anal.* 14(8–10):891–898, 1996.
364. P. H. Hindle, C. R. R. Smith. Ultra-fast web thickness and coating measurements with true, two-dimensional control. *Proc. Int. Conf. Near Infrared Spectrosc.*, 6th. 372–375, 1995.
365. A. B. Blakeney, G. D. Batten, L. A. Welsh. Leaf nitrogen determination using a portable near-infrared spectrometer. *Near-infrared Spectrosc.: Future Waves, Proc. Int. Conf. Near-infrared Spectrosc.*, 7th. 149–152, 1996.
366. Henrik Antti, Michael Sjostrom, Lars Wallbacks. Multivariate calibration models using near-IR reflectance (NIR) spectroscopy on pulp and paper industrial applications. *J. Chemom.* 10(5–6):591–603, 1996.
367. Zhaohui Sun, Amin Ibrahim, Philip B. Oldham, Tor P. Schultz, Terrance E. Conners. Rapid lignin measurement in hardwood pulp samples by near-infrared Fourier transform raman spectroscopy. *Agric. Food Chem.* 45(8):3088–3091, 1997.
368. Anthony J. Michell. Vibrational spectroscopy—A rapid means of estimating plantation pulpwood quality? *Appita J.* 47(1):29–37, 1994.
369. D. F. Leclerc, R. M. Hogikyan. Rapid determination of effective alkali and dead-load concentrations in kraft liquors by attenuated total reflectance infrared spectrometry. *J. Pulp Pap. Sci.* 21(7):J231–J237, 1995.
370. D. F. Leclerc, M. D. Ouchi. Rapid characterization of dirt specks in kraft pulp by infrared spectrometry. *J. Pulp Pap. Sci.* 22(3):J112–J117, 1996.
371. Dongsheng Wang, Yunping Yang, Jizuo Zou. New correction method for FT-IR on-line film thickness measurement. *Proc. SPIE-Int. Soc. Opt. Eng.* Vol. 2857: 12–20, 1996.
372. Brian D. Mindel. Process stream interfacing of in-line NIR analyzers. *Riv. Combust.* 49(11–12):467–470, 1995.
373. T. Ozpozan, B. Schrader, S. Keller. Monitoring of the polymerization of vinyl acetate by near-IR FT Raman spectroscopy. *Spectrochim. Acta* 53A(1):1–7, 1997.
374. R. Feldhoff, D. Wienke, K. Cammann, H. Fuchs. On-line post-consumer package identification by NIR spectroscopy combined with a FuzzyARTMAP classifier in an industrial environment. *Appl. Spectrosc.* 51:362–368, 1997.
375. W. H. A. M. van den Broek, D. Wienke, W. J. Melssen, R. Feldhoff, T. Huth-Fehre, T. Kantimm, L. M. C. Buydens. Application of a spectroscopic infrared focal-plane array sensor for on-line identification of plastic waste. *Appl. Spectrosc.* 51:856–865, 1997.
376. P. Damlin, C. Kvarnstroem, A. Ivaska. In situ external reflection Fourier-transform infrared spectroscopic study on the structure of the conducting polymer poly(paraphenylene). *Analyst* 121:1881–1884, 1996.
377. D. E. Pivonka, K. Russell, T Gero. Tools for combinatorial chemistry: in situ infrared analysis of solid-phase organic reactions. *Appl. Spectrosc.* 50:1471–1478, 1996.
378. G. Lachenal, I. Stevenson. Interaction and hydrogen-bonding studied by near-FT-IR spectroscopy. *NIR News* 7(6):10–12, 1996.
379. B. A. Kirsch, J. P. Chauvel, D. P. Denton, S. V. Lange, D. R. Lafevor, R. A. Bredeweg, W. K. Winnett, J. L. Dinaro. Improved process control in polyolefin polymerization by on-line Fourier transform infrared spectroscopy. *Process Control Qual.* 8:75–83, 1996.
380. M. Nielsen, P. Jurasek, J. Hayashi, E. Furimsky. Formation of toxic gases during pyrolysis of polyacrylonitrile and nylons. *J. Anal. Appl. Pyrolysis* 35(1):43–51, 1995.
381. H. W. Siesler. Near-infrared spectroscopy for polymer reaction and process control: synthesis, processing and recycling. *NIR News* 6(1):3–6, 1995.
382. P. Brush. The near-infrared analysis of polyols: Process monitoring in a hostile environment. *NIR News* 5(5):14–15, 1994.
383. D. M. Scott. A two-color near-infrared sensor for sorting recycled plastic waste. *Meas. Sci. Technol.* 6(2):156–159, 1995.
384. Lisheng Xu, Zongwei Shen, John R. Schlup. On-line remote monitoring of thermoset resin cure using near-infrared external reflection spectroscopy. *J. Adv. Mater.* 28(4):47–54, 1997.
385. Paul Dallin. NIR analysis in polymer reactions. *Process Control Qual.* 9(4):167–172, 1997.
386. Karin Stromberg. Challenge of introducing NIR on-line in polyol production. *Process Control Qual.* 9(4):179–183, 1997.
387. S. Jickells. Toxins tracked down in polymer packaging. *Lab. Equip. Dig.* 32(6):23, 1994.

388. Stephen W. Gaarenstroom, Patricia B. Coleman, Claudia M. Duranceau. Rapid identification of automotive plastics in dismantling operations. Evaluation of specular-reflectance infrared spectroscopy systems. *Soc. Automot. Eng.* SP-1263:47–54, 1997.
389. P. E. R. Mucci. Rapid identification of plastics using external beam mid-infrared spectroscopy. *R. Soc. Chem.* 199:53–70, 1997.
390. G. Zachmann, P. Turner. Rapid characterization of black polymeric material by mid-infrared reflectance spectroscopy. *R. Soc. Chem.* 199:71–76, 1997.
391. I. Rehman, C. Barnardo, R. Smith. Fourier transform infrared analysis of poly(ester-urethanes) at low temperature in-situ by using a newly constructed liquid nitrogen cooled sample stage. *J. Mater. Sci.* 32(10):2617–2621, 1997.
392. I. Linossier, F. Gaillard, M. Romand. Use of evanescent waves FT-IR spectroscopy for in-situ study of the degradation of polymer-substrate systems. *Proc. Annu. Meet. Adhes. Soc.* 18:86–88, 1995.
393. D. Fischer, T. Bayer, K. J. Eichhorn, M. Otto. In-line process monitoring on polymer melts by NIR spectroscopy. *Fresenius' J. Anal. Chem.* 359(1):74–77, 1997.
394. Marion G. Hansen, S. Vedula. In-line measurement of copolymer composition and melt index. *Polym. Process Eng.* 97:89–102, 1997.
395. B.-J. Niu, M. W. Urban. Surfactant exudation and film formation in 50%/59% Sty/n-BA latex films: Step-scan photoacoustic FT-IR spectroscopic studies. *Polym. Mater. Sci. Eng.* 75:41–42, 1996.
396. Atsushi Saito, Yoshie Urai, Koichi Itoh. Infrared and resonance Raman spectroscopic study on the photopolymerization process of the Langmuir-Blodgett films of a diacetylene monocarboxylic acid, 10,12-pentacosadiynoic acid. *Langmuir* 12(16):3938–3944, 1996.
397. Ann M. Brearley, Harvey S. Gold. Near-infrared measurement and control of polyamide manufacture. E. I. Du Pont de Nemours and Co., U.S. Patent No. 5532487 A, Issued: 07/02/96.
398. Thierry Buffeteau, Michel Pezolet. In situ study of photoinduced orientation in azopolymers by time-dependent polarization modulation infrared spectroscopy. *Appl. Spectrosc.* 50(7):948–955, 1996.
399. D. Chilling, H. Ritzmann. Rapid on-line identification of plastics using a novel ultrafast near infrared sensor. *Proc. Int. Conf. Near Infrared Spectrosc.*, 6th. 367–371, 1995.
400. B. A. Kirsch, J. P. Chauvel, D. P. Denton, S. V. Lange. Improved process control in polyolefin polymerization by on-line Fourier transform infrared spectroscopy. *Process Control Qual.* 8(2–3):75–83, 1996.
401. T. E. Long, R. D. Allen, L. J. Sorriero, B. A. Schell, D. M. Teegarden. Synthesis of polyester graft copolymers via the macromonomer technique: application of near-IR spectroscopy as a process monitor. *Polym. Prepr.* 37(2):678–679, 1996.
402. B.-J. Niu, L. R. Martin, L. K. Tebelius, Marek W. Urban. Latex film formation at surfaces and interfaces. Spectroscopic attenuated total reflectance and photoacoustic Fourier transform IR approaches. *ACS Symp. Ser.* 648:301–331, 1996.
403. K. W. Van Every, M. J. Elder. Polyolefin formulation validation using infrared spectrometric classification techniques. *J. Vinyl Addit. Technol.* 2(3):224–228, 1996.
404. Pia Damlin, Carita Kvarnstroem, Ari Ivaska. In situ external reflection Fourier transform infrared spectroscopic study on the structure of the conducting polymer poly(para-phenylene). *Analyst* 121(12):1881–1884, 1996.
405. G. Zachmann, P. Turner. Fast and reliable identification of black plastics. *Spectrosc. Eur.* 9(1):18, 20, 22, 1997.
406. F. A. DeThomas, P. J. Brush. Quality and process control during polyol production. *Proc. Int. Conf. Near Infrared Spectrosc.*, 6th. 382–386, 1995.
407. Richard Crosby Moessner. Process for measuring and controlling the neutralization of inorganic acids in an aromatic polyamide solution based on near-infrared spectroscopy. E. I. Du Pont de Nemours and Co., WO Patent No. 9607092 A1, Issued: 03/07/96.
408. Andre Martens, Bernard Descales, Didier Lambert, Jean Richard Llinas. Chemicals property determination in polymerization and organic reactions. BP Chemicals S. N.C., EP Patent No. 706041 A1, Issued: 04/10/96.
409. Norbert Eisenreich, Juergen Herz, Harald Kull, Wolfgang Mayer, Thomas Rohe. Fast on-line identification of plastics by near-infrared spectroscopy for use in recycling processes. *Annu. Tech. Conf. Soc. Plast. Eng.* 54(3):3131–3135, 1996.
410. O. Wollersheim, J. Hormes. Quantitative in situ FT-IR studies of polymers irradiated with synchrotron radiation. *Chem. Phys.* 204(1):129–134, 1996.
411. Caicai Wu, J. D. S. Danielson, James B. Callis, Mark Eaton, N. Lawrence Ricker. Remote in-line monitoring of emulsion polymerization of styrene by short-wavelength near-infrared spectroscopy Part I. Performance during normal runs. *Process Control Qual.* 8(1):1–23, 1996.
412. Caicai Wu, J. D. S. Danielson, James B. Callis, Mark T. Eaton, N. Lawrence Ricker. Remote, in-line monitoring of emulsion polymerization of styrene by short-wavelength near-infrared spectroscopy. Part II. Performance in the face of process upsets. *Process Control Qual.* 8(1):25–40, 1996.
413. Jovan Mijovic, Sasa Andjelic. Study of the mechanism and rate of bismaleimide cure by remote in-situ real time fiber optic near-infrared spectroscopy. *Macromolecules* 29(1):239–246, 1996.

REFERENCES

414. S. Sonja Sekulic, Howard W. Ward, II, Daniel R. Brannegan, Elizabeth D. Stanley, Christine L. Evans, Steven T. Sciavolino, Perry A. Hailey, Paul K. Aldridge. On-line monitoring of powder blend homogeneity by near-infrared spectroscopy. *Anal. Chem.* 68(3):509–513, 1996.
415. Hsin Her Yu, Leslie J. Fina. Electric field-induced dipole reorientation in oriented nylon 11 by in situ infrared spectroscopy. *J. Polym. Sci. B* 34(4):781–788, 1996.
416. Atul Khettry. In-line monitoring of molten polymeric processes (infrared spectroscopy, fiber optics). Doctoral dissertation, Univ. of Tennessee, Knoxville, TN, 1995, 167 pp.
417. Jovan Mijovic, Sasa Andjelic, Jose M. Kenny. In situ real-time monitoring of epoxy/amine kinetics by remote near infrared spectroscopy. *Polym. Adv. Technol.* 7(1):1–16, 1996.
418. John D. Kirsch, James K. Drennen. Near-infrared spectroscopic monitoring of the film coating process. *Pharm. Res.* 13(2):234–237, 1996.
419. Michael J. Elwell, Anthony J. Ryan, Henri J. M. Gruenbauer, Henry C. Van Lieshout. In-situ studies of structure development during the reactive processing of model flexible polyurethane foam systems using FT-IR spectroscopy, synchrotron SAXS, and rheology. *Macromolecules* 29(8):2960–2968, 1996.
420. Marion G. Hansen, Atul Khettry. In-line composition monitoring of molten poly(ethylene vinyl acetate). *Tappi J.* 78(9):129–134, 1995.
421. Amy L. Moe, Jiri D. Konicek. Process for measurement of the degree of cure and percent resin of glass fiber-reinforced epoxy resin prepreg. AlliedSignal Inc., U.S. Patent No. 5457319 A, Issued: 10/10/95.
422. Zhihong Ge, Richard Thompson, Sharon Cooper, Dean Ellison, Pat Tway. Quantitative monitoring of an epoxidation process by Fourier transform infrared spectroscopy. *Process Control Qual.* 7(1):3–12, 1995.
423. T. E. Long, H. Y. Liu, B. A. Schell, D. M. Teegarden, D. S. Uerz. Determination of solution polymerization kinetics by near-IR spectroscopy: real-time process monitoring. *Polym. Mater. Sci. Eng.* 71:146–147, 1994.
424. Keith L. Miller, Dave Curtin. On-line NIR monitoring of polyols. *Proc. Annu. Symp. Instrum. Process Ind.* 49:53–57, 1994.
425. Jin Cao, Xianliang Lu, Zhenli Zhang. The application of FT-IR spectrum method in photocuring process for polyester acrylate. *Proceedings of the 6th Japan-China Bilateral Symposium on Radiation Chemistry,* 1994:428–433, 1995.
426. Jovan Mijovic, Sasa Andjelic. In situ real-time monitoring of reactive systems by remote fiber-optic near-infrared spectroscopy. *Polymer* 36(19):3783–3786, 1995.
427. J. Graham, P. J. Hendra, P. Mucci. Rapid identification of plastics components recovered from scrap automobiles. *Plast., Rubber Compos. Process. Appl.* 24(2):55–67, 1995.
428. Marion G. Hansen, Atul Khettry. In-line monitoring of molten polymers. NIR spectroscopy, robust probes and rapid data analysis. *Annu. Tech. Conf. Soc. Plast. Eng.* 52(2):2220–2226, 1994.
429. S. H. Patel, D. B. Todd, M. Xanthos. Recent developments in on-line analytical techniques applicable to the polymer industry. *Annu. Tech. Conf. Soc. Plast. Eng.* 52(2):2214–2219, 1994.
430. J. W. Hall, F. A. DeThomas. Monitoring the production of polyurethanes with near-infrared spectroscopy. *Polyurethanes 94, Proc. Polyurethanes Conf.* 381–383, 1994.
431. Raymond Lew. Polyolefin extrusion processing: monitoring using size exclusion chromatography and in-line near-infrared spectroscopy. Doctoral dissertation, Univ. of Toronto, Toronto, ON, Canada, 1994, 211 pp.
432. Jitender Batra, Marion G. Hansen. In-line monitoring of titanium dioxide content in poly(ethylene terephthalate) extrusion. *Annu. Tech. Conf. Soc. Plast. Eng.* 52(2):2232–2235, 1994.
433. Amy Lynn Moe, Jiri Daniel Konicek. Process for measurement of the degree of cure and percent resin of fiberglass-reinforced epoxy resin prepreg. Alliedsignal Inc., WO Patent No. 9429698 A1, Issued: 12/22/94.
434. Marion G. Hansen, Atul Khettry. In-line monitoring of molten polymers: Near-infrared spectroscopy, robust probes, and rapid data analysis. *Polym. Eng. Sci.* 34(23):1758–1766, 1994.
435. Bert Lutz, Hendrik-Jan Luinge, John van der Maas, Rob van Agen. A low-cost external reflection unit for rapid analysis of carbon-filled rubbers. *Proc. SPIE-Int. Soc. Opt. Eng.* Vol. 2089:546–547, 1993.
436. Robert A. Fidler. Real-time analysis of PC/ABS blend composition via process-FT-IR spectroscopy. *Compalloy '93, Proc. Int. Congr. Compat. React. Polym. Alloying, 8th;* 119–131, 1993.
437. Masao Tamada, Hideki Omichi, Norimasa Okui. Real-time in-situ observation of vapor deposition polymerization of N-methylolacrylamide with IR-RAS. *Thin Solid Films* 260(2):168–173, 1995.
438. Marinus P. B. van Uum, Hans Lammers, Jaap P. de Kleijn. Process analysis: properties of poly(ethylene terephthalate) measured by near-infrared spectroscopy. 1. At-line analysis of poly(ethylene terephthalate) chips. *Macromol. Chem. Phys.* 196(6):2023–2028, 1995.
439. Hans Lammers, Marinus P. B. van Uum, Jaap P. de Kleijn. Process analysis: properties of poly(ethylene terephthalate) measured by near-infrared spectroscopy. 2. In-line analysis of poly(ethylene terephthalate) melt. *Macromol. Chem. Phys.* 196(6):2029–2034, 1995.
440. B. Ziegler, K. Herzog, R. Salzer. In-situ investigations of thermal processes in polymers by simultaneous differential scanning calorimetry and infrared spectroscopy. *J. Mol. Struct.* 348:457–460, 1995.

441. H. W. Siesler. Vibrational spectroscopy of polymers. Analysis, physics, and process control. *Adv. Chem. Ser.* 236:41–87, 1993.
442. Shaohua Liu, Peter Solomon, R. Carpio, B. Fowler, D. Simmons, J. Wang, R. Wise, G. Imper, N. B. Riley, et al. Modeling, simulation and control of single wafer process in cluster tool base on FT-IR-in-line sensor. *Mater. Res. Soc. Symp. Proc.* 389:269–274, 1995.
443. Qinbai Fan. In situ FT-IR studies of semiconductor/electrolyte interface (cadmium telluride, silicon). Doctoral dissertation, Cleveland State Univ., Cleveland, OH, 1995, 132 pp.
444. Zhao Ping, G. E. Nauer. In situ FT-IR-ATR spectroscopic investigations on the redox behavior of poly(thienylpyrrole) thin film electrodes in non-aqueous solutions. *Synth. Met.* 84(1–3):843–844, 1997.
445. V. Climent, A. Rodes, J. M. Orts, J. M. Feliu, J. M. Perez, A. Aldaz. On the electrochemical and in-situ Fourier transform infrared spectroscopy characterization of urea adlayers at Pt(100) electrodes. *Langmuir* 13(8):2380–2389, 1997.
446. Stuart Farquharson, Karen Kinsella, James R. Markham, Peter R. Solomon, Martin D. Carangelo, John R. Haigis, Nuggehalli M. Ravindra, Fei-Ming Tong, Malcolm J. Bevan, Glenn H. Westphal. Real-time process control of molecular beam epitaxial growth of mercury cadmium telluride films by Fourier transform infrared spectroscopy. *Transient Therm. Process. Tech. Electron. Mater. Proc. Symp.* 133–136, 1996.
447. R. P. Sperline, J. S. Jeon, S. Raghavan. FT-IR-ATR analysis of the silicon/aqueous solution interface using sputtered silicon thin films to access the 1550–1100 cm^{-1} spectral region. *Appl. Spectrosc.* 49(8):1178–1182, 1995.
448. Demetrius Papapanayiotou. In situ infrared spectroscopic studies of electrodeposition additive adsorption on copper. Doctoral dissertation, Univ. of Illinois, Urbana, IL, 1997, 184 pp.
449. A. Lee, S. R. Charagundla, A. Macek, R. Kanjolia, B. Hui. On-line spectroscopy of various metal organics used in compound semiconductors. *AT-PROCESS* 1(2):85–89, 1995.
450. Shaohua Liu, John R. Haigis, Marie B. DiTaranto, Karen Kinsella, James R. Markham, Qi Li, David B. Fenner, Peter R. Solomon, Stuart Farquharson, et al. Process monitoring and control of integrated circuit manufacturing using Fourier transform infrared spectroscopy. *Proc. SPIE-Int. Soc. Opt. Eng.* Vol. 2367:171–182, 1995.
451. Sateria Salim, K. F. Jensen, R. D. Driver. Fiber optics-based Fourier transform infrared spectroscopy for in-situ concentration monitoring in OMCVD. *Mater. Res. Soc. Symp. Proc.* 324:241–247, 1994.
452. Qinbai Fan, Lily M. Ng. In situ Fourier transform infrared-attenuated total reflection spectroscopy monitoring of polyaniline synthesis mechanism on the p-type silicon electrode. *J. Vac. Sci. Technol. A* 14(3, Pt. 2):1326–1329, 1996.
453. Feng Qin, Eduardo E. Wolf. Spatially resolved infrared spectroscopy: A novel technique for in situ study of spatial surface coverage during CO oxidation on supported catalysts. *Catal. Lett.* 39(1,2):19–25, 1996.
454. Ming-kai Lei, Teng-cai Ma, V. A. Emel'kin. Diffuse reflectance Fourier transform infrared spectrometry of BN films deposited on steels by remote microwave plasma CVD from borazine. *Chin. Phys. Lett.* 13(4):309–312, 1996.
455. L. Savary, J. Saussey, G. Costentin, M. M. Bettahar, J. C. Lavalley. Role of the nature of the acid sites in the oxydehydrogenation of propane on a VPO/TiO$_2$ catalyst. An in situ FT-IR spectroscopy investigation. *Catal. Lett.* 38(3,4):197–201, 1996.
456. Naohiko Fujino, Isamu Karino, Masashi Ohmori, Masatoshi Yasutake, Shigeru Wakiyama. Method and apparatus for analyzing minute foreign substances, and process for manufacturing semiconductor or LCD elements. Seiko Instruments Inc., Mitsubishi Denki Kabushiki Kaisha, EP Patent No. 727659A2, Issued: 08/21/96; EP Patent No. 727659A3, Issued: 03/04/98.
457. Glenn Ronald Howes. The application of Fourier transform infrared spectroscopy to the optimization of chemically-amplified photoresist processing. Doctoral dissertation, University of Wisconsin, Madison, WI, 1996, 192 pp.
458. Juock S. Namkung, Michael Hoke, Robert S. Rogowski, Sacharia Albin. Optical fiber FT-IR evanescent wave absorption spectroscopy of natural aluminum corrosion. *Proc. SPIE-Int. Soc. Opt. Eng.* Vol. 2883:655–669, 1996.
459. Qinbai Pan, Cong Pu, E. S. Smotkin. In situ Fourier transform infrared-diffuse reflection spectroscopy of direct methanol fuel cell anodes and cathodes. *Electrochem. Soc.* 143(10):3053–3057, 1996.
460. Jun Onoe, Kazuo Takeuchi. In situ high-resolution infrared spectroscopy of a photopolymerized C60 film. *Condens. Matter* 54(9):6167–6171, 1996.
461. J. J. Kellar, W. M. Cross, M. R. Yalamanchili, C. A. Young, J. D. Miller. Surface phase transitions of adsorbed collector molecules as revealed by in situ FT-IR/IRS spectroscopy. *Trans. Soc. Min. Metall. Explor.* 294:75–80, 1994.
462. D. Fornasiero, M. Montalti, J. Ralston. Kinetics of adsorption of ethyl xanthate on pyrrhotite: In situ UV and infrared spectroscopic studies. *J. Colloid Interface Sci.* 172(2):467–478, 1995.
463. Bernd R. Mueller, Gion Calzaferri. Thin Mo(CO)6-Y-zeolite layers: Preparation and in situ transmission FT-IR spectroscopy. *J. Chem. Soc. Faraday Trans.* 92(9):1633–1637, 1996.

REFERENCES

464. Ken-ichi Hanaoka, Hiroshi Ohnishi, Kunihide Tachibana. In situ monitoring of selective copper deposition processes in a metal-organic chemical vapor deposition using Fourier-transform infrared reflection-absorption spectroscopy. *Jpn. J. Appl. Phys.* 34(5A):2430–2439, 1995.
465. C. A. Melendres, G. A. Bowmaker, B. Beden, J. M. Leger. In-situ far infrared spectroscopy of electrode surfaces with a synchrotron source. *Proc. Electrochem. Soc.* 96–9:224–233, 1996.
466. Susan M. Stephens, Richard A. Dluhy. In-situ and ex-situ structural analysis of phospholipid-supported planar bilayers using infrared spectroscopy and atomic force microscopy. *Thin Solid Films* 284–285:381–386, 1996
467. G. L. J. Trettenhahn, G. E. Nauer, A. Neckel. In situ external reflection absorption FT-IR spectroscopy on lead electrodes in sulfuric acid. *Electrochim. Acta* 41(9):1435–1441, 1996.
468. Y. Vickie Pan, Ernesto Z. Barrios, D. Denice Denton. In situ infrared spectroscopy as a real time diagnostic for plasma polymer film deposition. *Appl. Phys. Lett.* 68(24):3386–3388, 1996.
469. J. A. Mielczarski, Z. Xu, J. M. Cases. Qualitative and quantitative evaluation of heterogeneous adsorbed monolayers on semiconductor electrode by infrared reflection spectroscopy. *J. Phys. Chem.* 100(17):7181–7184, 1996.
470. F. Ozanam, A. Djebri, J.-N. Chazalviel. The hydrogenated silicon surface in organic electrolytes probed through in situ IR spectroscopy in the ATR geometry. *Electrochim. Acta* 41(5):687–692, 1996.
471. Andrea E. Russell, Lavan Rubasingham, Patrick L. Hagans, Todd H. Ballinger. Cell and optics design for in situ far-infrared spectroscopy of electrode surfaces. *Electrochim. Acta* 41(5):637–640, 1996.
472. Sateria Salim. In situ Fourier transform infrared spectroscopy of chemistry and growth in chemical vapor deposition. Doctoral dissertation, Massachusetts Institute of Technology, Cambridge, MA, 1995.
473. Qinbai Fan, Cong Pu, Kevin L. Ley, E. S. Smotkin. In situ FT-IR-diffuse reflection spectroscopy of the anode surface in a direct methanol/oxygen fuel cell. *J. Electrochem. Soc.* 143(2):L21–L23, 1996.
474. Chan-Hwa Chung, Sang Heup Moon, Shi-Woo Rhee. In situ infrared spectroscopic study on the role of surface hydrides and fluorides in the silicon chemical vapor deposition process. *J. Vac. Sci. Technol. A* 13(6):2698–2702, 1995.
475. B. Beden. On the use of in situ UV-visible and infrared spectroscopic techniques for studying corrosion products and corrosion inhibitors. *Mater. Sci. Forum* 192–194:277–290, 1995.
476. I. T. Bae, M. Sandifer, Y. W. Lee, D. A. Tryk, C. N. Sukenik, D. A. Scherson. In situ fourier transform infrared spectroscopy of molecular adsorbates at electrode-electrolyte interfaces: A comparison between internal and external reflection modes. *Anal. Chem.* 67(24):4508–4513, 1995.
477. D. Olligs, U. Stimming, J. Stumper. On the mechanism of CO-electrooxidation on Pt-Ru alloys. An in-situ infrared spectroscopic study. *New Mater. Fuel Cell Syst. I, Proc. Int. Symp.,* 1st, 677–687, 1995.
478. C. Stuhlmann, I. Villegas, M. J. Weaver. Scanning tunneling microscopy and infrared spectroscopy as combined in-situ probes of electrochemical adlayer structure: Cyanide on Pt(111). Report 1994, TR-147, Order No. AD-A276715, 1995.
479. Toshimasa Wadayama, Yoshihisa Maiwa, Hironobu Shibata, Aritada Hatta. In situ IR spectroscopic study of the reaction of dimethylaluminum hydride with photochemically deposited amorphous silicon. *Jpn. J. Appl. Phys.* 34(6B):L779–L781, 1995.
480. S. Ghosh. At-line and on-line characterization of textile products using the NIR spectroscopy method. *Proc. Int. Conf. Near Infrared Spectrosc.,* 6th. 450–459, 1995.
481. Berend Jan Kip, Edo Augustinus Titus Peters, Jens Happel, Thomas Huth-Fehre, Frank Kowol. Identifying recyclable carpet materials using a hand-held infrared spectrometer. Dsm N. V.; Institut Für Chemo-und Biosensorik; WO Patent No. 9702481 A1, Issued: 01/23/97.
482. K. Kitagawa, S. Hayasaki, Y. Ozaki. In situ analysis of sizing agents on fiber reinforcements by near-infrared light-fiber optics spectroscopy. *Vib. Spectrosc.* 15(1):43–51, 1997.
483. M. Blanco, J. Coello, J. M. Garcia Fraga, H. Iturriaga, S. Maspoch, J. Pages. Determination of finishing oils in acrylic fibers by near infrared reflectance spectrometry. *Analyst* 122:777–781, 1997.

V.
NIR SPECTROSCOPY

13.

SHORT-WAVE NEAR-INFRARED SPECTROSCOPY

INTRODUCTION

Discussion occurs from time to time regarding the beginning and ending of the near-infrared (NIR) region as well as the precise band positions for vibrational information within the NIR spectral region. This chapter is an attempt to clarify this issue relative to a proposed starting wavelength for the NIR spectral region. The usual designation for the visible region includes 380–780 nm. The current "official" position defines the NIR region extending from 780 nm (12,800 cm^{-1}) to 2,500 nm (4,000 cm^{-1}), as specified by IUPAC and published in *Pure and Applied Chemistry* 57:105–120, 1985 [1]. It may be more appropriate to extend the NIR region to near 695 nm (14,388 cm^{-1}), and there is historical and experimental precedent for this claim. To give historical background on the reasoning behind this suggestion, a brief review of pertinent literature is presented here as a preface to the Experimental section. Current instrumentation provides high quality measurements of hydrocarbons with measurement pathlengths of 10 cm or more. What is considered the visible spectral region, often designated as a measurement region for electronic transitions, also contains vibrational information. The vibrational information occurs primarily as the fourth overtone for C-H stretching and is described in this chapter.

William Hershel is the recognized discoverer of the infrared region. His famous work, "Experiments on the Refrangibility of the Invisible Rays of the Sun," read April 24, 1800, at the Royal Society (*Phil. Transact. Roy. Soc.* 90:284–292) is described as the original discovery of infrared radiation. Herschel writes, "In that section of my former paper which treats of radiant heat, it was hinted, though from imperfect experiments, that the range of its refrangibility is probably more extensive than that of the prismatic colours; but, having lately had some favourable sunshine, and obtained a sufficient confirmation of the same, it will be proper to add the following experiments to those which have been given" [2]. Other early work comparing the effects of atomic grouping on infrared absorption of organic molecules was completed in 1882 by Captain Abney and Lt.-Col. Festing of the Royal Engineers. The work compared the "atomic groupings" of alcohols, halogens, aldehydes, ethers, nitrate, oxalate, sulphides [sic], nitric, carboxylic acids (and salts thereof), glycerine, benzines, anilines, turpentine, salicylate, oil, anhydride, inorganic acids, ammonia, and water, using an arbitrary empirical wavelength scale to compare band positions, strengths, and overall shapes [3].

DESCRIPTIVE ARTICLES OF HISTORICAL INTEREST

O.H. Wheeler [4] describes the near-infrared region as extending "from about 2 μ into the visible at about 0.7 μ" in a general discussion article (with 11 references). He also noted that "the term 'near infrared' formerly was used to denote the infrared spectrum to 23 μ, and to distinguish this region from that of the far-infrared above 23 μ." Wheeler credits Rawlins and Taylor [5] with this early use of the term *near-infrared*. Using a variety of instrumentation, the author identifies near-infrared bands (in microns) in tabular form as follows. For C-H stretch, the fundamental occurs at 3.5, the 1st overtone at 1.8, the 2nd overtone at 1.2, the 3rd overtone at 0.85, and the 4th overtone at 0.7 μ. The author identifies band positions for both O-H and N-H stretch as occurring near 2.8 for the fundamental, 1.4 for the 1st overtone, 0.95 for the 2nd overtone, and 0.7 μ for the 3rd overtone.

R. F. Goddu and D. A. Delker [6] provide two detailed tables showing (1) the spectra—structure correlations and average molar absorptivity for a number of functional groups for the NIR region, which they describe as extending from 1.0 to 3.1 μ, and (2) maximum recommended pathlengths for 12 solvents (useful for NIR spectroscopy) over the wavelength region 1.0–3.1 μ. The authors cite two references in this useful article.

EARLY REVIEWS

Professor J. W. Ellis [7] has reviewed work below 3 microns for absorption of organic liquids. The review cites 44 separate works related to studies made prior to June, 1929. In 1929 Professor Joseph W. Ellis wrote, "The region of the spectrum below 3 μ, although representing a relatively small portion of the total infra-red [sic] spectrum, is nevertheless quite significant in the study and interpretation of the absorption spectra of molecules. In particular, the absorption spectra of organic liquid molecules shows numerous bands in this region." He goes on to cite work by Coblentz, and Raman related to the observations of bands in the infrared region due to infrared absorptions and the Raman effect. Ellis goes on to refer to earlier work by Puccianti in 1900 with respect to the presence of specific near-infrared bands associated with "molecules [having] a hydrogen atom combined with a carbon atom." Ellis reports that bands relating to "the carbon-hydrogen linkage" were observed at 2.3–2.2, 1.7, 1.4, 1.2, 1.0, and 0.9 μ. With this work, and the work of other investigators, the modern science of near-infrared spectroscopy was under way.

W. Kaye [8] provides a summary review of the work in near-infrared spectroscopy from the late 1920s to April 1954. The author draws information from 106 references for this review. The author refers to the term "hydrogenic" stretching vibrations as CH, NH, and OH. Work involving measurements in the region of 0.7–3.5 μ is reviewed as pertaining to this "hydrogenic" stretch region. The author presents a Colthup-type chart of characteristic NIR bands and the accompanying references.

R. F. Goddu [9] provides an extensive review of near-infrared spectrophotometry prior to 1960. The author cites 110 literature sources of information for this review. The information within this review is organized into instrumentation (and methods), qualitative analysis, quantitative analysis (for C-H, N-H, O-H, thiols, P-H, carbonyls, nitriles, and miscellaneous groups). Also included in the work are inorganic applications, applications for solids and liquids, and future trends. The work provides a number of tables and spectra describing near-infrared absorption data.

K. B. Whetsel [10] reviews the significant work in near-infrared spectrophotometry prior to 1968. The review contains 336 references covering aspects of theory, instrumentation, and sampling techniques. Work for both inorganic and organic compounds is reviewed. Available pre-1968 instrumentation is described in tabular form, as is a list of NIR solvents and optical transparency from 1.0 to 3.0 microns. Other tables and figures illustrate such information as the effect of slit width on peak height for first- and second-overtone and combination N-H bands of aromatic amines, NIR spectra of organic compounds, NIR spectra of rare earth ions, and various band locations and assignments for a number of organic and inorganic compounds.

EXPERIMENTAL

SHORT-WAVE NEAR INFRARED AS A SEPARATE TOPIC

Schrieve et al. [11] discuss applications for the short-wave near-infrared (SW-NIR) region, referring to synonyms such as "the far-visible," the "near, near-infrared," and the "Herschel-infrared" to describe the range of approximately 700 to 1100 nm of the EMS (electromagnetic spectrum). The authors cite the increased interest of this spectral region to spectroscopists, particularly those involved with implementing process near-infrared measurements. The lower-cost components (sources and detectors) and the inherent reliability of these components are cited as advantages for process instrumentation. In the work, the authors state the major SW-NIR absorbances (in nm) for a variety of solvents: acetone (894, 908, and 1016); 1,1,1-trichloroethane (898, 1004, 1045); dichloromethane (884, 1024); dimethylformamide (914, 1014); glycerol (924, 1012); methanol (916, 1024); n-hexane (914, 928, 1020); tetrahydrofuran (906, 938, 1036); toluene (876, 910, 1020); and water (976). *Note:* This introductory material, in part, is repeated in Chapter 15.

EXPERIMENTAL

The aromatic C-H stretch fourth-overtone band from pure toluene occurs at approximately 695–725 nm (first-derivative zero crossover at 714.8 nm; peak rise above zero baseline at 694.5 nm); methyl C-H stretch fourth overtone near 745 nm; and methylene C-H stretch fourth overtone near 760 nm (Table 13.1 and Figure 13.1). These spectra closely resemble the second- and third-overtone spectra, with fewer features, as the harmonic increases, most likely attributable to fewer sum tone bands. The aromatic C-H stretch harmonic would clearly put 695 within the NIR region. This finding can be verified easily using a 10.0 cm pathlength with pure hydrocarbons or hydrocarbon mixtures. The measurements for this note were made at room temperature (25°C) using an FT-NIR instrument, 128 coadded scans per spectrum at 4-cm^{-1} resolution (Perkin-Elmer Model 2000 FT-IR spectrophotometer).

The third- and fourth-overtone measurements were taken using a 10.0 cm pathlength cell with the FT-NIR instrument. The first- and second-overtone measurements can be made using 2.0 cm and 1.0 mm-pathlength cells, respectively. The increased use of high-performance spectrophotometers and descriptions of the existence of C-H and O-H overtones near 0.7 microns (by O.W. Wheeler, W. Kaye, and G.D. Schrieve et al., and as clearly shown in this present note) are indications that an amendment moving the lower starting-wavelength region to 695 for NIR is arguably valid.

Table 13.1 Fourth-Overtone C-H Stretch Peak Positions for Three Hydrocarbons

Hydrocarbon	Aromatic C-H	Methyl C-H	Methylene C-H
Toluene	714.8 nm	744.5 nm	N/A
2,2,4-Dimethyl pentane	N/A	748.1 nm	766.0 nm
n-Decane	N/A	750.0 nm	761.4 nm

Measured Using a 10 cm Pathlength Liquid Cell.

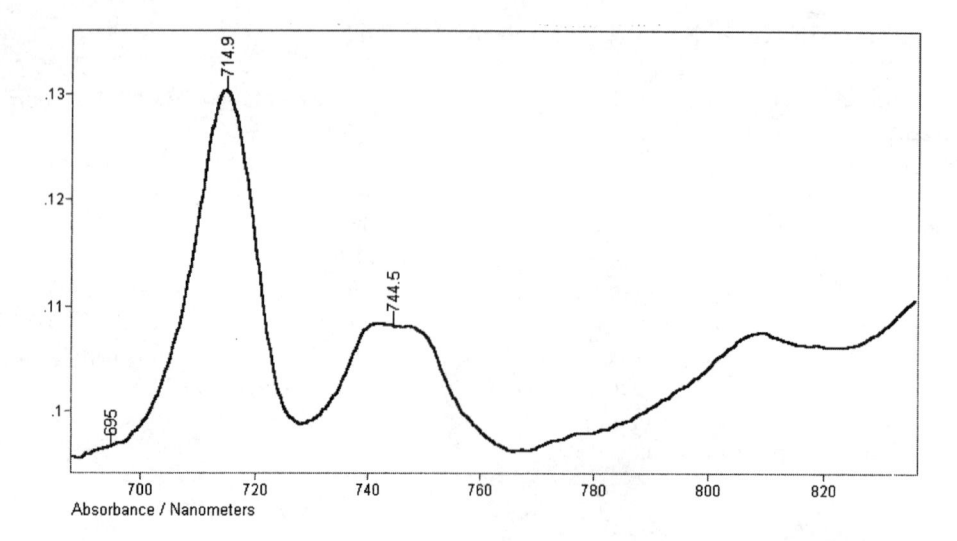

Fig. 13.1 The Fourth-overtone spectrum of pure toluene showing aromatic C-H stretch at 714.5 nm, methyl C-H stretch near 744.5 nm, and a combination band near 810 nm (10 cm pathlength cell).

ACKNOWLEDGEMENT

The author acknowledges *CAMPCLAN@prodigy.net,* the near-infrared users group, for their stimulating discussions and for raising this issue.

REFERENCES

1. CAMPCLAN@prodigy.net. Communication with Tony Davies (May 25, 1999).
2. William Hershel. Experiments on the refrangibility of the invisible rays of the sun. *Phil. Transact. Roy. Soc.* 90:284–292, 1800.
3. Capt. W. de W. Abney, Lt.-Col. Festing reported. On the influence of the atomic grouping in the molecules of organic bodies on their absorption in the infra-red region of the spectrum. *Phil. Transact.* 172:887–918, 1882.
4. O. H. Wheeler. Near infrared spectra. A neglected field of spectral study. *J. Chem. Education* 37:234–236, 1960.
5. F. I. G. Rawlins, A. M. Taylor. *Infrared Analysis and Molecular Structure.* Cambridge, U.K.: Cambridge University Press, 1929.
6. R. F. Goddu, D. A. Delker. Spectra–structure correlations for the near-infrared region. *Anal. Chem.* 32:140–141, 1960.
7. J. W. Ellis. Molecular absorption spectra of liquids below 3 µ. *Trans. Faraday Soc.* 25:888–898, 1928.
8. W. Kaye. Near-infrared spectroscopy: A review. I. Spectral identification and analytical applications. *Spectrochimica Acta* 6:257–287, 1954.
9. R. F. Goddu. Near-infrared spectrophotometry. *Advan. Anal. Chem. Instr.* 1:347–424, 1960.
10. K. B. Whetsel. Near-infrared spectrophotometry. *Appl. Spectrosc. Rev.* 2(1):1–67, 1968.
11. G. D. Schrieve, G. G. Melish, A. H. Ullman. The Herschel-infrared—A useful part of the spectrum. *Appl. Spectrosc.* 45:711–714, 1991.

14.

SW-NIR FOR ORGANIC COMPOSITION ANALYSIS

Table 14.1 demonstrates the basic functional group measurements that have useful signals in the short-wavelength near-infrared (SW-NIR) region (800–1100 nm) of the electromagnetic spectrum. As can be seen from the table, the SW-NIR region is used to measure molecular vibrations as combination bands for C-H groups, for second overtones of O-H and N-H groups, and for third-overtone C-H group measurements. All NIR spectroscopy is used to measure these basic organic functional groups resulting from molecular vibrations. The advantages of SW-NIR include high signal-to-noise ratios from readily available technologies, typically 25,000 : 1, as well as high throughput using fiber-optic cabling. An additional advantage of SW-NIR over other IR regions is the use of flow-cell pathlengths sufficiently large for industrial use (most often 5–10 cm). This range of pathlengths is useful in obtaining representative sample size measurements and in preventing fouling of internal cell optics. Figure 14.1 demonstrates the positions of third-overtone C-H stretch vibrations.

Table 14.2 on p. 139 shows the broad range of applications possible using the SW-NIR spectral region. The list of applications that have been successfully implemented using the NIR and SW-NIR techniques is vast. SW-NIR offers a possible technique for many analytical measurement requirements.

Table 14.3 on page 140 illustrates the potential for SW-NIR measurements. The approximate band locations shown are based on spectra of pure, undiluted materials measured with a low-resolution dispersive instrument. Most of the data was extracted from spectra shown in T. Hirschfeld and A. Zeev Hed, *The Atlas of Near-Infrared Spectra* (Philadelphia: Sadtler Research Laboratories, 1981). For the band assignments of Table 14.3 use the following:

1. Methyl C-H antisymmetrical stretch ($4v$) = 890–915 nm
2. Methylene C-H antisymmetrical stretch ($4v$) = 920–936 nm
3. Aromatic C-H stretch ($4v$) = 855–890 nm
4. O-H stretch ($3v$) = 950–965 nm
5. Combination bands or N-H = above 1000 nm

Table 14.1 Characteristic Second-, Third- and Fourth-Overtone
NIR Absorption Bands of Specified Functional Groups

Functional group structure	Band assignment(s)	Second overtone	Third overtone	Fourth overtone
ArCH . (aromatics)	C-H stretch 3rd overtone	—	875–881	710–720
. CH=CH (alkenes)	C-H stretch 3rd overtone	—	878	702
. CH_3 (methyl)	C-H stretch 3rd overtone	—	896–915	744–750
" "	C-H combination	—	1010–1025	—
. CH_2 (methylene)	C-H stretch 3rd overtone	—	911–936	760–766
R-OH (alcohol)	O-H stretch 2nd overtone	940–970	724	—
ArOH (phenol)	O-H stretch 2nd overtone	957–980	728	—
HOH (water)	O-H stretch 2nd overtone	960–990	729	—
Starch	O-H stretch 2nd overtone	967	725	—
Urea	Symmetric N-H stretch 2nd overtone	973	730	—
. $CONH_2$ (1° amide)	N-H stretch 2nd overtone	975	731	—
.CONHR′ (2° amide)	N-H stretch 2nd overtone	981	736	—
.CONHR′R″ (3° amide)	N-H stretch 2nd overtone	989	742	—
Cellulose	O-H stretch 2nd overtone	993	745	—
Urea (proteins & amines)	Symmetric N-H stretch 2nd overtone	993	745	—
$ArNH_2$ (aromatic amines)	N-H stretch 2nd overtone	995	746	—
. NH (amines/comb.)	N-H stretch 2nd overtone	1000	750	—
Protein	N-H stretch 2nd overtone	1007	755	—
Urea	N-H stretch 2nd overtone	1013	760	—
RNH_2	N-H stretch 2nd overtone	1020	765	—
Starch	O-H stretch 2nd overtone	1027	770	—
CONH	N-H stretch 2nd overtone	1047	785	—
=CH_2 (terminal)	C-H stretch 2nd overtone	1080	810	—

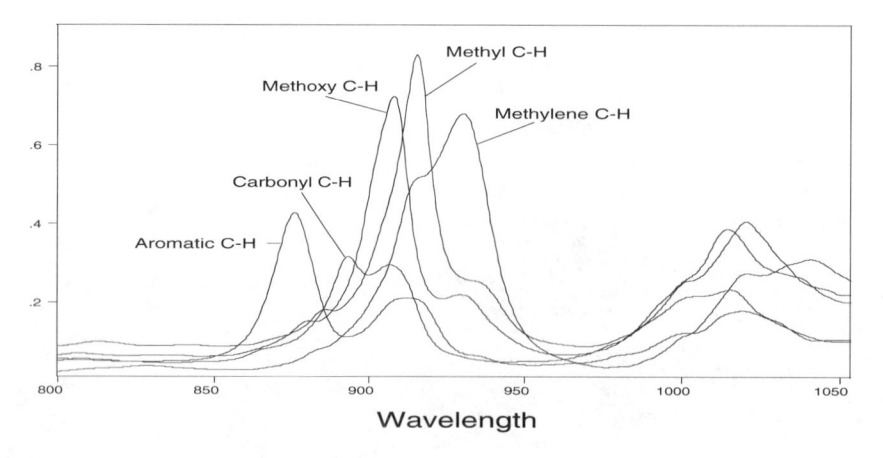

Fig. 14.1 Predominant short-wavelength near-infrared (SW-NIR) hydrocarbon spectral features (absorbance as a function of wavelength).

Table 14.2 Potential Applications for Short-Wavelength NIR

Matrix	Constituents	Matrix	Constituents
Aqueous electrolyte solutions	Cations and anions: $NaClO_4$, $NaCl$, $MgCl_2$, $NaBr$, NaI, LiX, KX; X=Cl, Br, I	Meat slurries	Fat, protein, moisture
		Milk	Lactose, fat, solids, protein
Beer	O.G., alcohol, sugars, water	Mineral oil	Polycyclic aromatic hydrocarbons (PAHs)
Blood	Cholesterol, glucose, hemoglobin	Organics	Water
		Oil-in-water emulsions	Oil and water
Caustic brine	Hydroxyl ion		
Cheese	Fat, protein, moisture	Petroleum products	Qualify and quantify hydrocarbon groups
Chocolate liquor	Fats, sugar, moisture	Pharmaceuticals	Constituents, alcohol, amino acids, sugars, microbial growth
Copolymers	Vinyl acetate, PVC, PVA		
Corn syrup	Saccharide distribution, dextrose equivalent, fructose actives, moisture	Polymers	Cure kinetics, chemical/physical properties, phase separation, hydroxyl number
Diesel fuels	Flash point, cloud point, cetane number, FIA		
Detergents	Actives and moisture	Polyolefins	Antioxidants
Egg products	Protein, moisture	Polyols	Hydroxyl number
Epoxy resins	Terminal CH_2 group	Polyurethane foams	Physical properties
Ethylene–propylene blends	Percentage polymer	Poly(ether urethane)	Bulk composition
Fermentation broth	Ethanol, cell density	Poly(ethylene terephthal)	Crystallinity/morphology
Fruit juices	Sugars, solids		
Gasoline	RON, MON, PON, oxygenates, aromatics, paraffins, olefins, RVP, MTBE, ethanol, distillation data	Proteins	Water binding
		Seawater	Electrolytes, salinity
		Shortening	I-value, SFI
		Soft drinks	Sugar, acidity
Gelatin	Water, protein	Solvents	Moisture
Glass	Hydroxyl groups	Sugar cane juice	Differential sugars, polarimetry, solids
Glycols	Water		
Heavy water	Protium	Toluene	Moisture
High-pressure natural gas	BTU content	Vinegar	Alcohol and acid
		Water in solvents	To less than 50 ppm
HPLC effluent	Multiple constituents	Wines	Alcohol
Jet fuel	See *Diesel fuel*	Yeast slurries	Solids, protein, cell density
Margarine	Fat, moisture, monoglycerides, glycerol		

Table 14.3 Approximate SW-NIR (third overtone) Band Locations for Common Organic Compounds

Neat material	Band locations	Neat material	Band locations
Acetic acid, ethyl ester	895, 995	Heptane	902, 1005
Acetic anhydride	890, 990	Heptanoic acid	915, 1030
Acetone	880, 890, 990	2-Heptanol	925, 960
Acetonitrile	875, 980, 1030	3-Heptanol	915, 930, 960, 1030
Benzene	865, 1005	Hexane	900, 1000
Benzoic acid	855	Hexanoic acid	942, 1010, 1060
Benzyl alcohol	875, 925, 975, 1020	Hexylamine	805, 930, 1010, 1020, 1040
Benzoyl chloride	866, 1010	Hexadecanol	930, 960, 1040
Butanol	912, 960, 1015	Isobutanol	915, 960, 1015
Butyl ether	900, 1010	Isooctane	910, 1010
Butyric acid	895, 1005	Isopropanol	905, 960, 1010
Chlorobenzene	868, 1015, 1050	Lauric acid	925, 1035
Chlorocyclohexane	895, 1010	Methyl cyclohexane	925, 933, 1020, 1033, 1066
2-Chloroethanol	885, 970	2-Methylfuran	840, 880, 900, 990, 1090
1-Chloropropane	895, 1005, 1040	Naphthalene	860
α-Chlorotoluene	870, 895, 1015	Nitrobenzene	860, 1005
p-Chlorotoluene	870, 910, 1010	Nitroethane	885, 995
p-Cresol	855, 890, 970	2-Nitropropane	860, 980
Cumene	870, 890, 1000	Nonane	930, 1005, 1030
Cyclohexane	860, 910, 1025	2-Nonanone	830, 910, 925, 1020, 1035
Cyclohexanol	917, 960, 1030	Octane	915, 1000, 1015
Cyclohexanone	905, 1020	1-Octanol	930, 960, 1040
Cyclohexene	880, 910, 1025	Pentane	900, 1000
Cyclopentanol	910, 960, 1030	Pentanol	915, 950, 1020
Cyclopentanone	890, 910, 1020	Propanol	915, 960, 1030
Decane	910, 1015	Propionic acid	905, 990, 1010, 1045
1-Decanol	930, 960, 1040	Pyridine	865, 990
1,2-Dichloroethane	880, 1010	Styrene	860, 1000
p-Dichlorobenzene	850	Toluene	860, 890, 995, 1045
N,N-Dimethylaniline	855, 885, 930, 995, 1040	Trichloroethylene	890, 1033
Diphenylamine	1015	α-3,4-Trichlorotoluene	860, 890, 1030
Ethanol	895, 995	Water	960
Ethylbenzene	870, 910, 1000, 1020, 1060	m-Xylene	860, 895, 1000, 1050
Formic acid	880, 1020	o-Xylene	860, 902, 1000, 1050
Glycerol	920, 990		

REFERENCES

1. D. M. Mayes, J. B. Callis. *Appl. Spectr.* 43:27, 1989.
2. J. J. Kelly, J. B. Callis. *Anal. Chem.* 62:1444, 1990.
3. J. J. Kelly, C. H. Barlow, T. M. Jinguji, J. B. Callis. *Anal. Chem.* 61:313, 1989.
4. J. Coates. *Hydrocarbon Technol. Int.* (1993).
5. J. Coates, S. Reber. *Am. Lab.* 24(18):20C, 1992.
6. J. Coates, T. Davidson, L. McDermot. *Spectroscopy* 7(9):40, 1992.
7. S. J. Swarin, C. A. Drumm. *Spectroscopy* 7(7):42, 1992.
8. A. Bonanno, P. Griffiths. *J. of Near-Infr Spectrosc* 1:13–23, 1993.
9. J. Workman, Jr., J. Coates, Spectroscopy 8(9): 36–42, 1993.
10. T. Hirschfeld, A. Zeev Hed. *The Altas of Near-Infrared Spectra.* Philadelphia: Sadtler Research Laboratories, 1981.

15.

INTERPRETIVE SPECTROSCOPY FOR NEAR-INFRARED

WHY INTERPRETIVE SPECTROSCOPY?

To begin this discussion a clear definition of two terms is required. First, a *spectrum* is the convolution of the measuring instrument function with the unique optical characteristics of the sample being measured (i.e., the sample is an active optical element of the spectrometer). Second, the *reference values* are those chemical or physical parameters *to be predicted using spectroscopic measurements;* a spectrum may or may not contain information related to the sample chemistry measured using a reference method. Interpretive spectroscopy provides a basis for the establishment of a known cause-and-effect relationship between instrument response and reference (analyte) data, in order to provide a more scientific basis for multivariate analytical spectrometry: *probability and statistics alone are a weak basis for analytical science.* When performing multivariate calibrations, analytically valid calibration models require a relationship between X (the instrument response data) and Y (the reference data); probability tells us only whether X and Y "appear" to be related. If no cause–effect relationship exists between X and Y, the model will have no true predictive importance. A knowledge of cause and effect creates a basis for decision making. Limitations of inferences derived from probability and statistics arise from limited knowledge of the characteristics and stability of:

- The set of samples used for calibration
- The set of measuring instrument(s) used for measurements
- The set of operators
- The set of measurement conditions

Probability can detect "alikeness" only in special cases, so cause–effect cannot be directly determined, only estimated. The requirement for a priori knowledge to provide a scientific basis for measurement poses several questions for the spectroscopist:

Question 1. Is X a true predictor of Y? Does cause–effect exist?

Question 2. If X is a true predictor of Y, what is the optimum mathematical relationship for describing the instrument response with respect to the reference data? (Such a relationship defines the optimum mathematical preprocessing and calibration algorithms for modeling.)

Question 3. If linear regression is to be used for calibration, do the five assumptions of regression apply? (1) Errors are independent of X (instrument response) or Y (reference) values. (2) The error distributions for both X and Y are known to be Gaussian. (3) The mean and variance of Y depend upon the absolute value of X. (4) The mean of each Y distribution is a straight-line function of X. (5) The variance of X is zero, and the variance of Y is exactly the same for all values of X.

Question 4. What are the effects of operator and measurement conditions on the change in instrument response relative to a sample?

Question 5. What are the effects of making measurements on multiple instruments with multiple operators?

Question 6. Does the instrument response represent a true measure of the analyte?

Question 7. What is the theoretical response for the measurement device with respect to the analyte?

Question 8. What is the limit of detection (LOD) relative to changes in X and Y? Is this limit acceptable for the intended measurement?

In routine analysis, at least three main examples of modeling problems are found:

The instrument response (X) is a predictor of reference values (Y). The limitation for modeling is in the representation of calibration set chemistry, sample presentation, and status of instrument and operator during measurement.

The instrument response does not change with a variation in the reference value. Therefore the cause for change in reference value (analyte) is not detected by the measurement instrument. Additional response information is required to describe the analyte.

The instrument response changes dramatically with no change in reference value. In this example additional clarification is required to define the relationship between the reference value and the spectroscopic/chemical composition of the sample, because factors other than chemical composition are affecting the reference data.

Factors affecting the integrity of teaching samples include the variations in sample chemistry, the variations in the physical condition of samples, and the variation in measurement conditions. Teaching sets must represent several sample "spaces," to include: compositional space, instrument space, and measurement condition (sample handling and presentation) space. Interpretive spectroscopy is a key intellectual process in approaching NIR measurements if one is to achieve an analytical understanding of these measurements.

INTRODUCTION TO NEAR-INFRARED SPECTROSCOPY

The discovery of the infrared region in 1800 is credited to William F. Hershel's famous work, "Experiments on the Refrangibility of the Invisible Rays of the Sun," read April 24, 1800, at the Royal Society (*Phil. Transact. Roy. Soc.* 90:284–292).

O. H. Wheeler [1] describes the near-infrared region as extending "from about 2 μ into the visible at about 0.7 μ" in a general discussion article (with 11 references). He also noted that "the term 'near infrared' formerly was used to denote the infrared spectrum to 23 μ, and to distinguish this region from that of the far infrared above 23 μ." Wheeler credits Rawlins and Taylor [2] with this early use of the term *near-infrared*. Using a variety of instrumentation, the author identifies near-infrared bands (in microns) in tabular form as follows. For C-H stretch, the fundamental occurs at 3.5, the 1st overtone at 1.8, the 2nd overtone at 1.2, the 3rd overtone at 0.85, and the 4th overtone at 0.7 μm. The author identifies band positions for both O-H and N-H stretch as occurring near 2.8 for the fundamental, 1.4 for the 1st overtone, 0.95 for the 2nd overtone, and 0.7 μ for the 3rd overtone.

R. F. Goddu and D. A. Delker [3] provide two detailed tables showing (1) the spectra–structure correlations and average molar absorptivity for a number of functional groups for the NIR region (1.0–3.1 μ), and (2) maximum recommended pathlengths for 12 solvents (useful for NIR spectroscopy) over the wavelength region 1.0–3.1 μ. The authors cite two references in this helpful article.

Early Reviews

Professor J. W. Ellis [4] has reviewed work below 3 microns for absorption of organic liquids. The review cites 44 separate works related to studies made prior to June 1929. In 1929 Professor Joseph W. Ellis wrote, "The region of the spectrum below 3 μ, although representing a relatively small portion of the total infra-red spectrum, is nevertheless quite significant in the study and interpretation of the absorption spectra of molecules. In particular, the absorption spectra of organic liquid molecules shows numerous bands in this region." He goes on to cite work by Coblentz, and Raman related to the observations of bands in the infrared region due to infrared absorptions and the Raman effect. Ellis goes on to refer to earlier work by Puccianti in 1900 with respect to the presence of specific near-infrared bands associated with, "molecules [having] a hydrogen atom combined with a carbon atom." Ellis reports that bands relating to "the carbon-hydrogen linkage" were observed at 2.3–2.2, 1.7, 1.4, 1.2, 1.0, and 0.9 μ. With this work, and the work of other investigators, the modern science of near-infrared spectroscopy was under way.

W. Kaye [5] provides a summary review of the work in near-infrared spectroscopy from the late 1920s to April 1954. The author draws information from 106 references for this review. The author refers to the term "hydrogenic" stretching vibrations as CH, NH, and OH. Work involving measurements in the region of 0.7–3.5 μ is reviewed. The author presents a Colthup-type chart of characteristic NIR bands, and the accompanying references. R. F. Goddu [6] provides an extensive review of near-infrared spectrophotometry prior to 1960. The author cites 110 literature sources of information for this review. K. B. Whetsel [7] reviews the significant work in near-infrared spectrophotometry prior to 1968. The review contains 336 references covering aspects of theory, instrumentation, and sampling techniques. Stark et al. [8] review work related to near-infrared analysis (NIRA) prior to 1986. The review, containing 300 references, describes the history, instrumentation, computations related to qualitative and quantitative analysis, and applications for NIRA. The authors cite the classical reviews of W. Kaye, O. Wheeler, R. Goddu, R. Goddu and D. Delker, K. Whetsel, and C. Tossi and A. Pinto, briefly describing each work.

SW-NIR as a Separate Topic

Schrieve et al. [9] discuss applications for the short-wave near-infrared (SW-NIR) region, referring to synonyms such as "the far-visible," the "near, near-infrared," and the "Herschel-infrared" to describe the range of approximately 700 to 1100 nm of the EMS (electromagnetic spectrum). The authors cite the increased interest of this spectral region to spectroscopists, particularly those involved with implementing process near-infrared measurements.

SPECTROSCOPY IN THE NEAR-INFRARED REGION

Since 1912, research investigations into the molecular structures of organic compounds using infrared spectroscopy has grown. This early work by W. W. Coblentz reported on the IR absorption of water. Coblentz used the spectral region from 1 to 8 microns. The decade of the

1960s brought about a prolific series of papers related to "direct determination" and to the measurement of light transmittance and reflectance properties of intact biological materials [10–22]. A summary of the most notable review papers describing the history of NIR interpretive spectroscopy would include Refs. 8 and 23–34.

Flour and Grain Analysis

In 1973, P. Williams reported the use of a commercial NIR grain analyzer for analyses of cereal products, following the pioneer work of Norris and others. Later, Williams and Karl Norris would edit a comprehensive text [35] on the subject of NIR analysis for commercially important biological materials. The book includes an exhaustive set of references (986) covering the early aspects of NIR in food and agricultural products analysis. Additional key references are given in Refs. 36–40. In September 1980, at the 65th AACC meeting in Chicago, B. G. Osborne et al. presented a detailed description of research into NIR analysis of wheat flour. He reported measurements for protein, moisture, ash, alpha-amylase, starch damage, water absorption, grain texture, particle size, loaf score, and pentosans in wheat flour used in the baking industry. The information in this presentation was later included in a textbook written by B. Osborne and T. Fearn [41]. References 42 and 43 describe the early work, most of which used multiple linear regression to identify key calibration wavelengths relating NIR instrument response to reference analytical data. P. Williams and coworkers [44] described the work in flour milling using near-infrared spectroscopy for the determination of moisture in cereals and cereal grains. The authors describe the uses of NIR for protein and moisture analysis in hard and soft wheat flours. Additional references describing wavelength selection and methodology for flour applications are presented in Refs. 45–47.

Forage Analysis

Forage analysis using NIR measurement has been a major application of the technique, largely due to the work of J. S. Shenk, M. Westerhaus, W. Barton, G. Marten, N. Martin, and a host of others who improved upon the technique and worked toward its widespread use and acceptance among scientists as a valid analytical technique. One could not mention NIR and Forage analysis without listing the primary reference source in the field since August 1985. The handbook edited by G.C. Marten, then of the University of Minnesota, J. S. Shenk of The Pennsylvania State University, and F. E. Barton of the Richard Russel Research Center, USDA, has become the most used handbook for NIR forage analysis [48].

A second comprehensive information source for NIR analysis of forages is a book chapter, containing 106 references, by Shenk, Workman, and Westerhaus [49]. Other key initial references for this application are found in Refs. 50–61.

Hydrocarbon Analysis

The following text points to only a few of the myriad references available that give key or important wavelengths for the designation of hydrocarbon groups. It also outlines references giving the general procedures that are universally followed for multivariate calibration of near-infrared instruments. Infrared spectroscopy, including the narrower term *near-infrared spectroscopy,* does not directly measure hydrocarbon classes such as olefins or napthenes, as such, but rather it measures functional group absorptions such as methyl, methylene, methine, and aromatic stretching and deformation vibrations. The ratios of these various absorptions, when correlated (using a variety of well-described multivariate calibration techniques) to known physical or compositional parameters, for a learning or teaching set, provides a correlation estimate of the hydrocarbon class composition from various unknown complex mixtures. References 62–84 describe classic and multivariate spectroscopic measurements of hydrocarbons.

NEAR-INFRARED MEASUREMENT TECHNIQUE

Infrared energy is the electromagnetic energy of molecular vibration. The energy band is defined for convenience as the near-infrared (0.78–2.50 microns), the infrared (or mid-infrared) (2.50–40.0 microns), and the far-infrared (40.0–1000 microns). Table 15.1 illustrates the region of the EMR (electromagnetic radiation) spectrum referred to as the near-infrared region. The table shows the molecular interactions associated with the energy frequencies (or corresponding wavelengths) of the various regions. Dominant near-infrared spectral features include: methyl C-H, methylene C-H, methoxy C-H, carbonyl C-H, aromatic C-H, hydroxyl O-H; N-H from primary amides, secondary amides (both alkyl, and aryl group associations), N-H from primary, secondary, and tertiary amines; and N-H from amine salts. Table 15.2 compares the band intensities for mid-infrared (MIR) versus near-infrared (NIR) for the C-H stretch region.

Properties of Infrared–Near-Infrared Energy

Light has both particle and wave properties. Quantum theory tells us that the energy of a light "particle," or photon, E_p, is given by

$$E_p = h\nu \qquad (1)$$

where h = Planck's constant (or 6.6256×10^{-27} erg-s), and ν is the frequency of light (or the number of vibrations per second, or in units of s^{-1}). Thus the energy for any specific photon can be quantified, and it is this energy that interacts with the vibrating bonds within *infrared-active molecules*. The subsequent values for wavelength, wavenumber, and frequency for both the visible and the extended near-infrared regions are shown in Table 15.3.

Table 15.1   Spectroscopic Regions of Interest for Organic Analysis

Region	Wavelength	Characteristic Measured
Ultraviolet	190–360 nm	Transitions of delocalized pi electrons
Visible	360–780 nm	Electronic transitions: color measurements
NIR (near-infrared)	780–2500 nm	Overtone and combination bands of fundamental molecular vibrations
IR (infrared)	2500–40,000 nm	Fundamental molecular vibrations: stretching, bending, wagging, scissoring
FIR (far-infrared)	4×10^4 to 10^6 nm	Molecular rotation

Table 15.2   Relative Band Intensities: MIR vs. NIR for C-H Stretch

Band	Wavelength region	Relative Intensity	Recommended sample cell pathlength (for liquid hydrocarbons)
Fundamental (ν)	3380–3510 nm	1	0.1–4 mm
First overtone (2ν)	1690–1755 nm	0.01	0.1–2 cm
Second overtone (3ν)	1127–1170 nm	0.001	0.5–5 cm
Third overtone (4ν)	845–878 nm	0.0001	5–10 cm
Fourth overtone (5ν)	N/A	0.00005	10–20 cm

Table 15.3 Equivalent Wavelength, Wavenumber, and Frequency values for the Visible and Near-Infrared Spectral Regions

Region	Wavelength (cm)	Wavelength (microns)	Wavelength (nanometers)	Wavenumber (cm^{-1})	Frequency (hertz)
Visible	3.5×10^{-5} to 7.8×10^{-5}	0.35–0.78	350–780	28,571–12,821	8.563×10^{14} to 3.842×10^{14}
Extended near-infrared	7.8×10^{-5} to 3.0×10^{-4}	0.78–3.0	780–3000	12,821–3333	3.842×10^{14} to 9.989×10^{13}

Ideal Harmonic Oscillator

Light from a spectrophotometer is directed to strike a matrix consisting of one or more types of molecules. If the molecules do not interact with the light, then the light passes through the matrix with no interaction whatsoever. If molecules interact with the light in a very specific way (i.e., molecular absorption) we refer to them as *active* or *infrared active*. For NIR energy, the overtones of X–H bonds, i.e., N-H, C-H, and O-H hydrogenic stretching and bending, are of the greatest interest. Infrared-active molecules can be seen as consisting of mechanical models with vibrating dipoles. Each dipole model vibrates with a *specific frequency* and *amplitude,* as shown using a simple model (Figure 15.1).

Note that the term *frequency* refers to the number of vibrations per unit of time, designated by the Greek letter v (nu) and generally specified in units of s^{-1} or Hertz (Hz). *Amplitude* is defined by the interatomic distance covered at the extremes of the vibrating dipole, and is dependent upon the amount of energy absorbed by the infrared-active bond. When incoming photons from a spectrophotometer source lamp (after they have passed through the monochromator or interferometer) strike different molecules in a sample, one of two direct results may occur: (1) the disturbing energy does not match the natural vibrational frequency of the molecule, or (2) the disturbing frequency does match the vibrational frequency of the molecule. When there is a match between the disturbing frequency of the illumination energy and the natural vibrational frequency of a molecule in the sample, the molecule absorbs this energy, which in turn increases the vibrational amplitude of the absorbing dipoles. However, regardless of the increase in amplitude, the frequency of the absorbing vibration remains constant.

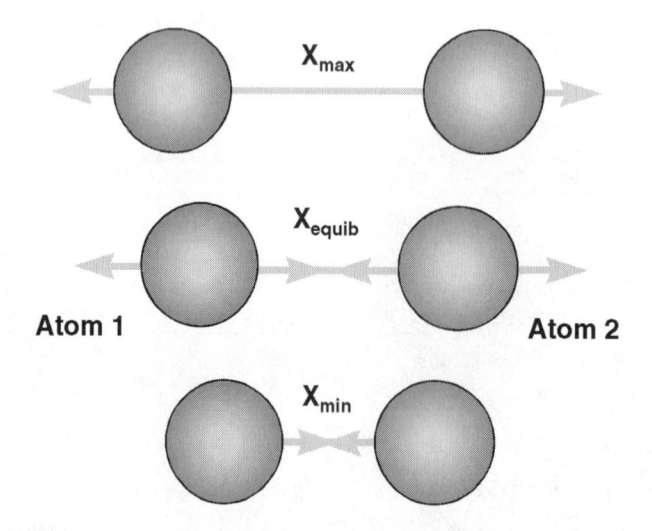

Fig. 15.1 Model of infrared-active molecule as a vibrating dipole between two atoms. (See color plate 4).

Another name for the dipole model from Figure 15.1 is *ideal harmonic oscillator*. The frequency at which the dipole (or ideal harmonic oscillator) vibrates (stretches or bends) is dependent upon the bond strength and the masses of the atoms bonded together. When the harmonic oscillator (HO) vibrates, the vibrational energy is continuously changing from kinetic to potential and back again. The total energy in the bond is proportional to the frequency of the vibration. Hooke's law (in our case referring to the elasticity properties of the HO) is applied to illustrate the properties of the two atoms with a well-behaved springlike bond between them. The natural frequency of vibration for a bond (or any two masses connected by a spring) is given by the well-known relationship

$$v = \frac{1}{2\pi} \sqrt{\frac{K}{\left(\frac{1}{m_1} + \frac{1}{m_2}\right)}} \qquad (2)$$

where K is a force constant that varies from one bond to another, m_1 = the mass of atom 1, and m_2 = the mass of atom 2. *Note:* As a first approximation, the force constant (K) for single bonds is ½ that of a double bond and ⅓ that of a triple bond. Also note that as the mass of the atoms increases, the frequency of the vibration decreases.

The effects of quantum mechanics on a simple HO indicate that we cannot treat the bond between two atoms quite as simply as two masses connected by a spring. This is no surprise, since quantum mechanical evidence has shown that vibrational energy between atoms in a molecule is quantized into discrete energy levels. When the conditions are right, vibrational energy in a molecule "jumps" from one energy level to another. The discrete vibrational energy levels for any molecule E_{VIB} are given by:

$$E_{VIB} = hv\left(v + \frac{1}{2}\right) \qquad (3)$$

where h = Plank's constant, v (Greek nu) = vibrational frequency of the bond, and v = vibrational quantum number (which can only have the integer values of 0, 1, 2, 3, and so on).

Anharmonic Oscillator

The concept of an *anharmonic oscillator* allows for the more realistic calculations of the positions of the allowed overtone transitions. The energy levels for these overtones are not found as the product of exact integer multiples and the fundamental frequency. In fact, the following expression defines the relationships between wavenumber (for a given bond) and the vibrational energy of that bond. The relationship is calculated from the Schroedinger equation to yield

$$\bar{v} = \left(\frac{E_{VIB}}{hc}\right) = \bar{v}_1 v - x_1 \bar{v}_1 (v + v^2) \qquad (4)$$

where v = an integer number, i.e., 0, 1, 2, 3, ..., n, and $x_1 \bar{v}_1$ = the unique anharmonicity constant for each bond.

Calculations of band positions using Eq. (4) will more closely approximate observed band positions than those calculated from the ideal harmonic oscillator expression found in Eq. (2). For a rule of thumb, the first overtone for a fundamental can be calculated as a 1% to 5% shift due to anharmonicity. Thus the expression

$$\lambda = \left[\frac{\lambda_1}{\kappa} + \frac{\lambda_1}{\kappa}(0.01, 0.02, \ldots, 0.05)\right] \qquad (5)$$

is used to illustrate the occurrence of overtone absorption using a fundamental absorption occurring at approximately 2632 cm^{-1}. Converting to nanometers, using $10^7 \div v$ (in cm^{-1}) = 3800 nm, the first overtone would occur at

$$v = \left[\frac{3800}{2} + (3800 \cdot 0.01)\right] = 1938 \text{nm} \quad \text{to} \quad v = \left[\frac{3800}{2} + (3800 \cdot 0.05)\right] = 2090 \text{nm}$$

And thus one would expect the first overtone to occur somewhere between 1938 and 2090 nm rather than the 1900 nm calculated using a simple harmonic oscillator model. Calculations for wavelength positions for the first overtone ($2v$), second overtone ($3v$), and third overtone ($4v$) within this chapter are computed with the assumption of a 1% to 5% frequency shift due to anharmonicity.

Molecules that absorb NIR energy vibrate in two fundamental modes: (1) *stretching* and (2) *bending*. Stretching is defined as a continuous change in the interatomic distance along the axis between two atoms, and *bending* is defined as a change in the bond angle between two atoms. Figure 15.2 illustrates the often-repeated stretching and bending interactions defining infrared-active species within infrared-active molecules.

Note that near-infrared (0.78–2.5 microns) spectral features arise from the molecular absorptions of the overtones (0.78–1.8 microns) and combination (1.8–2.5 microns) bands from fundamental vibrational bands found in the mid-infrared region. For fundamental vibrations there exists a series of overtones with decreasing intensity as the transition number (overtone) increases. Combination bands arise as the summation of fundamental bands, their intensity decreasing with an increase in the summation frequency. Most near-infrared absorptions result from the harmonics and overtones of X–H fundamental stretching and bending vibrational modes. Other important functional groups relative to near-infrared spectroscopy include hydrogen bonding, carbonyl carbon-to-oxygen stretch, carbon-to-nitrogen stretch, carbon-to-carbon stretch, and metal halides.

Molecular vibrations in the near-infrared regions consist of stretching and bending combination and overtone bands. Stretching vibrations occur at higher frequencies (lower wavelengths) than bending vibrations. Stretching vibrations are either symmetric or asymmetric; bending vibrations are either in-plane or out-of-plane. In-plane bending consists of scissoring and rocking; out-of-plane bending consists of wagging and twisting. From highest frequency to lowest, the vibrational modes occur as stretching, in-plane bending (scissoring), out-of-plane bending (wagging), twisting, and rocking. The bands observed most often in the near-infrared include the combination bands, second or third harmonics of O-H, N-H, and C-H fundamentals.

Fundamental and overtone absorptions arise when molecules are excited from the ground

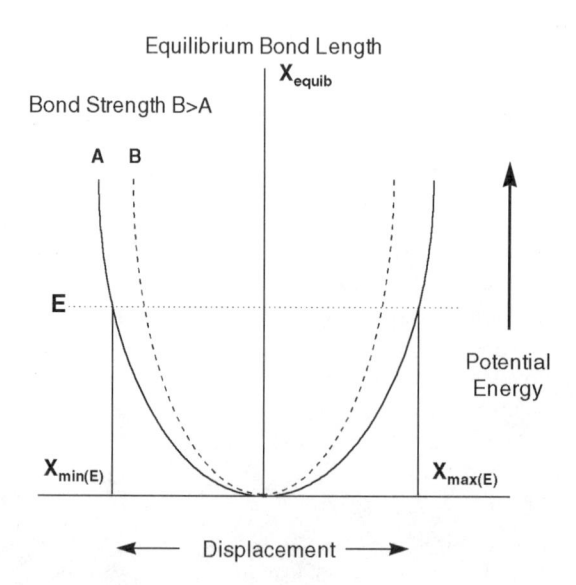

Fig. 15.2 Potential energy curve showing minimum and maximum amplitude for the harmonic oscillator model as a continuum. (See color plate 4).

state to a higher-energy excited state. Fundamental vibrations will change in amplitude when absorbing energy of the same frequency from an outside source that strikes the vibrating bond(s).

Variations in hydrogen bonding manifest themselves as changes in the force constants of the X–H bonds. Generally bands will shift in frequency and broaden due to the formation of hydrogen bonding. Since combination bands result as the summation of two or more fundamental vibrations, and overtones occur as the result of the multiples of fundamental vibrations, frequency shifts related to hydrogen bonding have a greater relative effect on combination and overtone bands than on their corresponding fundamentals. This feature of the near-infrared region alerts one to the importance of the relative hydrogen bonding effects brought about by solvent and temperature variations.

Precise band assignments are difficult in the near-infrared region due to the fact that a single band may be attributable to several possible combinations of fundamental and overtone vibrations all severely overlapped. The influence of hydrogen bonding results in band shifts to lower frequencies (higher wavelengths), a decrease in hydrogen bonding due to dilution and higher temperatures result in band shifts to higher frequencies (lower wavelengths). Band shifts of the magnitude of 10–100 cm^{-1}, corresponding to from a few to 50 nm, may be observed. The substantial effect of hydrogen bonding should be kept in the forefront of thought when composing calibration sample sets and experimental designs for near-infrared experiments.

Near-infrared spectra contain information relating to differences in bond strengths, chemical species, electronegativity, and hydrogen bonding. For solid samples, information with respect to scattering, diffuse reflectance, specular reflectance, surface gloss, refractive index, and polarization of reflected light are all superimposed on the near-infrared vibrational information. Aspects related to hydrogen bonding and hydronium ion concentration are included within the spectra.

Light can interact with the sample as reflection, refraction, absorption, scattering, diffraction, and transmission. Signal losses from the sample can occur as specular reflection, internal scattering, refraction, complete absorption, transmission loss during reflectance measurements, and trapping losses. Spectral artifacts can also arise as offset or multiplicative errors due to coloration of the sample, variable particle sizes and resultant variability in apparent pathlength, refractive index changes in clear liquids relative to temperature changes, and pathlength differences due to temperature-induced density changes.

During measurement of a sample, the light energy entering the sample will be attenuated to some extent; the light entering the sample is able to interact with the sample emerging as attenuated transmitted or reflected light. The frequency and quantity of light absorbed yield information regarding both the physical and compositional information of the sample. For reflectance spectra the diffuse and specular energy fractions are superimposed. The intensity of reflected energy is a function of multiple factors, including the angles of incidence and observation, the sample packing density, crystalline structure, refractive index, particle size and distribution, and absorptive and scattering qualities.

The types of vibrations found in near-infrared and infrared spectroscopy are designated by v (Greek letter nu), with a subscript designating whether the vibration is stretching (S) or bending (B); e.g., stretching is designated as v_S and bending vibrations as v_B. Combination bands resulting from the sum of stretching and bending modes are designated as $v_S + v_B$, and harmonics are designated as kv_S, where k is an integer number, such as 2 (first overtone), 3 (second overtone), ..., k(k-1th overtone) and v_S is the frequency of the fundamental stretch vibration for a specific functionality.

Illustration of the Harmonic Oscillator

The classical harmonic oscillator model is often illustrated using a potential energy curve (Figure 15.2). The larger the mass, the lower the frequency of the molecular vibration, yet the potential energy curve does not change. The relationship for a simple diatomic molecule model is

$$\text{Frequency} = v \text{ (nu)} = \frac{1}{2\pi}\sqrt{\frac{k}{\mu}}$$

where k is the force constant for the bond holding the two atoms together and μ is the reduced mass from the individual atoms.

For any bond there is a limit to the amplitude of motion away from and toward the opposing atom (X_{max} and X_{min}, respectively); the potential energy increases at either extreme, i.e., either the minimum or maximum distance between atoms. The total potential energy (E_τ) for a bond following the harmonic oscillator model is given by $E_\tau = (1/2)k \cdot x^2$, where k is the force constant for the bond holding the two atoms together and x is the change in displacement from the center of equilibrium. For any specific bond there is a limit to the total energy and maximum displacement of the atoms (or dipoles). Quantum mechanical theory is used to describe more closely the behavior of these atoms connected by bonds with varying force constants. Using quantum theory it is possible to describe a mathematical relationship for the behavior of two oscillating atoms as following not a continuous change in potential energy states for the vibrating system, but actually a set of discrete energy levels (E_n) given by $E_n = (n + 1/2)h\nu$, where n is the oscillating dipole quantum number for discrete potential energy levels, i.e., 0, 1, 2, 3, . . . , n. The potential energy curve that describes the behavior of oscillating dipoles better is given in Figure 15.3.

Selection Rule

The selection rule states that in the quantum mechanical model a molecule may only absorb (or emit) light of an energy equal to the spacing between two levels. For the harmonic oscillator (dipole involving two atoms) the transitions can only occur from one level to the next higher (or lower).

The infrared spectrum for any molecule will exhibit a band corresponding to the frequency of the energy at which light can absorb, namely, E_0, E_1, E_3, and so forth. Figure 15.3 illustrates the discrete quanta potential energy curves for both strong and weakly bonded dipoles. The energy levels are not equally spaced for the harmonic oscillator as shown in the figure. Note that the figure illustrates the energy levels for the fundamental vibrations (or oscillations) only. In addition to the fundamentals, the overtone bands also occur at less than 2 times, 3 times, and 4 times the frequency of the fundamental vibrations, and at much less intensity; later discussions in this chapter will describe the factors involving the infrared and near-infrared spectroscopic activity of molecules.

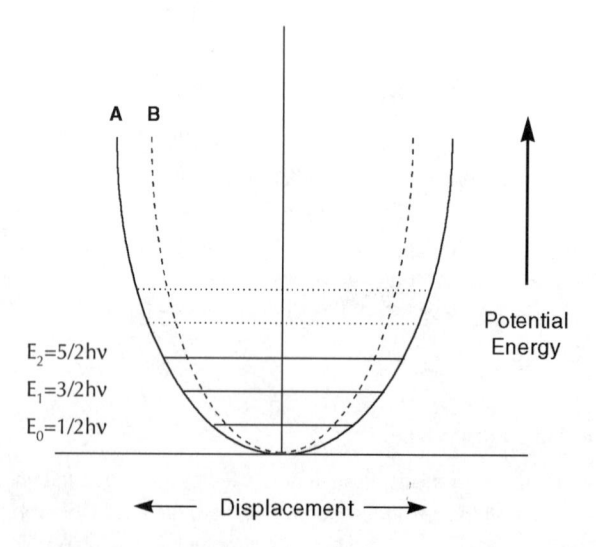

Fig. 15.3 Illustration of the discrete quanta potential energy curves (E_0, E_1, E_2, etc.) for both strong and weakly bonded dipoles as $E_n = [n + (1/2)]h\nu$. (See color plate 5).

Illustration of the Anharmonic Oscillator

The model of the anharmonic oscillator more closely follows the actual condition for molecular absorption than does that of the harmonic oscillator. Figure 15.4 illustrates the differences between the ideal harmonic oscillator case, which has been discussed in detail in this chapter, and the anharmonic oscillator model (better representing the actual condition of molecules). Unlike the ideal model illustrated by the harmonic oscillator expression, the anharmonic oscillator involves considerations such that when two atoms are in close proximity (minimum distance) they repel one another; when two atoms are separated by too large a distance, the bond breaks. The anharmonic oscillator potential energy curve is most useful to predict behavior of real molecules.

INTERPRETIVE NIR SPECTROSCOPY

The energy absorbed by a matrix consisting of organic compounds depends upon the chemical composition of the matrix, defined by the species (or type) of molecule present, the concentration of these individual species, and the interactions between the molecules in the matrix. In order for NIR (or any other vibrational spectroscopic measurement technique) to be valid, one must be absolutely assured that different types of molecules absorb at unique frequencies. Due to the broad band nature of NIR spectra, consisting of overlapping combination and overtone bands, the individual species are not as well resolved as they are in the mid-infrared region. In addition, many compounds absorb NIR energy throughout the entire wavelength region, making it difficult, if not impossible, to clearly resolve a usable baseline for simple peak-height or peak-area quantitative methods. This brings one to the same conclusion drawn by early investigators that the NIR region is not especially useful as a quantitative measurement technique. At the very least, novel techniques for spectral manipulation were required to "interpret" the poorly resolved bands and "compensate" for background interferences.

Group Frequencies

Frequencies that are characteristics of groups of atoms (termed *functional groups*) are given common group names and can be assigned general near-infrared locations (in nanometers) for

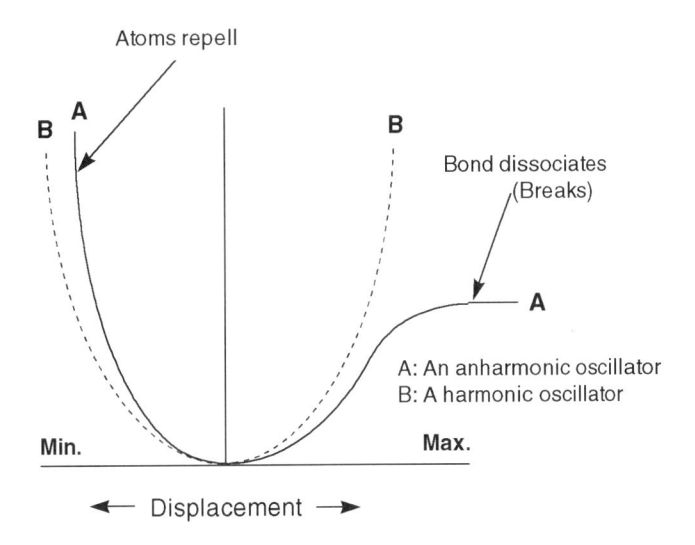

Fig. 15.4 Illustration for the differences in potential energy curves between the ideal harmonic oscillator model and the harmonic oscillator model (better representing the actual condition of molecules). (See color plate 6).

major groups associated with stretching vibrational bands. The locations for the fundamental frequencies, overtones, and combination bands is provided within the text of this chapter. References 85–88 give a rich source of information as to locations of fundamental frequencies for the major function groups (see Table 15.4 for examples) within the infrared and near-infrared regions.

Fingerprint Frequencies

Fingerprint frequencies are due to molecular vibrations of the entire molecule rather than being specific to a particular functional group. Since the infrared spectrum is unique to an individual molecule it is termed the *fingerprint* of that molecule, much as the fingerprint of a person indicates the specific identity of that person [87]. However, in identifying the unique "fingerprint vibrations" of a molecule, information is given as to the approximate molecular formula and composition of the molecule.

Dividing the Near-Infrared Spectrum

The group frequencies are given as those bands occurring below 6667 nm, and the fingerprint frequencies are those above 6667 nm. Thus, for the near-infrared region (below 2500 nm), the fingerprint frequencies do not occur. The group frequencies below 6667 nm usually demonstrate bands of reasonable intensity, with somewhat reliable interpretation based upon frequency (wavelength) position.

Coupling of Vibrations

Coupling indicates that the oscillators, or molecular vibrations of two or more molecules, are interactive, so the original vibrational energy states (if the vibrations could occur independent of one another) result in split energy states due to the interaction of the vibrations. Coupling is divided into two basic orders, first and second (Fermi resonance). First-order coupling can be involved in several important infrared group frequencies. For example, CO_2 has the two separate (uncoupled oscillators) of C=O, each occurring at approximately 5405 nm (v at ~1850 cm^{-1}), 2756.6–2972.8 nm ($2v$), 1855.7–2071.9 nm ($3v$), 1405.3–1621.5 ($4v$). The interactive (coupled oscillators) energy states occur at 4257 nm (v at 2349 cm^{-1}, asymmetric stretch), 2171.1–2341.4 nm ($2v$), 1461.6–1631.9 nm ($3v$), and 1106.8–1277.1 nm ($4v$); and 7463 nm (v at 1340 cm^{-1}, symmetric stretch), 3806.1–4104.7 nm ($2v$), 2562.3–2860.8 nm ($3v$), and 1940.4–2238.9 nm ($4v$). First-order coupling is involved for multiple infrared group frequencies, including [88, p. 33]:

- The stretches for all cumulated double bonds, X=Y=Z, e.g., C=C=N
- The stretches in XY$_2$, including –CH$_2$–, and H$_2$O
- The stretches in XY$_3$ groups, including –CH$_3$
- The deformations of XY$_3$ groups, including –CH$_3$, CCl$_3$
- The N-H in-plane bend of secondary amides, e.g., R–CO–NH–R′

Fermi Resonance (or Second-Order Coupling)

Fermi resonance [89] is the interaction or coupling of two vibrational energy states, with resultant separation of the states, where one of the states is an overtone or a sum tone. An overtone vibration occurs as the integer multiple of a fundamental vibration in frequency space (intensity falls off rapidly with the higher multiples):

- The first overtone of a fundamental vibration (v) is equal to $2 \times v$.
- The second overtone of a fundamental vibration is equal to $3 \times v$.
- The third overtone is equal to $4 \times v$.

Table 15.4 Example IR-Active Functional Groups with Assigned Group Frequencies

Hydrocarbons, etc.

C–C, paraffinics (or alkanes)

CH_3–$(CH_2)_N$–CH_3, normal paraffins

R–CH_3, methyl C–H

R–CH_2R′, methylene C–H

R–CHR′R″, methine C–H

C=C, olefinic group (or alkenes)

–CH=CH_2, vinyls

R′R″C=CH_2, vinylidenes

>C=C=CH_2, allenes

–C≡C–, acetylenes (or acetylinics)

C_NH_N, aromatics

C_NH_{2N}, naphthalenes (or cycloalkanes)

Carbonyl compounds

R–C=O–H, aldehydes

R–C=O–R′, ketones

C=C=O–R′, ketenes

R–C=O–OR′, esters

S–C=O–O–R′, thiol esters

R–C=O—C=O—R, anhydrides

–C–O–O–C–, peroxides

–O–C=O–C=O–O–, oxalates

–O–C=O–, carboxy-

–C=O–, carbonyl group

–C=O–NH_2, primary amides

–C=O–NHR, secondary amides

–C=O–NR′R″, tertiary amides

Ethers

—OCH_3, methoxy (or ether group)

C—O—C, ethers

Hydroxy

–OH, hydoxyl- [O-H]

R–OH, alcohols [O-H]

–CH_2–OH, primary alcohols

R′R″CH–OH, secondary alcohols

R′R″R‴C–OH, tertiary alcohols

Ar–OH, phenolics (or phenols) [O-H]

Nitrogen compounds

—C≡N, nitriles

R—C≡N–, nitriles

S–C≡N⁻, thiocyanate

—C≡N⁻, cyanide

C–NO, nitroso- group

R′R″NNO, nitrosamines

R–NO⁻$_2$, nitro- group (or nitrite)

R–NH_2, primary amines

R′R″NH, secondary amines

R–NH⁺$_3$, R′R″NH⁺$_2$, amine salts

N=N, azo- group

–N=N⁺=N⁻, –N⁻–N⁺≡N; azides

NNO, azoxy group

R–O–N–N=O, organic nitrites

–NO_3, nitrates

ON=NO, nitroso- group

H_3N^+–CH–COO⁻, amino acids

Note: overtones are the special case of sumtones where the frequencies are identical. A sum tone is the general case of an overtone where the frequencies are not equal and where a variety of vibrational energy states can occur:

A binary sum tone is equal to the sum of two fundamentals, e.g., $v_i + v_k$.

A ternary sum tone is equal to the sum of three fundamental vibrations, e.g., $v_i + v_k + v_m$.

Other sum tones can occur, such as the sum of an overtone and a fundamental vibration, e.g., $2v_i + v_k$.

Three requirements are stated for Fermi resonance [88, p. 34]:

> The zero-order frequencies must be close together (typically within 30 cm^{-1}).
>
> The fundamental and the overtone or sumtone must have the same symmetry.
>
> There must be a mechanism for the interaction of the vibrations. *Note:* (1) The vibrations cannot be separate (or localized) in distinctly different parts of the molecule. (2) The vibrations must be mechanically interactive in order that the interaction of one vibration affects another.

The results of Fermi resonance are important for infrared, and near-infrared spectroscopy. Fermi resonance causes the following effects on spectral bands:

> The resultant bands are moved in position from their expected frequencies.
>
> Overtone bands are more intense than expected.
>
> There may exist doublet bands where only singlets were expected.
>
> Solvent changes can bring about slight shifts in the frequency location of a band, and intensities can be greatly changed.

GROUP FREQUENCIES OF HYDROCARBONS

Aliphatic Hydrocarbons (*n*-Alkanes)

The number of normal vibrations occurring within the infrared region for saturated hydrocarbons is $3N-6$, where N is the sum of carbon plus hydrogen atoms; e.g., *n*-hexane (C_6H_{14}) exhibits $3(20) - 6 = 20$ normal vibrational modes. Internal rotation and overlapping of bands complicate the interpretation of alkane (parafinnic) group spectra. C-H stretching vibrations can be expressed to a first approximation by the straightforward use of Hooke's law describing the vibration between two masses connected by a vibrating spring, as previously presented.

The first consideration in the description of alkane group spectral interpretation would include the C-H stretching vibrations for methyl (–CH$_3$) and methylene (–CH$_2$–) groups. The *methyl C-H stretching vibrations* (Figure 15.5) are found in two basic modes:

> The asymmetric (degenerate or out-of-phase) methyl C-H stretch at 3365–3388 nm (v at 2962 ± 10 cm^{-1}), 1716.2–1863.4 nm (2v), 1155.3–1298.7 nm (3v), 874.9–1016.4 nm (4v)
>
> The symmetric (in-phase) methyl C-H stretch at 3470–3494 nm (v at 2872 ± 10 cm^{-1}), 1769.7–1921.7 nm (2v), 1191.4–1339.37 nm (3v), 902.2–1048.2 nm (4v); and 3406–3429 nm (v at 2926 ± 10 cm^{-1}), 1737.1–1886.0 nm (2v), 1169.4–1314.5 nm (3v), 885.6–1028.7 nm (4v)

Three C-H stretching modes correspond to three coupled C-H stretching vibration oscillators. The molecular extinction coefficient for methyl C-H stretch remains constant irrespective of increasing alkane chain length.

The *methylene C-H stretching vibrations* (Figure 15.6) are also found in two basic modes:

> The asymmetric (degenerate or out-of-phase) methylene C-H stretch at 3406–3429 nm (v at 2926 ± 10 cm^{-1}), 1737.1–1886.0 nm (2v), 1169.4–1314.5 nm (3v), and 885.6–1028.7 nm (4v)

The symmetric (in-phase) methylene C-H stretch at 3493–3517 nm (v at 2853 ± 10 cm^{-1}), 1781.4–1934.4 nm ($2v$), 1199.3–1348.2 nm ($3v$), and 908.2–1055.1 nm ($4v$)

The two methylene stretching modes result from two coupled C-H stretching vibration oscillators. Increasing the chain length within alkane groups brings about a linear increase in the molecular extinction coefficient of methylene C-H stretching bands.

The carbon-hydrogen vibrational modes for alkanes also contain both methyl and methylene bending vibrations. The *methyl C-H bending vibrations* are found in two modes:

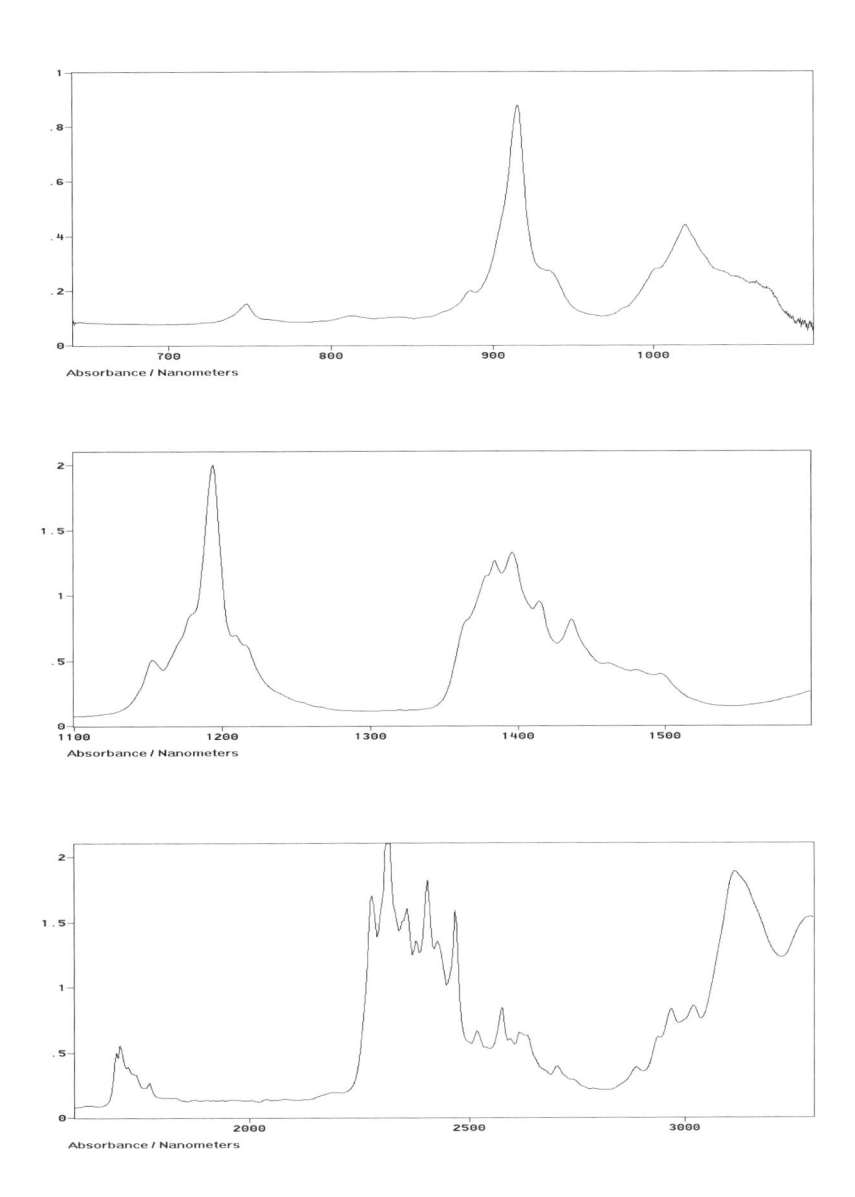

Fig. 15.5   Illustration of aliphatic C-H (as predominantly methyl C-H stretch) from 2,2,4-dimethyl pentane (iso octane). The top spectrum (from left to right) shows the 4th overtone ($5v$), the 3rd overtone ($4v$), and combination bands. The center spectrum illustrates the 2nd overtone ($3v$) and combination bands. The bottom spectrum presents the 1st overtone ($2v$) and combination bands and a portion of the fundamental.

The asymmetric (degenerate or out-of-phase deformation) methyl C-H bending at 6803–6897 nm (v at 1460 ± 10 cm^{-1}), 3469.5–3793.4 nm ($2v$), 2335.7–2643.9 nm ($3v$), and 1768.8–2069.1 nm ($4v$).

The symmetric (in-phase or "umbrella") methyl C-H bending at 7220–7326 nm (v at 1375 ± 10 cm^{-1}), 3682.2–4029.3 nm ($2v$), 2478.9–2808.3 nm ($3v$), and 1877.2–2197.8 nm ($4v$). This functional group vibrational frequency is dependent upon the atom to which the methyl group is attached.

There is no coupling of methyl C-H bending modes to the rest of the system.

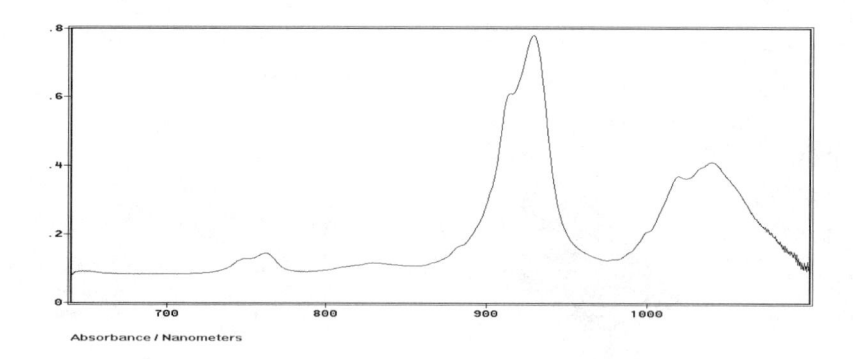

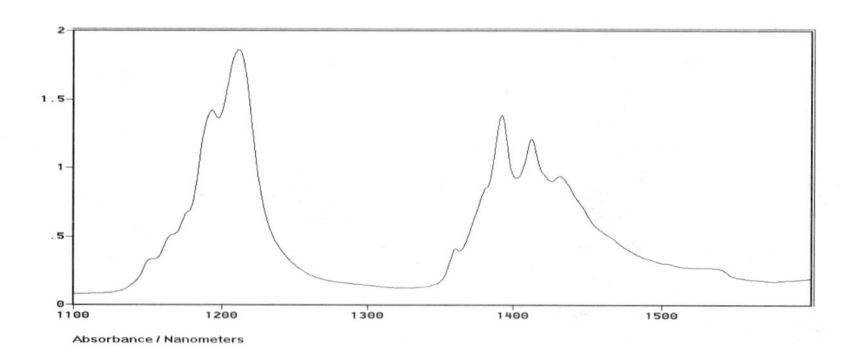

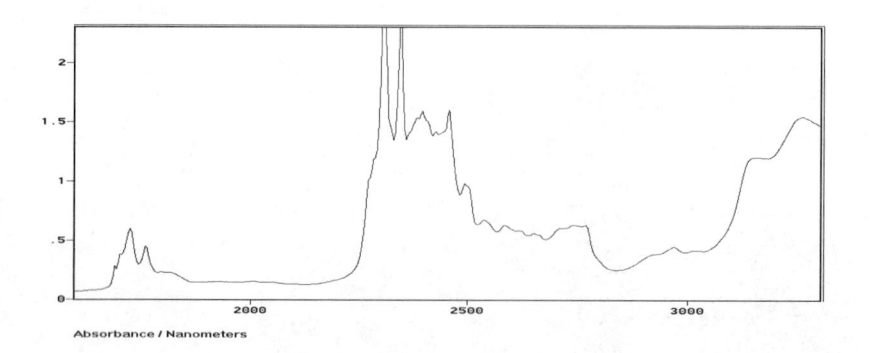

Fig. 15.6 Illustration of aliphatic C-H (as predominantly methylene C-H stretch) from *n*-decane. The top spectrum (from left to right) shows the 4th overtone ($5v$), the 3rd overtone ($4v$), and combination bands. The center spectrum illustrates the 2nd overtone ($3v$) and combination bands. The bottom spectrum presents the 1st overtone ($2v$) and combination bands and a portion of the fundamental.

The *methylene C-H bending vibrations* are found in four basic modes:

The symmetric (scissoring) methylene C-H bending at 6826–6920 nm (*v* at 1455 ± 10 cm^{-1}), 3481.3–3806.0 nm (2*v*), 2343.6–2652.7 nm (3*v*), and 1774.8–2076.0 nm (4*v*); no coupling is observed for this vibration.

The in-phase twisting vibration at 7634–7752 nm (*v* at 1300 ± 10 cm^{-1}), 3893.3–4198.7 nm (2*v*), 2621.0–2971.6 nm (3*v*), and 1984.8–2325.6 nm (4*v*); coupling is observed for this vibration.

The coupled wagging vibrations at 7692–10,000 nm (*v* at 1300–1000 cm^{-1}), 3922.9–5500 nm (2*v*), 2640.9–3833.3 nm (3*v*), 1999.9–3000 nm (4*v*).

The in-phase rocking vibration for alkanes with four or more carbons is not found within the near-infrared region.

In addition, other bending modes of methyl and methylene can be observed. The alkane carbon-to-carbon stretching and carbon-to-carbon bending vibrations also occur as weak group frequencies.

Alkenes (Olefins) and Alkynes (Acetylenes)

The olefins include bands for saturated hydrocarbons as well as the olefinic C=C stretching and olefinic C-H stretching and bending vibrations.

Olefinic C-H Stretch

Fundamental olefinic C-H stretch occurs at 3226–3333 nm (*v* at 3100–3000 cm^{-1}), 1645.3–1833.2 nm (2*v*), 1107.6–1277.7 nm (3*v*), and 838.8–999.9 nm (4*v*). C-H stretch for an unsaturated carbon occurs below 3333 nm (*v* above 3000 cm^{-1}), 1699.8–1833.2 nm (2*v*), 1144.3–1277.7 nm (3*v*), and 866.6–999.9 (4*v*). The saturated carbon C-H bond stretching occurs above 3333 nm (*v* below 3000 cm^{-1}).

Olefinic C-H stretch can occur as an end group, e.g. R–R–C=C, or as an imbedded group associated with other carbons on either end, e.g., R–C=C–R. The absorption bands for these two olefinic types occur for the end-group type from 3226 nm to 3268 nm (*v$_i$* at 3100–3060 cm^{-1}), 1645.3–1797.4 nm (2*v$_i$*), 1107.6–1252.7 nm (3*v$_i$*), 838.8–980.4 nm (4*v$_i$*). The imbedded group stretch vibration can occur from 3279 to 3322 nm (*v$_i$* at 3050–3010 cm^{-1}), 1672.3–1827.1 nm (2*v$_i$*), 1125.8–1273.4 nm (3*v$_i$*), and 852.5–996.6 nm (4*v$_i$*). Note that compounds containing the vinyl group (–CH=CH$_2$) have weak absorption bands at approximately 3247 nm to 3300 nm (3080–3030 cm^{-1}), 1656.0–1815.0 nm (2*v$_i$*), 1114.8–1265.0 nm (3*v$_i$*), and 844.2–990.0 nm (4*v$_i$*) [88, p. 53].

Alkyne C-H Stretch

The sp-type C-H stretch for ≡C-H occurs at approximately 3030 nm (*v* at 3300 cm^{-1}), 1545.3–1666.5 nm (2*v*), 1040.3–1161.5 nm (3*v*), and 787.8–909.0 nm (4*v*). The sp^1-type C-H stretch associated with aromatic and olefinic carbons occurs at 3226–3333 nm (*v* at 3100–3000 cm). The sp^1-type stretch for saturated carbons is observed at the high-frequency ranges between 3344 and 3571 nm (*v* at 2990 and 2800 cm^{-1}), 1705.4–1964.1 nm (2*v*), 1148.1–1368.9 nm (3*v*), and 869.4–1071.3 nm (4*v*).

Olefinic C-H Bend

Two modes of bending occur in olefinic hydrogens, in-plane and out-of-plane. These bands do not appear with any importance within the near-infrared region, and the following two paragraphs are for informational purposes only. In-plane bending represents a change in the C-H bond angle along the same plane, occurring at 6920–8333 nm (v at 1455–1200 cm^{-1}) as a weak and not very useful group frequency for the presence of olefinic hydrogens. Out-of-plane bending occurs as the C-H bonds vibrate outside the flat olefinic group plane. Out-of-plane bending occurs at approximately 10,000–15,385 nm (1000–650 cm^{-1}) and represents excellent infrared bands for identifying the presence of olefinic groups; however, the overtones do not occur within the near-infrared spectral region. The out-of-plane C-H bend–related bands are group frequencies that are distinct, not being affected by conjugation or coupling with other vibrations in the olefinic molecule.

There is only one infrared-active C-H bend of the four possible modes for olefinic bending vibrations. The four bending modes include (1) *twisting or rotational motion* and (2) opposite twisting at the individual olefinic carbons producing a *torsion* at the C=C bond. *Wagging,* or a motion of the hydrogen bonds, takes two forms: (3) one where the two groups of two hydrogens move in opposite planes, causing a rotational motion around a common center of axis, and (4) the second, where all four hydrogens move in or out of the plane together. The first type of wag is Raman-only active, and the second type of wag is the only infrared-active type, though not observed in the near-infrared region.

AROMATIC SYSTEMS

Aromatic systems produce strong and sharp infrared bands due to the relatively rigid molecular structures; see Figure 15.7. There are multiple vibrational modes present in aromatic systems including: (1) carbon-hydrogen stretching, (2) in-plane bending, and (3) out-of-plane bending. Carbon-carbon interactions include (4) ring-stretching, and (5) ring-bending modes. Internal vibrations caused by groups associated with the aromatic system also exhibit vibrational group frequencies. The total vibrational modes for aromatic systems is equal to $3N-6$ vibrations, where N is the number of carbons in the ring structure. These vibrational modes are shown in Ref. 90, p. 262. Carbon-hydrogen stretching vibrations occur from approximately 3226 nm to 3333 nm (v at 3100–3000 cm^{-1}), 1645.3–1833.2 nm ($2v$), 1107.6–1277.7 nm ($3v$), and 838.8–999.9 nm ($4v$). Infrared spectra of aromatics using high-resolution instrumentation shows a pair of composite bands near 3257 nm (v at 3070 cm^{-1}), 1661.1–1791.4 nm ($2v$), 1118.2–1248.5 nm ($3v$), 846.8–977.1 nm ($4v$); and 3300 nm (v at 3030 cm^{-1}), 1683.0–1815.0 nm ($2v$), 1133.0–1265.0 nm ($3v$), 858.0–990.0 nm ($4v$). These composite bands arise as the combined interactions of several stretching modes discussed briefly earlier.

The carbon-carbon ring stretching occurs as a vibrational pair near 6250 nm (v at 1600 cm^{-1}), 3187.5–3437.5 nm ($2v$), 2145.8–2395.8 nm ($3v$), 1625–1875.0 nm ($4v$). This pair of frequencies, termed a *degenerate* pair, results from two stretching modes of the carbon-carbon ring stretching. However, due to Fermi resonance, one of the vibrations (near 6270 nm, or 1595 cm^{-1}) is not observed in the infrared spectrum. When there is substitution on the aromatic ring, the degeneracy is lost and the pair of bands occur near 6250 nm (v at 1600 cm^{-1}) and 6329 nm (1580 cm^{-1}). The 6329 nm band will be more intense if there is conjugation with the aromatic system by an electron withdrawing group or if substitution by O or N has occurred. Note the occurrence of overtones for this band at 3227.8–3481.0 nm ($2v$), 2173.0–2426.1 nm ($3v$), and 1645.5–1898.7 nm ($4v$). Para-disubstitution of a benzene ring by two identical groups yields to the center-of-symmetry rule, and thus the carbon-to-carbon ring stretching is infrared forbidden. It follows that if the para-disubstituted groups are not identical, some weak bands will occur near the 6250 nm (1600 cm^{-1}) region (see overtone positions in the preceding text).

The carbon-hydrogen bending (wag) vibrations occur as in-plane bending and out-of plane bending. The in-plane bending vibrations are much less consistent than the out-of-plane bending vibrations. The in-plane bending vibrational frequency depends upon the number of C-H groups present and the substituent groups.

AROMATIC SYSTEMS

Out-of-plane bending is much more consistent and intense in the infrared region. Out-of-plane bending is not particularly influenced by the substituents present, but the frequency for the bending vibrations depends upon the number of adjacent hydrogens. The carbon-hydrogen bending vibrations give rise to a series of combination bands termed *sumtones* or *summation bands.* These sumtones, summation bands, or combination bands occur from 5000 nm to 5882 nm (v at 2000–1700 cm^{-1}), 2550–3235.1 nm (2v), 1716.7–2254.8 nm (3v), and 1300.0–1764.6 nm (4v).

The patterns are used in infrared spectroscopy to determine ring substituents. The pattern for the sumtones is more important than their exact position. These composite band patterns are given in most infrared texts, such as Ref. 88 (p. 82), and those found in the Additional References section at the end of this chapter.

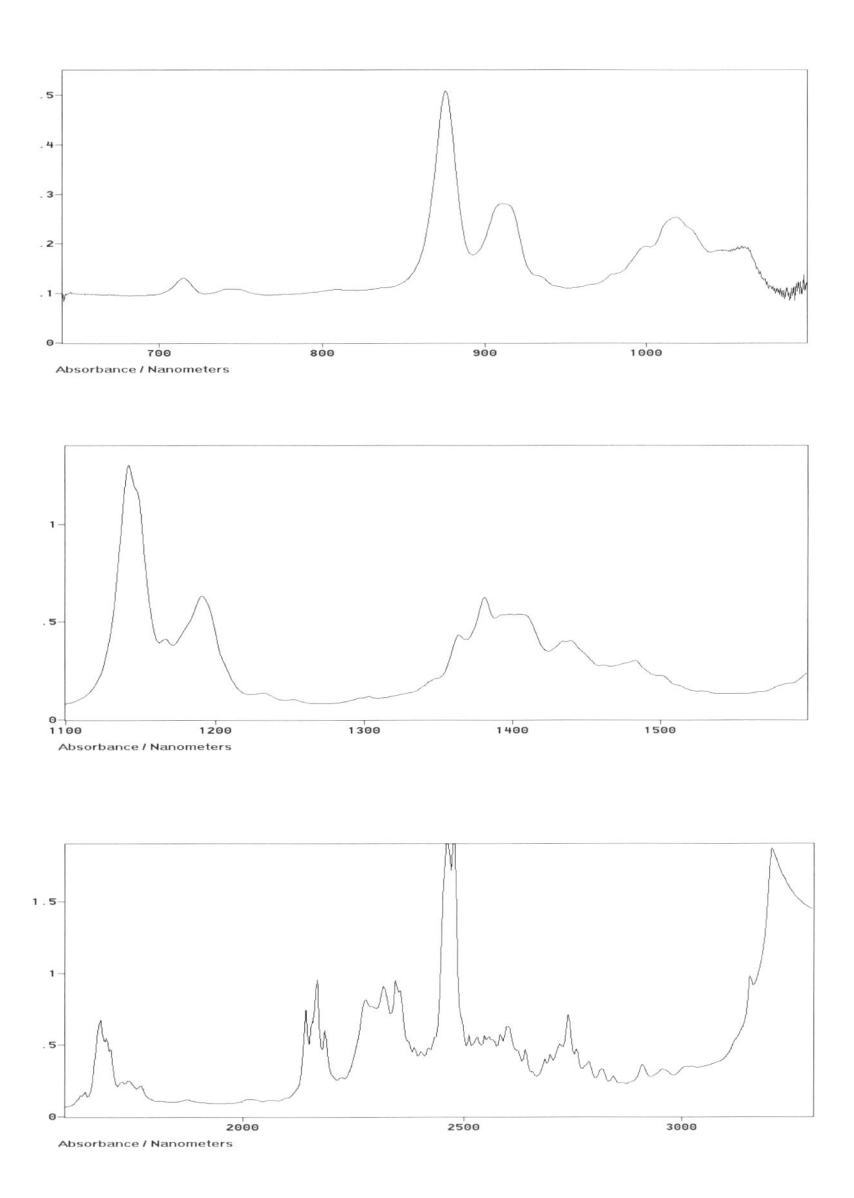

Fig. 15.7 Aromatic C-H stretch as illustrated using toluene NIR spectra. The top spectrum (from left to right) shows the 4th overtone (5v), the 3rd overtone (4v), and combination bands. The center spectrum illustrates the 2nd overtone (3v) and combination bands. The bottom spectrum presents the 1st overtone (2v) and combination bands and a portion of the fundamental vibrational band.

CARBONYL COMPOUNDS (>C=O)

Compounds containing carbonyl groups exhibit frequencies over a broad range from 5128 nm to 6452 nm (v at 1950–1550 cm^{-1}), 2615.3–3548.6 nm ($2v$), 1760.6–2473.3 nm ($3v$), 1333.3–1935.6 nm ($4v$). The position or frequency of the carbonyl C=O stretching vibration is affected by (1) the isotope effects and mass change of substituted groups, (2) the bond angles of substituted groups, (3) electronic (resonance and inductive) effects, and (4) interactions of these effects. Substituents with higher mass decrease the C=O stretch frequency; increasing the mass or bond angles of substituents also decreases the frequency by up to 40 cm^{-1} at 1715 cm^{-1}; this is equivalent to an increase of 139 nm at 5831 nm. Similarly, decreasing the substituent mass or bond angles between the carbonyl carbon and its substituents increases the frequency by 25 cm^{-1} above the nominal 1715 cm^{-1} carbonyl C=O stretch frequency, equivalent to a decrease in wavelength of 84 nm at 5831 nm. More electronegative (electron-withdrawing) substituents will increase the carbonyl carbon-oxygen stretch frequency by up to 100 cm^{-1} above the 1715-cm^{-1} nominal frequency, a decrease of 321 nm at 5831 nm. Conjugation of the carbonyl group to aromatic or olefinic groups tends to lower the frequency for both C=O and C=C by 30–40 cm^{-1}, dependent upon the ring size of the substituent. For example, carbonyl conjugation with aromatic ring size of 6 carbons produces C=O stretch frequencies at 5935 nm (v at 1685 cm^{-1}), 3026.9–3264.3 nm ($2v$), 2037.7–2275.1 nm ($3v$), and 1543.1–1780.5 nm ($4v$). For carbonyl conjugation with aromatic ring size of 5, a stretch occurs at 5928 nm (v at 1687 cm^{-1}), 3023.3–3260.4 nm ($2v$), 2035.3–2272.4 nm ($3v$), and 1541.3–1778.4 nm ($4v$). For aromatic ring size of 4, a stretch occurs at 5931 nm (v at 1686 cm^{-1}), 3024.8–3262.1 ($2v$), 2036.3–2273.6 nm ($3v$), and 1542.1–1779.3 nm ($4v$). For aromatic ring size of 3, a stretch occurs at 5963 nm (v at 1677 cm^{-1}), 3041.4–3279.7 nm ($2v$), 2047.3–2285.8 nm ($3v$), and 1550.4–1788.9 nm ($4v$). Electronic effects, such as resonance and inductive type, produce a double-bond character and increase the frequency of the carbonyl stretch.

Amides (R–C=ONH–X)

Amides are a special case of carbonyl compounds that are subject to frequency shifts due to hydrogen bonding. The frequencies for dilute solutions of amides can vary by as much as 250–275 nm (75 cm^{-1}) at ~5880 nm as compared to the same molecule with hydrogen bonding (such as a Nujol mull preparation). Dilute solutions of amides exhibit carbonyl frequencies as listed in the following text. Primary amides (R–C=O–NH$_2$) occur at 5764 nm (v at 1735 cm^{-1}), 2939.6–3170.2 nm ($2v$), 1979.0–2209.5 ($3v$), 1498.6–1729.2 ($4v$). Secondary amides (R–C=O–NHR) occur at 5865 nm (v at 1705 cm^{-1}), 2991.2–3225.8 nm ($2v$), 2013.7–2248.3 nm ($3v$), 1524.9–1759.5 nm ($4v$). Tertiary amides, R–(C=O)–NR$_2$, are found at 6042 nm (v at 1655 cm^{-1}), 3081.4–3323.1 nm ($2v$), 2074.4–2316.1 nm ($3v$), and 1570.9–1812.6 nm ($4v$).

These respective hydrogen-bonded vibrations occur at frequencies shifted from those listed above. For primary amides, vibration occurs at 5882–6061 nm (v at 1700–1650 cm^{-1}), 2999.8–3333.6 nm ($2v$), 2019.5–2323.4 nm ($3v$), and 1529.3–1818.3 nm ($4v$). For secondary amides, vibrations occur at 5952–6116 nm (v at 1680 to 1635 cm^{-1}), 3035.5–3363.8 nm ($2v$), 2043.5–2344.5 nm ($3v$), and 1547.5–1834.8 nm ($4v$). For tertiary amides, vibrations occur at 6024 nm (v occurs at 1660 cm^{-1}), 3072.2–3313.2 nm ($2v$), 2068.2–2309.2 nm ($3v$), and 1566.2–1807.2 nm ($4v$).

The N-H stretch for amides occurs as two bands between 2976 nm and 3150 nm (v_i at 3360 and 3175 cm^{-1}). Geometrically, two forms of amides exist as either the trans or cis forms, with the trans form demonstrating a frequency of 3008–3053 nm (v at 3325–3275 cm^{-1}), 1534.1–1679.2 nm ($2v$), 1032.7–1170.3 nm ($3v$), and 782.1–915.9 nm ($4v$). The cis form is from 3140 nm to 3190 nm (v at 3185–3135 cm^{-1}), 1601.4–1754.5 nm ($2v$), 1078.1–1222.8 nm ($3v$), and 816.4–957.0 nm ($4v$). The geometries included are as R–(C=O)–(N–R')–H where R and R' are either in the trans or the cis configuration.

The term *amide I* band refers to the C=O carbonyl stretch, and the nominal C=O stretch frequency is 5831 nm (v at 1715 cm^{-1}), 2973.8–3207.1 nm ($2v$), 2002.0–2235.2 ($3v$), and

1516.1–1749.3 nm (4ν). Since the amide form of the carbonyl has the electronegative N atom, lowered by resonance with the nitrogen electrons and lowered by hydrogen bonding, the C=O nominal frequency for the amide I stretch is approximately 6061 nm (ν at 1650 cm^{-1}), 3091.1–3333.6 nm (2ν), 2080.9–2323.4 nm (3ν), and 1575.9–1818.3 nm (4ν).

The term *amide II* refers to the NH$_2$ scissoring associated with primary amides, R–(C=O)–NH$_2$. The amide II frequency results from the combined interactions of the C-N stretch and the N-H in-plane bend; these vibrations interact to form bands at the 6494 nm (ν at 1540 cm^{-1}) region, with overtones at 3311.9–3571.7 nm (2ν), 2229.6–2489.4 nm (3ν), and 1688.4–1948.2 nm (4ν). In addition, bands are formed at the 7634–8197-nm (ν at 1310–1220 cm^{-1}) region, with overtones at 3893.3–4508.4 nm (2ν), 2621.0–3142.2 nm (3ν), and 1984.8–2459.1 nm (4ν). A band at 6410–6515 nm (ν at 1560–1535 cm^{-1}) indicates a trans substituted amide; this band is lost in the cis-amide configuration. Overtones for this band occur at 3269.1–3583.3 nm (2ν), 2200.8–2497.4 nm (3ν), and 1666.6–1954.5 nm (4ν).

X–H FUNCTIONAL GROUPS (O-H AND N-H)

X–H frequencies occur below 5000 nm (above 2000 cm^{-1}), except for Boron compounds, which demonstrate bands from 4762 nm to 6173 nm (ν at 2100–1620 cm^{-1}). Interaction of X–H with other groups does not usually occur, due to the high frequency of these bands and the low mass of hydrogen, so the frequency of X–H is specific and reliable for infrared spectroscopy. The X–H stretch follows the two-body harmonic oscillator approximation (discussed in an earlier section of this chapter). The relationship of X–H to the harmonic oscillator allows the frequency of any specific X–H group to be related to the periodic table; the frequency of an X–H group increases as one goes upward in any column or to the right within any row of the periodic table. The first row of the periodic table gives H-H with a nominal fundamental stretch at 2404 nm (4160 cm^{-1}) [88, p. 136]. The second row yields:

ν(Li-H) = 7353 nm (1360 cm^{-1})

ν(Be-H) = 5089 nm (1965 cm^{-1})

ν(B-H) = 4000 nm (2500 cm^{-1})   and   4762–6173 nm (2100–1620 cm^{-1})

ν(C-H) = 3384 nm (2955 cm^{-1})

ν(N-H) = 2941 nm (3400 cm^{-1})

ν(O-H) = 2770 nm (3610 cm^{-1})

ν(F-H) = 2532 nm (3950 cm^{-1})

In the third row the more important ν(X–H) include:

ν(Si-H) = 4577 nm (2185 cm^{-1})

ν(P-H) = 4255 nm (2350 cm^{-1})

ν(S-H) = 3876 nm (2580 cm^{-1})

The most important of these for the discussion of near-infrared spectroscopy include O-H and C-H.

Hydrogen Bonding (X–H · · · Y)

A hydrogen-bonded molecular system has four possible modes:

O-H stretch found at 2941–4000 nm (ν at 3400 to 2500 cm^{-1}), with overtones occuring at 1499.9–2200 nm (2ν), 1009.7–1533.3 (3ν), and 764.7–1200 nm (4ν).

R-O-H in-plane bend at 7143–7692 nm (v at 1400–1300 cm^{-1}), with overtones occuring at 3642.9–4230.6 nm ($2v$), 2452.4–2948.6 nm ($3v$), and 1857.2–2307.6 nm ($4v$).

The R-O-H out-of-plane bend at 15,152 nm (v at 660 cm^{-1}); no vibrations are found in the near-infrared region.

The –O-H $\cdots$ O stretch at 57,143 nm (v at 175 cm^{-1}); no vibrations are found within the near-infrared region.

Specific frequencies for molecules containing O-H with hydrogen bonding include:

Alcohols (R-O-H) with v(O-H) = 2981 nm (v at 3355 cm^{-1}), with overtones at 1520.3–1639.6 nm ($2v$), 1023.5–1142.7 nm ($3v$), and 775.1–894.3 nm ($4v$)

Carboxylic acids (–COOH) with v(O-H) = 3333 nm (v at 3000 cm^{-1}), with overtones at 1699.8–1833.2 nm ($2v$), 1144.3–1277.7 nm ($3v$), and 866.6–999.9 nm ($4v$)

Amines (R–NH–X) with v(N-H) = 3000 nm (v at 3300 cm^{-1}), with overtones occuring at 1530–1650 nm ($2v$), 1030–1150 nm ($3v$), and 780–900 nm ($4v$)

In general, hydrogen bonding has several main effects on the frequency of X–H bands: (1) It lowers the frequency of the X–H stretch as compared to the gas phase and dilute liquid phase of a molecule; (2) it raises the X–H bend frequency; (3) it broadens all O–H stretch bands by up to 300–475 nm (300–500 cm^{-1}); (4) it intensifies all bending and stretching X–H bands.

O-H Stretch

It is common knowledge among infrared spectroscopists that any fundamental band occurring above 2857 nm (v at 3500 cm^{-1}) can be confidently assigned to the O-H stretch group. A sharp band exists for the stretching frequency of O-H for the infrared spectrum of O-H containing the following systems:

Primary alcohols (–CH$_2$–OH) exhibit v_S(O-H) = 2740 nm (v at 3650 cm^{-1}), 1397.4–1507.0 nm ($2v$), 940.7–1050.3 nm ($3v$), and 712.4–822.0 nm ($4v$).

Secondary alcohols (R–CHOH–R′) exhibit v_S(O-H) = 2762 nm (v at 3620 cm^{-1}), 1408.6–1519.1 nm ($2v$), 948.3–1058.8 nm ($3v$), and 718.1–828.6 nm ($4v$).

Tertiary alcohols (R–CR′OH–R″) demonstrate v_S(O-H) = 2770 nm (v at 3610 cm^{-1}), 1412.7–1523.5 ($2v$), 951.0–1061.8 nm ($3v$), and 720.2–831.0 nm ($4v$).

Phenols have v_S(O-H) = 2762–2793 nm (v at 3620–3580 cm^{-1}), 1408.6–1519.1 nm ($2v$), 948.3–1070.7 nm ($3v$), and 718.1–837.9 nm ($4v$).

Carboxylic acids have v_S(O-H) = 2817–2857 nm (3550–3500 cm^{-1}), 1436.7–1571.4 nm ($2v$), 967.2–1095.2 nm ($3v$), and 732.4–857.1 nm ($4v$).

To distinguish between primary, secondary, and tertiary alcohols using infrared spectra requires the use of these bands as well as the v_S(C-O) = 8547–9615 nm (v at 1170–1040 cm^{-1}), 4903.7–5288.3 nm ($2v$), 3301.2–3685.8 nm ($3v$), and 2499.9–2884.5 nm ($4v$); this band is not observed in the near-infrared region. The O-H in-plane bend occurs as v_B(O-H) =

X–H FUNCTIONAL GROUPS (O-H AND N-H)

7143–7692 nm (v at 1400–1300 cm^{-1}), 3642.9–4230.6 nm ($2v$), 2452.4–2948.6 nm ($3v$), and 1857.2–2307.6 nm ($4v$). The C-O-H out-of-plane bend v_B (O-H) = 13,158–15385 nm (760–650 cm^{-1}) is not observed in the near-infrared.

Carboxylic acids (–COOH) form dimers (hydrogen bonding between molecules as (–C=O $\cdots$ H–O–C–) in condensed states that exhibit four distinct frequencies:

A very broad, very strong O-H stretch at approximately 3333 nm (v at 3000 cm^{-1}; at a lower frequency than alcohols), 1699.8–1833.2 ($2v$), 1144.3–1277.7 ($3v$), and 866.6–999.9 nm ($4v$).

Summation tones as one to three weak bands near 3846 nm (v at 2600 cm^{-1}), 1961.5–2115.3 nm ($2v$), 1320.5–1474.3 nm ($3v$), and 1000–1153.8 nm ($4v$).

Carbonyl (C=O) stretch vibration near 5797 nm (v at 1725 cm^{-1}), 2956.5–3188.4 nm ($2v$), 1990.3–222.2 nm ($3v$), 1507.2–1739.1 nm ($4v$).

A weak, broad band indicating a dimer condition near 10,471 nm (955 cm^{-1}) is not observed in the near-infrared.

N-H Group Frequencies and Nitrogen-Containing Compounds

The average frequencies for free N-H groups include the following:

Primary amides, –(C=O)–NH$_2$, v_S (N-H) = 2857 nm (v at 3500 cm^{-1}), 1457.1–1571.4 nm ($2v$), 980.9–1095.2 nm ($3v$), 742.8–857.1 nm ($4v$). For primary amides a band also occurs at 2933 nm (v at 3410 cm^{-1}), 1495.8–1613.2 nm ($2v$), 1007.0–1124.3 nm ($3v$), and 762.6–879.9 nm ($4v$).

Secondary amides, –(C=O)–NHC, v_S (N-H) = 2899 nm (v at 3450 cm^{-1}), 1478.5–1594.5 nm ($2v$), 995.3–111.3 nm ($3v$), 753.7–869.7 nm ($4v$).

Alkyl (aliphatic-NH-X) groups include:

Alkyl-NH$_2$, v_S (N-H) = 2946 nm (v at 3395 cm^{-1}), 1502.5–1620.3 nm ($2v$), 1011.5–1129.3 nm ($3v$), and 766–883.8 nm ($4v$); and 3003 nm (3330 cm^{-1}), 1531.5–1651.7 nm ($2v$), 1031.0–1151.2 nm ($3v$), and 780.8–900.9 nm ($4v$).

Alkyl-NH-X, v_S (N-H) = 2981 nm (v at 3355 cm^{-1}), 1520.3–1639.6 nm ($2v$), 1023.5–1142.7 nm ($3v$), 775.1–894.3 nm ($4v$).

For aryl (aromatic-NH-X), the N-H stretch fundamentals occur as:

Aryl-NH$_2$, v_S (N-H) = 2869 nm (v at 3485 cm^{-1}), 1463.2–1578.0 nm ($2v$), 985.0–1099.8 nm ($3v$), and 745.9–860.7 nm ($4v$); and a second band is found near 2946 nm (v at 3395 cm^{-1}), 1502.5–1620.3 nm ($2v$), 1011.5–1129.3 nm ($3v$), and 766–883.8 nm ($4v$).

Aryl-NH-X, v_S (N-H) = 2911 nm (v at 3435 cm^{-1}), 1484.6–1601.1 nm ($2v$), 999.4–1115.9 nm ($3v$), 756.9–873.3 nm ($4v$). *Note:* The N-H frequency is lower and less intense than the O-H frequency; the hydrogen bonding with N-H is weaker and of less effect than in O-H.

Primary amines (R–NH$_2$) demonstrate two N-H stretching bands due to coupling, the in-phase (or symmetric) and the out-of-phase (or asymmetric):

Aliphatic amines, (C_nH_{2n+1})–NHX, demonstrate medium-strength in-phase frequencies from 2999 to 3053 nm (v at 3335–3275 cm^{-1}), 1529.5–1613.2 nm ($2v$), 1029.7–1124.3 nm ($3v$), and 779.7–879.9 nm ($4v$). The out-of-phase medium-strength frequency occurs from 2933–2985 nm (v at 3410–3350 cm^{-1}), 1495.8–1641.8 nm ($2v$), 1007.0–1144.3 nm ($3v$), 762.2–895.5 nm ($4v$).

Aromatic amines exhibit two medium-strength frequencies: an in-phase frequency of 2920–3003 nm (v at 3425–3330 cm^{-1}), 1489.2–1651.7 nm ($2v$), 1002.5–1151.2 nm ($3v$), and 759.2–900.9 nm ($4v$); and an out-of-phase frequency at 2849–2950 nm (v at 3510–3390 cm^{-1}), 1453.0–1622.5 nm ($2v$), 978.2–1130.8 nm ($3v$), 740.7–885.0 nm ($4v$).

The scissoring for primary amines occurs as a medium-to-strong, broad band from 6042 to 6289 nm (v at 1655–1590 cm^{-1}), 3081.4–3459.0 nm ($2v$), 2074.4–2410.8 nm ($3v$), and 1570.9–1886.7 nm ($4v$). The NH$_2$ wagging is a strong, broad band near-12,500 nm (v at 800 cm^{-1}) and is not observed in the near-infrared region.

Secondary amines (R'R''–NH) exhibit a single, broad N-H stretch vibrational frequency near 3021 nm (v at 3310 cm^{-1}), 1540.7–1661.6 nm ($2v$), 1037.2–1158.1 nm ($3v$), and 785.5–906.3 nm ($4v$). An N-H bend occurs as a weak-to-medium-strength band near 6667 nm (v at 1500 cm^{-1}), 3400.2–3666.9 nm ($2v$), 2289.0–2555.7 nm ($3v$), 1733.4–2000.1 nm ($4v$).

To separate spectral features for the various amines, i.e., primary, secondary, and tertiary, it is helpful to point to obvious differences between the groups. For example, the number of N-H stretches for the three types are: primary (2), secondary (1), and tertiary (0). For N-H bending vibrations, primary amines have a band near 6042–6289 nm (v at 1655–1590 cm^{-1}) and secondary amines near 6667 nm (1500 cm^{-1}); this feature is absent in tertiary amines.

The C-H stretch vibration for CH$_2$ is to be found for all three types of amines from 3503 to 3683 nm (2855–2715 cm^{-1}), 1786.5–2025.7 nm ($2v$), 1202.7–1411.8 nm ($3v$), and 910.8–1104.9 nm ($4v$). For primary amines, the C-H stretch will be found at higher frequencies nearer to 3448 nm (v at 2900 cm^{-1}), 1758.5–1896.4 nm ($2v$), 1183.8–1321.7 nm ($3v$), and 896.5–1034.4 nm ($4v$).

Amine salts as R–NH$_3^+$ and R'R''–NH$_2^+$ exhibit several weak bands near 3571–4545 nm (v at 2800–2200 cm^{-1}), 1821.2–2499.8 nm ($2v$), 1226.0–1742.3 nm ($3v$), 928.5–1363.5 nm ($4v$). These salts exhibit a weak-to-medium, very broad band centered near 4000 nm (v at 2500 cm^{-1}), 2040–2200 nm ($2v$), 1373.3–1533.3 nm ($3v$), and 1040.0–1200 nm ($4v$). This broad band is lower in frequency than the C-H stretch and up to 314 nm (500 cm^{-1}) in width at the baseline [88, p. 147].

Amines (R–NH–X)

Amines are polar compounds, and both primary and secondary amines can form intermolecular hydrogen bonding. All classes of amines are water soluble (up to 6 carbon atoms) and will form hydrogen bonds in aqueous solutions. Near-infrared absorptions for amines occur, in decreasing wavelength order, as:

2020 nm	Primary N-H combination band
2000–2050 nm	N-H stretch and N-H bend combination band
1990–2020 nm	Free and bonded primary amines
1960 nm	Combination of N-H stretch and NH$_2$ bending, for example, meprobamate
1500–1550 nm	Primary amines first overtone aliphatic N-H stretch
1530 nm	N-H stretch overtone for aliphatic-NH system

X–H FUNCTIONAL GROUPS (O-H AND N-H)

1450 nm	Cyclic amines, e.g., pyrroles, indoles, and carbazoles, show intense first-overtone N-H stretch
1010–1020 nm	Aliphatic (alkyl-) N-H stretch second overtone
995–1015 nm	Aromatic (aryl-) N-H stretch second overtone

Whetsel et al. [91] provided near-infrared spectra for primary aromatic amines. The authors provide spectra for aniline and 39 of its ring-substituted derivatives, giving special attention to N-H bands located between 1000 and 2000 nm. The band positions are reported in Table 15.5.

Amides (–C=O–NX'X", also shown as –NHCO–)

2817–3279 nm	Fundamental N-H for acyl-group of amides (v at 3550–3050 cm^{-1}) (R–C=O–)
5917–6061 nm	The C=O stretch fundamental (v at 1690–1650 cm^{-1})
2000 nm (5000 cm^{-1})	Combination band
1500 nm (6667 cm^{-1})	First overtone

Primary amides (–C=O–NH$_2$)

1000–1050 nm

1450–1500 nm

2850–2950 nm

Other Nitrogen-Containing Compounds

Reference 7 delineates locations of multiple bands for nitrogen-containing compound (in nanometers):

Ring-substituted derivatives of aniline (Ar-NH$_2$), N-H stretch

1981 nm	Combination
1503 nm	1st-overtone symmetrical stretch
1460 nm	1st-overtone asymmetric stretch

Para substitution with –SCN

1966 nm	Combination
1487 nm	1st-overtone symmetric stretch
1141 nm	1st-overtone asymmetric stretch

Table 15.5 Band Positions and Intensities for Primary Aromatic Amines in references [91–97].

Band designation	Location	Intensity (as ε)
Combination bands	1970 nm	1.75
First-overtone symmetric stretch	1490 nm	1.35
First-overtone asymmetric stretch	1450 nm	0.20
Second-overtone symmetric stretch	1020 nm	0.04

Source: Ref. 91

Para substitution with –CN

1961 nm	Combination
1484 nm	1st-overtone symmetric stretch
1438 nm	1st-overtone asymmetric stretch

Para substitution with –NO$_2$

1959 nm	Combination
1484 nm	1st-overtone symmetric stretch
1436 nm	1st-overtone antisymmetric stretch

Meta substitution with –NO$_2$

1963 nm	Combination
1487 nm	1st-overtone symmetric stretch
1441 nm	1st-overtone antisymmetric stretch

Ortho substitution with –NO$_2$

1967 nm	Combination
1495 nm	1st-overtone symmetric stretch
1432 nm	1st-overtone antisymmetric stretch

NIR MEASUREMENTS OF H$_2$O, HDO, D$_2$O

In their classic paper, Buijs and Choppin [98] reported on the NIR absorption spectrum of water near 1150 nm. These workers proposed that free water is the result of three main molecular species with differing numbers of hydrogen bonds. The three species include zero-, one-, and two-bonded groups.

Vand and Senior [99] later expanded on the work of Buijs and Choppin by concluding that the three component bands do exist as S_0, S_1, and S_2 but with a distribution of hydrogen bond energies with a continuous distribution of hydrogen bond lengths and bond angles associated with each of the three species. The authors conclude that the degree of orderliness for the water structure is $S_2 > S_1 > S_0$.

Other workers, such as Fornes and Chaussidon [100], and Philip and Jolicoeur [101] confirmed the work of Vand and Senior and resolved the spectrum of pure water into its three component band distributions, assuming a Gaussian distribution of hydrogen energies for each hydrogen-bonded species in free water. The three component bands are centered at: S_0, 955–957 nm (10,460 ± 10 cm^{-1}); S_1, 977–978 nm (10,230 ± 10 cm^{-1}); and S_2, 1016–1018 nm (9830 ± 10 cm^{-1}). Baly et al. [102] described the spectroscopy in detail for the near-infrared and infrared absorption spectra of natural water (H$_2$O), heavy water (containing 99.6 weight % D$_2$O), and a deduced HDO spectrum (from H$_2$O-HDO-D$_2$O mixtures). Absorption band positions, vibration mode assignments, molecular extinction coefficients, and some half-band widths are identified for the liquid samples in an extensive table.

The near infrared spectra (909–1053 nm) are given for pure water and aqueous solutions of alkali halides, MgCl$_2$, NaClO$_4$, and R$_4$NBr. The measurements were made at temperatures between 10°C and 55°C and at pressures to 500 MPa [103]. The authors report that for O-H, v (max) shifts to a higher wavenumber (lower wavelength) with increasing temperature (given constant pressure). The authors also noted that v (max) shifts to a higher wavenumber (lower wavelength) with increasing pressure (given constant temperature). By varying temperature and pressure, the authors reported isobestic points for pure water at 980–1000 nm (10,000–10,200 cm^{-1}) for temperature variation (at fixed pressure) and 971–990 nm (10,100–10,300 cm^{-1}) for pressure variation (at fixed temperature).

The use of near-infrared spectroscopy for moisture analysis has been used extensively, with a variety of key wavelengths used in calibrations (Table 15.6).

Table 15.6 Bands Associated with "Moisture," O-H Hydrogen Bonding

"Moisture"-associated band	Band location
Combination of O-H stretch/HOH deformation: O-H bend 2nd overtone	1920–1940 nm
O-H stretch 1st overtone	1440–1470 nm
O-H stretch 2nd overtone	950–980 nm

MEASUREMENT OF ALCOHOLS USING NIR

Pathlengths used for O-H–based samples in the NIR:

800–1100 nm (7–10 cm)

1100–2500 nm (0.1–1.0 cm)

2500–4000 nm (0.5–1 mm)

Alcohol product–related band assignments:

C-H 3rd overtone-methyl stretch at 890–910 nm

C-H 3rd overtone-methylene stretch at 905–940 nm

HOH 2nd overtone combination (symmetric OH stretch ($2v_1$) + v_3 bend) at 960 nm

C-H stretch 2nd overtone at 1080–1130 nm

ROH 1st overtone at 1410 nm

HOH 1st overtone at 1440–1450 nm

Davenel et al. [104] reported measurements of alcoholic fermentation in wines using filter near-infrared (FIR) spectrometry and Fourier-transform near-infrared (FT-NIR) spectrometry. The alcohol and total sugar contents were monitored during the fermentation process for wines. The authors report SEP (standard error of prediction) for the FIR method as 2.5 g/L for total sugars and 0.15% (v/v) absolute for ethanol determination. Table 15.7 shows the reported wavelengths used in regression equations for each of the parameters mentioned.

Kuehl and Crocombe [105] described a method of quantitative analysis of a four-component mixture using FT-IR and ATR. The mixture containing methanol, ethanol, and acetone in an aqueous solution was studied. The authors reported using the analytical bands at 8051 nm (1242 cm^{-1}) for acetone, 9524 nm (1050 cm^{-1}) for methanol, and 9804 nm (1020

Table 15.7 Wavelengths Used for Regression Equations (Demonstrating Smallest SEC) for Determination of Ethanol and Total Sugars in Wines

Ethanol (volume %)	Total sugars (g/L)
2129, 2269, 2289 nm (FT-NIR)	1595, 1692, 2129, 2243, 2269, 2281, 2289, 2309, 2338 nm (FT-NIR)
2207, 2270 (FIR)	1445, 1680, 1722, 1735, 1818, 2139, 2230, 2270, 2345 nm (FIR)

Source: Ref. 104.

cm^{-1}) for ethanol. Average relative errors reported for the determinations were 1.57 % for acetone, 0.47 % for methanol, and 0.67 % for ethanol (for composition ranges from 0.1000 to 0.5000 volume % for each component).

Buchanan et al. [106] describe the use of a 500 µm multimode step-index fiber-optic coupled to a PbS detector for near-infrared (981–1731 nm) wavelength measurements; the spectrometer was tested for the determination of ethanol in wine. The authors reported prediction of ethanol to ±0.33 volume % in the range of 4.61–18.61 vol. %. The calibration sample set consisted of 66 total samples: 57 wine, 6 champagne, 2 port, and 1 beer. A nonlinear iterative partial least squares (NI-PALS) algorithm was used with full spectral information for calibration. The authors explain that the use of telecommunications-grade optical fiber limited the success of this experiment in terms of achievable signal-to noise ratio.

Cavinato et al. [107] describe the use of a 2-mm-inside-diameter fiber-optic bundle combined with a silicon photodiode array spectrophotometer for measurements of ethanol during the time course of fermentation. The SW-NIR instrument (700–1100 nm) was used to measure through the fiber-optic bundle probe placed outside the wall of a glass fermentation vessel. Data was pretreated using a second derivative, and a MLR (multiple linear regression) algorithm was used for calibration. Results obtained were reported as standard error of prediction (SEP). For ethanol in ethanol/water mixtures, an SEP of 0.2% (w/w) was achieved over the range of 0–15% (w/w) ethanol; for ethanol in ethanol/yeast/water solutions, an SEP of 0.27% (w/w) was obtained. The authors describe basic spectroscopy for the mixtures as shown in Table 15.8.

APPLICATION OF NIR TO NATURAL PRODUCTS

Natural product analysis includes the analysis of food and feedstuffs for nutritional quality relative to health and profitability. The major parameters of interest using NIR spectroscopy include: protein, fiber, moisture, fats and oils, starches and sugars, and other minor constituents.

Food and Agricultural Products

The application of NIR to food and agriculture occurred largely as a result of the work of Norris et al. in 1968 [17,20,21]. Norris recognized the potential for diffuse-reflectance measurements in the NIR region for the rapid and routine quantitative analysis of major constituents (such as oil and moisture) in grain and forage materials. Food and forage materials are found to exhibit identifiable C-H, N-H, and O-H absorption bands in the 1400–2500 nm region [49, 108]. The figures show typical NIR spectra for forage products. The bands associated with free water include the 960 nm, 1270 nm, 1410 nm, 1940 nm, and above-2200 nm bands. Functional groups associated with amino acids and protein are found near 2150 nm and 2180 nm.

Table 15.8 Major Analytical Bands, Locations, and Assignments for Ethanol in Ethanol/Water and Ethanol/Yeast/Water Solutions

Parameter	Analytical band location	Band assignment
Water	960 nm	Overtone combination ($2v_1 + v_3$), where v_1 is the symmetric O-H stretch and v_3 is the O-H bending mode
	829 and 730 nm	Overtone ($5v$)
Ethanol	905 nm	Third-overtone C-H stretch of methyl group
	935 nm	Third-overtone C-H stretch of methylene group

Source: Ref. 107.

Functional groups associated with lignin, cellulose, and hemicellulose relative to fiber determinations are known to occur at 1500 nm, 2200 nm, and 2348 nm.

Norris recognized [51,52] the more important absorption peaks in 1976. He then proposed that analytical NIR instruments be capable of measuring absorbances at several wavelengths for foodstuffs and grain analysis, to include: 1680 nm (reference), 1940 nm (free water), 2100 nm (O-H bend + C-O stretch combination band resulting from starch and simple carbohydrate associations), 2180 nm (protein functional group associations), 2310 nm (C-H bend 2nd overtone relative to oil methylene groups), and 2336 nm (starch and cellulose functionalities). For forage constituent work he proposed 1672 nm (reference), 1700 nm (C-H stretch associations with methyl groups from cellulose or lignin), 1940 nm (free water), 2100 nm (carbohydrate association), and 2336 nm (C-H stretch and C-H deformation combination associated with cellulose and starch). Occasionally 2230 nm was added as a reference wavelength. Norris' work generated an increased interest and use of near-infrared reflectance analysis as a rapid nondestructive technique for assessing the quality of grain and feedstuffs. Traditional NIR wavelengths associated with major natural product constituents include: 2270 nm (lignin), 2310 nm (oil), 2230 nm (reference region), 2336 nm (cellulose), 2180 nm (protein), 2100 nm (carbohydrate), 1940 nm (water), and 1680 nm (reference region).

Fats, Oils, and Waxes

See Table 15.9 for reference near-infrared spectral data for fats and oils.

Starches and Sugars

See Table 15.10 for reference near-infrared spectral data for starches and sugars.

Table 15.9 Bands Associated with "Fats and Oils" as C-H and O-H Related Bands

"Fats and oils"–associated band	Band location
Combination C-H stretch/C-C stretch	2380 nm
C-H bend 2nd overtone	2310 nm
Combination C-H stretch/C=O stretch and C-H deformation	2140 nm
Combination O-H stretch/HOH deformation; O-H bend 2nd overtone	2070 nm
O-H stretch 1st overtone	1415 nm
Combination C-H band	1010–1025 nm
Methylene C-H stretch 3rd overtone	911–935 nm
Methyl C-H stretch 3rd overtone	896–915 nm

Table 15.10 Bands Associated with "Starches and Sugars" as C-H and O-H Related Bands

"Starches and sugars"–associated bands	Band location
Combination C-H stretch/C-C stretch and C-O-C stretch	2500 nm
Combination C-H stretch/CH_2 deformation	2280–2330 nm
Combination O-H stretch/HOH deformation	2100 nm
O-H stretch 1st overtone	1450 nm
O-H stretch 2nd overtone	1010–1030 nm
C-H stretch 3rd overtone	850–900 nm

Cellulose and Cellulosic Compounds

An excellent reference article for background reading is supplied by Blackwell [109]. The article reviews the work in infrared and Raman spectroscopy of cellulosic materials up to 1977, including a detailed table giving calculated and observed band assignments for cellulose I. Table 15.11 shows observed fundamental positions for cellulose I functional groups with corresponding first- and second-overtone *calculated* positions between 1100 and 2500 nm. Bands thought to associate with fiber parameters, such as cellulose, and lignin are presented in Tables 15.12 and 15.13, respectively. Figures 15.8a and 15.8b demonstrate the occurrence of wavelength selections in fiber calibrations using NIR.

Wingfield [110] discusses the possibilities for NIR detection of cellulose in flour. He basically summarizes work up until the publication and cites early workers in the field. Marton and Sparks [111] have reported measurements of lignocellulose by using simple linear regression of lignin content vs. the absorbance ratio of 1510/1310 cm^{-1}; these frequencies correspond to a ratio of second overtones at 2208/2545 nm. Gould et al. [112] reported using a frequency shift of the 2900-cm^{-1} band (3448 nm), corresponding to a frequency shift at a first-

Table 15.11 Cellulose I–Associated Bands.

"Cellulose I"– associated band	Fundamental band location, absorption strength	1st overtone (calculated)	2nd overtone (calculated)
O-H stretch	(3025 nm, weak)	1513 nm	1008 nm
C-H stretch	(3372 nm, weak)	1686 nm	1124 nm
CH$_2$ antisymmetric stretch	(3399 nm, weak)	1700 nm	1133 nm
C-H stretch	(3435 nm, very weak)	1718 nm	1145 nm
CH$_2$ symmetric stretch	(3505 nm, weak)	1753 nm	1168 nm
C-O-H deformation	(6748 nm, weak)	(3374 nm)	2249 nm
CH$_2$ deformation	(7013 nm, medium)	(3506 nm)	2338 nm
C-O-H, C-C-H deformation	(7117 nm, weak)	(3559 nm)	2372 nm
C-C-H deformation	(7369 nm, medium)	(3685 nm)	2456 nm
CH$_2$ and C-O-H deformation	(7496 nm, medium)	(3748 nm)	2499 nm

Bands not visible in the NIR region are shown in parentheses.

Table 15.12 Bands Associated with "Fiber" as Cellulose Listed in Table 15.10

"Fiber"-associated band	Band location
Combination of C-H stretch/C-C stretch	2488 nm
CH$_2$ deformation 2nd overtone	2352 nm
Combination of CH$_2$ symmetric stretch/=CH$_2$ deformation	2347 nm
Combination of C-H stretch/C-H deformation	2335 nm
Combination of O-H stretch/C-O stretch 2nd overtone	1820 nm
C-H stretch 1st overtone	1780 nm
O-H stretch 1st overtone	1490 nm
O-H stretch 2nd overtone	985–995 nm

overtone band near 1724 nm. Schultz et al. [113] reported prediction of percent glucose, xylose, and lignin using DRIFTS (diffuse-reflectance infrared Fourier-transform spectroscopy). The authors report r-squared values of 0.949 for lignin, 0.921 for glucose, and 0.973 for xylose. Equations were generated using ratios of mid-infrared wavelengths and stepwise elimination regression. Mitchell and coworkers [114] determined acetyl content of cellulose acetate using NIR from 35 to 44.8% acetyl. The spectra were measured in 5 cm cells against a reference solution of pyrrole containing 5% carbon tetrachloride. The maximum absorbance at 1445 nm was plotted against percent acetyl to obtain a calibration line. A standard deviation of 0.22% was indicated by the NIR method.

Protein

Protein is measured in the NIR region as its associated functional groups. The locations of these absorptions are at 973–1020 nm (symmetrical N-H stretch 2nd overtone), 1500–1530 nm (symmetrical N-H stretch 1st overtone), 2050–2060 nm (N-H stretching vibrations), 2060 nm (the carbonyl stretch of the primary amide, Maillard effect–bound protein is also observed here), and the 2168–2180-nm region (N-H bend second overtone + C=O stretch/N-H in-plane bending/C-N stretch combination bands). Hermans and Scheraga [115] made measurements of the backbone peptide hydrogen-bond system (NH group) using NIR (700–3500 nm) measurements. The authors describe the use of NIR to distinguish between hydrogen-bonded and non-hydrogen-bonded N-H and O-H groups. Molecules examined include methanol, aniline, and poly-γ-benzyl-L-glutamate.

Table 15.13 Bands Associated with "Fiber" as Lignin Listed in Table 15.10

"Fiber"-associated band	Band location
C-H stretch 1st overtone	1685 nm
O-H stretch 1st overtone	1410–1420 nm
Combination of C-H stretch modes	1417–1440 nm
C-H stretch 2nd overtone	1170 nm
C-H stretch 3rd overtone	875–900 nm

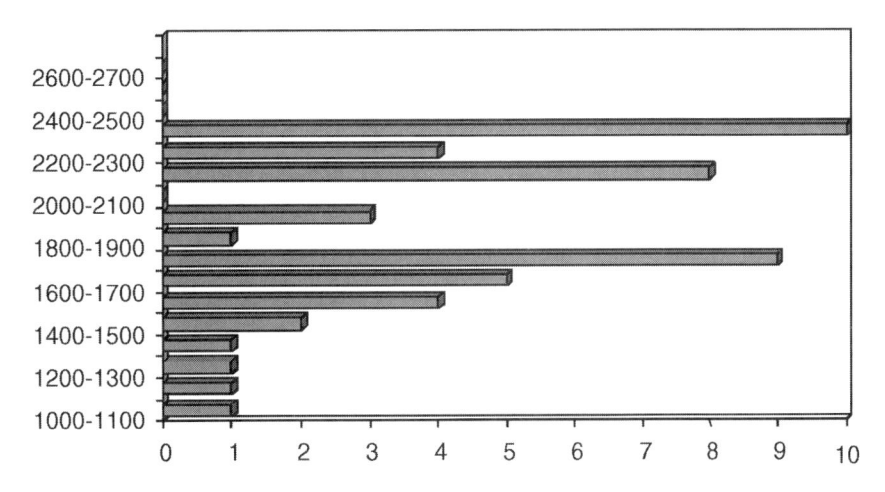

Fig. 15.8a. Occurrence for specific NIR wavelength selections in plant material (forage) multivariate "fiber" calibrations.

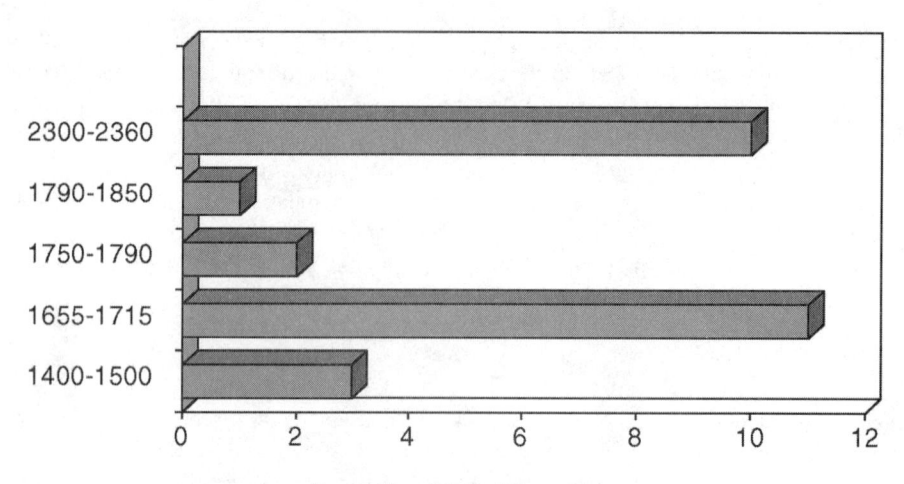

Fig. 15.8b. Detail of Figure 15.9a.

Elliott and Ambrose [116] identify absorption bands for polypeptides and proteins at approximately 3505 cm^{-1} (2853 nm) as the N-H stretching band, and 4825 cm^{-1} (2073 nm) in the overtone region. The main result of their work was to demonstrate that a band at 4840 cm^{-1} (2066 nm) is useful for distinguishing the presence of extended versus folded configurations for polypeptides and proteins, even in the presence of liquid-phase water. The band near 2073 nm was classified earlier by Glatt and Ellis [117] on work in nylon as a combination band of the N-H deformation and stretching modes. Krikorian and Mahpour [118] have described overtone and combination bands in the near-infrared region for primary and secondary amides. Their work includes several tables, with band assignments found in a summarized form in Table 15.14.

K. T. Hecht and D. L Wood [119] describe in a detailed study the band assignments for peptides, as shown in Table 15.15. The authors identify the bands for a porcupine quill spectrum as in the 2500-nm region, CH stretching and deformation; 2170-nm region, carbonyl 2× (first overtone) stretching plus peptide group; 2070 NH stretching plus peptide group mode; 1982 nm as OH stretching and deformation; 1724 nm as 2× (first overtone) CH stretching; 1524 nm as 2× NH stretching; and 1495 nm as H$_2$O 2× (first overtone) stretching plus deformation. Bands thought to associate with protein as Kjeldahl nitrogen are presented in Table 15.16. Figures 15.9a and 15.9b demonstrate the occurrence of wavelength selections in protein calibrations using NIR.

Fraser and MacRae [120] have reported important absorption bands for natural product proteins such as nylon 2053 nm (4870 cm^{-1}) resulting from a combination of the peptide absorptions at 3026 nm (3305 cm^{-1}) and 6494 nm (1540 cm^{-1}). For Feather Rachis, the authors report absorp-

Table 15.14 N-H–Related Band Assignments for Primary Amides

Reported band assignments	Fundamental	First overtone	Anharmonicity
Symmetric N-H stretching vibration	2924 nm, 3420 cm^{-1}	1484 nm, 6737 cm^{-1}	103 cm^{-1}
Asymmetric N-H stretching vibration	2833 nm, 3530 cm^{-1}	1430 nm, 6995 cm^{-1}	65 cm^{-1}
Combination of two N-H stretching vibrations (asymmetric and symmetric)	2878 nm, 3475 cm^{-1}	1470 nm, 6805 cm^{-1}	145 cm^{-1}

tion bands at 2012 nm (4970 cm^{-1}) and 1984 nm (5040 cm^{-1}). The shoulder at 1984 nm (5040 cm^{-1}) was reported as resulting from side-chain amide groups. For beta-keratin, the paper reports important absorption bands at 2174 nm (4600 cm^{-1}), 2062 nm (4850 cm^{-1}), and 2012 nm (4970 cm^{-1}). The authors are interested in studying the amorphous components of naturally occurring protein structures. Protein associated bands are given in Table 15.16.

Table 15.15 Band Assignments for Peptide Groups as Delineated by K. T. Hecht and D. L. Wood

Reported band assignments	Fundamental	First overtone (calculated, est. 100 cm^{-1} anharmonicity)	Second overtone (calculated, est. 100 cm^{-1} anharmonicity)
NH stretching (hydrogen bonded)	3060 nm, 3280 cm^{-1}	1548 nm	1032 nm
CH$_2$ antisymmetric stretching	3420 nm, 2925 cm^{-1}	1739 nm	1159 nm
CH$_2$ symmetric stretching	3510 nm, 2853 cm^{-1}	1784 nm	1189 nm
C=O stretching (H-bonded)	6150 nm, 1650 cm^{-1}	3125 nm	2083 nm
–NCHO– group vibration	6450 nm, 1550 cm^{-1}	3333 nm	2222 nm

Calculated positions for first- and second-overtone positions are by the author.

Table 15.16 Bands Associated with "Proteins"

"Protein"-associated band	Band location
Combination bands consisting of: N-H bend 2nd overtone; C-H stretch/C=O stretch combination; C=O stretch/N-H in-plane bend/C-N stretch combination bands	2148–2200 nm
C=O carbonyl stretch 2nd overtone of primary amide ('amide I' band)	2030–2080 nm
Aromatic C-H stretch 1st overtone	1620–1700 nm
N-H stretch 1st overtone	1480–1550 nm
N-H stretch 2nd overtone	975–1015 nm

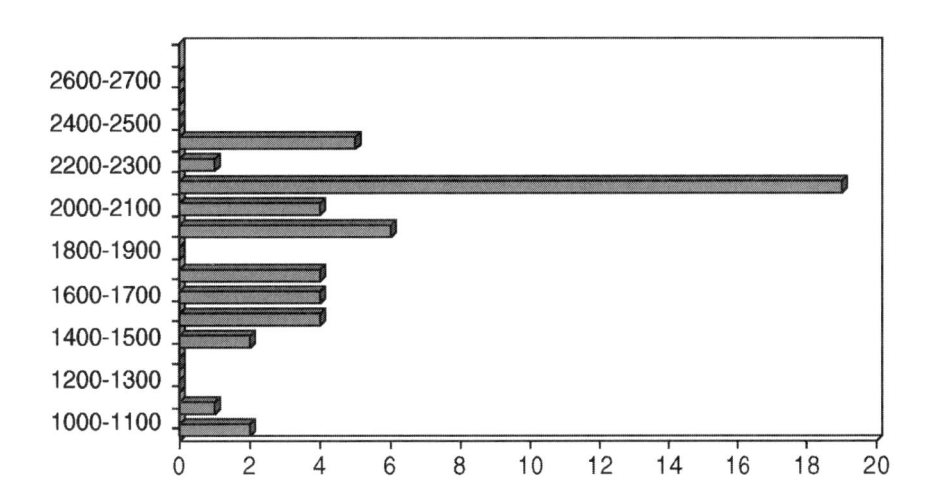

Fig. 15.9a. Occurrence for specific NIR wavelength selections in plant material (forage) multivariate "protein" calibrations.

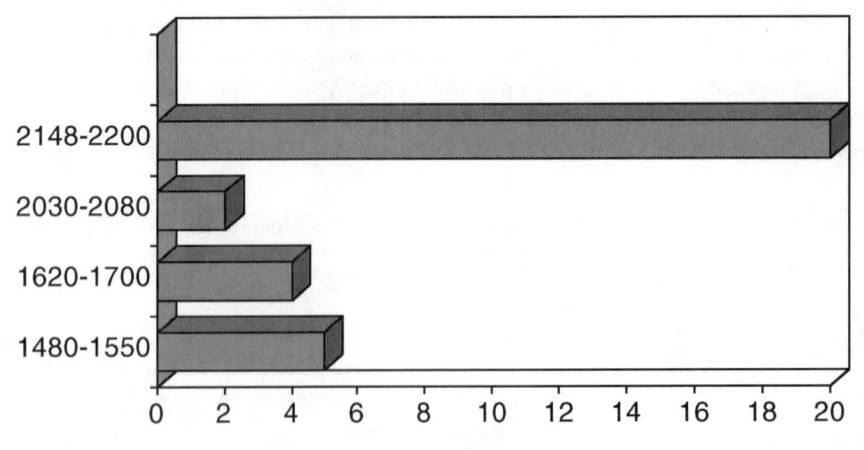

Fig. 15.9b. Detail of Figure 15.9a.

ACKNOWLEDGMENT

The author would like to thank Dr. John Coates for his editorial assistance with the manuscript and help in the production of the illustrations for this review.

REFERENCES

1. O. H. Wheeler. Near-Infrared Spectra. A neglected field of spectral study. *J. Chem. Education* 37:234–236, 1960.
2. F. I. G. Rawlins, A. M. Taylor, *Infrared Analysis and Molecular Structure.* Cambridge, U.K.: Cambridge University Press, 1929.
3. R. F. Goddu, D. A. Delker. Spectra–structure correlations for the near-infrared region. *Anal. Chem.* 32:140–141, 1960.
4. J. W. Ellis. Molecular absorption spectra of liquids below 3μ. *Trans. Faraday Soc.* 25:888–898, 1928 [sic].
5. W. Kaye. Near-infrared spectroscopy: A review. I. Spectral identification and analytical applications. *Spectrochimica Acta* 6:257–287, 1954.
6. R. F. Goddu. Near-Infrared Spectrophotometry. *Advan. Anal. Chem. Instr.* 1:347–424, 1960.
7. K. B. Whetsel. Near-Infrared Spectrophotometry. *Appl. Spectrosc. Rev.* 2(1):1–67, 1968.
8. E. Stark, K. Luchter, M. Margoshes. Near-infrared analysis (NIRA): A technology for quantitative and qualitative analysis. *Appl. Spectrosc. Rev.* 22(4):335–399, 1986.
9. G. D. Schrieve, G. G. Melish, A. H. Ullman. The Herschel-infrared—A useful part of the spectrum. *Appl. Spectrosc.* 45:711–714, 1991.
10. G. S. Birth. A nondestructive technique for detecting internal discolorations in potatoes. *Am. Potato J.* 37(2):53–60, 1960.
11. G. S. Birth. Measuring the smut content of wheat. *Trans. ASAE* 3(2):19–21, 1960.
12. W. L. Butler, K. H. Norris. The spectrophotometry of dense light-scattering material. *Arch. Biochem. Biophys.* 87,(1): 31–40, 1960.
13. K. H. Norris, W. L. Butler. Techniques for obtaining absorption spectra on intact biological samples. *IRE Trans. Bio-Med. Electron.* 8(3): 153–157, 1961.
14. K. H. Norris, J. D. Rowan. Automatic detection of blood in eggs. *Ag. Eng.* 43(3):154–159, March 1962.
15. J. R. Hart, K. H. Norris, C. Golumbic. Determination of the moisture content of seeds by near-infrared spectrophotometry of their methanol extracts. *Cereal Chem.* 39(2):94–99, March 1962.
16. G. Birth, K. Olsen. Nondestructive detection of water core in delicious apples. *Am. Soc. Horticul. Soc.* 85:74–84, 1964.
17. K. H. Norris, J. R. Hart. Direct spectrophotometric determination of moisture content of grain and seeds. *Proc. of the 1963 International Symposium on Humidity and Moisture.* Vol. 4. New York: Reinhold, pp. 19–25, 1965

REFERENCES

18. G. S. Birth, K. H. Norris. The difference meter for measuring interior quality of foods and pigments in biological tissues. *United States Department of Agriculture, Agricultural Research Service,* Tech. Bulletin No. 1341, Sept. 1965.
19. D. R. Massie, K. H. Norris. Spectral reflectance and transmittance properties of grain in the visible and near-infrared. *Trans. ASAE* 8:588–600, 1965.
20. I. Ben-Gera, K. H. Norris. Direct spectrophotometric determination of fat and moisture in meat products. *J. Food Sc.* 33(1):64–67, 1968.
21. I. Ben-Gera, K. H. Norris. Determination of moisture content in soybeans by direct spectrophotometry. *Israel J. Ag. Res.* 18(3):125–132, July 1968.
22. I. Ben-Gera, K. H. Norris. Influence of fat concentration on the absorption spectrum of milk in the near-infrared region. *Israel J. Ag. Res.* 18(3):117–124, July 1968.
23. W. Gordy, P. C. Martin. *J. Chem. Phys.* 7:99, 1939.
24. W. Gordy, S. C. Stanford. *J. Chem Phys.* 9:204, 1941.
25. G. B. B. M. Sutherland. *Discuss. Faraday Soc.* 9:274, 1950.
26. W. Kaye, C. Canon, R. G. Devaney. *J. Opt. Soc. Am.* 41:658, 1951.
27. W. Kaye, R. G. Devaney. *J. Opt. Soc. Am.* 42:567, 1952.
28. W. Kaye. *Spectrochim. Acta* 6:257–287, 1954.
29. W. Kaye. *Spectrochim. Acta* 1:181–204, 1955.
30. L. Weyer. Near-infrared spectroscopy of organic substances. *Appl. Spectrosc. Rev.* 21:1, 1985.
31. C. A. Watson. Near-infrared reflectance spectrophotometric analysis of agricultural products. *Analyt. Chem.* 49(9):835A–840A, 1977.
32. D. L. Wetzel. Near-infrared reflectance analysis, sleeper among spectroscopic techniques. *Anal. Chem.* 55(12):1165A–1176A, 1983.
33. W. Fred McClure. Near-infrared spectroscopy, the giant is running strong. *Anal. Chem.* 66(1): 43A–53A, 1994.
34. J. Workman, Jr. A review of process near-infrared spectroscopy: 1980–1994. *J. Near-Infra. Spectrosc.* 1:221–245, 1993.
35. P. Williams, K. Norris, eds. *Near-Infrared Technology in the Agricultural and Food Industries.* St. Paul, MN: American Association of Cereal Chemists, 1987.
36. T. Hymowitz, J. W. Dudley, F. I. Collins, C. M. Brown. Estimations of protein and oil concentration in corn, soybean, and oat seed by near-infrared light reflectance. *Crop Sci.* 14:713–715, Sept.–Oct. 1974.
37. R. A. Stermer, Y. Pomeranz, R. J. McGinty. Infrared reflectance spectroscopy for estimation of moisture of whole grain. *Cereal Chem.,* 54(2):345–351, 1977.
38. W. H. Hunt, B. Elder, K. Norris. Collaborative study on infrared reflectance devices for determination of protein in hard red winter wheat, and for protein and oil in soybeans. *Cereal Foods World.* 22(10):534–536,1977.
39. D. L. Wetzel. Near infrared (diffuse) reflectance analysis: State of the art. *Cereal Foods World* 26, 306 (1981); D. L. Wetzel. Near-infrared (diffuse) reflectance analysis: Grating and interferometric instruments. *Cereal Foods World* 27:415, 1982.
40. W. H. Hunt, D. W. Fulk, T. Thomas. Effect of type of grinder on protein values of hard red winter wheat when analyzed by infrared reflectance devices. *Cereal Foods World.* 23(3):143–144, 1978.
41. B. Osborne, T. Fearn. *Near-Infrared Spectroscopy in Food Analysis.* New York: Wiley, 1986.
42. B. Osborne. A near-infrared research composition analyzer. *ESN-European Spectrosc. News* 35, 1981.
43. B. G. Osborne, S. Douglas, T. Fearn. Assessment of wheat texture by near-infrared reflectance measurements in Buhler-milled flour. *J. Sci. Food Agric.* 32:200–202, 1981.
44. P. C. Williams, B. N. Thompson, D. Wetzel, G. W. McLay, D. Loewen. Near-infrared instruments in flour mill quality control. *Cereal Foods World* 26(5):234–237, 1981.
45. R. Tkachuk. Protein analysis of whole wheat kernels by near-infrared reflectance. *Cereal Foods World* 26(10):584–587, 1981.
46. R. Tkachuk. Oil and protein analysis of whole rapeseed kernels by near-infrared reflectance spectroscopy. *JAOCS,* 819–822, August, 1981.
47. P. C. Williams, K. H. Norris, C. W. Gehrke, K. Bernstein. Comparison of near-infrared methods for measuring protein and moisture in wheat. *Cereal Foods World* 28(2):149–152, 1983.
48. G. C. Marten, J. S. Shenk, F. E. Barton II. Near-infrared reflectance spectroscopy (NIRS): Analysis of forage quality. *United States Department of Agriculture—Agricultural Research Service, Agriculture Handbook* No. 643, August 1985.
49. J. S. Shenk, J. J. Workman, Jr., M. O. Westerhaus. Application of NIR spectroscopy to agricultural products. In D. Burns, E. Ciurczak, eds., *Near-Infrared Analysis.* New York: Marcel Dekker, 1992.
50. T. Hymowitz, J. W. Dudley, F. I. Collins, C. M. Brown. Estimation of protein and oil concentration in corn, soybean, and oat seed by near-infrared light reflectance. *Crop Sci.* 14:713–715, 1974.
51. K. H. Norris, R. F. Barnes. Infrared reflectance analysis of nutritive value of feedstuffs. In *Proc. First Int. Symp. Feed Comp.,* Utah Agr. Exp. Sta., Utah State Univ., Logan, pp. 237–241, 1976.
52. K. H. Norris, R. F. Barnes, J. E. Moore, J. S. Shenk. Predicting forage quality by infrared reflectance spectroscopy. *J. Anim. Sci.* 43(4):889–897, 1976.

53. J. S. Shenk, M. R. Hoover. Infrared reflectance spectro—computer design and application. *Proc. 7th Technicon Int. Cong.,* 2, Tarrytown, NY, 122, 1976.
54. J. S. Shenk, K. H. Norris, R. F. Barnes, G. W. Fissel. Forage and feedstuff analysis with infrared reflectance spectro/computer system. *13th International Grassland Congress,* Leipzig, Germany, pp. 1439–1442, May 18–27, 1977.
55. J. S. Shenk, M. O. Westerhaus, M. R. Hoover. Analysis of forages by infrared reflectance. *J. Dairy Sci.* 62:807–812, 1979.
56. D. Burdick, F. E. Barton II, B. D. Nelson. Rapid determination of forage quality with a near infrared filter spectrometer. *Proc. 36th Southern Pasture and Forage Crop Improvement Conf.,* Beltsville, MD, pp. 81–86, May 13, 1979.
57. F. E. Barton II, D. Burdick. Preliminary study on the analysis of forages with a filter-type near-infrared reflectance spectrometer. *J. Agr. Food Chem.* 27:1248, 1979.
58. F. E. Barton II, D. Burdick. Prediction of forage quality with NIR reflectance spectroscopy. *Proc. International Grassland Congress,* Vol. 16, pp. 532–534, June 15–24, 1981.
59. S. W. Coleman. Fast grass analysis. *Oklahoma State University Agricultural Experiment Station Research Report,* pp. 77–80, May 1982.
60. F. E. Barton II, S. W. Coleman. Potential for near-infrared reflectance spectroscopy for measuring forage quality. *Oklahoma State University, Agricultural Experimental Station,* pp. 73–76, April 1981.
61. J. E. Winch, H. Major. Predicting nitrogen and digestibility of forages using near-infrared reflectance photometry. *Can. J. Plant Sci.* 61:45–51, Jan. 1981.
62. F. W. Rose. Quantitative analysis with respect to the component structural groups, of the infrared (1 to 2 microns) molal absorptive indices of 55 hydrocarbons. *J. Natl. Bur. Standards* 20:129, 1938.
63. D. L. Fry, R. E. Nusbaum, H. M. Randall. The analysis of multicomponent mixtures of hydrocarbons in the liquid phase by means of infrared absorption spectroscopy. *J. Appl. Phys.* 17:150, 1946.
64. S. C. Fulton, J. J. Heigl. Spectroscopy in the petroleum industry. *Instruments.* 20:35, 1947.
65. E. K. Plyler, R. Stair, C. J. Humphreys. Infrared absorption spectra of seven cyclopentanes and five cyclohexanes. *J. Res. Natl. Bur. Standards* 38:211, 1947.
66. J. W. Kent, J. Y. Beach. Infrared spectrometric quantitative analysis of multicomponent liquid hydrocarbon mixtures. *Anal. Chem.* 19:290, 1947.
67. J. J. Heigl, M. F. Bell, J. U. White. Application of infrared spectroscopy to the analysis of liquid hydrocarbons. *Anal. Chem.* 19:293, 1947.
68. R. R. Hibbard, A. P. Cleaves. Carbon-hydrogen groups in hydrocarbons. *Anal. Chem.* 21:486, 1949.
69. M. F. Bell. Analysis of East Texas virgin naphtha fractions boiling up to 270 degrees F. *Anal. Chem.* 22:1005, 1950.
70. A. Evans, R. R. Hibbard, A. S. Powell. Determination of carbon-hydrogen groups in high molecular weight hydrocarbons. *Anal. Chem.* 23:1604, 1951.
71. O. W. Wheeler. Near-infrared spectra of organic compounds. *Chem. Rev.* 59:629, 1959.
72. A. S. Ahmadi. Infrared spectroscopic analysis of aromatic hydrocarbons in petroleum distillates. *Iranian Petroleum Institute* 30:164, 1968.
73. J. S. Mattson. 'Fingerprinting' of oil by infrared spectrometry. *Anal. Chem.* 43:1872, 1971.
74. U. Bernhard, P. H. Berthold. Application of NIR spectrometry for the structural group analysis of hydrocarbon mixtures. *Jena Rev.* 248, 1975.
75. C. W. Brown, P. F. Lynch, and M. Ahmadjian. Applications of infrared spectroscopy in petroleum analysis and oil spill identification. *Appl. Spectrosc. Rev.* 9:223, 1975.
76. T. Hirschfeld, A. Z. Hed, eds. *The Atlas of Near-Infrared Spectra.* Philadelphia: Sadtler Research Laboratories, 1981.
77. D. M. Mayes, J. B. Callis. A photodiode-array—based near-infrared spectrophotometer for the 600–1100 nm wavelength region. *Appl. Spectrosc.* 43:27, 1989.
78. J. Kelly, C. H. Barlow, T. M. Jinguji, J. B. Callis. Prediction of gasoline octane numbers from near-infrared spectral features in the range 660–1215 nm. *Anal. Chem.* 61:313, 1989.
79. J. J. Kelly, J. B. Callis. Fast, non-destructive analytical procedure for the simultaneous estimation of major classes of hydrocarbon constituents of finished gasolines. *Publication Announcement #67, Center for Process Analytical Chemistry,* Univ. of Wash., Seattle, WA, 44 pp. (personal communication), Feb. 9, 1989.
80. S. J. Foulk, V. J. Catalano. Determination of octane number using remote sensing NIR spectroscopy. *Am. Lab.* Nov. 1989.
81. J. J. Kelly, J. B. Callis. Nondestructive analytical procedure for simultaneous estimation of the major classes of hydrocarbon constituents of finished gasolines. *Anal. Chem.* 62:1444, 1990.
82. S. J. Swarin, C. A. Drumm. Prediction of gasoline properties with near-infrared spectroscopy and chemometrics. *International Fuels and Lubricants Meeting and Exposition,* Toronto, CA, pp. 1–10, Oct. 7–10, 1991.
83. S. J. Swarin, C. A. Drumm. Predicting gasoline properties using near-IR spectroscopy. *Spectroscopy* 7:42, 1992.

REFERENCES

84. R. DiFoggio, M. Sadhukhan, M. L. Ranc. Near-infrared offers benefits and challenges in gasoline analysis. *Oil Gas J.* 87, May 1993.
85. J. Workman, J. Coates. Interpretive spectroscopy for SW-NIR. *The Pittsburgh Conference,* No. 126, 1995.
86. R. F. Goddu, D. A. Delker. Spectra–structure correlations for near-infrared. *Anal. Chem.* 32:140–141, 1960.
87. W. J. Hershel. *Nature* 23, 76, Nov. 25, 1880.
88. D. W. Mayo, ed. *Infrared Spectroscopy: I. Instrumentation, and II. Instrumentation, Raman Spectra, Polymer Spectra, Sample Handling, and Computer-Assisted Spectroscopy* (Vol. 1). Bowdoin College, ME, 1994.
89. Von E. Fermi. Über den Ramaneffekt des Kohlendioxyds. *Z. Physik* 71:250–259, 1931
90. N. B. Colthup, L. H. Daly, S. E. Wiberly. *Introduction to Infrared and Raman Spectroscopy.* San Diego, CA: Academic Press, pp. 31–33, 1990.
91. K. B. Whetsel, W. E. Roberson, M. W. Krell. Near-infrared spectra of primary aromatic amines. *Anal. Chem.* 30(10):1598–1604, 1958.
92. T. Hirschfeld, A. Z. Hed. *The Atlas of Near-Infrared Spectra.* In Books No. 95, 483. Philadelphia: Sadtler Research Laboratories, 1981.
93. R. F. Goddu, ed. *Near-Infrared Spectra* In Volume 1, *Advances in Analytical Chemistry and Instrumentation.* pp. 364–366, 1960.
94. R. R. Hibbard, A. P. Cleaves. *Anal. Chem.* 21(4):486–492, 1949.
95. A. Evans, R. R. Hibbard. *Anal. Chem.* 23(11):1604–1610, 1951.
96. F. Rose, Jr. *J. Res. Natl. Bur. Standards* 20:129–157, 1938.
97. O. H. Wheeler. *Chem. Rev.* 59:129–666, 1959.
98. K. Buijs, G. R. Choppin. *J. Chem. Phys.* 39:2035, 1963.
99. V. Vand, W. A. Senior. *J. Chem. Phys.* 43:1878, 1965.
100. V. Fornes, J. Chaussidon. *J. Chem. Phys.* 68:4667, 1978.
101. P. R. Philip, C. Jolicoeur. *J. Chem. Phys.* 77:3071, 1973.
102. J. G. Bayly, V. B. Kartha, W. H. Stevens. The absorption spectra of liquid phase H_2O, HDO, and D_2O from 0.7 to 10 μm. In *Infrared Physics.* Vol. 3, pp. 211–223. London: Pergamon Press, 1963.
103. A. Inoue, K. Kojima, Y. Taniguchi, K. Suzuki. Near-infrared spectra of water and aqueous electrolyte solutions at high pressures. *J. Solution Chemistry,* 13(11):811–823, 1984.
104. A. Devenel, P. Grenier, B. Foch, J. C. Bouvier, P. Verlaque, J. Pourcin. Filter, Fourier transform infrared, and areometry, for following alcoholic fermentation in wines. *J. Food Sc.* 56(6):1635–1638, 1991.
105. D. Kuehl, R. Crocombe. The quantitative analysis of a model fermentation broth. *Appl. Spectros.,* 38(6):907–909, 1984.
106. B. R. Buchanan, D. E. Honigs, C. J. Lee, W. Roth. Detection of ethanol in wines using optical-fiber measurements and near-infrared analysis. *Appl. Spectrosc.* 42(6):1106–1111, 1988.
107. A. G. Cavinato, D. M. Mayes, Z. Ge, J. B. Callis. Noninvasive method for monitoring ethanol in fermentation processes using fiber-optic near-infrared spectroscopy. *Anal. Chem.* 62:1977–1982, 1990.
108. J. J. Workman, Jr. Doctoral dissertation, Columbia Pacific University, San Rafael, CA, 1984.
109. J. Blackwell. Infrared and Raman spectroscopy of cellulose. In J. C. Arthur, Jr., ed. *Cellulose Chemistry and Technology.* ACS Symposium Series No. 48, American Chemical Society, Chapter 14, p. 206, 1977.
110. J. Wingfield. NIR detection of cellulose as a milling control parameter. *Assoc. Operative Millers—Bulletin,* pp. 3769–3770, Nov. 1979.
111. J. Marton, H. E. Sparks. *TAPPI* 50:363–368, 1967.
112. J. M. Gould, R. V. Greene, S. H. Gordon. Book of abstracts. *188th National Meeting of the American Chemical Society,* Philadelphia, Aug. 1984.
113. T. P. Shultz, M. C. Templeton, G. D. McGinnis. Rapid determination of lignocellulose by diffuse reflectance Fourier transform infrared spectrometry. *Anal. Chem.* 57:2867–2869, 1985.
114. J. A. Mitchell, C. D. Bockman, Jr., A. V. Lee. Determination of acetyl content of cellulose acetate by near-infrared spectroscopy. *Anal. Chem.* 29:499–502, 1957.
115. J. Hermans, H. A. Scheraga. Structural studies of ribonuclease. IV. The near-infrared absorption of the hydrogen-bonded peptide NH group. *138th Meeting of the American Chemical Society,* New York, 1960.
116. A. Elliott, J. Ambrose. Evidence of chain folding in polypeptides and proteins. *Disc. Faraday Soc.* 9:246–251, 1950.
117. Glatt, Ellis. *J. Chem. Physics.* 16:551, 1948.
118. S. E. Krikorian, M. Mahpour. The identification and origin of N-H overtone and combination bands in the near-infrared spectra of simple primary and secondary amides. *Spectrochim. Acta* 29A:1233–1246, 1973.
119. K. T. Hecht, D. L. Wood. The near infra-red spectrum of the peptide group. *Disc. Faraday Soc.* 14:174, 1955.

120. R. B. D. Fraser, T. P. MacRae. Hydrogen-deuterium reaction in fibrous proteins. I. *J. Chem. Physics* 29(5):1024–1028, 1958.

ADDITIONAL REFERENCES

Infrared Spectroscopy—General

Griffiths, P. R., de Haseth, J. A. (1986). *Fourier Transform Infrared Spectrometry.* New York: Wiley, 656 pp.
Smith, A. L. (1979). *Applied Infrared Spectroscopy.* New York: Wiley, 322 pp.
Colthup, N. B., Daly, L. H., Wiberley, S. E. (1990). *Introduction to Infrared and Raman Spectroscopy,* 3d ed. Boston: Academic Press, 547 pp.
Ferraro, J. R., Krishnan, K. (1990). *Practical Fourier Transform Infrared Spectroscopy.* Boston: Academic Press, 534 pp.

Short-Wave NIR

Bonanno, A. S., Griffiths, P. R. (1993). Short-wave near infrared spectra of organic liquids. *J. Near-Infrared Spectrosc.* 1:13–23.

Long-Wave NIR

Ellis, J. W. (1928). Molecular absorption spectra of liquids below 3 microns. *Trans. Faraday Soc.* 25:888–898.
Goddu, R. F. (1960). Near-infrared spectrophotometry. *Advan. Anal. Chem. Instr.* 1:347–424.
Goldstein, M. (1979). Near-infrared diffuse reflectance analysis—Are we sure we know what we are measuring? *Proc. Soc. Photo-Opt. Instr. Eng.* (SPIE) 197:256–264.
Kaye, W. (1954). Near-infrared spectroscopy. *Spectrochimica Acta,* 6:257–287.
Wheeler, O. W. (1960). Near-infrared spectra. *J. Chem. Educ.* 37:234–236,.
Whetsel, K. B. (1968). Near-infrared spectrophotometry. *Appl. Spectrosc. Rev.* 2:1–67.

General Vibrational Frequency Locations

de W. Abney, W., Festing, Lt. Col. (1882). On the influence of the atomic grouping in the molecules of organic bodies on their absorption in the infra-red region of the spectrum. *Phil. Transact.* 172:887–918.
Bellamy, L. J., Mayo, D. W. (1976). Infrared frequency effects of lone pair interactions with antibonding orbitals on adjacent atoms. *J. Phys. Chem.* 80:1217–1220.
Fang, H. L., et al. (1984). Overtone spectroscopy of nonequivalent methyl C-H oscillators. Influence of conformation on vibrational overtone energies. *J. Chem Phys.* 88:410–416.

Hydrocarbons

Tosi, C., Pinto, A. (1970). Near-infrared spectroscopy of hydrocarbon functional groups. *Spectrochim. Acta.* 28A:585–597.

Water

Darling, B. T., Dennison, D. M. (1940). The water vapor molecule. *Phys. Rev.* 57:128–139.

Hydrogen Bonding

Finch, J. N., Lippincott, E. R. (1956). Hydrogen bond systems: Temperature dependence of OH frequency shifts and OH band intensities. *J. Chem. Phys.* 24:908–909.
Finch, J. N., Lippincott, E. R. (1957). Hydrogen bond systems—temperature dependence of OH frequency shifts and OH band intensities. *J. Chem. Phys.* 61:894–902.
Lippincott, E. R., Schroeder, R. (1955). One-dimensional model of the hydrogen bond. *J. Chem. Phys.* 23:1099–1106.
Lippincott, E. R., et al. (1959). Potential-function model of hydrogen bond systems, hydrogen bonding. *Papers Symposium Ljubljana* 1957, 361–374.

Murthy, A. S. N., Rao, C. N. R. (1968). Spectroscopic studies of the hydrogen bond. *Appl. Spectrosc. Rev.* 2:69–191.

Schroeder, R., Lippincott, E. R. (1957). Potential function model of hydrogen bonds. II. *J. Chem. Phys.* 61:921–928.

Stutman, J. M., Posner, A. S. (1962). *Nature* 193:368–369.

Ammonia and Methane

Hayward, R. J., Henry, B. R. (1974). Anharmonicity in polyatomic molecules: A local-mode analysis of the XH-stretching overtone spectra of ammonia and methane. *J. Mol. Spectrosc.* 50:58–67.

Cyclic Alkanes and Alkenes

Henry, B. R., et al. (1980). A local mode description of the CH-stretching overtone spectra of the cycloalkanes and cycloalkenes. *J. A. C. S.* 102:515–519.

Aromatic Amines

Whetsel, K., et al. (1957). Near-infrared analysis of *N*-alkyl and *N*-alkyl-*N*-hydroxyalkyl aromatic amine mixtures. *Anal. Chem.* 29:1006–1009.

Whetsel, K., et al. (1958). Near-infrared analysis of mixtures of primary and secondary aromatic amines. *Anal. Chem.* 30:1594–1597.

Whetsel, K. B., et al. (1958). Near-infrared spectra of primary aromatic amines. *Anal. Chem.* 30:1598–1604.

The N-H Group

Krikorian, S. E., Mahpour, M. (1973). The identification and origin of N-H overtone and combination bands in the near-infrared spectra of simple primary and secondary amides. *Spectrochim. Acta* 29A, 1233–1246.

Amino Acids

Leifer, A., Lippincott, E. R. (1957). The infrared spectra of some amino acids. *J.A.C.S.* 79:5098–5101.

Proteins

Ellis, J. W., Bath, J. (1938). Modifications in the near infra-red absorption spectra of protein and of light and heavy water molecules when water is bound to gelatin. *J. Chem. Phys.* 6:723–729.

Fatty Acids

Holman, R. T., Edmondson, P. R. (1956). near-infrared spectra of fatty acids and some related substances. *Anal. Chem.* 28:1533–1538.

Alkenes

Fang, H. L., Compton, D. A. C. (1988). Overtone spectroscopy of nonequivalent C-H oscillators in 1-alkenes and dienes. *J. Chem. Phys.* 92:7185–7192.

Goodu, R. F. (1957). Determination of unsaturation by near-infrared spectrophotometry. *Anal. Chem.* 29:1790–1794.

Epoxides

Goddu, R. F., Delker, D. A. (1958). Determination of terminal epoxides by near-infrared spectrophotometry. *Anal. Chem.* 30:2013–2016.

Phenolic Hydroxyl

Goddu, R. F. (1958). Determination of phenolic hydroxyl by near-infrared spectrophotometry. *Anal. Chem.* 30:2009–2013.

Group Frequencies—General

Bellamy, L. J. (1979). Some wider uses of group frequencies. *Appl. Spectrosc.* 33:439–443.

Vibrational States in Benzene

Bray, R. G., Berry, M. J. (1979). Intramolecular rate processes in highly vibrationally excited benzene. *J. Chem. Phys.* 71(12):4909–4922.

Henry, B. R., Siebrand, W. (1968). Anharmonicity in polyatomic molecules. The CH-stretching overtone spectrum of benzene. *J. Chem Phys.* 49(12):5369–5376.

Spears, K. G., Rice, S. A. (1971). Study of the individual vibronic states of the isolated benzene molecule. *J. Chem. Phys.* 55(12):5561–5581.

Coupling in C-H Frequencies

Henry, Bryan R. (1977). Use of local modes in the description of highly vibrationally excited molecules. *Acc. Chem. Res.* 10:207–213.

Local Modes in C-H Stretch Combation Bands

Burberry, M. S., Albrecht, A. C. (1979). Local mode combination bands and local mode mixing. *J. Chem. Phys.* 71(11):4631–4640.

Local Modes in Overtones

Child, M. S. (1985). Local mode overtone spectra. *Acc. Chem. Res.* 18:45–50.

Heller, D. F., Mukamel, S. (1979). Theory of vibrational overtone line shapes of polyatomic molecules. *J. Chem. Phys.* 70(1):463–472.

Normal Modes

Duncan, J. L. (1991). The determination of vibrational anharmonicity in molecules from spectroscopic observations. *Spectrochim. Acta* 47A:1, 1–27.

C-H Stretch in Benzene

Hayward, R. J., Henry, B. R. (1973). Anharmonicity in polyatomic molecules. *J. Mol. Spectrosc.* 46:207–213.

Carbon Number Prediction

Schrieve, G. D., Ullman, A. H. (1991). Carbon number prediction from Herschel-infrared spectra using partial least-squares regression. *Appl. Spectrosc.* 45:713–714.

Solids at High Pressure

Weir, C. E. et al. (1959). Infrared studies in the 1- to 15-micron region to 30,000 atmospheres. *J.N.B.S.* 63A:55–62.

Weir, C. E. et al. (1959). Studies of infrared absorption spectra of solids at high pressures. *Spectrochim. Acta* 16:58–73.

Methyl Iodide

Herzberg, G., Herzberg, L. (1949). Absorption spectrum of methyl iodide in the near infrared. *Can. J. Res.* 270:332–338.

16.

FUNCTIONAL GROUPINGS AND CALCULATED LOCATIONS IN NANOMETERS (NM) FOR NIR SPECTROSCOPY*

*Numbers in parentheses are outside of the near-infrared region.

Group		Molecular structure	Relative intensity	1st Overtone	2nd Overtone	3rd Overtone	4th Overtone
I. C-H stretch							
a. Alkane, C_nH_{2n+2}		$CH_3-(CH_2)_N-CH_3$, normal paraffins	m-s	1690–1755	1120–1212	890–936	720–766
	Methyl	$-C-H_3$	m-s				
	Antisymmetric stretch			(typically 1682–1695)	(typically 1190–1194)	(typically 890–915)	(typically 744–750)
	Symmetric stretch			1735–1747	1157–1165	917–924	734–740
	Methylene	$-C-H_2$	m-s	1703–1759	1135–1215	924–936	760–770
	Antisymmetric stretch			1703–1715	1135–1200	901–907	721–726
	Symmetric stretch			1746–1759	1164–1172	924–936	738–753
b. Alkene, C_nH_{2n}		>C=C<, olefinic group					
	Monosubstituted	H >C=C<^H_H R	m m	1645–1660 1615–1625	1097–1107	823–830	~662
	Cis, disubstituted	H >C=C<^H_R R	m	1645–1660	1097–1107	823–830	~662
	Trans, disubstituted	H >C=C<^R_H R	m	1645–1660	1097–1107	823–830	~662
	Gem, disubstituted	R >C=C<^H_H R	m	1615–1625	1077–1083	808–813	~649
	Trisubstituted	R >C=C<^R_H R	m	1645–1660	1097–1107	823–830	~662
c. Alkyne, C_nH_{2n-2}		$-C≡C-H$	s	~1515	~1010	~758	~606
d. Aromatic, C_nH_n		Ar(C)–H	v	1650–1680	1100–1142	857–890	710–725
II. C-H bend							
a. Alkane, C_nH_{2n+2}							
	Methine	R–CHR′R″	w	(~3730)	2487	~1865	1492
	Methylene	R–CH$_2$R′	m	(3371)	2247–2307	1685–1730	1348
	Methyl	R–CH$_3$	m s	(3401) (3625–3650)	2267–2333 2417–2433	1700–1750 1813–1825	1360 1450–1460
	Gem, dimethyl	CH$_3$–CHR–CH$_3$	s s	(3610–3625) (3650–3665)	2407–2413 2433–2443	1805–1813 1825–1833	1444–1450 1460–1466
	Tert-butyl	(CH$_3$)$_3$–CH	m s	(3585–3610) (~3650)	2390–2407 ~2433	1793–1805 ~1833	1434–1444 1460
b. Alkene, C_nH_{2n}		>C=C<					
	Monosubstituted	H >C=C<^H_H R	s s s	(5075–5250) (5465–5525) (3520–3545)	(3350–3383) (3643–3683) 2347–2363	2513–2538 2733–2763	2030–2100 2186–2210 1408–1418
	Cis, disubstituted	H >C=C<^H_R R	s	(~7250)	(~4833)	(~3625)	2900
	Trans, disubstituted	H >C=C<^R_H R	s m	(5155–5210) (3820–3860)	(3437–3473) (2547–2573)	(2578–2605) 1910–1930	2062–2084 1528–1544

Group		Molecular structure	Relative intensity	1st Overtone	2nd Overtone	3rd Overtone	4th Overtone
	Gem, disubstituted	R >C=C<^H_H R	s s	(5585–5650) (3520–3545)	(3723–3767) 2347–2363)	(2793–2825) 1760–1773	2234–2260 1408–1418
	Trisubstituted	R >C=C<^R_H R	s	(5950–6330)	(3967–4220)	(2975–3165)	2380–2532
c.	Alkyne, C_nH_{2n-2}	–C≡C–H	s	(~7950)	(~5300)	(~3975)	3180
d.	Aromatic, C_nH_n	Ar(C)–H					
	Monosubstituted (5 hydrogens)	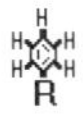	v, s	(~6650) (~7150)	(~4433) (~4767)	(~3325) (~3575)	(2660) (2860)
	Disubstituted (4 hydrogens)	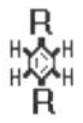	v, s	(~6650)	(~4433)	(~3325)	(2660)
	Trisubstituted (3 hydrogens)		v, s	(~6400)	(~4267)	(~3200)	(2560)
	Tetrasubstituted (2 hydrogens)		v, m	(~6000)	(~4000)	(~3000)	2400
	Pentasubstituted (1 hydrogen)		v, w	(~5650)	(~3767)	(~2825)	2260
III. C-C stretch							
a.	Alkene, C_nH_{2n}	>C=C<	v	(2975–3085)	1983–2057	1488–1543	1190–1234
	Monosubstituted	As shown above	m	(~3040)	~2027	~1520	1216
	Cis, disubstituted	As shown above	m	(~3015)	~2010	~1508	1206
	Trans, disubstituted	As shown above	m	(~2985)	~1990	~1493	1194
	Gem, disubstituted	As shown above	m	(~3025)	~1660	~1513	1210
	Trisubstituted	As shown above	m	(~2995)	~1997	~1498	1198
	Tetrasubstituted	As shown above	w	(~2995)	~1997	~1498	1198
	Diene	>C=C< ... >C=C<	w w	(~3030) (~3125)	~2020 ~2083	~1515 ~1563	1212 1250
b.	Alkyne, C_nH_{2n-2}						
	Monosubstituted		m	(2335–2380)	1557–1587	1168–1190	934
	Disubstituted		v, w	(2210–2285)	1473–1523	1105–1143	884–914
c.	Allene	>C=C=C<	m m	(~2550) (~4700)	~1837 (~3133)	~1275 ~2350	1020 1880
d.	Aromatic	containing	v v m m	(~3125) (~3165) (~3335) (~3450)	~2083 ~2110 ~2223 ~2300	~1563 ~1583 ~1668 ~1725	1250 1266 1334 1380
IV. Carbonyl		>C=O					
a.	Ketones	R >C=O R′					
	Saturated, acyclic	alkane (R) >C=O (R′) alkane	s	(2900–2935)	1933–1957	1450–1468	1160–1174

Group		Molecular structure	Relative intensity	1st Overtone	2nd Overtone	3rd Overtone	4th Overtone
	Saturated, cyclic						
	6- or more membered ring	O=C (or) AR >C=O R'	s	(2900–2935)	1933–1957	1450–1468	1160–1174
	5-membered ring	O=C	s	(2855–2825)	1903–1917	1450–1438	1142–1130
	4-membered ring	O=C	s	(~2815)	~1877	~1408	1126
	Alpha-, beta-unsaturated, acyclic		s	(2970–3005)	1980–2003	1485–1503	1188–1202
	Alpha-, beta-unsaturated, cyclic						
	6- or more membered ring		s	(2970–3005)	1980–2003	1485–1503	1188–1202
	5-membered ring		s	(2900–2925)	1933–1950	1450–1463	1160–1170
	Alpha-, beta-, alpha'-, beta'-, unsaturated, acyclic		s	(2995–3005)	1997–2003	1498–1503	1198–1202
	Aryl		s	(2940–2975)	1960–1983	1470–1488	1176–1190
	Diaryl		s	(2995–3010)	1997–2007	1498–1505	1198–1204
	Alpha-diketones		s	(2890–2925)	1927–1950	1445–1463	1156–1170
	Beta-diketones (enolic)		s	(3050–3250)	2033–2167	1525–1625	1220–1300
	1,4-Quinones			(2960–3010)	1973–2007	1480–1505	1184–1204
	Ketenes			(~2325)	~1500	~1163	930
b.	Aldehydes, carbonyl stretch	>C=O					
	Saturated, aliphatic		s	(2875–2905)	1917–1937	1438–1453	1150–1162
	Alpha-, beta-unsaturated, aliphatic		s	(2935–2975)	1957–1983	1468–1488	1174–1190
	Alpha-, beta-, gamma-, delta-unstaturated, aliphatic		s	(2975–3010)	1983–2007	1488–1505	1190–1204
	Aryl		s	(2915–2950)	1943–1967	1485–1475	1166–1180
	C-H stretching, two bands		w w	1725–1775 1800–1850	1150–1183 1200–1233	863–888 900–925	690–710 720–740
c.	ESTERS	R-C=O-OR'					
	Saturated, acyclic		s	(2855–2880)	1903–1920	1428–1440	1142–1152
	Saturated, cyclic: delta- and larger lactones		s	(2855–2880)	1903–1920	1428–1440	1142–1152
	Gamma-lactones		s	(2810–2840)	1873–1893	1405–1420	1124–1136
	Beta-lactones		s	(~2750)	~1833	~1375	1100
	Unsaturated, vinyl ester		s	(2780–2825)	1853–1883	1390–1413	1112–1130
	Alpha-, beta-unsaturated and aryl		s	(2890–2910)	1927–1940	1445–1455	1156–1164
	Alpha, beta-unsaturated delta-lactone		s	(2790–2910)	1927–1940	1445–1455	1116–1164
	Alpha, beta-unsaturated gamma-lactone		s	(2840–2875)	1893–1917	1420–1438	1136–1150

Group		Molecular structure	Relative intensity	1st Overtone	2nd Overtone	3rd Overtone	4th Overtone
1112	Beta, gamma-unsaturated gamma-lactone		s	(~2780)	~1853	~1390	
	Alpha-ketoesters		s	(2850–2875)	1900–1917	1425–1438	1140–1150
	Beta-ketoesters (enolic)		s	(~3030)	~2020	~1515	1212
	Carbonates		s	(2810–2875)	1873–1917	1405–1438	1124–1150
V. Carboxylic acids		—O—C=O—					
	Carbonyl stretching: saturated aliphatic		s	(2900–2940)	1933–1960	1450–1470	1160–1176
	Alpha-, beta-unsaturated aliphatic aryl		s	(2915–2960) 2440–2975	1943–1973 1960–1983	1458–1480 1470–1488	1166–1184 976–1190
	Hydroxyl stretching (bonded), several bands		w	1850–2000	1233–1333	925–1000	740–800
	Carboxylate anion stretching		s	(3105–3225)	2070–2150	1553–1613	1242–1290
VI. Anhydrides							
	Stretching		s	(3575–3845)	2383–2563	1788–1923	1350–1538
	Saturated, acyclic		s s	(2705–2780) (2795–2875)	1803–1853 1863–1917	1353–1390 1398–1438	1082–1112 1118–1150
	Alpha-, beta-unsaturated and aryl, acyclic		s s	(2735–2810) (2825–2905)	1823–1873 1883–1937	1368–1405 1413–1453	1094–1124 1130–1162
	Saturated, 5-member ring		s s	(2675–2745) (2830–2855)	1783–1830 1887–1903	1338–1373 1415–1428	1070–1098 1132–1142
	Alpha-, beta-unsaturated, 5-member ring		s s	(2705–2780) (2735–2810)	1803–1853 1823–1973	1353–1390 1368–1405	1082–1112 1094–1124
VII. Acyl halides							
	Stretch: acyl fluorides		s	(~2705)	~1803	~1353	1082
	Acyl chlorides		s	(~2785)	~1857	~1393	1114
	Acyl bromides		s	(~2765)	~1843	~1383	1106
	Alpha-, beta-unsaturated and aryl		s m	(2855–2860) (2860–2910)	1870–1907 1907–1940	1403–1430 1430–1455	1142–1144 1144–1164
	COF_2		s	(2595)	1730	1298	1038
	$COCl_2$		s	(2735)	1823	1368	1094
	$COBr_2$		s	(2735)	1823	1368	1094
VIII. Amides							
a. Carbonyl stretch							
	Primary, concentrated solutions and solids		s	(~3030)	~2020	~1515	1212
	Primary, dilute solutions		s	(~2910)	~1973	~1480	1164
	Secondary, concentrated solutions and solids		s	(2975–3070)	1983–2047	1488–1535	1190–1228
	Secondary, dilute solutions		s	(2940–2995)	1960–1997	1470–1498	1176–1198
	Tertiary, solutions and solids		s	(2995–3070)	1997–2047	1498–1535	1198–1228
	Cyclic, delta-lactams, dilute solutions		s	(~2975)	~1983	~1480	1190
	Cyclic, gamma-lactams, dilute solutions		s	(~2940)	~1960	~1470	1176

Group		Molecular structure	Relative intensity	1st Overtone	2nd Overtone	3rd Overtone	4th Overtone
		Cyclic, gamma-lactams fused to second ring, dilute solution	s	(2855–2940)	1903–1960	1428–1470	1142–1176
		Cyclic, beta-lactams, dilute solution	s	(2840–2890)	1893–1927	1420–1445	1136–1156
		Cyclic, beta-lactams fused to second ring, dilute solutions	s	(2810–2825)	1873–1883	1405–1413	1124–1130
		Urea, acyclic	s	(~3010)	~2007	~1505	1204
		Urea, cyclic, 6-membered ring	s	(~3050)	~2033	~1525	1220
		Ureas, cyclic 5-member ring	s	(~2905)	~1937	~1453	1162
		Urethanes	s	(2875–2960)	1917–1973	1438–1480	1150–1184
		Imides, acyclic	s	(~2925)	~1950	~1463	1170
		Imides, cyclic, 6-membered ring	s s	(~2925) (~2940)	~1950 ~1960	~1463 ~1470	1170 1176
		Imides, cyclic, alpha-, beta-unsaturated, 6-membered ring	s s	(~2890) (~2995)	~1927 ~1997	~1445 ~1490	1156 1198
		Imides, cyclic, 5-membered ring	s s	(~2825) (~2940)	~1883 ~1960	~1413 ~1470	1130 1176
		Imides, cyclic, alpha-, beta-unsaturated, 5-membered ring	s s	(~2790) (~2925)	~1863 ~1950	~1398 ~1463	1116 1170
	b.	N-H stretch					
		Primary, free; two bands	m m	~1430 ~1470	953–989	~715 ~735	(572) (588)
		Primary, bonded; two bands	m m	~1495 ~1575	~997 ~1050	~748 ~788	(598) (630)
		Secondary, free; one band	m	~1460	~975	~730	(584)
		Secondary, bonded; one band	m	1500–1600	1000–1067 981	750–800	(600–640)
	c.	N-H Bending					
		Primary amides, dilute solution	s	(3085–3145)	2057–2097	1543–1573	1234–1258
		Secondary amides, dilute solution	s	(3225–3310)	2150–2207	1613–1655	1290–1324
IX. Alcohols and phenolics, O-H stretch							
	a.	O-H Stretching	v, sh	1370–1395	947–980	685–698	(548)
		With hydrogen bonding, O–H···H– single bridge compounds	v, sh	1410–1450	955–990	705–725	(564)
		Polymeric O-H	s,b	1470–1565	980–1043 993 (cellulose)	735–783	(588)
		Chelated compounds	w,b	1550–2000	1034–1334	775–1000	(620)
	b.	O-H bend and C-O stretch combination					
		Primary alcohols	s s	(~4750) (3700–3950)	(~3167) 2467–2633	~2375 1850–1975	1900 1480–1580

Group		Molecular structure	Relative intensity	1st Overtone	2nd Overtone	3rd Overtone	4th Overtone
	Secondary alcohols		s	(~4550)	(~3033)	~2275	1820
			s	(3700–3950)	2467–2633	1850–1975	1480–1580
	Tertiary alcohols		s	(~4350)	(~2900)	~2175	1740
			s	(3550–3300)	2367–2533	1775–1900	1420–1320
	Phenols		s	(~4175)	(~2767)	~2075	1670
			s	(3559–3300)	2367–2533	1775–1900	1424–1320
X. Amines, N-H stretch							
a.	Primary, free; two bands		m	~1430	953–980	~715	(572)
			m	~1470		~735	
	Secondary, free; one band		m	1430–1510	953–1007	715–755	(572)
	Imines; one band	=N-H	m	1470–1515	980–1010	735–758	(588)
	Amine salts		m	1600–1650	1067–1100	800–825	(640)
b.	N-H bend						
	Primary		s–m	(3030–3145)	2020–2097	1515–1573	1212–1258
	Secondary		w	(3030–3225)	2020–2150	1515–1613	1212–1290
	Amine salts		s	(3125–3175)	2083–2117	1563–1588	1250–1270
			s	(~3335)	~2223	~1668	1334
c.	C-N stretch						
	Aromatic, primary		s	(3730–4000)	2487–2667	1865–2000	1492–1600
	Aromatic, secondary		s	(3705–3905)	2470–2603	1853–1953	1482–1562
	Aromatic, tertiary		s	(3680–3820)	2453–2547	1840–1913	1472–1528
	Aliphatic		w	(4100–4900)	2673–3267	2050–2450	1640–1960
			w	(~3505)	~2367	~1775	1402
XI. Unsaturated N compounds, C=N stretch							
a.	Alkyl nitriles		m	2210–2230	1473–1487	1105–1115	884–892
	Alpha-, beta-, unsaturated alkyl nitriles		m	2235–2255	1490–1503	1118–1128	894–902
	Aryl nitriles		m	2230–2250	1487–1500	1115–1125	892–900
	Isocyanates		m	2200–2240	1467–1487	1100–1115	880–896
	Isocyanides		m	2250–2415	1500–1610	1125–1208	900–966
b.	Imines and oximes						
	Alkyl compounds		v	(2960–3033)	1973–2033	1480–1525	1184–1213
	Alpha-, beta-unsaturated		v	3010–3070	2070–2047	1505–1535	1204–1228
	–N=N— stretch, azo- compounds		v	3070–3175	2047–2117	1535–1588	1228–1270
	–N=C=N–, diimides		s	2320–2350	1547–1557	1160–1168	928–940
	–N$_3$ stretch, azides		s	2315–2340	1543–1567	1158–1180	926–936
			w	(3730–4240)	2487–2827	1865–2120	1492–1700
	C-NO$_2$, nitro- compounds						
	Aromatic		s	(3185–3335)	2185–2223	1593–1668	1274–1334
			s	(3650–3850)	2433–2567	1825–1925	1460–1540
	Aliphatic		s	(3185–3225)	2123–2150	1593–1613	1274–1290
			s	(3625–3650)	2417–2433	1813–1825	1450–1460
	O-NO$_2$, nitrates		s	(3030–3125)	2020–2083	1515–1563	1212–1250
			s	(3850–4000)	2567–2667	1925–2000	1540–1600
	C-NO, nitroso		s	(3125–3335)	2083–2223	1563–1668	1250–1334

Group	Molecular structure	Relative intensity	1st Overtone	2nd Overtone	3rd Overtone	4th Overtone
O-NO, nitrites		s	(2975–3125)	1983–2083	1488–1515	1190–1250
		s	(3075–3105)	2050–2070	1538–1553	1230–1242
XII. Halogen, C-X stretch						
C-F		s	(3550–5000)	2366–3333	1775–2500	1420–2000
C-Cl		s	(6250–8300)	(4167–5533)	(3125–4150)	(2500–3320)
C-Br		s	(8300–10000)	(5533–6667)	(4150–5000)	(3320–4000)
C-I		s	(~10000)	(~6667)	(~5000)	(~4000)
XIII. Sulfur compounds						
a. S-H stretch		w	1925–1960	1283–1307	963–980	770–784
b. C=S stretch		s	(4165–4760)	(2777–3173)	2083–2380	1666–1904
c. S=O stretch						
Sulfoxides		s	(4675–4855)	(3175–3237)	2338–2428	1870–1942
Sulfones		s	(4310–4385)	(2873–2923)	2155–2193	1724–1754
		s	(3705–3845)	2470–2563	1853–1923	1482–1538
Sulfites		s	(4065–4350)	(2710–2900)	2033–2175	1626–1740
		s	(3500–3705)	2333–2470	1750–1853	1400–1482
Sulfonyl chlorides		s	(4220–4295)	(2813–2863)	2110–2148	1688–1718
		s	(3650–3730)	2433–2487	1825–1865	1460–1492
Sulfonamides		s	(4240–4385)	(2827–2923)	2120–2193	1696–1754
		s	(3705–3845)	2470–2563	1853–1923	1482–1538
Sulfonic acids		s	(4135–4350)	(2757–2900)	2068–2175	1654–1740
		s	(4715–4855)	(3143–3237)	2358–2428	1886–1942
		s	(~7700)	(~5133)	(~3850)	(3080)

Group	Molecular structure	Relative intensity	1st Overtone	2nd Overtone	3rd Overtone	4th Overtone
XIV. Specific materials						
Alcohols				RO-H at 913–967		
Aqeous electrolytes				Changes in hydrogen bonding at 960–980		
Cellulose						
Epoxy resins				Terminal CH$_2$ at 1100	Terminal CH$_2$ at 825	
Fats					C-H stretch at 845	
Feeds						
Forages						
Lignin						
Oils					C-H stretch at 840–910	
Polymers					–CH=CH$_2$, C-H at 820	
Polyols				O-H at 973		
Proteins				N-H stretch at 1007		
Silica				SiO-H at 966		
Starches				O-H stretch at 973		
Water/moisture				O-H at 960–980		

Group	Molecular structure	Relative intensity	1st Overtone	2nd Overtone	3rd Overtone	4th Overtone
Acetone					907 (methyl C-H stretch)	
Cyclohexane					935 (methyl C-H stretch) 924 (methylene C-H)	
Ethylbenzene					911 (methyl C-H stretch) 876 (aromatic C-H)	
Isopropanol				O-H at 964	908 (methyl C-H stretch)	
MTBE					907 (methyl C-H stretch) 929 (methylene C-H)	
n-Decane					914 (methyl C-H stretch) 930 (methylene C-H)	
n-Heptane					914 (methyl C-H stretch) 929 (methylene C-H)	
Pentane					913 (methyl C-H stretch) 929 (methylene C-H)	
p-Xylene					913 (methyl C-H stretch) 881 (aromatic C-H)	
tert-Butanol				O-H at 967	908 (methyl C-H stretch)	
Toluene					912 (methyl C-H stretch) 876 (aromatic C-H)	
Trimethylpentane					915 (methyl C-H stretch) 936 (methylene C-H	
Water				O-H at 972		
Isooctane		S m	1695 (methyl C-H) 1704–1775 (methylene C-H)	1194 (methyl) 1209 (methylene)	915 (methyl) 935 (methylene)	748 (methyl) 766 (methylene)
n-Decane		S m	1692 (methyl C-H) 1727–1770 (methylene C-H)	1193 (methyl) 1212 (methylene)	914 (methyl) 930 (methylene)	748 (methyl) 761 (methylene)
		S m	1703 (methyl C-H) 1680 (aromatic C-H)	1191 (methyl) 1142 (aromatic)	911 (methyl) 875 (aromatic)	748 (methyl) 715 (aromatic)

17.

NIR SPECTRAL CORRELATION CHARTS

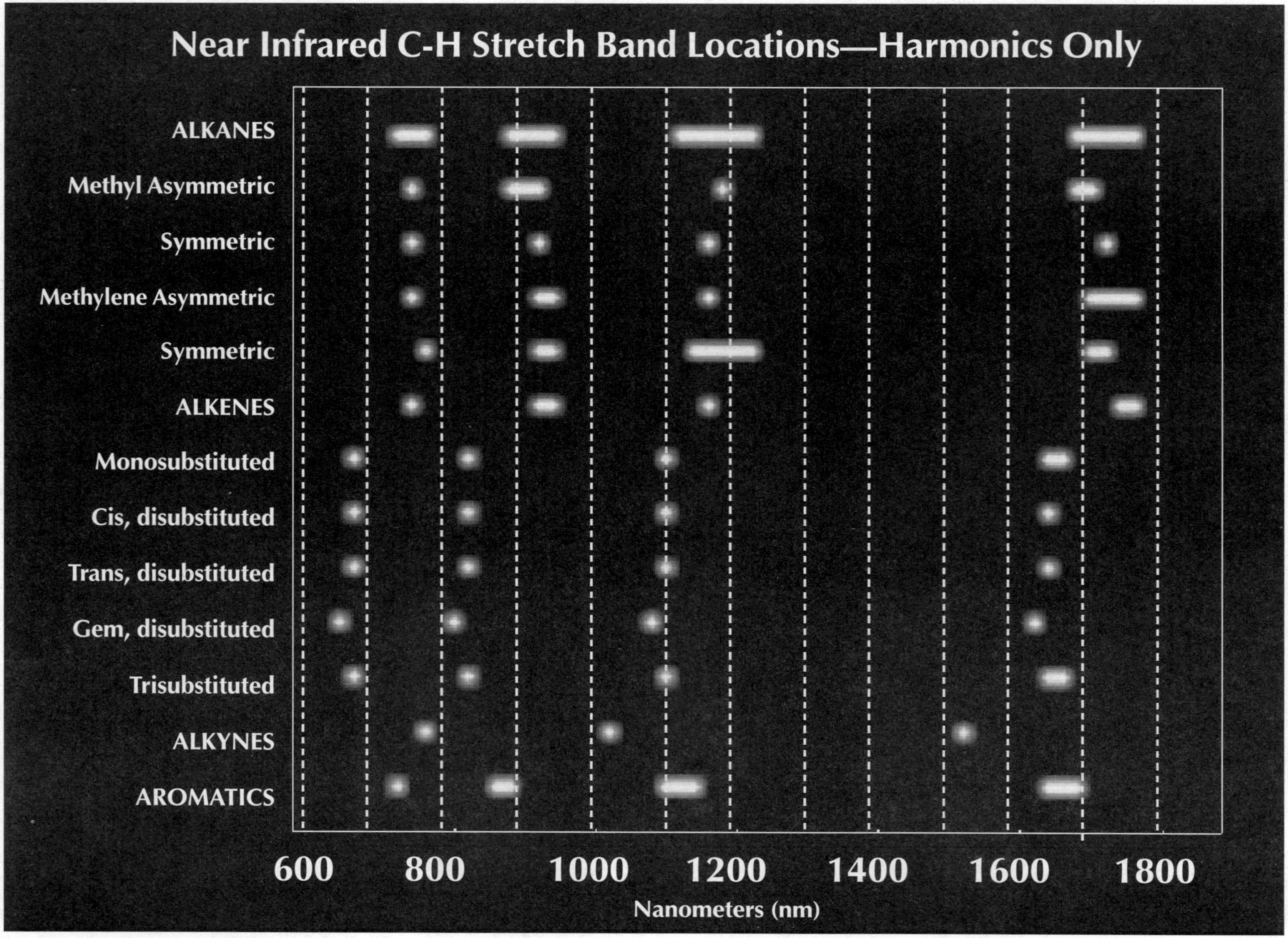

Near Infrared C-H Stretch Band Locations—Harmonics Only

Nanometers (nm)

ALKANES
Methyl Asymmetric
Symmetric
Methylene Asymmetric
Symmetric
ALKENES
Monosubstituted
Cis, disubstituted
Trans, disubstituted
Gem, disubstituted
Trisubstituted
ALKYNES
AROMATICS

(See color plate 7.)

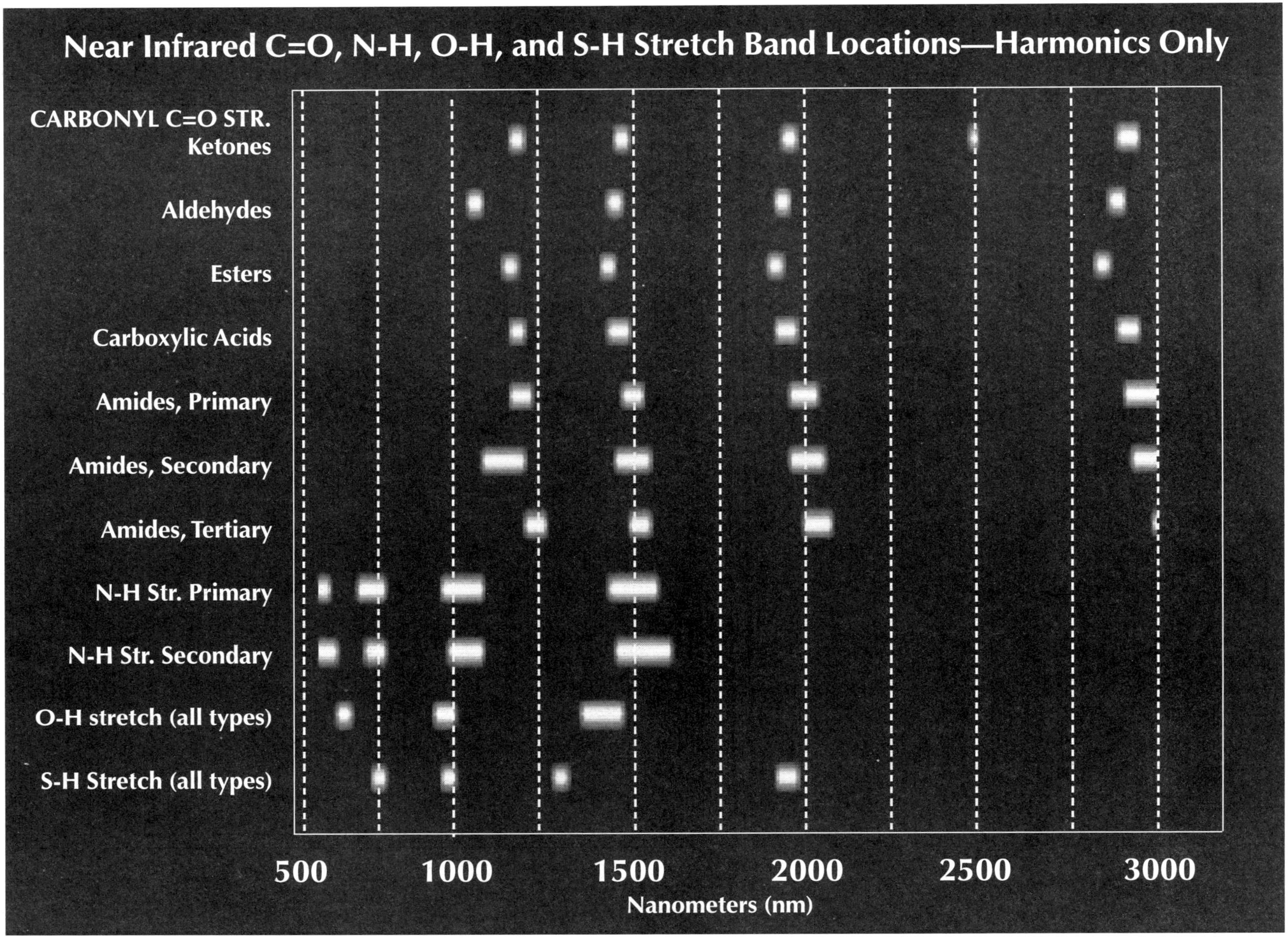

Near Infrared C=O, N-H, O-H, and S-H Stretch Band Locations—Harmonics Only

18.

NIR BAND ASSIGNMENTS FOR ORGANIC COMPOUNDS, POLYMERS, AND RUBBERS

Table 18.1   C-H, N-H, and O-H Stretch Absorption Bands for Specific Long Wavelength NIR (1100–2500 nm) Functional Groups (1st to 4th Overtones)

Structure	Bond vibration	Location (in nm) of 1st overtone	Location (in nm) of 2nd overtone	Location (in nm) of 3rd overtone	Location (in nm) of 4th overtone
ArCH (Aromatics)	C-H stretching	1680–1684	1143	875	715
.CH$_2$-CH$_2$. (Methylene)	C-H stretching	1726–1727	1212	927–930	761–764
.CH$_3$ (Methyl)	C-H stretching	1743–1762	1191–1194	911–915	745–748
.CH$_3$ (Methyl)	C-H combination	2020–2050	1347–1367		
>C=O (carbonyl)	>C=O stretching	2915	1920–1945	1420–1460	1140–1170
R-OH (Alcohols)	O-H stretching	1410–1455	940–970	—	—
ArOH (Phenols)	O-H stretching	1421–1470	940–980	—	—
HOH (Water)	O-H stretching	1440–1485	960	—	—
Starch	O-H stretching	1451	967	—	—
Urea	Sym. N-H stretching	1460	973–993	—	—
.CONH$_2$ (Primary amides)	N-H stretching	1463–1484	975–989	—	—
.CONHR (Secondary amides)	N-H stretching	1472	981	—	—
Cellulose	O-H stretching	1490	993	—	—
ArNH$_2$ (Aromatic amines)	N-H stretching	1493	995	—	—
.NH (Amines, general)	N-H stretching	1500	1000	—	—
Protein	N-H stretching	1511	1007	—	—
				—	—

Pure organic compounds and materials exhibit similar NIR harmonic bands and positions, represented in Table 18.2.

Table 18.2   Spectral Correlation Chart for Selected Organic Compounds and Materials

Material	Band location (in nm)	Band assignment
Acetic acid, ethyl ester	725	Asymmetric methyl (C-H) stretch—4th overtone
	740	Asymmetric methylene (C-H) stretch—4th overtone
	870	Asymmetric methyl (C-H) stretch—3rd overtone
	890	Asymmetric methylene (C-H) stretch—3rd overtone
	1130	Asymmetric methyl (C-H) stretch—2nd overtone
	1170	Asymmetric methylene (C-H) stretch—2nd overtone
	1410	C=O stretch—3rd overtone
	1660	Asymmetric methyl (C-H) stretch—1st overtone
	1705	Asymmetric methylene (C-H) stretch—1st overtone
	1910	C=O stretch—2nd overtone
Benzoyl chloride	866	Aromatic C-H stretch—3rd overtone
	1150	Aromatic C-H stretch—2nd overtone
	1660–1675	Aromatic C-H stretch—1st overtone
	2140–2150	C-H stretch/C=O stretch combination band
o-Xylene	860	Aromatic C-H stretch—3rd overtone
	905	Asymmetric methyl (C-H) stretch—3rd overtone
	1130	Aromatic C-H stretch—2nd overtone
	1180	Asymmetric methyl (C-H) stretch—2nd overtone

Material	Band location (in nm)	Band assignment
	1680–1700	Aromatic C-H stretch—1st overtone/asymmetric methyl (C-H) stretch—1st overtone
Nitrobenzene	860	Aromatic C-H stretch—3rd overtone
	1110	Aromatic C-H stretch—2nd overtone
	1640	Aromatic C-H stretch—1st overtone
Cyclohexanol	915	Methylene (C-H) stretch—3rd overtone
	960	O-H stretch—2nd overtone
	1200	Methylene (C-H) stretch—2nd overtone
	1410	O-H stretch—1st overtone
	1715–1740	Cyclic methylene (C-H) stretch—1st overtone
Pyridine	860	Aromatic C-H stretch—3rd overtone
	1120	Aromatic C-H stretch—2nd overtone
	1670	Aromatic C-H stretch—1st overtone
Formic acid	890	Carbonyl associated C-H stretch—3rd overtone
	1165	Carbonyl associated C-H stretch—2nd overtone
	1705–1750	Carbonyl associated C-H stretch—1st overtone
Propyl alcohol	870	Asymmetric methyl (C-H) stretch—3rd overtone
	910	Asymmetric methylene (C-H) stretch—3rd overtone
	960	O-H stretch—2nd overtone
	1190	Asymmetric methyl (C-H) stretch—2nd overtone
	1205	Asymmetric methylene (C-H) stretch—2nd overtone
	1410	O-H stretch—1st overtone
	1695	Asymmetric methyl (C-H) stretch—1st overtone
	1710	Asymmetric methylene (C-H) stretch—1st overtone
p-Cresol	860	Aromatic C-H stretch—3rd overtone
	890	Asymmetric methyl (C-H) stretch—3rd overtone
	1120	Aromatic C-H stretch—2nd overtone
	1170	Asymmetric methyl (C-H) stretch—2nd overtone
	1670	Aromatic C-H stretch—1st overtone
	1690	Asymmetric methyl (C-H) stretch—1st overtone
4-Methyl-2-pentanol	860	Asymmetric methyl (C-H) stretch—3rd overtone
	895	Asymmetric methylene (C-H) stretch—3rd overtone
	960	O-H stretch—2nd overtone
	1120	Asymmetric methyl (C-H) stretch—2nd overtone
	1170	Asymmetric methylene (C-H) stretch—2nd overtone
	1410	O-H stretch—1st overtone
	1690	Asymmetric methyl (C-H) stretch—1st overtone
	1705	Asymmetric methylene (C-H) stretch—1st overtone
1,2-Dichloroethane	890	Asymmetric methylene (C-H) stretch—3rd overtone
	1150	Asymmetric methylene (C-H) stretch—2nd overtone
	1700	Asymmetric methylene (C-H) stretch—1st overtone
Diphenylamine	890	Aromatic C-H stretch—3rd overtone
	1140	Aromatic C-H stretch—2nd overtone

Material	Band location (in nm)	Band assignment
	1490	Aromatic C-H stretch—1st overtone
	1670	N-H stretch—1st overtone
Butyl alcohol	910	Asymmetric methyl (C-H) stretch—3rd overtone
	915	Asymmetric methylene (C-H) stretch—3rd overtone
	960	O-H stretch—2nd overtone
	1180	Asymmetric methyl (C-H) stretch—2nd overtone
	1204	Asymmetric methylene (C-H) stretch—2nd overtone
	1415	O-H stretch—1st overtone
	1702	Asymmetric methyl (C-H) stretch—1st overtone
	1715	Asymmetric methylene (C-H) stretch—1st overtone
1-Nitropropane	890	Asymmetric methyl (C-H) stretch—3rd overtone
	905	Asymmetric methylene (C-H) stretch—3rd overtone
	1130	Asymmetric methyl (C-H) stretch—2nd overtone
	1170	Asymmetric methylene (C-H) stretch—2nd overtone
	1685	Asymmetric methyl (C-H) stretch—1st overtone
	1715	Asymmetric methylene (C-H) stretch—1st overtone
Pentyl alcohol	905	Asymmetric methyl (C-H) stretch—3rd overtone
	910	Asymmetric methylene (C-H) stretch—3rd overtone
	945	O-H stretch—2nd overtone
	1185	Asymmetric methyl (C-H) stretch—2nd overtone
	1195	Asymmetric methylene (C-H) stretch—2nd overtone
	1400	O-H stretch—1st overtone
	1695	Asymmetric methyl (C-H) stretch—1st overtone
	1715	Asymmetric methylene (C-H) stretch—1st overtone
Morpholine	908	Methylene (C-H) stretch—3rd overtone
	966	N-H stretch—2nd overtone
	1185	Methylene (C-H) stretch—2nd overtone
	1490–1530	N-H stretch—1st overtone
	1730	Methylene (C-H) stretch—1st overtone
Iodoethane	880	Asymmetric methyl (C-H) stretch—3rd overtone
	905	Asymmetric methylene (C-H) stretch—3rd overtone
	1145	Asymmetric methyl (C-H) stretch—2nd overtone
	1165	Asymmetric methylene (C-H) stretch—2nd overtone
	1690	Asymmetric methyl (C-H) stretch—1st overtone
	1700	Asymmetric methylene (C-H) stretch—1st overtone
2-Ethyl-1-butanol	910	Asymmetric methyl (C-H) stretch—3rd overtone
	925	Asymmetric methylene (C-H) stretch—3rd overtone
	960	O-H stretch—2nd overtone
	1180	Asymmetric methyl (C-H) stretch—2nd overtone
	1210	Asymmetric methylene (C-H) stretch—2nd overtone
	1410	O-H stretch—1st overtone
	1700	Asymmetric methyl (C-H) stretch—1st overtone
	1715	Asymmetric methylene (C-H) stretch—1st overtone

Material	Band location (in nm)	Band assignment
Nitromethane	870	Asymmetric methyl (C-H) stretch—3rd overtone
	1130	Asymmetric methyl (C-H) stretch—2nd overtone
	1650–1700	Asymmetric methyl (C-H) stretch—1st overtone
2-Heptanol	908	Asymmetric methyl (C-H) stretch—3rd overtone
	930	Asymmetric methylene (C-H) stretch—3rd overtone
	960	O-H stretch—2nd overtone
	1185	Asymmetric methyl (C-H) stretch—2nd overtone
	1205	Asymmetric methylene (C-H) stretch—2nd overtone
	1410	O-H stretch—1st overtone
	1700	Asymmetric methyl (C-H) stretch—1st overtone
	1720	Asymmetric methylene (C-H) stretch—1st overtone
2-Chloroethanol	890	Asymmetric methylene (C-H) stretch—3rd overtone
	965	O-H stretch—2nd overtone
	1155	Asymmetric methylene (C-H) stretch—2nd overtone
	1420	O-H stretch—1st overtone
	1700	Asymmetric methylene (C-H) stretch—1st overtone
4-Methyl-2-pentanol acetate	885	Asymmetric methyl (C-H) stretch—3rd overtone
	910	Asymmetric methylene (C-H) stretch—3rd overtone
	1190	Asymmetric methyl (C-H) stretch—2nd overtone
	1215	Asymmetric methylene (C-H) stretch—2nd overtone
	1410	C=O stretch—3rd overtone
	1690	Asymmetric methyl (C-H) stretch—1st overtone
	1720	Asymmetric methylene (C-H) stretch—1st overtone
	1910	C=O stretch—2nd overtone
Benzyl alcohol	870	Aromatic C-H stretch—3rd overtone
	925	Asymmetric methylene (C-H) stretch—3rd overtone
	975	O-H stretch—2nd overtone
	1140	Aromatic C-H stretch—2nd overtone
	1210	Asymmetric methylene (C-H) stretch—2nd overtone
	1390	O-H stretch—1st overtone
	1675	Aromatic C-H stretch—1st overtone
	1685	Asymmetric methylene (C-H) stretch—1st overtone
o-Methoxyphenol	860	Aromatic C-H stretch—3rd overtone
	900	Asymmetric methyl (ROC-H) stretch—3rd overtone
	975	O-H stretch—2nd overtone
	1100	Aromatic C-H stretch—2nd overtone
	1166	Asymmetric methyl (ROC-H) stretch—2nd overtone
	1420	O-H stretch—1st overtone
	1645	Aromatic C-H stretch—1st overtone
	1690	Asymmetric methyl (ROC-H) stretch—1st overtone
1-Decanol	915	Asymmetric methyl (C-H) stretch—3rd overtone
	933	Asymmetric methylene (C-H) stretch—3rd overtone
	960	O-H stretch—2nd overtone

Material	Band location (in nm)	Band assignment
	1190	Asymmetric methyl (C-H) stretch—2nd overtone
	1210	Asymmetric methylene (C-H) stretch—2nd overtone
	1405	O-H stretch—1st overtone
	1725	Asymmetric methyl (C-H) stretch—1st overtone
	1760	Asymmetric methylene (C-H) stretch—1st overtone
1-Octanol	915	Asymmetric methyl (C-H) stretch—3rd overtone
	930	Asymmetric methylene (C-H) stretch—3rd overtone
	960	O-H stretch—2nd overtone
	1190	Asymmetric methyl (C-H) stretch—2nd overtone
	1210	Asymmetric methylene (C-H) stretch—2nd overtone
	1405	O-H stretch—1st overtone
	1725	Asymmetric methyl (C-H) stretch—1st overtone
	1760	Asymmetric methylene (C-H) stretch—1st overtone
n-Decane	905	Asymmetric methyl (C-H) stretch—3rd overtone
	915	Asymmetric methylene (C-H) stretch—3rd overtone
	1180	Asymmetric methyl (C-H) stretch—2nd overtone
	1190	Asymmetric methylene (C-H) stretch—2nd overtone
	1705	Asymmetric methyl (C-H) stretch—1st overtone
	1740	Asymmetric methylene (C-H) stretch—1st overtone
Acetoacetic acid, ethyl ester	860	Asymmetric methyl (C-H) stretch—3rd overtone
	890	Asymmetric methylene (C-H) stretch—3rd overtone
	1125	Asymmetric methyl (C-H) stretch—2nd overtone
	1166	Asymmetric methylene (C-H) stretch—2nd overtone
	1410	C=O stretch—3rd overtone
	1665	Asymmetric methyl (C-H) stretch—1st overtone
	1705	Asymmetric methylene (C-H) stretch—1st overtone
	1910	C=O stretch—2nd overtone
	2100	Asymmetric C-O-O stretch—3rd overtone
Isopropyl alcohol	910	Asymmetric methyl (C-H) stretch—3rd overtone
	960	O-H stretch—2nd overtone
	1180	Asymmetric methyl (C-H) stretch—2nd overtone
	1410	O-H stretch—1st overtone
	1680–1695	Asymmetric methyl (C-H) stretch—1st overtone
Methylcyclohexane	750	Asymmetric methyl (C-H) stretch—4th overtone
	760	Methylene (C-H) stretch—4th overtone
	920	Asymmetric methyl (C-H) stretch—3rd overtone
	930	Methylene (C-H) stretch—3rd overtone
	1198	Asymmetric methyl (C-H) stretch—2nd overtone
	1215	Methylene (C-H) stretch—2nd overtone
	1705	Asymmetric methyl (C-H) stretch—1st overtone
	1725	Methylene (C-H) stretch—1st overtone
Hexadecanol	910	Asymmetric methyl (C-H) stretch—3rd overtone
	930	Methylene (C-H) stretch—3rd overtone

Material	Band location (in nm)	Band assignment
	960	O-H stretch—2nd overtone
	1160	Asymmetric methyl (C-H) stretch—2nd overtone
	1210	Methylene (C-H) stretch—2nd overtone
	1405	O-H stretch—1st overtone
	1725	Asymmetric methyl (C-H) stretch—1st overtone
	1760	Methylene (C-H) stretch—1st overtone
p-Dichlorobenzene	850	Aromatic C-H stretch—3rd overtone
	1110	Aromatic C-H stretch—2nd overtone
	1640	Aromatic C-H stretch—1st overtone
Cyclohexanone	905	Asymmetric methylene (C-H) stretch—3rd overtone
	1185	Asymmetric methylene (C-H) stretch—2nd overtone
	1702	Asymmetric methylene (C-H) stretch—1st overtone
	2125	C-H stretch/C=O stretch combination
	2170	Asymmetric C-H stretch/C-H deformation combination
Glycerol	760	O-H stretch—3rd overtone
	905	Asymmetric methylene (C-H) stretch—3rd overtone
	995	O-H stretch—2nd overtone
	1195	Asymmetric methylene (C-H) stretch—2nd overtone
	1450–1580	O-H stretch—1st overtone
	1695	Asymmetric methylene (C-H) stretch—1st overtone
Cyclohexane	715	Asymmetric methyl (C-H) stretch—4th overtone
	755	Methylene (C-H) stretch—4th overtone
	860	Asymmetric methyl (C-H) stretch—3rd overtone
	905	Methylene (C-H) stretch—3rd overtone
	1120	Asymmetric methyl (C-H) stretch—2nd overtone
	1190	Methylene (C-H) stretch—2nd overtone
	1660	Asymmetric methyl (C-H) stretch—1st overtone
	1705	Methylene (C-H) stretch—1st overtone
Ethanol	745	O-H stretch—3rd overtone
	890	Asymmetric methyl (C-H) stretch—3rd overtone
	915	Asymmetric methylene (C-H) stretch—3rd overtone
	995	O-H stretch—2nd overtone
	1180	Asymmetric methyl (C-H) stretch—2nd overtone
	1190	Asymmetric methylene (C-H) stretch—2nd overtone
	1500–1600	O-H stretch—1st overtone
	1685	Asymmetric methyl (C-H) stretch—1st overtone
	1715	Asymmetric methylene (C-H) stretch—1st overtone
Acetone	740	Asymmetric methyl (C-H) stretch—4th overtone
	875–890	Asymmetric methyl (C-H) stretch—3rd overtone
	1150	Asymmetric methyl (C-H) stretch—2nd overtone
	1450	C=O stretch—3rd overtone
	1660–1720	Asymmetric methyl (C-H) stretch—1st overtone
	1940	C=O stretch—2nd overtone

Material	Band location (in nm)	Band assignment
Hexylamine	2100–2140	C-H stretch/C=O stretch combination
	760	Asymmetric methyl (C-H) stretch—4th overtone
	805	Asymmetric methylene (C-H) stretch—4th overtone
	915	Asymmetric methyl (C-H) stretch—3rd overtone
	925	Asymmetric methylene (C-H) stretch—3rd overtone
	1045	N-H stretch—3rd overtone
	1190	Asymmetric methyl (C-H) stretch—2nd overtone
	1210	Asymmetric methylene (C-H) stretch—2nd overtone
	1520	N-H stretch—2nd overtone
	1730	Asymmetric methyl (C-H) stretch—1st overtone
	1760	Asymmetric methylene (C-H) stretch—1st overtone
2,5-Diethoxyaniline	760	Asymmetric methyl (C-H) stretch—4th overtone
	805	Asymmetric methylene (C-H) stretch—4th overtone
	915	Asymmetric methyl (C-H) stretch—3rd overtone
	925	Asymmetric methylene (C-H) stretch—3rd overtone
	1035	N-H stretch—3rd overtone
	1190	Asymmetric methyl (C-H) stretch—2nd overtone
	1210	Asymmetric methylene (C-H) stretch—2nd overtone
	1495	N-H stretch—2nd overtone
	1730	Asymmetric methyl (C-H) stretch—1st overtone
	1760	Asymmetric methylene (C-H) stretch—1st overtone
	1970	N-H stretch and N-H bend combination
Acetonitrile	720	Asymmetric methyl (C-H) stretch—4th overtone
	875	Asymmetric methyl (C-H) stretch—3rd overtone
	1140	Asymmetric methyl (C-H) stretch—2nd overtone
	1670, 1700	Asymmetric methyl (C-H) stretch—1st overtone
	2250	C≡N stretch—1st overtone
2-Butanone	725	Asymmetric methyl (C-H) stretch—4th overtone
	740	Asymmetric methylene (C-H) stretch—4th overtone
	870	Asymmetric methyl (C-H) stretch—3rd overtone
	890	Asymmetric methylene (C-H) stretch—3rd overtone
	1120	Asymmetric methyl (C-H) stretch—2nd overtone
	1160	Asymmetric methylene (C-H) stretch—2nd overtone
	1400	C=O stretch—3rd overtone
	1680	Asymmetric methyl (C-H) stretch—1st overtone
	1710	Asymmetric methylene (C-H) stretch—1st overtone
	1900	C=O stretch—2nd overtone
Heptanal	740	Asymmetric methyl (C-H) stretch—4th overtone
	760	Asymmetric methylene (C-H) stretch—4th overtone
	900	Asymmetric methyl (C-H) stretch—3rd overtone
	905	Asymmetric methylene (C-H) stretch—3rd overtone
	1180	Asymmetric methyl (C-H) stretch—2nd overtone
	1195	Asymmetric methylene (C-H) stretch—2nd overtone

Material	Band location (in nm)	Band assignment
	1410	C=O stretch—3rd overtone
	1690	Asymmetric methyl (C-H) stretch—1st overtone
	1710	Asymmetric methylene (C-H) stretch—1st overtone
	1930	C=O stretch—2nd overtone
Phenanthrene	870	Aromatic C-H stretch—3rd overtone
	1140	Aromatic C-H stretch—2nd overtone
	1680	Aromatic C-H stretch—1st overtone
Chloromethane	880	Asymmetric methyl (C-H) stretch—3rd overtone
	1160	Asymmetric methyl (C-H) stretch—2nd overtone
	1660	Asymmetric methyl (C-H) stretch—1st overtone
Dimethylamine	760	Asymmetric methyl (C-H) stretch—4th overtone
	915	Asymmetric methyl (C-H) stretch—3rd overtone
	1035	N-H stretch—3rd overtone
	1180	Asymmetric methyl (C-H) stretch—2nd overtone
	1515	N-H stretch—2nd overtone
	1720	Asymmetric methyl (C-H) stretch—1st overtone
N,N-Dimethylacetamide		
	725	Asymmetric methyl (C-H) stretch—4th overtone
	740	Asymmetric methylene (C-H) stretch—4th overtone
	890	Asymmetric methyl (C-H) stretch—3rd overtone
	905	Asymmetric methylene (C-H) stretch—3rd overtone
	1166	Asymmetric methyl (C-H) stretch—2nd overtone
	1180	Asymmetric methylene (C-H) stretch—2nd overtone
	1460	C=O stretch—3rd overtone
	1680	Asymmetric methyl (C-H) stretch—1st overtone
	1700	Asymmetric methylene (C-H) stretch—1st overtone
	1930	C=O stretch—2nd overtone
N-Butylacetamide	725	Asymmetric methyl (C-H) stretch—4th overtone
	740	Asymmetric methylene (C-H) stretch—4th overtone
	890	Asymmetric methyl (C-H) stretch—3rd overtone
	910	Asymmetric methylene (C-H) stretch—3rd overtone
	1030	N-H stretch—3rd overtone
	1185	Asymmetric methyl (C-H) stretch—2nd overtone
	1205	Asymmetric methylene (C-H) stretch—2nd overtone
	1430	C=O stretch—3rd overtone
	1500	N-H stretch—2nd overtone
	1680–1700	Asymmetric methyl (C-H) stretch—1st overtone
	1715	Asymmetric methylene (C-H) stretch—1st overtone
	1940	C=O stretch—2nd overtone
Methanol	750	Asymmetric methyl (C-H) stretch—4th overtone
	905	Asymmetric methyl (C-H) stretch—3rd overtone
	1000	O-H stretch—2nd overtone
	1185	Asymmetric methyl (C-H) stretch—2nd overtone

Material	Band location (in nm)	Band assignment
	1400–1600	O-H stretch 1st overtone
	1695	Asymmetric methyl (C-H) stretch—1st overtone
Carbon tetrachloride	1380	Asymmetric Cl-C-Cl stretch
	1680	Asymmetric Cl-C-Cl stretch
	1870	Asymmetric Cl-C-Cl stretch
	2360	Asymmetric Cl-C-Cl stretch
Carbon disulfide	1380	Asymmetric S-C-S stretch
	1880	Asymmetric S-C-S stretch
	1925	Asymmetric S-C-S stretch
	2205	Asymmetric S-C-S stretch
Formamide	765	C=O-H stretch—4th overtone
	905	C=O-H stretch—3rd overtone
	1000	N-H stretch—3rd overtone
	1190	C=O-H stretch—2nd overtone
	1460	C=O stretch—3rd overtone
	1500	N-H stretch—2nd overtone
	1760	C=O-H stretch—1st overtone
	1930	C=O stretch—2nd overtone

Table 18.3 C-H, N-H, and O-H Stretch Absorption Bands for Specific Long Wavelength NIR (1100–2500 nm) Functional Groups (1st to 4th overtones).

Structure	Bond vibration	Location (in nm) of 1st overtone	Location (in nm) of 2nd overtone	Location (in nm) of 3rd overtone	Location (in nm) of 4th overtone
ArCH (Aromatics)	C-H stretching	1680–1684	1143	875	715
.CH$_2$-CH$_2$. (Methylene)	C-H stretching	1726–1727	1212	927–930	761–764
.CH$_3$ (Methyl)	C-H stretching	1743–1762	1191–1194	911–915	745–748
.CH$_3$ (Methyl)	C-H combination	2020–2050	1347–1367		
>C=O (carbonyl)	>C=O stretching	2915	1920–1945	1420–1460	1140–1170
R-OH (Alcohols)	O-H stretching	1410–1455	940–970	—	—
ArOH (Phenols)	O-H stretching	1421–1470	940–980	—	—
HOH (Water)	O-H stretching	1440–1485	960	—	—
Starch	O-H stretching	1451	967	—	—
Urea	Sym. N-H stretching	1460	973–993	—	—
.CONH$_2$ (Primary amides)	N-H stretching	1463–1484	975–989	—	—
.CONHR (Secondary amides)	N-H stretching	1472	981	—	—
Cellulose	O-H stretching	1490	993	—	—
ArNH$_2$ (Aromatic amines)	N-H stretching	1493	995	—	—
.NH (Amines, general)	N-H stretching	1500	1000	—	—
Protein	N-H stretching	1511	1007	—	—
				—	—

Source: From J. Workman and A. Springsteen (eds.), *Applied Spectroscopy: A Compact Reference for Practitioners.* Boston: Academic Press, 1998.

COLOR PLATE 1

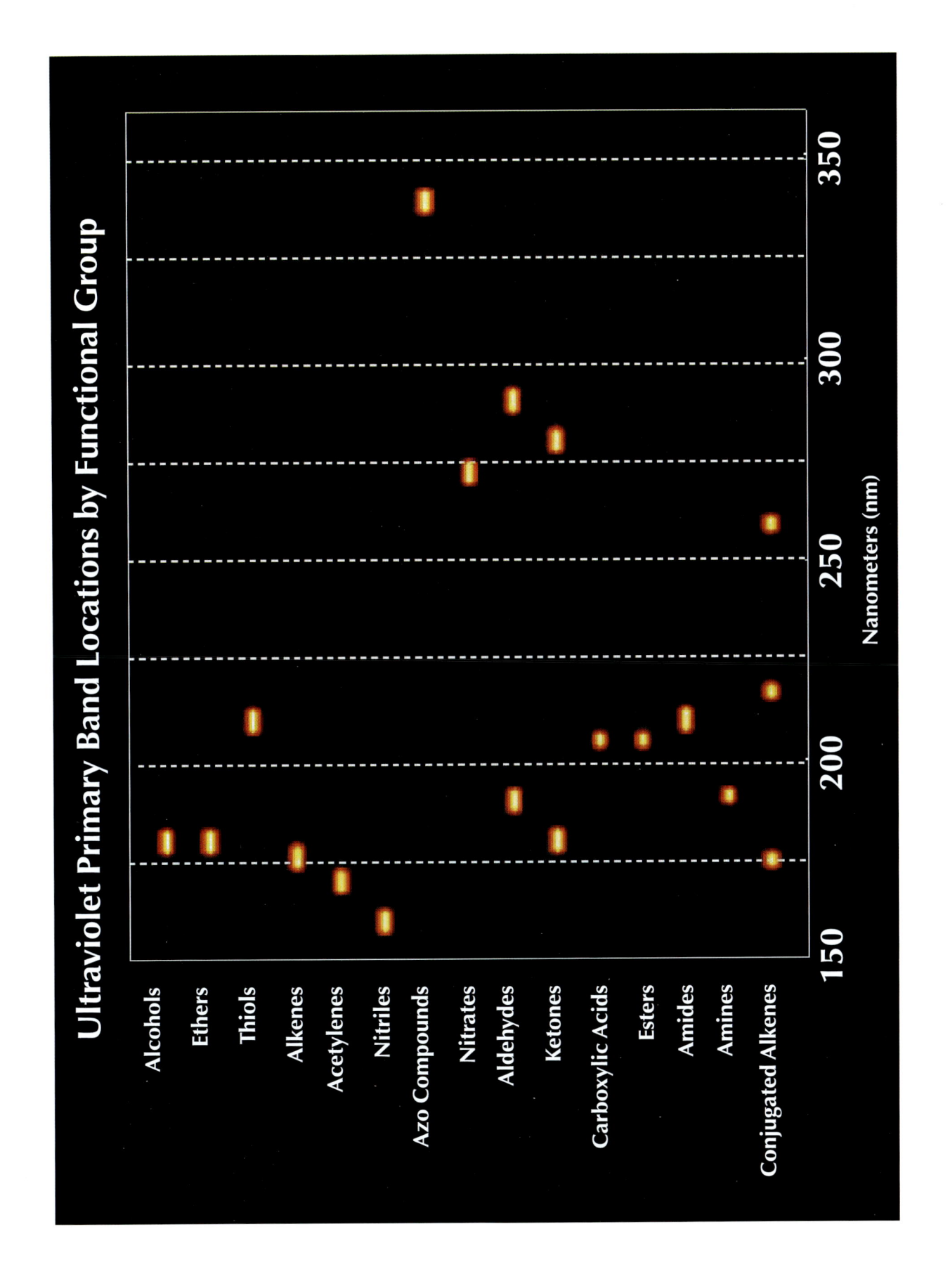

Ultraviolet Primary Band Locations by Functional Group

COLOR PLATE 2

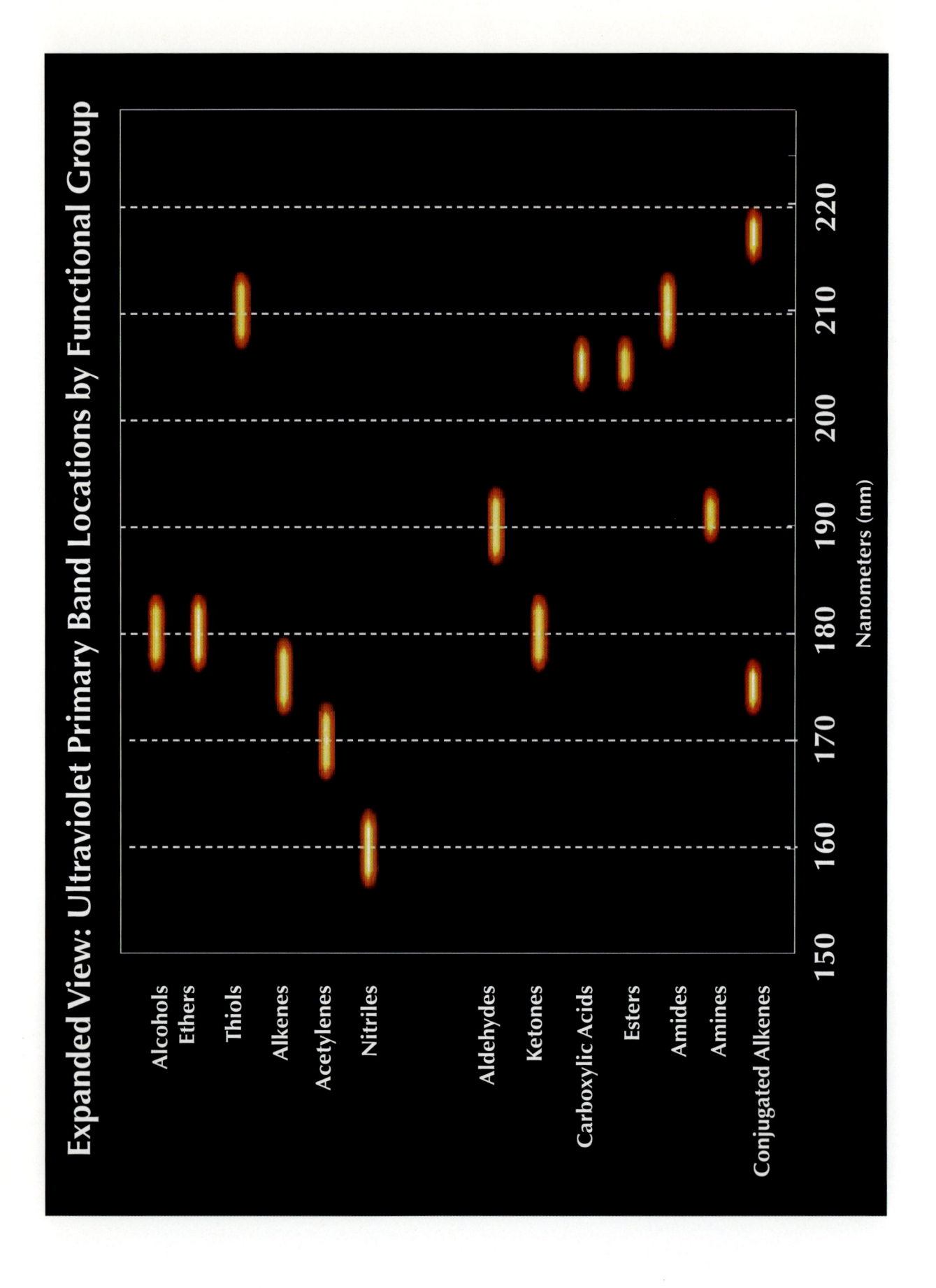

Expanded View: Ultraviolet Primary Band Locations by Functional Group

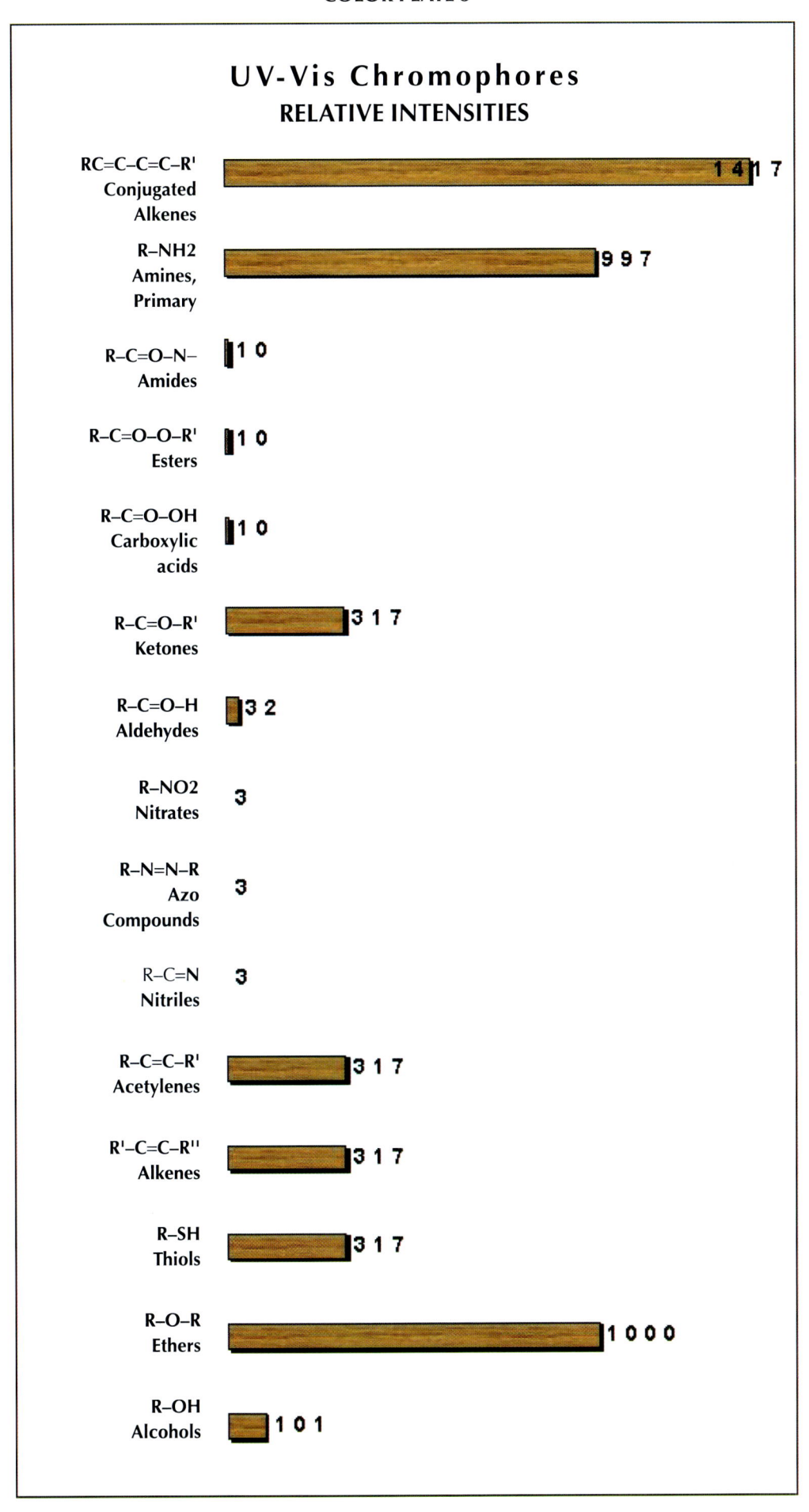

COLOR PLATE 3

UV-Vis Chromophores
RELATIVE INTENSITIES

Chromophore	Relative Intensity
RC=C–C=C–R' Conjugated Alkenes	1417
R–NH2 Amines, Primary	997
R–C=O–N– Amides	10
R–C=O–O–R' Esters	10
R–C=O–OH Carboxylic acids	10
R–C=O–R' Ketones	317
R–C=O–H Aldehydes	32
R–NO2 Nitrates	3
R–N=N–R Azo Compounds	3
R–C=N Nitriles	3
R–C=C–R' Acetylenes	317
R'–C=C–R" Alkenes	317
R–SH Thiols	317
R–O–R Ethers	1000
R–OH Alcohols	101

COLOR PLATE 4

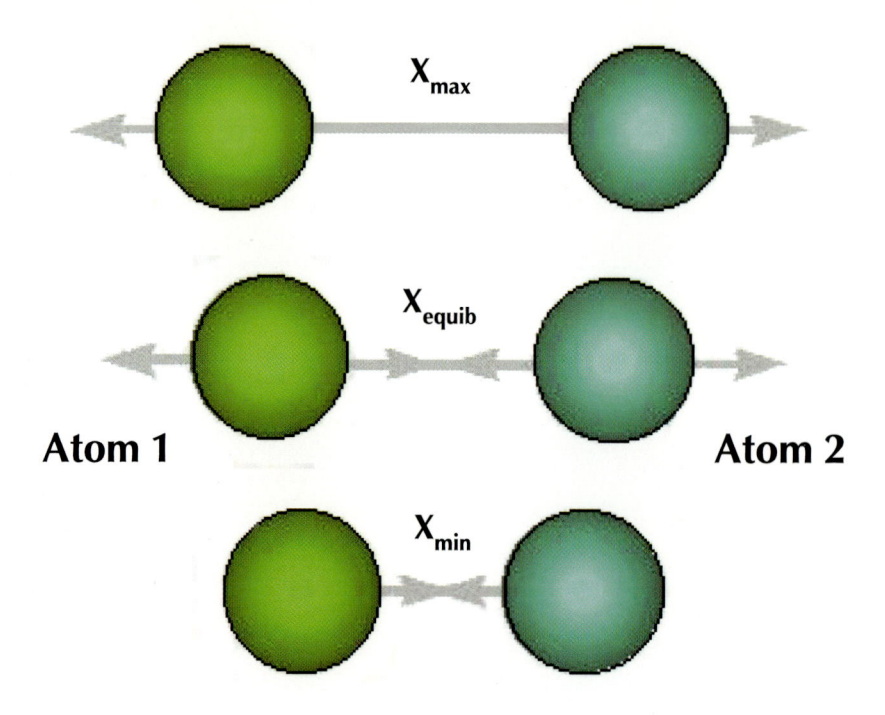

Fig. 15.1 Model of infrared-active molecule as a vibrating dipole between two atoms.

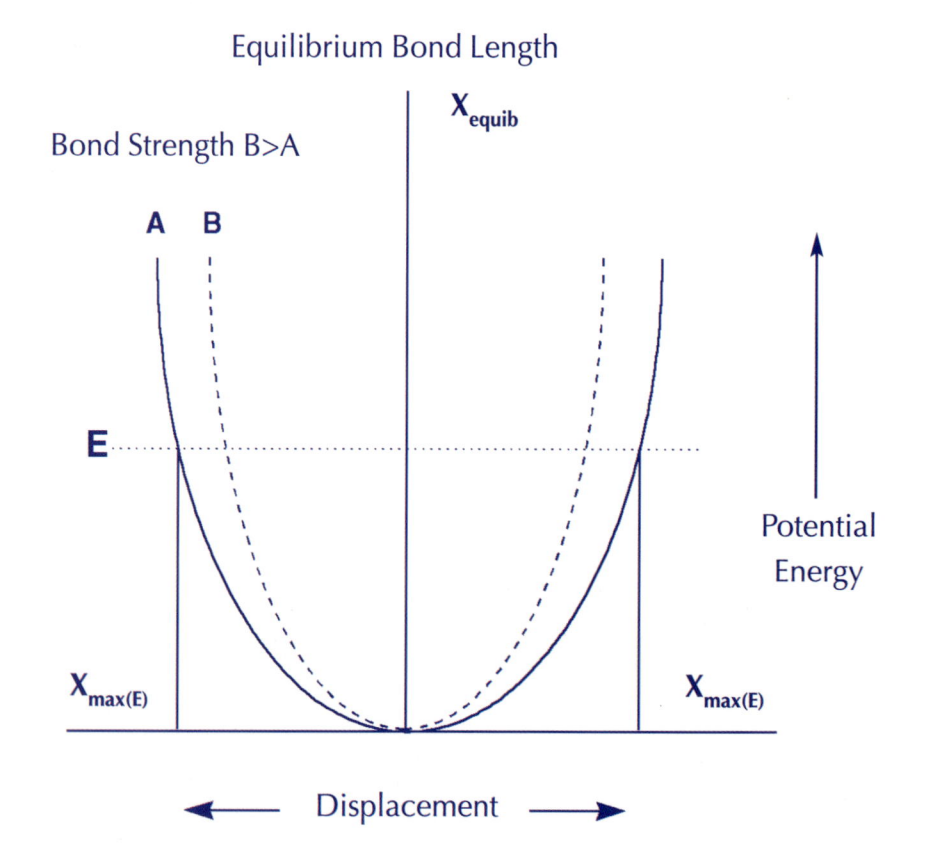

Fig. 15.2 Potential energy curve showing minimum and maximum amplitude for the harmonic oscillator model as a continuum.

COLOR PLATE 5

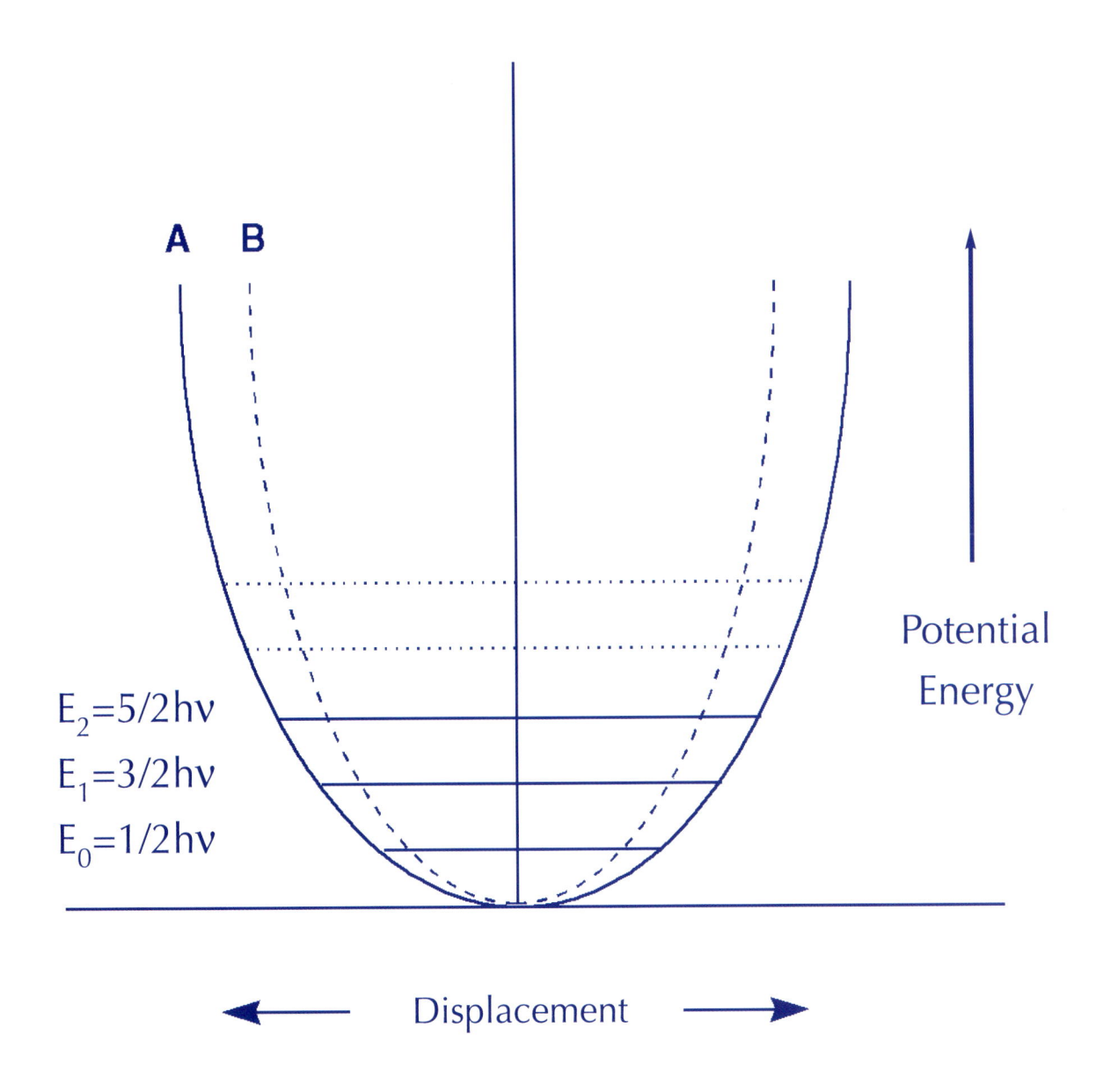

Fig. 15.3 Illustration of the discrete quanta potential energy curves (E_0, E_1, E_2, etc.) for both strong and weakly bonded dipoles as $E_n = [n + (1/2)]h\nu$.

COLOR PLATE 6

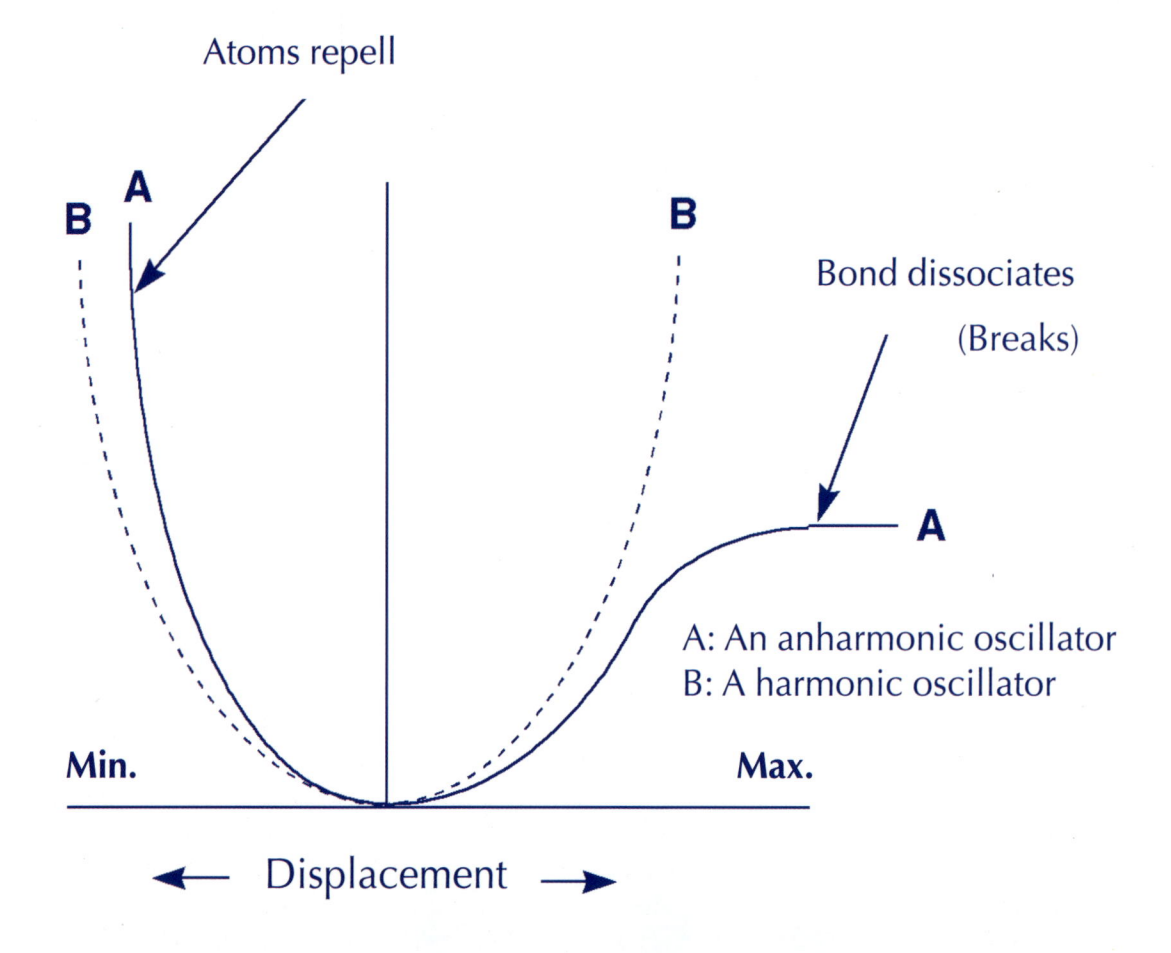

Fig. 15.4 Illustration for the differences in potential energy curves between the ideal harmonic oscillator model and the harmonic oscillator model (better representing the actual condition of molecules).

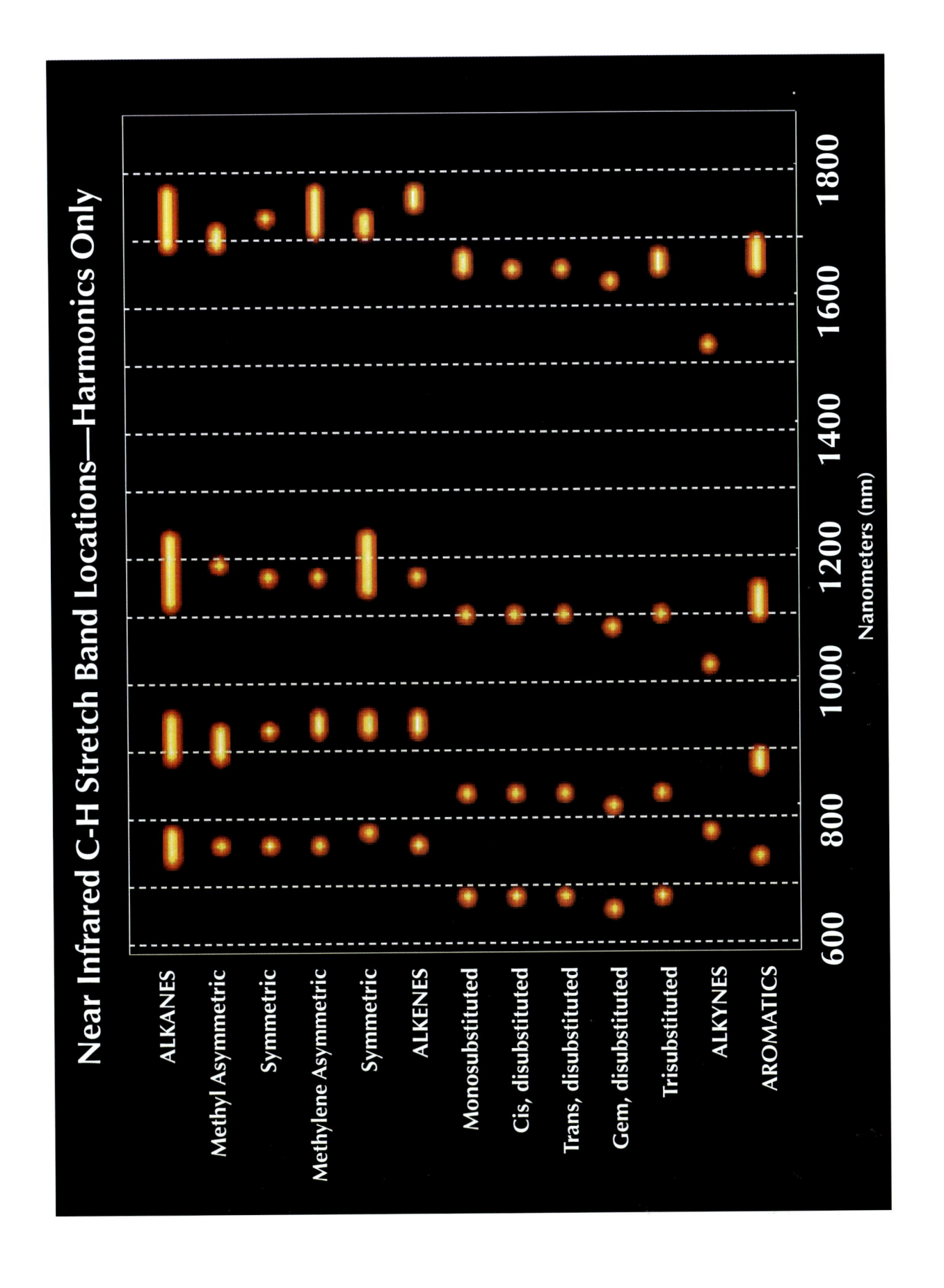

COLOR PLATE 7

Near Infrared C-H Stretch Band Locations—Harmonics Only

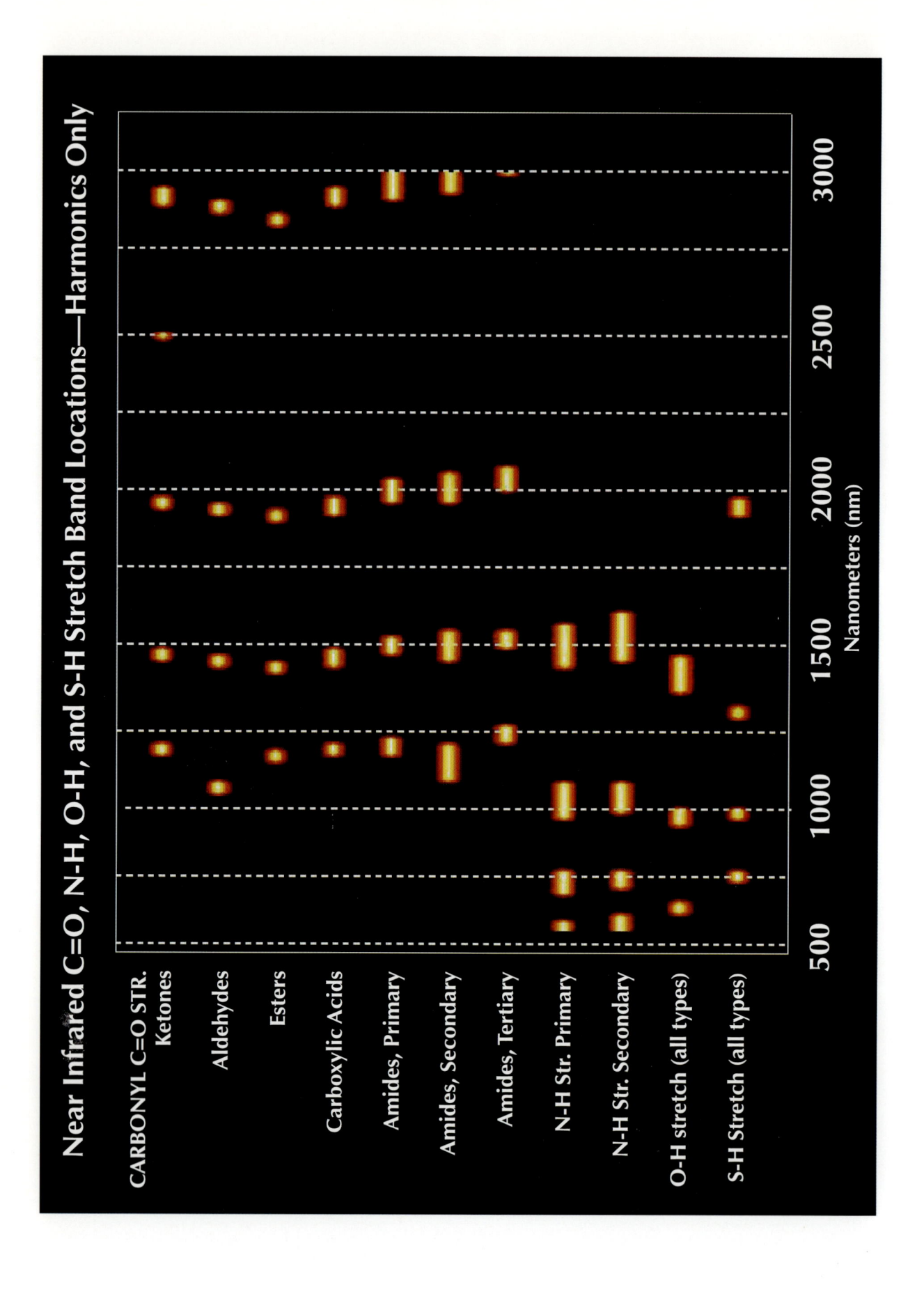

COLOR PLATE 8

COLOR PLATE 9

Infrared band locations for C-H stretching and bending modes

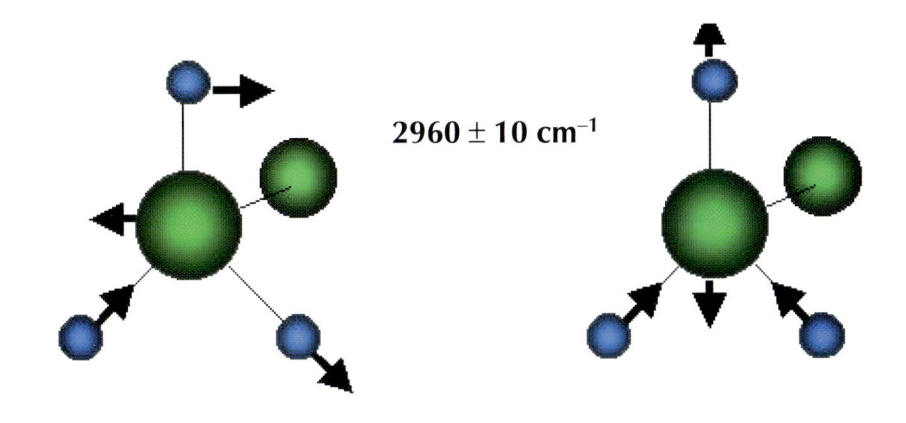

2960 ± 10 cm^{-1}

methyl C-H asymmetric stretch

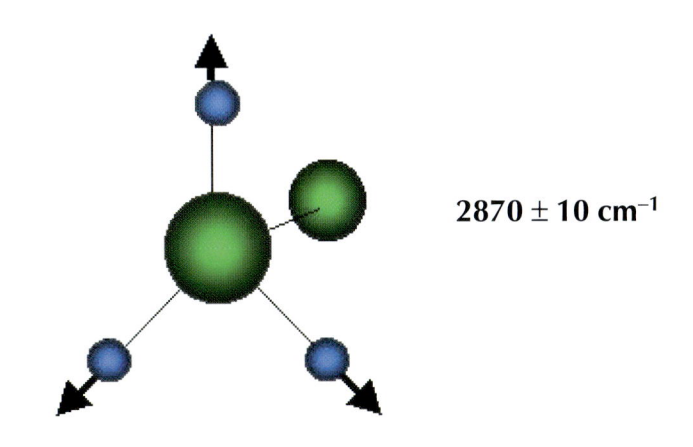

2870 ± 10 cm^{-1}

methyl C-H symmetric stretch

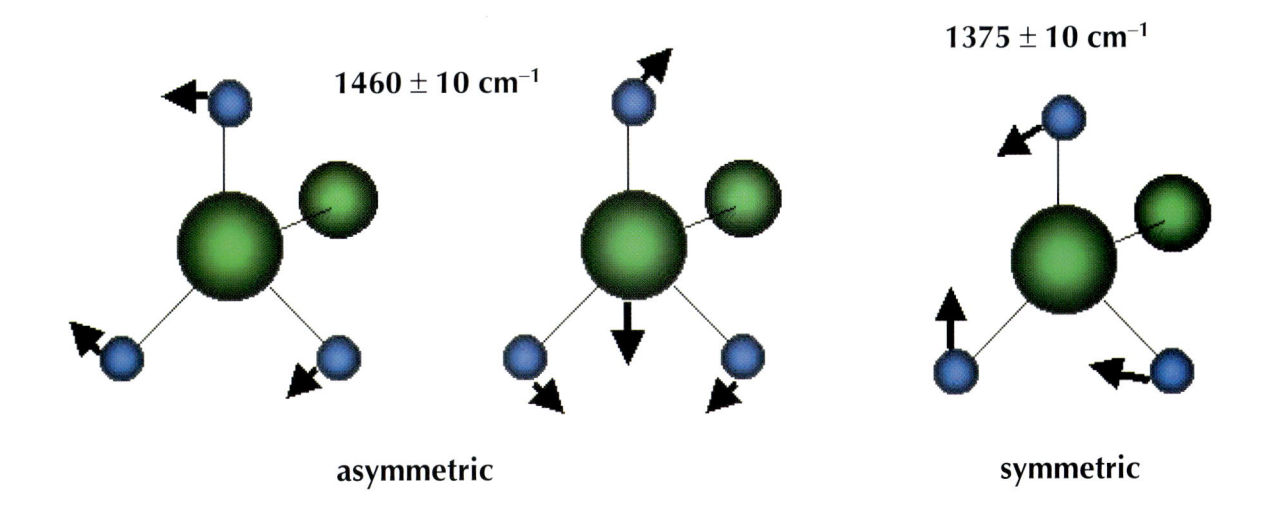

1460 ± 10 cm^{-1}

1375 ± 10 cm^{-1}

asymmetric

symmetric

methyl C-H bending mode

COLOR PLATE 10

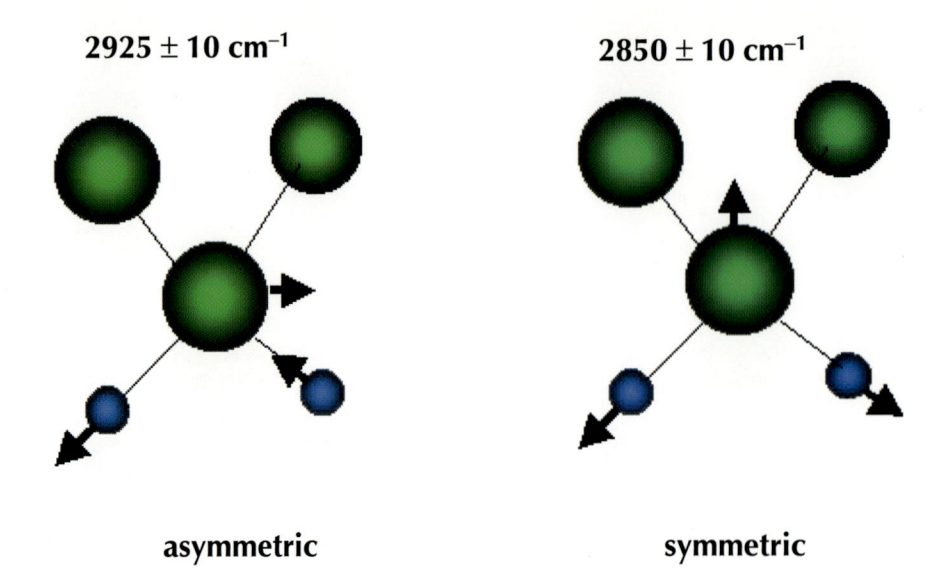

2925 ± 10 cm^{-1} 2850 ± 10 cm^{-1}

asymmetric   symmetric

methylene C-H stretching modes

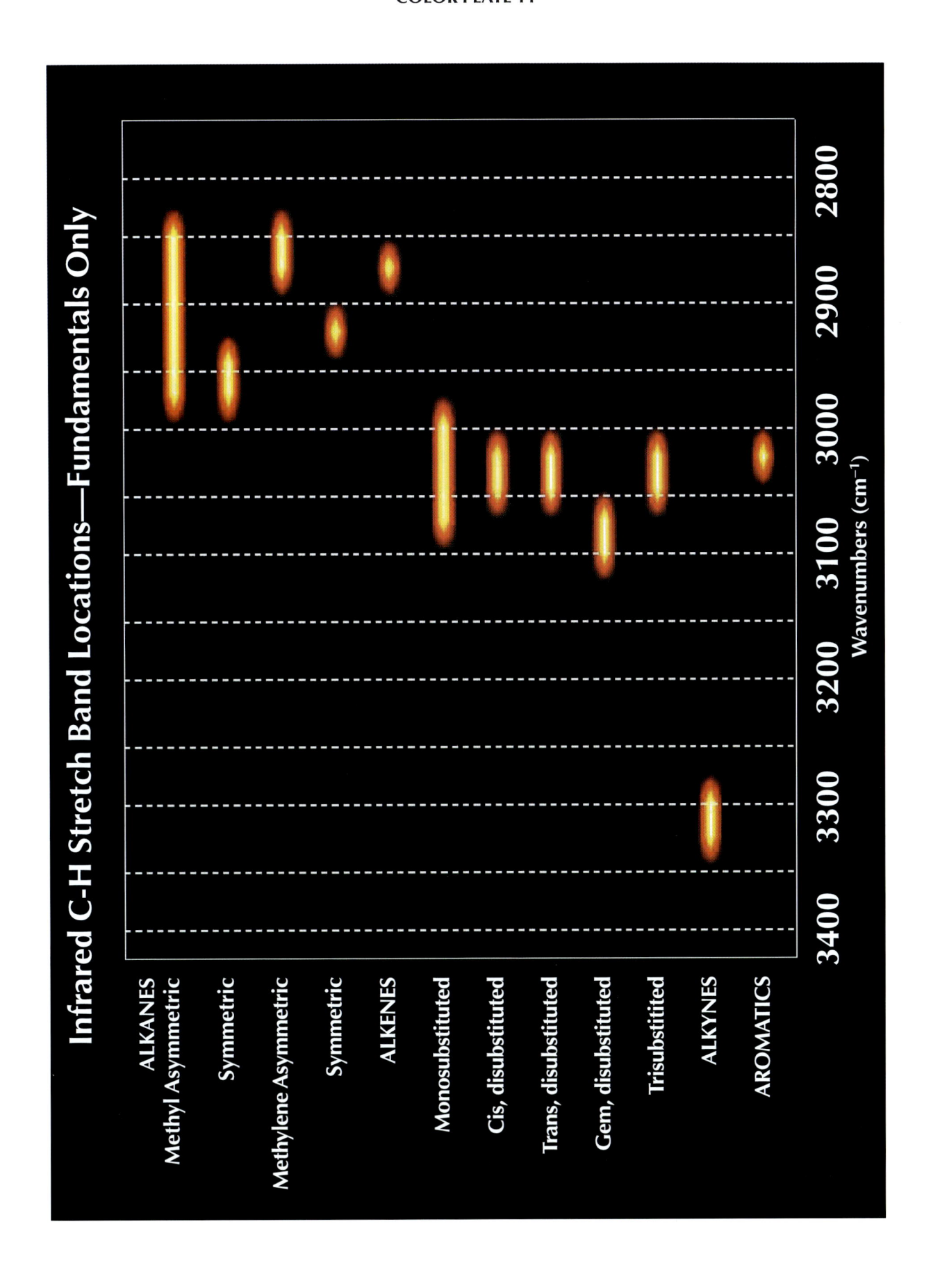

COLOR PLATE 11

Infrared C-H Stretch Band Locations—Fundamentals Only

COLOR PLATE 12

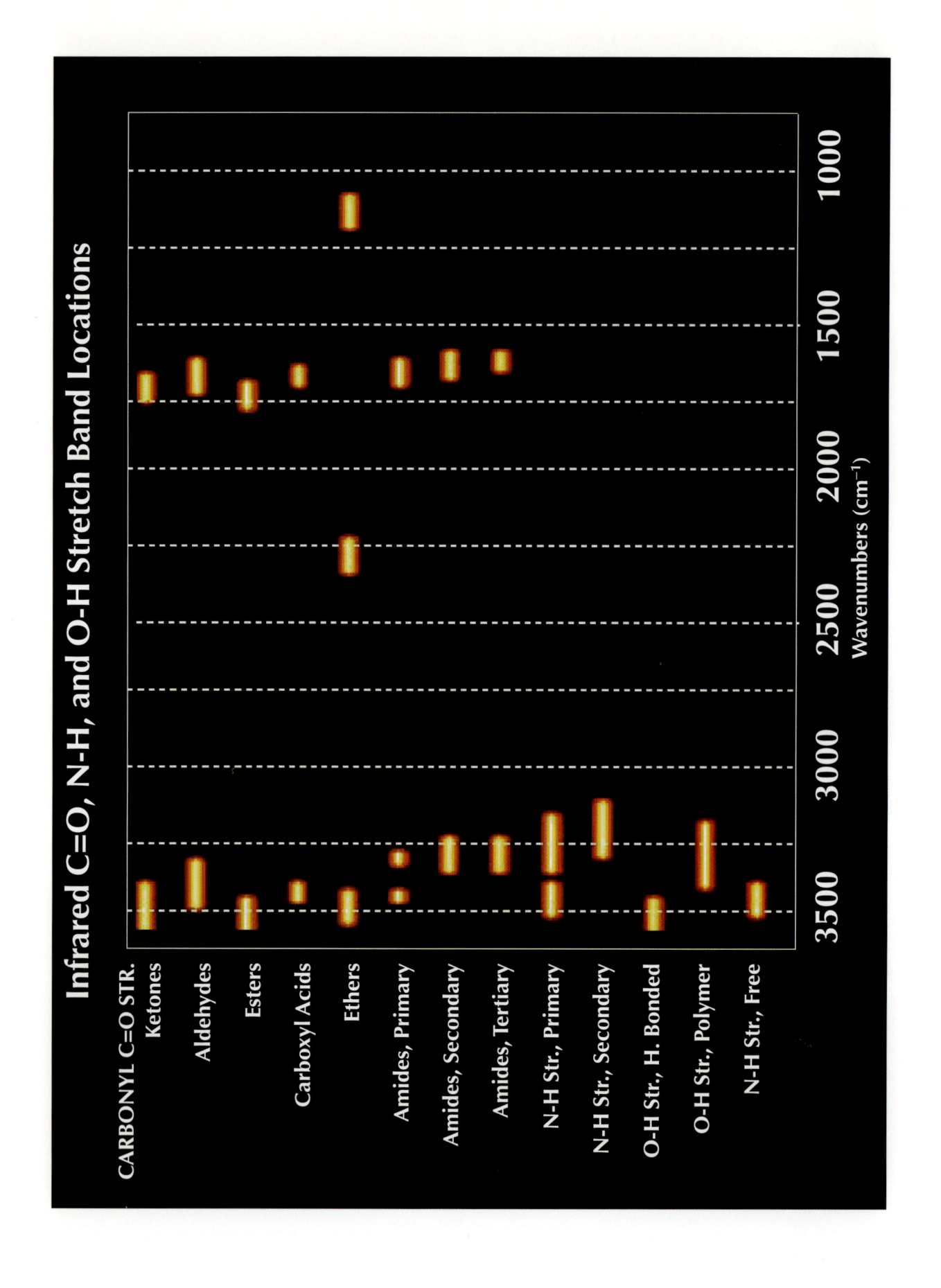

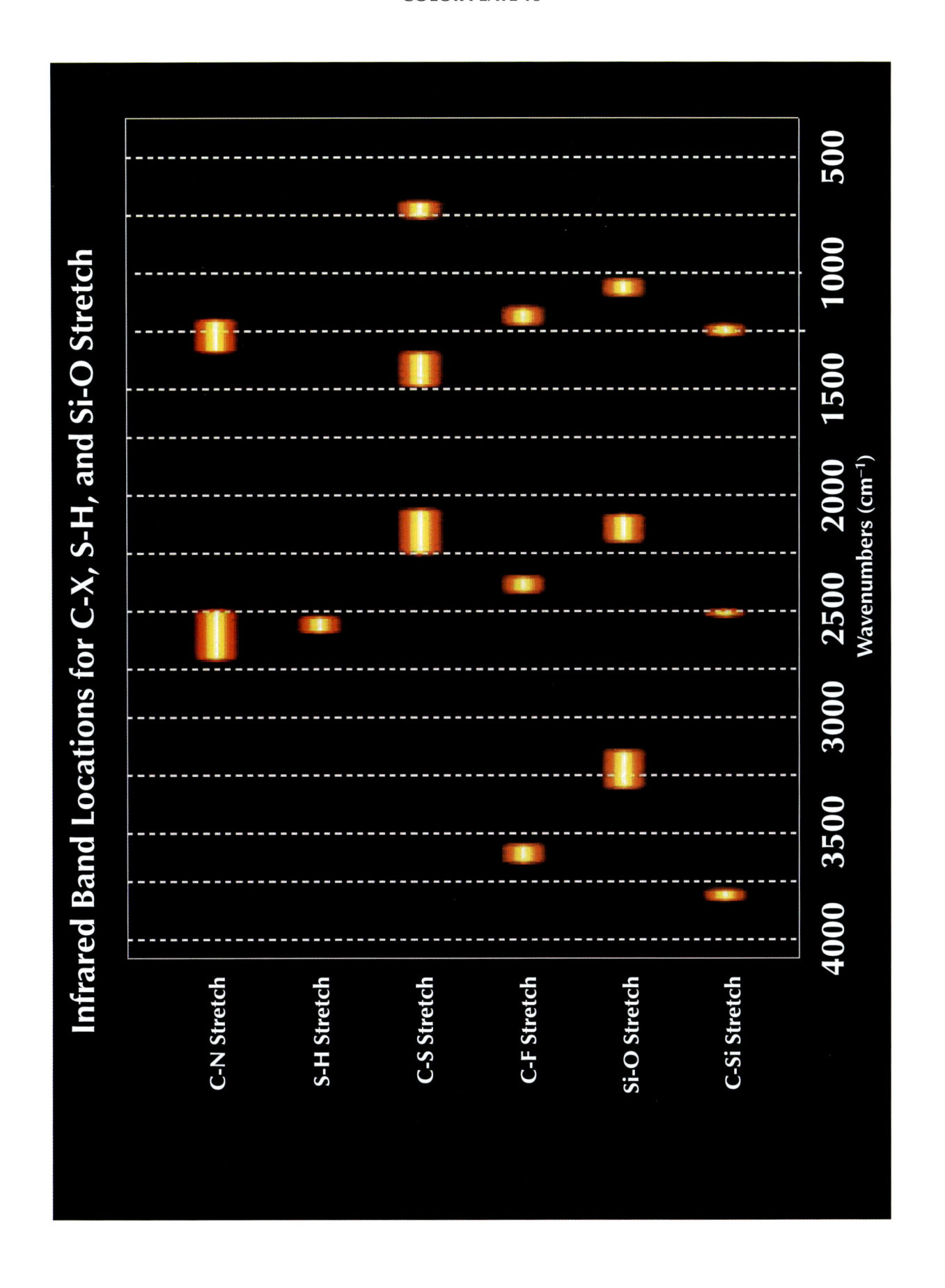

COLOR PLATE 13

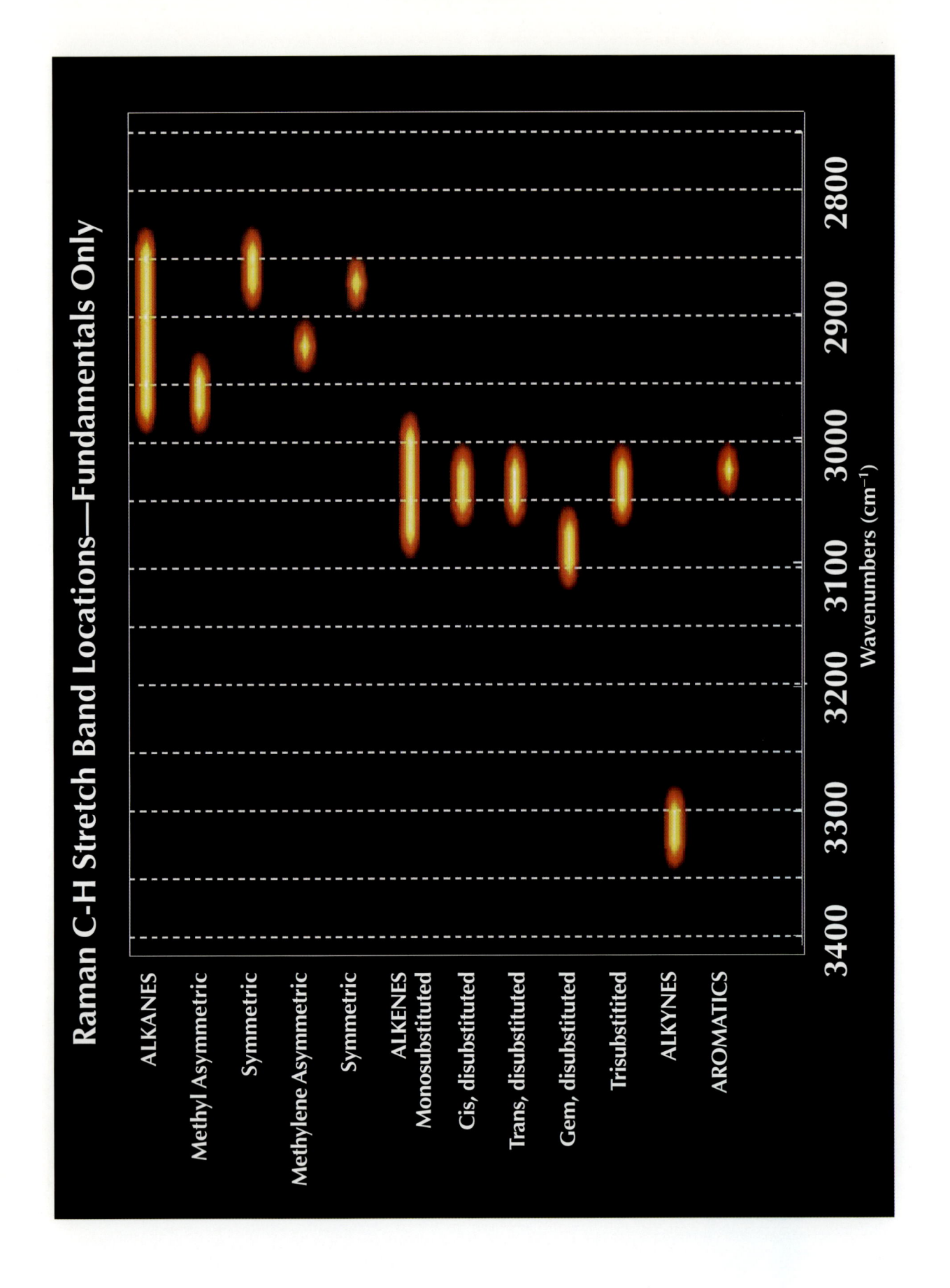

COLOR PLATE 14

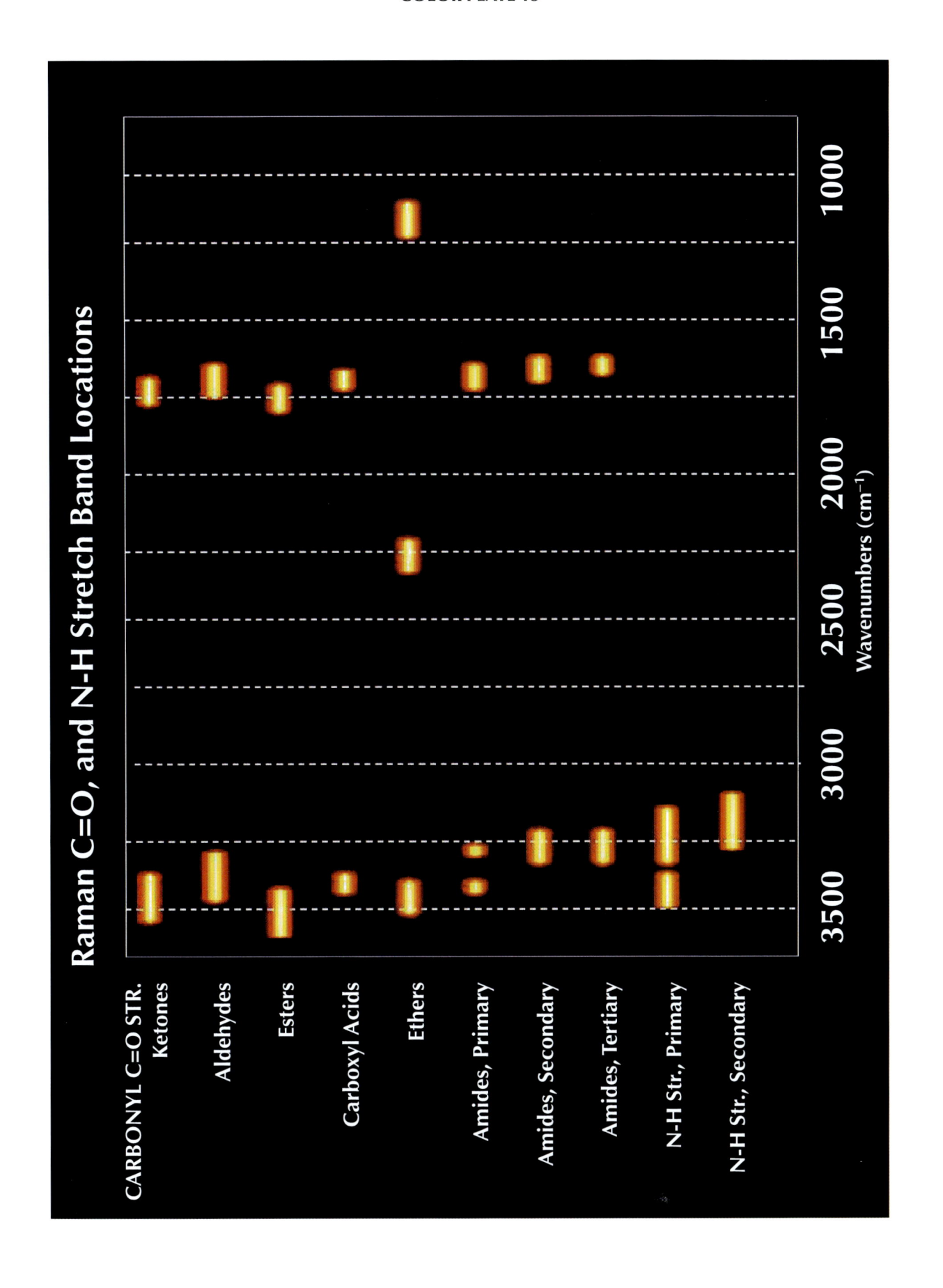

COLOR PLATE 15

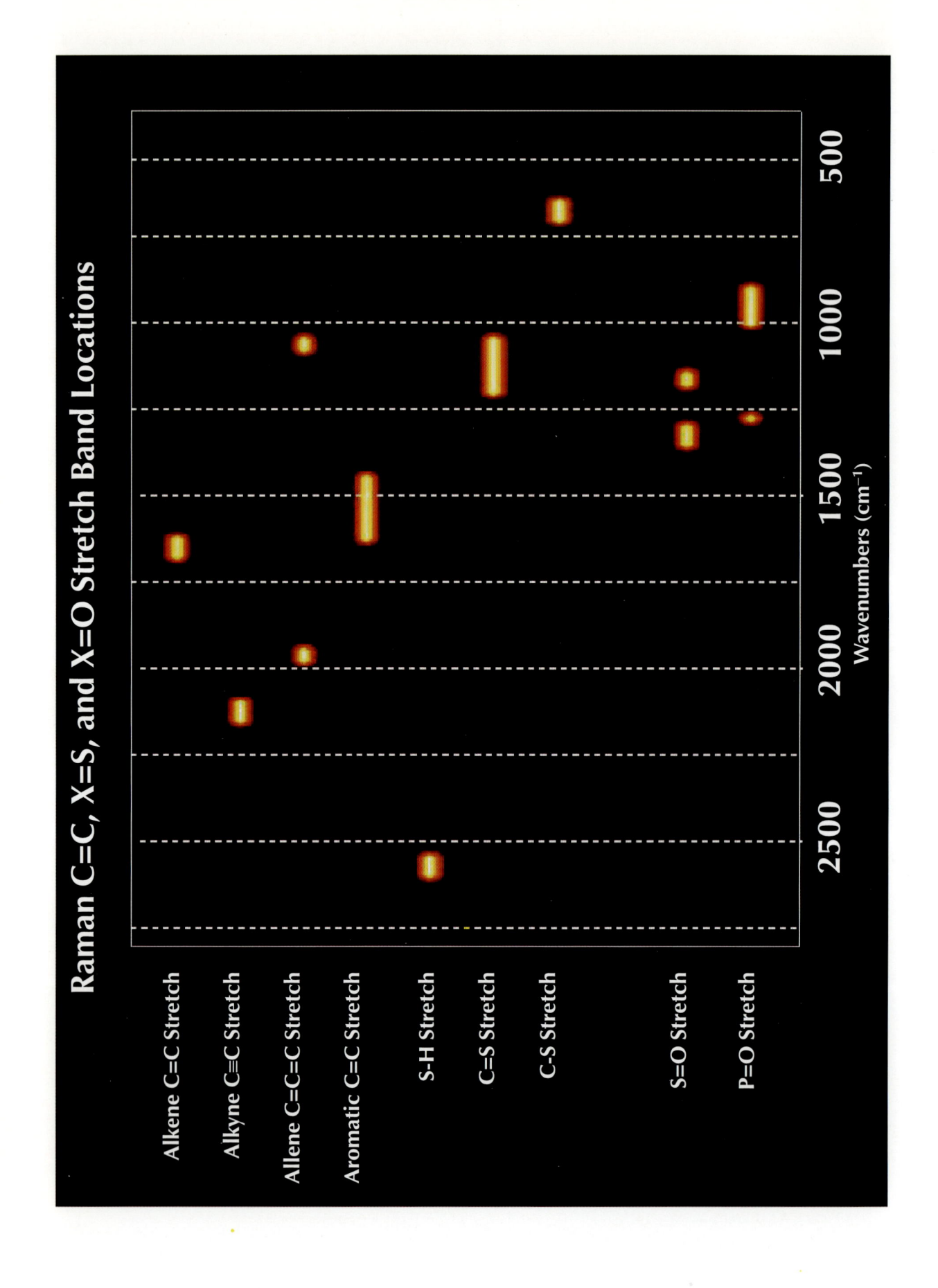

COLOR PLATE 16

Typical polymer and rubber materials exhibit similar NIR bands and positions, represented in Tables 18.4 and 18.5.

Table 18.4   Spectral Correlation Chart for Polymer Materials

Polymer	Band location (in nm)	Band assignment
Poly(propylene)	1192	Asymmetric methyl (C-H) stretch—2nd overtone
	1220	Asymmetric methylene (C-H) stretch—2nd overtone
	1394	Methyl and methylene (C-H) combination
	1700	Asymmetric methyl (C-H) stretch—1st overtone
	1726	Asymmetric methylene (C-H) stretch—1st overtone
	1820	Symmetric methyl (C-H) stretch—1st overtone
	2323	C-H bend—2nd overtone
	2383	C-H stretch and C-C stretch combination
	2454	C-H combination band
Poly(propylene) + poly(ethylene) copolymer	1728	Asymmetric C-H stretch—1st overtone
	2314	C-H bend—2nd overtone
	2356	C-H stretch and C-C stretch combination
Poly(styrene)	1142	Aromatic (C-H) stretch—2nd overtone
	1684	Aromatic (C-H) stretch—1st overtone
Starch	1456	O-H stretch—1st overtone
	1700	Asymmetric methyl (C-H) stretch—1st overtone
	1739	Asymmetric methylene (C-H) stretch—1st overtone
	1927	O-H stretch + HOH deformation + O-H bend comb.
	2105	O-H stretch + HOH deformation combination
	2291	C-H stretch + CH_2 deformation combination
	2488	C-H stretch + C-C stretch + C-O-C stretch comb.
Poly(acrylic acid)	1185	Asymmetric methylene (C-H) stretch—2nd overtone
	1428	O-H stretch—1st overtone
	1695	Asymmetric methyl (C-H) stretch—1st overtone
	1735	Asymmetric methylene (C-H) stretch—1st overtone
	1924	O-H stretch—1st overtone + C=O stretch—2nd overtone + O-H stretch/HOH deformation combination + O-H bend—2nd overtone
	2149	C-H stretch/C=O stretch combination + symmetric C-H deformation
	2286	C-H stretch + CH_2 deformation + C-H bend—2nd overtone
	2489	C-H stretch + C-C stretch + C-O-C stretch comb

Table 18.5 Spectral Correlation Chart for Rubber Materials

Rubber	Band location (in nm)	Description
Styrene, ethylene butylene, styrene copolymer	1193	Methyl (C-H) stretch—2nd overtone
	1391	C-H stretch mode combination
	1724	Asymmetric methylene (C-H) stretch—1st overtone + aromatic C-H stretch—1st overtone
	2180	C-H bend—2nd overtone
	2283	C-H stretch + CH_2 deformation combination
	2324	C-H stretch + CH_2 deformation combination
	2386	C-H stretch + C-C stretch + C-O-C stretch comb.
Styrene-isoprene-styrene	1195	Methyl (C-H) stretch—2nd overtone
	1400	C-H stretch mode combination
	1719	Asymmetric methylene (C-H) stretch—1st overtone + aromatic C-H stretch—1st overtone
	2171	C-H bend—2nd overtone
	2291	C-H stretch + CH_2 deformation combination
	2336	C-H stretch + CH_2 deformation combination
	2470	C-H stretch + C-C stretch + C-O-C stretch comb.
Silicone (dimethyl siloxane)	1452	Si-O stretch—1st overtone
	1748	Methyl (C-H) stretch—1st overtone
	1933	Si-O-H stretch + Si-O-Si deformation combination
	2295	C-H bend—2nd overtone

VI.

INFRARED AND RAMAN SPECTROSCOPY

19.

REVIEW OF INTERPRETIVE SPECTROSCOPY FOR RAMAN AND INFRARED

For infrared and Raman spectroscopic theory the reader is referred to the References and Additional Sources sections at the conclusion of this chapter. This review is for the express purpose of assisting the spectroscopist in band assignments for infrared and Raman spectra.

GROUP FREQUENCIES

Fundamental frequencies that are characteristics of groups of atoms (termed *functional groups*) with common group names and general infrared locations (in wavenumbers) are discussed in this review. Functional groups include the following [1]:

–OH, hydoxyl- [O-H]

R-OH, alcohols [O-H]

–CH$_2$–OH, primary alcohols

R'R"CH-OH, secondary alcohols

R'R"R'"C-OH, tertiary alcohols

Ar-OH, phenolics (or phenols) [O-H]

–OCH$_3$, methoxy (or ether group)

C-O-C, ethers

R-C=O-H, aldehydes

R-C=O-R', ketones

C=C=O-R', ketenes

R-C=O-OR', esters

S-C=O-O-R', thiol esters

R-C=O-C=O-R, anhydrides

–C-O-O-C-, peroxides

–O-C=O-C=O-O-, oxalates

–O-C=O–, carboxy-

–C=O-, carbonyl group

–C=O-NH$_2$, primary amides

–C=O-NHR, secondary amides

–C=O-NR'R", tertiary amides

–C≡N, nitriles

C-C, paraffinics (or alkanes)

CH$_3$-(CH$_2$)$_N$-CH$_3$, normal paraffins

R-CH$_3$, methyl C-H

R-CH$_2$-R', methylene C-H

R-CHR'R", methine C-H

C=C, olefinic group (or alkenes)

–CH=CH$_2$, vinyls

R'R"C=CH$_2$, vinylidenes

>C=C=CH$_2$, allenes

–C≡C-, acetylinics

R-C≡N, nitriles

C$_N$H$_N$, aromatics

C$_N$H$_{2N}$, naphthalenes (or cycloalkanes)

C-NO, nitroso- group

CONO, nitrites

R'R"NNO, nitrosamines

R-NO$_2$, nitro- group

R-NH$_2$, primary amines

R'R"NH, secondary amines

N=N, azo-group

–N=N$^+$=N$^-$, -N$^-$-N$^+$≡N, azides

NNO, azoxy group

R-O-N-N=O, organic nitrites

–NO$_3$, nitrates

ON=NO, nitroso-group

H$_3$N$^+$-CH-COO$^-$, amino acids

FINGERPRINT FREQUENCIES

Fingerprint frequencies are due to interactions of the molecular vibrations of the entire molecule rather than specific functional groups. Since the infrared spectrum is unique to an individual molecule, it is termed the *fingerprint* of that molecule, much as the fingerprint of a person indicates the specific identity of that person. However, in identifying the unique "fingerprint vibrations" of a molecule, information is given as to the approximate molecular formula and composition of the molecule. (Incidentally, William James Hershel, grandson of F. William Hershel, was the first to identify the uniqueness of individual human fingerprints [2]).

DIVIDING THE INFRARED SPECTRUM

The group frequencies are given as those bands occurring above 1500 cm^{-1}, and the fingerprint frequencies are those below 1500 cm^{-1}. The group frequencies above 1500 cm^{-1} usually demonstrate bands of reasonable intensity, where reliable interpretation is possible, based upon frequency position, whereas the fingerprint frequencies contain bands for group and fingerprint frequencies. This mixed (or overlapping) band region below 1500 cm^{-1} may give inaccurate interpretation based only upon frequency position. In the fingerprint region the most useful interpretive information is based upon shape (e.g., broad, very broad, sharp, very sharp); or intensity (e.g., weak, medium, intense). The fingerprint region is also useful for interpretation relative to the absence of a particular band. In summary, the group frequency region is used to derive a first assumption about the identity of the sample based on its infrared spectrum, and then the fingerprint region is used to verify the assumptions by the presence or absence of bands as well as their shapes and intensities. Group frequencies are used with general rules and comparative known spectra to make positive identification of a sample spectrum. The ideal group frequencies have the characteristics of: 1) always being found above 1500 cm^{-1}, (2) having intense amplitude, (3) having identifiable and consistent position in frequency, (4) being reliable and always appearing when a particular group is present, (5) being isolated without interferences, (6) being narrow (in the range of ±25 cm^{-1}), and (7) having groups with more than one characteristic frequency (used for further verification of the group's presence) [3–5].

COUPLING OF VIBRATIONS

Coupling indicates that the oscillators, or molecular vibrations of two or more molecules, are interactive, so the original vibrational energy states (if the vibrations could occur independent of one another) result in split energy states due to the interaction of the vibrations. Coupling is divided into two basic orders, first and second (Fermi resonance). First-order coupling can be involved in several important infrared group frequencies. For example, CO_2 has the two separate (uncoupled oscillators) of C=O, each occurring at approximately 1850 cm^{-1}. The interactive (coupled oscillators) energy states occur at 2349 cm^{-1} (asymmetric stretch) and 1340 cm^{-1} (symmetric stretch). First-order coupling is involved for multiple infrared group frequencies [6], such as: (1) the stretches for all cumulated double bonds, X=Y=Z, e.g., C=C=N; (2) the stretches in XY_2, including $-CH_2-$, and H_2O; (3) the stretches in XY_3 groups, including $-CH_3$; (4) the deformations of XY_3 groups, including $-CH_3$, CCl_3; and (5) the N-H in-plane bend of secondary amides, e.g., R–CO–NH–R'.

FERMI RESONANCE (OR SECOND ORDER COUPLING)

Fermi resonance is the interaction or coupling of two vibrational energy states, with resultant separation of the states, where one of the states is an overtone or a sumtone. An overtone vibration occurs as the integer multiple of a fundamental vibration in frequency space (intensity falls off rapidly with the higher multiples):

The first overtone of a fundamental vibration (v_i) is equal to $2v_i$.

The second overtone of a fundamental vibration is equal to $3v_i$.

The third overtone is equal to $4v_i$.

Note: overtones are the special case of sumtones where the frequencies are identical. A sumtone is the general case of an overtone where the frequencies are not equal and where a variety of vibrational energy states can occur:

A binary sumtone is equal to the sum of two fundamentals, e.g., $v_i + v_k$.

A ternary sumtone is equal to the sum of three fundamental vibrations, e.g., $v_i + v_k + v_m$.

Other sumtones can occur, such as the sum of an overtone and a fundamental vibration, e.g., $2v_i + v_k$.

Three requirements are stated for Fermi resonance [7]:

The zero-order frequencies must be close together (typically within 30 cm^{-1}).

The fundamental and the overtone or sumtone must have the same symmetry.

There must be a mechanism for the interaction of the vibrations. *Note:* (1) The vibrations cannot be separate (or localized) in distinctly different parts of the molecule. (2) The vibrations must be mechanically interactive in order that the interaction of one vibration affects another.

The results of Fermi resonance are important for infrared, Raman, and near-infrared spectroscopy. Fermi resonance causes the following effects on spectral bands:

The resultant bands are moved in position from their expected frequencies.

Overtone bands are more intense than expected.

There may exist doublet bands where only singlets were expected.

Solvent changes can bring about slight shifts in the frequency location of a band, and intensities can be greatly changed [5, 8–9].

GROUP FREQUENCIES OF ALKANES (PARAFFINS)

Normal Hydrocarbons (*n*-alkanes)

The number of normal vibrations occurring within the infrared region for saturated hydrocarbons is $3N - 6$, where N is the sum of carbon plus hydrogen atoms; e.g., *n*-hexane (C_6H_{14}) exhibits $3(20) - 6 = 20$ normal vibrational modes. Internal rotation and overlapping of bands complicate the interpretation of alkane (paraffin) group spectra. C-H stretching vibrations can be expressed to a first approximation by the straightforward use of Hooke's law describing the vibration between two masses connected by a vibrating spring.

The first consideration in the description of alkane group spectral interpretation would include the C-H stretching vibrations for methyl (–CH$_3$) and methylene (–CH$_2$–) groups. The *methyl C-H stretching vibrations* are found in two basic modes:

The antisymmetric (degenerate or out-of-phase) methyl C-H stretch at 2962 ± 10 cm^{-1}.

The symmetric (in-phase) methyl C-H stretch at 2872 ± 10 cm^{-1}.

Three C-H stretching modes correspond to three coupled C-H stretching vibration oscillators. The molecular extinction coefficient for methyl C-H stretch remains constant irrespective of increasing alkane chain length.

The *methylene C-H stretching vibrations* are also found in two basic modes:

The antisymmetric (degenerate or out-of-phase) methylene C-H stretch at 2926 ± 10 cm^{-1}.

The symmetric (in-phase) methylene C-H stretch at 2853 ± 10 cm^{-1}.

The two methylene stretching modes result from two coupled C-H stretching vibration oscillators. Increasing the chain length within alkane groups brings about a linear increase in the molecular extinction coefficient of methylene C-H stretching bands.

The carbon-hydrogen vibrational modes for alkanes also contain both methyl and methylene bending vibrations. The *methyl C-H bending vibrations* are found in two modes:

The antisymmetric (degenerate or out-of-phase deformation) methyl C-H bending at 1460 ± 10 cm^{-1}.

The symmetric (in-phase or "umbrella") methyl C-H bending at 1375 ± 10 cm^{-1}. This functional group vibrational frequency is dependent upon the atom to which the methyl group is attached.

Examples from the literature illustrating band frequency location relative to the atom attached to methyl group (X-CH$_3$) are presented in Table 19.1 [10].

There is no coupling of methyl C-H bending modes to the rest of the system.

The *methylene C-H bending vibrations* are found in four basic modes:

The symmetric (scissoring) methylene C-H bending at 1455 ± 10 cm^{-1}, where no coupling is observed.

Table 19.1 Examples of Band Frequency Locations Relative to the Atom Attached to Methyl Group (X–CH$_3$)

Group	Fundamental frequency (±10 cm^{-1})	Fundamental wavelength (in nm)
Pb-CH$_3$	1165 cm^{-1}	8511–8658 nm
Sn-CH$_3$	1190	8333–8475
Sb-CH$_3$	1200	8264–8403
Hg-CH$_3$	1200	8264–8403
Ge-CH$_3$	1235	8032–8163
As-CH$_3$	1250	7937–8065
I-CH$_3$	1252	7924–8052
Se-CH$_3$	1280	7752–7874
Br-CH$_3$	1305	7605–7722
Si-CH$_3$	1255 ± 5 cm^{-1}	7937–8000
P-CH$_3$	1300	7634–7752
B-CH$_3$	1320	7519–7634
S-CH$_3$	1325	7491–7605
Cl-CH$_3$	1355	7326–7435
Carboxyl C-CH$_3$	1355	7326–7435
C-CH$_3$	1378 ± 5 cm^{-1}	7231–7283
N-CH$_3$	1425	6969–7067
O-CH$_3$	1455	6826–6920
F-CH$_3$	1475	6734–6826

Source: Ref. 10.

The in-phase twisting vibration at 1300 ± 10 cm^{-1} with observed coupling.

The coupled wagging vibrations at 1300–1000 cm^{-1}.

The in-phase rocking vibration for alkanes with four or more carbons is located at 720 ± 10 cm^{-1}, where coupling is observed.

In addition, other bending modes of methyl and methylene can be observed. The alkane carbon-to-carbon stretching and carbon-to-carbon bending vibrations also occur as weak group frequencies.

Branched Chain Hydrocarbons (Iso-, Gem-Dimethyl, Tertiary Butyl, and Cyclic Groups)

Methine groups (tertiary methyl, R$_3$-C-H) are not important for the infrared region and do not exhibit unique infrared spectral characteristics.

Isopropyl and gem-dimethyl groups exhibit a symmetric bending vibration doublet at 1370 ± 10 cm^{-1} with coupling absent.

Tertiary butyl substitution exhibits a symmetric bending vibration doublet from 1390 to 1365 cm^{-1} that does not show mechanical coupling.

Cyclic hydrocarbons demonstrate methylene C-H stretching modes based upon ring size. In addition there are a number of spectral effects based upon lone electron pair interactions with carbon-hydrogen molecular orbitals. The reader is referred to a number of references describing the unique spectra of cyclic hydrocarbons with various substituted functional groups. Primary references are listed under the Reference Textbooks section at the end of this review chapter.

GROUP FREQUENCIES OF ALKENES (OLEFINS)

C-H Stretch of Alkenes

The olefins include bands for saturated hydrocarbons as well as the olefinic C=C stretching and olefinic C-H stretching and bending vibrations.

Olefinic C-H Stretch

Olefinic C-H stretch occurs at 3100–3000 cm^{-1}. C-H stretch for an unsaturated carbon occurs above 3000 cm^{-1}, whereas the saturated carbon C-H bond stretching occurs below 3000 cm^{-1}. The sp-type C-H stretch for ≡C-H occurs at approximately 3300 cm^{-1}; the sp^2-type C-H stretch associated with aromatic and olefinic carbons occurs at 3100–3000 cm^{-1}; and the sp^3-type stretch for saturated carbons is observed at the high-frequency ranges between 2990 and 2800 cm^{-1}.

Olefinic C-H stretch can occur as an end group, e.g. R–R–C=C, or as an imbedded group associated with other carbons on either end, e.g., R–C=C–R. The absorption bands for these two olefinic types occur from 3060 to 3100 cm^{-1} for the end-group type and from 3010 to 3050 cm^{-1} for the imbedded type. Note that compounds containing the vinyl group (–CH=CH$_2$) have weak absorption bands at approximately 3080 cm^{-1} and 3030 cm^{-1}.

Olefinic C=C Stretch

Olefinic C=C stretch is infrared active, but the intensity is quite variable, depending upon atoms associated to the C=C group. Table 19.2 below summarizes type appearance of absorption bands associated with olefinic C=C stretch for vinyl-type molecules.

Factors affecting the frequencies for the various olefinic groups include *conjugation,* which is the sharing of electrons between the C=C olefinic bond and atoms or molecules attached to

GROUP FREQUENCIES OF ALKENES (OLEFINS)

Table 19.2 Locations of C=C Stretch Vibrations–Related Olefinic-Containing Molecules

Olefinic molecule type (R-type substituents)*	C=C Stretch band location	Relative band intensity
Trans, tri- and tetrasubstituted	1685–1665 cm^{-1} (5935–6006 nm)	Weak
Cis olefins	1665–1625 cm^{-1} (6006–6154 nm)	Medium to strong
Monosubstituted	1660–1630 cm^{-1} (6024–6135 nm)	Medium to strong
Trans, disubstituted	1680–1685 cm^{-1} (5952–5935 nm)	Weak
Cis, disubstituted	1669–1630 cm^{-1} (5992–6135 nm)	Medium to strong
Disubstitution on a single carbon	1669–1630 cm^{-1} (5992–6135 nm)	Medium to strong
Trisubstitution	1680–1685 cm^{-1} (5952–5935 nm)	Weak
Tetrasubstitution	1680–1685 cm^{-1} (5952–5935 nm)	Weak

*Substituents other than R-groups will change the frequencies of absorption.

either of the carbons in the C=C group. The associated groups lower the frequency by 20–30 cm^{-1}. The conjugated groups that lower the frequency of the –C=C– stretch include: nitrile (–C≡N), occurring at 1609 cm^{-1}; carbonyl (–CO–R), occurring at 1617 cm^{-1}; phenyl (–C$_6$H$_5$), at 1630 cm^{-1}; vinyl (–CH=CH$_2$), at 1621 cm^{-1}; and methyl (–CH$_3$), at 1651 cm^{-1}. In the Raman measurement technique, conjugated alkenes demonstrate strong absorption bands.

The attachment of =C to a variety of ring sizes affects the frequency of the C=C olefinic stretch. A six-member phenyl group attached to a (=C) group yields a 1651-cm^{-1} band; a five-member ring yields a 1657-cm^{-1} band; a four-member ring exhibits a band at 1678 cm^{-1}, a three-member ring exhibits a band at 1781 cm^{-1}; and the alyne two-carbon C=C olefinic group (H$_2$C=C=CH$_2$) occurs at 1974 cm^{-1}.

Carbonyl groups attached to olefinic groups yield a variety of frequency shifts, depending on the conjugation between carbonyl, the olefinic group, and the other attached groups. The section on carbonyls discusses this in greater detail.

Cyclic olefins exhibit unique frequencies for –C=C– stretch depending on the number of carbons within the cyclic olefin group and the type of molecule or atom attached to the olefinic carbons. Table 19.3 illustrates the frequency for olefinic C=C stretch dependent upon the size of the cyclic olefin and the attached group types. An increase in the electronegativity of the attached group raises the frequency of the olefinic C=C stretch; resonance lowers the frequency of the C=C stretch. And there are interactions when the attached groups exhibit both electronegative and resonance features.

Table 19.3 C=C Stretch Frequency with Respect to Carbon Size of Cyclic Olefin and Type of Attached Group

Atom or molecule attached to both sites of olefinic carbons (–C=C–)	Number of carbons in cyclic olefin			
	6	5	4	3
–H	1653 cm^{-1} (6050 nm)	1611 cm^{-1} (6207 nm)	1565 cm^{-1} (6390 nm)	1648 cm^{-1} (6068 nm)
–CH$_3$	1683 cm^{-1} (5942 nm)	1686 cm^{-1} (5931 nm)	1695 cm^{-1} (5900 nm)	1887 cm^{-1} (5299 nm)

In the IR region the intensity of the fundamental olefinic carbon-carbon stretch, designated by $\tilde{v}$ (C=C) is highly variable. In some cases the band is completely absent; in other cases the band may be very strong. If there is little change in the dipole moment of an olefinic containing molecule, the C=C stretching mode will be infrared inactive. Thus if the C=C group is across the center of symmetry for a molecule the C=C stretch will be forbidden in the infrared. A center of symmetry is where identical atoms or molecules are attached coaxially across the C=C olefinic group. As a general rule, the more symmetrically substituted the olefin is the weaker is the infrared band for the olefinic molecule. Bands present in the infrared spectrum for an olefin can indicate the presence of an olefinic group, but the absence of such bands does not give conclusive evidence for the absence of an olefin.

Olefinic C-H Bend

Two modes of bending occur in olefinic hydrogens, in-plane and out-of-plane. In-plane bending represents a change in the C-H bond angle along the same plane, occurring at 1455–1200 cm^{-1} as a weak and not very useful group frequency for the presence of olefinic hydrogens. Out-of-plane bending occurs as the C-H bonds vibrate outside the flat olefinic group plane. Out-of-plane bending occurs at approximately 1000–650 cm^{-1} and represents excellent bands for identifying the presence of olefinic groups. The out-of-plane C-H bend–related bands are group frequencies that are distinct, not being affected by conjugation or coupling with other vibrations in the olefinic molecule.

There is only one infrared-active C-H bend of the four possible modes for olefinic bending vibrations. The four bending modes include (1) *twisting or rotational motion* and (2) opposite twisting at the individual olefinic carbons producing a *torsion* at the C=C bond. *Wagging,* or a motion of the hydrogen bonds, takes two forms: (3) one where the two groups of two hydrogens move in opposite planes, causing a rotational motion around a common center of axis, and (4) the second, where all four hydrogens move in or out of the plane together. The first type of wag is Raman-only active, and the second type of wag is the only infrared-active type. Table 19.4 summarizes the frequencies for C-H out-of plane bending given the number of R-groups attached to the two olefinic carbons.

MOLECULES WITH TRIPLE BONDS (–C≡)

Acetylenes (–C≡C–)

Triple-bonded carbon-to-carbon functional groups are termed acetylinic (or acetylenes). Due to symmetry about the C≡C bond, the absorptions for C≡C are weak in the infrared. For acety-

Table 19.4 Locations of C-H Out-of-Plane Bending Vibrations for Olefinic Molecules

Olefinic molecule type (R-type substituents)*	C-H Bend frequencies	Relative band intensity/shape
Monosubstituted	995–985 cm^{-1} (10,050–10,152 nm)	Very strong
	915–905 cm^{-1} (10,929–11,050 nm)	Very strong
Trans, disubstituted cis olefins	970–960 cm^{-1} (10,309–10,417 nm)	Very strong
Cis, disubstituted	730–650 cm^{-1} (13,699–15,385 nm)	Weak and broad
Disubstitution on a single carbon	895–885 cm^{-1} (11,173–11,299 nm)	Very strong
Trisubstitution	840–790 cm^{-1} (11,905–12,658 nm)	Strong
Tetrasubstitution	N/A	N/A

* Substituents other than R-groups will change the frequencies of absorption for the C-H out-of plane bend [4].

lene (H-C≡C-H), the C≡C stretch is forbidden in the infrared and thus not observed. For R-C≡C-H (monosubstituted), the C≡C stretch occurs at 2140–2100 cm^{-1} and is weak in intensity. For R-C≡C-R' (disubstituted acetylenes), the C≡C stretch occurs from 2260 to 2190 cm^{-1}, with very weak, sharp bands. The ≡C-H stretch band for monosubstituted acetylenes is medium and sharp in the infrared, occurring at 3310–3290 cm^{-1}.

The bending vibrations for ≡C-H occur as medium-strong bands at approximately 650 cm^{-1}.

Nitriles (R-C≡N)

The frequency for unconjugated nitriles, where –C≡N is not associated with electron-withdrawing groups, occurs at 2260–2240 cm^{-1}. Where conjugation occurs, such as aromatic-C≡N, or C=C-C≡N, the frequency is found from 2240 to 2210 cm^{-1}.

Hydrogen cyanide produces the only differences to the above rules. In the case of HCN, frequencies occur at 2089 cm^{-1} for gas phase and 2062 cm^{-1} for liquid phase. In the case of acetonitrile, the frequency for C≡N stretch is found at 2250 cm^{-1}; for RCN the frequency of 2260–2240 cm^{-1} is observed.

Related groups have different frequencies, depending upon electron sharing of the associated atom or group with the C≡N bond. As examples, nitrile oxide –C≡NO occurs at 2350 cm^{-1}, approximately 100 cm^{-1} higher than C-C≡N; and thiocyanate RS-C≡N occurs at 2150 cm^{-1}, approximately 100 cm^{-1} lower than C-C≡N.

CUMULATED DOUBLE-BOND SYSTEMS (A=B=C)

As examples, Table 19.5 illustrates the frequencies for a linear and a nonlinear system in the order of frequency. Note that the linear system, carbon dioxide, has a much greater frequency differential between the antisymmetric and symmetric modes, due to the high interaction between two colinear oscillators sharing a common central atom. This first-order interaction between two (hypothetical) fundamental vibrations is termed *coupling*.

AROMATIC SYSTEMS

Aromatic systems produce strong and sharp infrared bands due to the relatively rigid molecular structures. There are multiple vibrational modes present in aromatic systems, including: (1) carbon-hydrogen stretching, (2) in-plane bending, and (3) out-of-plane bending. Carbon-carbon interactions include (4) ring stretching and (5) ring bending modes. Internal vibrations caused by groups associated with the aromatic system also exhibit vibrational group frequencies. The total vibrational modes for aromatic systems is equal to $3N - 6$ vibrations, where N is the number of carbons in the ring structure. These vibrational modes are shown in Ref. 11.

Table 19.5 Coupling for Cumulated Double Bond Systems

Wavelength (Frequency) (cm^{-1})	Functional group	Name	Bond angle*
2349 cm^{-1} (4257 nm)	→O=C=O→	Asymmetric	180°, linear
1340 cm^{-1} (7463 nm)	←O=C=O→	Symmetric	
1362 cm^{-1} (7342 nm)	→O=S=O→	Asymmetric	119°, nonlinear
1151 cm^{-1} (8688 nm)	←O=S=O→	Symmetric	

*Bond angle is only one consideration for the frequency difference for interacting oscillators. Another important consideration is the mass of the central atom.

Carbon-hydrogen stretching vibrations occur from approximately 3100 to 3000 cm^{-1}. Infrared spectra of aromatics using high-resolution instrumentation show a pair of composite bands near 3070 cm^{-1} and 3030 cm^{-1}. These composite bands arise as the combined interactions of several stretching modes discussed briefly earlier.

The carbon-carbon ring stretching occurs as a vibrational pair near 1600 cm^{-1}. This pair of frequencies is considered a degenerate pair and results from two stretching modes of the carbon-carbon ring stretching. However, due to Fermi resonance, one of the vibrations near 1595 cm^{-1} is not observed in the infrared spectrum. When there is substitution on the aromatic ring, the degeneracy is lost and the pair of bands occur near 1600 cm^{-1} and 1580 cm^{-1}. The 1580-cm^{-1} band will be more intense if there is conjugation with the aromatic system by an electron-withdrawing group or if substitution by O or N has occurred. Paradisubstitution of a benzene ring by two identical groups yields to the center-of-symmetry rule, and thus the carbon-to-carbon ring stretching is infrared forbidden. It follows that if the paradisubstituted groups are not identical, some weak bands will occur near the 1600-cm^{-1} region.

The carbon-hydrogen bending (wag) vibrations occur as in-plane bending and out-of plane bending vibrations. The in-plane bending vibrations are much less consistent than the out-of-plane bending vibrations. The in-plane bending vibrational frequency depends upon the number of C-H groups present and the substituent groups. Out-of-plane bending is much more consistent and intense in the infrared region. Out-of-plane bending is not particularly influenced by substituents present, but the frequency for the bending vibrations depends upon the number of adjacent hydrogens. The typical presentation for the out-of-plane bending vibrations for benzene is given in Table 19.6.

The carbon-hydrogen bending vibrations give rise to a series of combination bands termed *sumtones* or *summation bands*. These sumtones, summation bands, or combination bands occur from 2000 to 1700 cm^{-1}. The patterns are used in infrared spectroscopy to determine ring substituents. The pattern for the sumtones is more important than their exact position. These composite band patterns are given in most infrared texts, such as Ref. 12.

Vibrational States in Benzene

Hayward and Henry [13] calculate the frequencies for all eight members of the C-H stretching overtone spectrum in benzene. The paper provides a limited generalization method for calculating the anharmonicity constants for molecules where nearly identical oscillators are involved and when these oscillators have vibrational motion that is approximately independent of the rest of a polyatomic molecule. The authors report the overtone frequencies (in cm^{-1}) of v = 1–8 as 3053, 5940, 8737, 11,411, 14,003, 16,458, 18,807, and 21,025, respectively. The equivalent wavelength positions (in nm) for these overtone bands are 3275, 1684, 1145, 876, 714, 608, 532, and 476.

Heller and Mukamel [14] describe a general theory for estimating the overtone absorption line shapes for polyatomic molecules. The authors expand on the molecular Hamiltonian and

Table 19.6 Typical Out-of-Plane Bending Vibrations for Benzene

Substitution on Ring	Adjacent hydrogens	C-H bend vibration Wavelength (in nm)	C-H bend vibration Frequency (in cm^{-1})
None	6	13072–13605	765–735
Mono-	5	12987–13605	770–735
Ortho- (di-)	4	12903–13514	775–740
Tri-	3	12346–13333	810–750
Tetra-	2	11628–12500	860–800
Penta-	1	10989–11765	910–850

local-mode theory and provide details for overtone line-shape estimation. Numerical estimation of benzene overtone line-shapes are given along with the derivation and detailed explanation of both theory and methodology.

A description of the aspects related to local mode combination bands for aromatic and aliphatic hydrocarbons is given in a paper by Burberry and Albrecht [15]. The authors explain the origin of side bands in the local-mode (LM) overtone vibrations as resulting from LM–LM combination bands of various C-H oscillators. The local-mode overtones are described by theory and observation for methyl groups in tetramethylsilane (TMS) as well as benzene.

Child [16] discusses issues on the use of local modes for spectroscopic assignment and interpretation. Examples discussed include water, acetylene, methane, and silane. The author generally reviews the topic for pre-1985 literature, including 42 references.

CARBONYL COMPOUNDS (>C=O)

Compounds containing carbonyl groups exhibit frequencies over a broad range of 1950–1550 cm^{-1}. The position or frequency of the carbonyl C=O stretching vibration is affected by (1) the isotope effects and mass change of substituted groups, (2) bond angles of substituted groups, (3) electronic (resonance and inductive) effects, and (4) interactions of these effects. Substituents with higher mass decrease the C=O stretch frequency. Increasing the mass or bond angles of substituents also decreases the frequency by up to 40 cm^{-1} at 1715 cm^{-1}; this is equivalent to an increase of 139 nm at 5831 nm. Similarly, decreasing the substituent mass or the bond angles between the carbonyl carbon and its substituents increases the frequency by 25 cm^{-1} above the nominal 1715-cm^{-1} carbonyl C=O stretch frequency, equivalent to a decrease in wavelength of 84 nm at 5831 nm. More electronegative (electron-withdrawing) substituents will increase the carbonyl carbon-oxygen stretch frequency by up to 100 cm^{-1} above the 1715-cm^{-1} nominal frequency, a decrease of 321 nm at 5831 nm. Conjugation of the carbonyl group to aromatic or olefinic groups tends to lower the frequency for both C=O and C=C by 30–40 cm^{-1}, dependent upon the ring size of the substituent. For example, carbonyl conjugation with aromatic ring sizes of 6, 5, 4, and 3 carbons produces C=O stretch frequencies at 1685 cm^{-1}, 1687 cm^{-1}, 1686 cm^{-1}, and 1677 cm^{-1}, respectively. Electronic effects such as resonance and inductive type produce a double-bond character and increase the frequency of the carbonyl stretch.

Special Cases of the Carbonyl

The group called anhydrides (R–C=O–O–C=O–R) consists of two carbonyls connected by an oxygen. The frequency for the carbonyl stretch occurs from 1820 to 1750 cm^{-1}; the in-phase (←C=O→ and ←C=O→) stretch for the two carbonyl groups in the anhydride occurs at 1820 cm^{-1}. In the out-of-phase carbonyl pair (←C=O→ and →C=O←), the frequency for the carbonyl is 1750 cm^{-1}. In cyclic anhydrides, the in-phase frequency is 1870–1850 cm^{-1}, and the out-of-phase occurs at 1800–1780 cm^{-1}. Solvent effects like hydrogen bonding to the carbonyl oxygen lowers the carbonyl C=O frequency. As in other carbonyl-containing molecules, smaller mass substituents or decreasing bond angles of carbonyl carbon substituent bonds increase the frequency for the carbonyl. Large inductive effects from the addition of largely electronegative substituents also increase the frequency for the carbonyl. When a halogen is attached to the alpha carbon attached to the carbonyl carbon, the frequency is increased by 20–30 cm^{-1} (i.e., 60–90 nm).

Acid dimers interact by strong molecular hydrogen bonding and lower the carbonyl frequency by 25–60 cm^{-1}. The interaction is between the carbonyl oxygen and the hydrogen associated with the acid molecules:

R–C=O–OH · · · O=C–R–OH

AMIDES (R–C=ONH–X)

Amides are the special case of carbonyl compounds that are subject to frequency shifts due to hydrogen bonding. The frequencies for dilute solutions of amides can vary by as much as 75 cm^{-1} at ~1701 cm^{-1} as compared to the same molecule with hydrogen bonding (such as a Nujol mull preparation). Dilute solutions of amides exhibit carbonyl frequencies at 1735 cm^{-1} (primary, R–C=O–NH$_2$), 1705 cm^{-1} (secondary, R–C=O–NHR), and 1655 cm^{-1} (tertiary, R–C=O–NR$_2$). The respective hydrogen-bonded vibrations occur at 1700–1650 cm^{-1} for primary, 1680–1635 cm^{-1} for secondary, and 1660 cm^{-1} for tertiary amides [17].

The N-H stretch for amides occurs as two bands between 3360 and 3175 cm^{-1}. Geometrically, two forms of amides exist as either the trans or cis form with the trans form demonstrating a frequency of 3325–3275 cm^{-1} and the cis form 3185–3135 cm^{-1}. The geometries are as R–C=O–N–R'–H, where R and R' are either in the trans or the cis configuration.

The term *amide I* band refers to the C=O carbonyl stretch [18]. In the section on carbonyls it was mentioned that the nominal C=O stretch frequency is 1715 cm^{-1}, since the amide form of the carbonyl has the electronegative N atom, lowered by resonance with the nitrogen electrons and lowered by hydrogen bonding. The C=O nominal frequency for the amide I stretch is approximately 1650 cm^{-1}.

The term *amide II* refers to the NH$_2$ scissoring associated with primary amides (R–C=O–NH$_2$). The amide II frequency results from the combined interactions of the C-N stretch and the N-H in-plane bend; these vibrations interact to form bands at the 1540 cm^{-1} region and the 1310–1220 cm^{-1} region. A band at 1560–1535 cm^{-1} indicates a trans-substituted amide; this band is lost in the cis-amide configuration [18].

The carboxylate ion (O=C–O$^-$) has two identical C$\cdots$O, C$\cdots$O bonds, consisting of the in-phase ←C$\cdots$O→, ←C$\cdots$O→; and out-of-phase ←C$\cdots$O→ →C$\cdots$O← bands. The broad out-of-phase band occurs at to 1615–1550 cm^{-1}, whereas the broad in-phase band has a frequency of 1400–1300 cm^{-1}. The in-plane band is not particularly useful in spectroscopy due to its variable intensity and position.

ETHERS (C–O–C)

The ether group can be distinguished by the degree of saturation for the ether carbons. A saturated ether has the maximum number of hydrogens (6) associated with the ether carbons. Unsaturated ethers can be either single unsaturated or double unsaturated, depending upon the number of unsaturated methoxy carbons (=C–O–) within the ether group. Saturated ethers exhibit a very strong C-O stretch at a frequency of 1160–1050 cm^{-1}; doubly unsaturated ethers exhibit a strong, slightly broadened C-O stretch frequency at 1275–1230 cm^{-1}. Epoxides C–O–C ↑ ↑ demonstrate a symmetric stretch near 1255 cm^{-1} and an antisymmetric stretch of 860 cm^{-1} (11,628 nm).

ESTERS (R-CO-OR')

The ester group exhibits the frequencies of a mixed ether. It shows a very strong band at 1260–1150 cm^{-1} and a strong band at 1200–1000 cm^{-1}. These two bands correspond to the C–O for RCO–OR' and the O–R' of RCOO–R'.

ANHYDRIDES (R–C=O–C=O–R)

Cyclic anhydrides show bands for doubly unsaturated ethers at approximately 1260 cm^{-1}. Open-chain anhydrides show the saturated ether band at 1060 cm^{-1}.

PEROXIDES (–C–O–O–C–)

Infrared-active bands are extremely difficult to identify using infrared spectroscopy. There is an ether C-O stretch vibration near 1100 cm^{-1} that can be useful for identification.

X–H FUNCTIONAL GROUPS (O-H AND N-H)

These bands are very weak to absent in Raman spectra. For the infrared region X–H frequencies occur above 2000 cm^{-1}, except for boron compounds, which demonstrate bands from 2100 to 1620 cm^{-1}. Interaction of X–H with other groups does not usually occur, due to the high frequency of these bands and the low mass of hydrogen. Therefore the frequency of X–H is specific and reliable for infrared spectroscopy. The X–H stretch follows the two-body harmonic oscillator approximation (discussed in another section of this book). The relationship of X–H to the harmonic oscillator allows the frequency of any specific X–H group to be related to the periodic table. The frequency of an X–H group increases as one goes upward in any column or to the right within any row of the periodic table. The first row of the periodic table gives H-H with a nominal fundamental stretch at 4160 cm^{-1}. The second row yields v(Li-H) = 1360 cm^{-1}, v(Be-H) = 1965 cm^{-1}, v(B-H) = 2500 cm^{-1} and 2100–1620 cm^{-1}, v(C-H) = 2955 cm^{-1}, v(N-H) = 3400 cm^{-1}, v(O-H) = 3610 cm^{-1}, and v(F-H) = 3950 cm^{-1}. In the third row the more important v(X–H) include: v(Si-H) = 2185 cm^{-1}, v(P-H) = 2350 cm^{-1}, and v(S-H) = 2580 cm^{-1}. These bands are weak to nonexistent using Raman.

Hydrogen Bonding (X–H $\cdots$ Y)

A hydrogen-bonded molecular system has four possible modes: (1) O-H stretch found at 3400–2500 cm^{-1}, (2) R-O-H in-plane bend at 1400–1300 cm^{-1}, (3) the R-O-H out-of-plane bend at 660 cm^{-1}, and (4) the –O-H $\cdots$ O stretch at 175 cm^{-1}.

Specific frequencies for molecules containing O-H with hydrogen bonding include: alcohols (R-O-H), with v(O-H) = 3355 cm^{-1}; carboxylic acids (–COOH), with v(O-H) = 3000 cm^{-1}; and amines (R–NH–X), with v(N-H) = 3300 cm^{-1}. In general, hydrogen bonding has several main effects on the frequency of X–H bands: (1) it lowers the frequency of the X–H stretch as compared to the gas phase and dilute liquid phase of a molecule; (2) it raises the X–H bend frequency; (3) it broadens all O-H stretch bands by up to 300–500 cm^{-1}; (4) it intensifies all bending and stretching X–H bands.

O-H Stretch

It is common knowledge among infrared spectroscopists that any fundamental band occurring above 3500 cm^{-1} can be confidently assigned to the O-H stretch group. A sharp band exists for the stretching frequency of O-H for the infrared spectrum of O-H–containing systems. Primary alcohols (–CH$_2$–OH) exhibit v_s(O-H) = 3650 cm^{-1}; secondary alcohols (R–CHOH–R') exhibit v_s(O-H) = 3620 cm^{-1}; tertiary alcohols (R–CR'OH–R'') demonstrate v_s(O-H) = 3610 cm^{-1}; phenols have v_s(O-H) = 3620–3580 cm^{-1}; and carboxylic acids have v_s(O-H) = 3550–3500 cm^{-1}. To distinguish between primary, secondary, and tertiary alcohols using infrared spectra requires the use of these bands as well as the v_s(C-O) = 1170–1040 cm^{-1}.

The O-H in-plane bend occurs as v_B(O-H) = 1400–1300 cm^{-1}. The C-O-H out-of-plane bend v_B(O-H) = 760–650 cm^{-1}.

Carboxylic acids (–COOH) form dimers (hydrogen bonding between molecules as (–C=O $\cdots$ H–O–C–) in condensed states that exhibit four distinct frequencies: (1) a very broad, very strong O-H stretch at approximately 3000 cm^{-1}, at a lower frequency than alcohols); (2) summation tones as one to three weak bands near 2600 cm^{-1}; (3) a carbonyl (C=O) stretch vibration near 1725 cm^{-1}; and (4) a weak, broad band indicating a dimer condition near 955 cm^{-1}.

N-H Group Frequencies and Nitrogen-Containing Compounds

The average frequencies for free N-H groups include primary amides (–C=O–NH$_2$), v_s(N-H) = 3500 cm^{-1} and 3410 cm^{-1}; and secondary amides (–C=O–NHC), v_s(N-H) = 3450 cm^{-1}. Alkyl (aliphatic-NH-X) groups include alkyl-NH$_2$, v_s(N-H) = 3395 cm^{-1} and 3330 cm^{-1}; for alkyl-NH-X, v_s(N-H) = 3355 cm^{-1}. In the case of aryl (aromatic-NH-X), the N-H stretch fundamentals occur as v_s(N-H) = 3485 cm^{-1} and 3395 cm^{-1} for aryl-NH$_2$, and v_s(N-H) = 3435 cm^{-1} for aryl-NH-X. *Note:* The N-H frequency is lower and less intense than the O-H frequency; hydrogen-bonding with N-H is weaker and of less effect than with O-H.

Primary amines (R–NH$_2$) demonstrate two N-H stretching bands due to coupling: the in-phase (or symmetric) and the out-of-phase (or antisymmetric). Aliphatic amines, (C$_n$H$_{2n+1}$)–NHX, demonstrate in-phase frequencies from 3335 to 3275 cm^{-1}, with medium strength; out-of-phase frequency occurs from 3410 to 3350 cm^{-1}, also with medium strength. Aromatic amines exhibit in-phase and out-of-phase frequencies of 3425–3330 cm^{-1} and 3510–3390 cm^{-1}, both with medium strength. The scissoring for primary amines occurs from 1655 to 1590 cm^{-1}, as a medium-to-strong, broad band; the NH$_2$ wagging is a strong, broad band near 800 cm^{-1}.

Secondary amines (R'R''–NH) exhibit a single broad N-H stretch vibrational frequency near 3310 cm^{-1}; an N-H bend occurs as a weak-to-medium-strength band near 1500 cm^{-1}.

To separate spectral features for the various amines, i.e., primary, secondary, and tertiary, it is helpful to point to obvious differences between the groups. For example, the number of N-H stretches for the three types are primary (2), secondary (1), and tertiary (0). For N-H bending vibrations, primary amines have a band near 1655–1590 cm^{-1} and secondary amines near 1500 cm^{-1}. This feature is absent in tertiary amines. Lastly, the C-H stretch vibration for CH$_2$ is to be found for all three types of amines from 2855 to 2715 cm^{-1}; for primary amines, the C-H stretch will be found at higher frequencies nearer to 2900 cm^{-1}.

Amine salts as R–NH$^+_3$ and R'R''–NH$^+_2$ exhibit several weak bands near 2200 cm^{-1}. These salts exhibit a weak-to-medium, very broad band band, centered near 2500 cm^{-1}, lower in frequency than the C-H stretch and up to 500 cm^{-1} (i.e., 314 nm) in width at the baseline.

AMINES (R–NH–X)

Amines are polar compounds, and both primary and secondary amines can form intermolecular hydrogen bonding. All classes of amines are water soluble (up to 6 carbon atoms) and will bond with hydrogen in aqueous solutions.

AMIDES (–C=O–NX'X'', ALSO SHOWN AS –NHCO–)

3550–3050 cm^{-1} as fundamental N-H for acyl-group of amide (R–C=O–)

1690–1650 cm^{-1} as the C=O stretch fundamental

5000 cm^{-1} as combination band

6667 cm^{-1} as first overtone

Raman spectral features are dominated by bands representing the vibrational information from acetylenic C≡C stretching, olefinic C=C stretch; N=N (azo-) stretch, S-H (thio-) stretch, C=S stretch, C-S stretch, and S-S stretch. Raman spectra also contain such molecular vibrational information as CH$_2$ twist and wagging, carbonyl C=O stretch associated with esters, acetates, and amides; C-Cl (halogenated hydrocarbons) stretching, and –NO$_2$ (nitro-/nitrite) stretch. In addition, Raman yields information content of phenyl-containing compounds at 1000 cm^{-1}.

REFERENCES

1. R. F. Goddu, D. A. Delker. Spectra–structure correlations for near-infrared. *Anal. Chem.* 32:140–141, 1960.
2. William James Hershel. *Nature* 23:76, Nov. 25, 1880.
3. Lionel J. Bellamy. *The Infrared Spectra of Complex Molecules* (2 vols.). 2nd ed. New York: Chapman and Hall, 1980.
4. Norman B. Colthup, Lawrence H. Daly, Stephen E. Wiberley. *Introduction to Infrared and Raman Spectroscopy.* 1st ed. New York: Academic Press, 1975.
5. Daimay Lin-Vien, Norman B. Colthup, William G. Fateley, Jeanette G. Grasselli. *The Handbook of Infrared and Raman Characteristic Frequencies of Organic Molecules.* New York: Academic Press, 1991.
6. Jeanette Grasselli, Robert Hannah, Dana Mayo, Foil Miller. *Infrared Spectroscopy I. Interpretation of Spectra and II. Instrumentation, Raman Spectra, Polymer Spectra, Sample Handling and Computer-Assisted Spectroscopy.* Brunswick, ME: Bowdoin College, p. 33, 1994.
7. Jeanette Grasselli, Robert Hannah, Dana Mayo, Foil Miller. *Infrared Spectroscopy I. Interpretation of Spectra and II. Instrumentation, Raman Spectra, Polymer Spectra, Sample Handling and Computer-Assisted Spectroscopy.* Brunswick, ME: Bowdoin College, p. 34, 1994.
8. J. Bertran, L. Ballestare. *Spectrochim. Acta* 39A:123, 1983.
9. Jeanette Grasselli, Robert Hannah, Dana Mayo, Foil Miller. *Infrared Spectroscopy I. Interpretation of Spectra and II. Instrumentation, Raman Spectra, Polymer Spectra, Sample Handling and Computer-Assisted Spectroscopy.* Brunswick, ME: Bowdoin College, pp. 34–35, 1994.
10. Jeanette Grasselli, Robert Hannah, Dana Mayo, Foil Miller. *Infrared Spectroscopy I. Interpretation of Spectra and II. Instrumentation, Raman Spectra, Polymer Spectra, Sample Handling and Computer-Assisted Spectroscopy.* Brunswick, ME: Bowdoin College, p. 44, 1994.
11. Norman B. Colthup, Lawrence H. Daly, Stephen E. Wiberley. *Introduction to Infrared and Raman Spectroscopy.* 1st ed. New York: Academic Press, p. 262, 1975.
12. Norman B. Colthup, Lawrence H. Daly, Stephen E. Wiberley. *Introduction to Infrared and Raman Spectroscopy.* 1st ed. New York: Academic Press, p. 262, 1975.
13. R. J. Hayward, B. R. Henry. Anharmonicity in polyatomic molecules. *J. Mol. Spectrosc.* 46:207–213, 1973.
14. D. F. Heller, S. Mukamel. Theory of vibrational overtone line shapes of polyatomic molecules. *J. Chem. Phys.* 70:463–472, 1979.
15. M. S. Burberry, A. C. Albrecht. Local Mode Combination Bands and Local Mode Mixing. *J. Chem. Phys.* 71:4631–4640, 1979.
16. M. S. Child. Local Mode Overtone Spectra. *Acc. Chem. Res.* 18:45–50, 1985.
17. Jeanette Grasselli, Robert Hannah, Dana Mayo, Foil Miller. *Infrared Spectroscopy I. Interpretation of Spectra and II. Instrumentation, Raman Spectra, Polymer Spectra, Sample Handling and Computer-Assisted Spectroscopy.* Brunswick, ME: Bowdoin College, p. 117, 1994.
18. R. T. Morrison, R. N Boyd. *Organic Chemistry.* Boston: Allyn and Bacon, 1973.

ADDITIONAL SOURCES

Reference Textbooks

For a comprehensive early work listing 2701 references related to the details on infrared spectroscopy up to 1944 see:

Barnes, Robert B., Gore, Robert C., Liddel, Urner, Williams, Van Zandt. *Infrared Spectroscopy Industrial Applications and Bibliography.* New York: Reinhold, 1944.

Other solid reference textbooks covering aspects of band characterization and sampling methods include the following.

Bellamy, Lionel J. *The Infrared Spectra of Complex Molecules* (2 Vol.s). 2nd ed. New York: Chapman and Hall, 1980.
Colthup, Norman B., Daly, Lawrence H., Wiberley, Stephen E. *Introduction to Infrared and Raman Spectroscopy.* 1st ed. New York: Academic Press, 1975.
Ferraro, John R., Krishnan, K. eds. *Practical Fourier Transform Infrared Spectroscopy.* New York: Academic Press, 1990.
Grasselli, Jeanette, Hannah, Robert, Mayo, Dana, Miller, Foil. *Infrared Spectroscopy I. Interpretation of Spectra and II. Instrumentation, Raman Spectra, Polymer Spectra, Sample Handling and Computer-Assisted Spectroscopy.* Brunswick, ME: Bowdoin College, 1994.

Griffiths, Peter R., de Haseth, James A. *Fourier Transform Infrared Spectrometry.* New York: Wiley, 1986.
Hendra, Patrick, Jones, Catherine, Warnes, Gavin. *Fourier Transform Raman Spectroscopy.* New York: Ellis Horwood, 1991.
Lin-Vien, Daimay, Colthup, Norman B., Fateley, William G., Grasselli, Jeanette G. *The Handbook of Infrared and Raman Characteristic Frequencies of Organic Molecules.* New York: Academic Press, 1991.
Scheuing, David R. *Fourier Transform Infrared Spectroscopy in Colloid and Interface Science.* Washington, DC: American Chemical Society, 1990.
Socrates, George. *Infrared Characteristic Group Frequencies.* 2nd ed. New York: Wiley, 1997.

Fermi

Fermi, Von E. Über den Ramaneffekt des Kohlendioxyds. *Z. Physik* 71:250–259, 1931.

General Vibrational Locations

Abney, W. de W., Festing, Lt. Col. On the influence of the atomic grouping in the molecules of organic bodies on their absorption in the infra-red region of the spectrum. *Phil. Transact.* 172:887–918, 1882.
Bellamy, L. J., Mayo, D. W. Infrared frequency effects of lone pair interactions with antibonding orbitals on adjacent atoms. *J. Phys. Chem.* 80:1217–1220, 1976.
Fang, H. L., et al. Overtone spectroscopy of nonequivalent methyl C-H oscillators. Influence of conformation on vibrational overtone energies. *J. Chem. Phys.* 88:410–417, 1984.

Hydrocarbons

Tosi, C., Pinto, A. Near-infrared spectroscopy of hydrocarbon functional groups. *Spectrochim. Acta.* 28A:585–597, 1970.

Water

Darling, B. T., Dennison, D. M. The water vapor molecule. *Phys. Rev.* 57:128–139, 1940.

Hydrogen Bonding

Finch, J. N., Lippincott, E. R. Hydrogen bond systems: Temperature dependence of OH frequency shifts and OH band intensities. *J. Chem. Phys.* 24:908, 1956.
Lippincott, E. R., Schroeder, R. One-dimensional model of the hydrogen bond. *J. Chem. Phys.* 23:1099–1106, 1955.
Lippincott, E. R., et al. Potential-function model of hydrogen bond systems, hydrogen bonding. *Papers Symposium Ljubljana* 1957:361–374, 1959.
Murthy, A. S. N., Rao, C. N. R. Spectroscopic studies of the hydrogen bond. *Appl. Spectrosc. Rev.* 2:69–191, 1968.
Schroeder, R., Lippincott, E. R. Potential function model of hydrogen bonds. II. *J. Chem. Phys.* 61:921–928, 1957.
Stutman, J. M., Posner, A. S. *Nature* 193:368–369, 1962.

Ammonia and Methane

Hayward, R. J., Henry, B. R. Anaharmonicity in polyatomic molecules: A local-mode analysis of the XH-stretching overtone spectra of ammonia and methane. *J. Mol. Spectrosc.* 50:58–67, 1974.

Cyclic Alkanes and Alkenes

Henry, B. R., et al. A local mode description of the CH-stretching overtone spectra of the cycloalkanes and cycloalkenes. *J.A.C.S.* 102:515–519, 1980.

Aromatic Amines

Whetsel, K., et al. Near-infrared analysis of N-alkyl and N-alkyl-N-hydroxyalkyl aromatic amine mixtures. *Anal. Chem.* 29:1006–1009, 1957.
Whetsel, K., et al. Near-infrared analysis of mixtures of primary and secondary aromatic amines. *Anal. Chem.* 30:1594–1597, 1958.

Whetsel, K., et al. Near-infrared spectra of primary aromatic amines. *Anal. Chem.* 30:1598–1604, 1958.

N-H

Krikorian, S. E., Mahpour, M. The identification and origin of N-H overtone and combination bands in the near-infrared spectra of simple primary and secondary amides. *Spectrochim. Acta* 29A:1233–1246, 1973.

Amino Acids

Leifer, A., Lippincott, E. R. The infrared spectra of some amino acids. *J.A.C.S.* 79:5098–5101, 1957.

Proteins

Ellis, J. W., Bath, J. Modifications in the near infra-red absorption spectra of protein and of light and heavy water molecules when water is bound to gelatin. *J. Chem. Phys.* 6:723–729, 1938.

Fatty Acids

Holman, R. T., Edmondson, P. R. Near-infrared spectra of fatty acids and some related substances. *Anal. Chem.* 28:1533–1538, 1956.

Alkenes

Fang, H. L., Compton, D. A. C. Overtone spectroscopy of nonequivalent C-H oscillators in 1-alkenes and dienes. *J. Chem. Phys.* 92:7185–7192, 1988.

Goodu, R. F. Determination of unsaturation by near-infrared spectrophotometry. *Anal. Chem.* 29:1790–1794, 1957.

Epoxides

Goddu, R. F., Delker, D. A. Determination of terminal epoxides by near-infrared spectrophotometry. *Anal. Chem.* 30:2013–2016, 1958.

Phenolic Hydroxyl

Goodu, R. F. Determination of phenolic hydroxyl by near-infrared spectrophotometry. *Anal. Chem.* 30:2009–2013, 1958.

Group Frequencies

Bellamy, L. J. Some wider uses of group frequencies. *Appl. Spectrosc.* 33:439–443, 1979.

Vibrational States in Benzene

Bray, R. G., Berry, M. J. Intramolecular rate processes in highly vibrationally excited benzene. *J. Chem. Phys.* 71(12):4909–4922, 1979.

Henry, B. R., Siebrand, W. Anharmonicity in polyatomic molecules. The CH-stretching overtone spectrum of benzene. *J. Chem. Phys.* 49(12):5369–5376, 1968.

Spears, K. G., Rice, S. A. Study of the individual vibronic states of the isolated benzene molecule. *J. Chem. Phys.* 55(12):5561–5581, 1971.

Coupling in CH

Henry, Bryan R. Use of local modes in the description of highly vibrationally excited molecules. *Acc. Chem. Res.* 10:207–213, 1977.

Local Modes in CH Stretch Combination Bands

Burberry, M. S., Albrecht, A. C. Local mode combination bands and local mode mixing. *J. Chem. Phys.* 71(11):4631–4640, 1979.

Local Modes in Overtones

Child, M. S. Local mode overtone spectra. *Acc. Chem Res.* 18:45–50, 1985.
Heller, D. F., Mukamel, S. Theory of vibrational overtone line shapes of polyatomic molecules. *J. Chem. Phys.* 70(1):463–472, 1979.

Normal Modes

Duncan, J. L. The determination of vibrational anharmonicity in molecules from spectroscopic observations. *Spectrochim. Acta,* 47A:1, 1–27, 1991.

CH Stretch in Benzene

Hayward, R. J., Henry, B. R. Anharmonicity in polyatomic molecules. *J. Mol. Spectrosc.* 46:207–213, 1973.

Carbon Number Prediction

Schrieve, G. D. Ullman, A. H. Carbon number prediction from Herschel-infrared spectra using partial least-squares regression. *Appl. Spectrosc.* 45:713–714, 1991.

Solids at High Pressure

Weir, C. E. et al. Infrared studies in the 1- to 15-micron region to 30,000 atmospheres. *J.N.B.S.* 63A:55–62, 1959.
Weir, C. E. et al. Studies of infrared absorption spectra of solids at high pressures. *Spectroschim. Acta* 16:58–73, 1959.

Methyl Iodide

Herzberg, G., Herzberg, L. Absorption spectrum of methyl iodide in the near infrared. *Can. J. Res.* 270:332–338, 1949.

20.

FUNCTIONAL GROUPINGS AND CALCULATED LOCATIONS IN WAVENUMBERS (cm^{-1}) FOR IR SPECTROSCOPY*

*Numbers in parentheses are outside of the infrared region.

Group			Molecular structure	Relative intensity	Fundamental	1st Overtone	2nd Overtone
I.	C-H stretch						
	a.	Alkane, C_nH_{2n+2}	$CH_3-(CH_2)_N-CH_3$, normal paraffins	m-s	2959–2849	5917–5698	8873–8547
		Methyl	$-C-H_3$				
		Antisymmetric stretch		s	2973–2950	5945–5900	8913–8850
		Symmetric stretch		s	2882–2862	5964–5724	8643–8584
		Methylene	$-C-H_2$				
		Antisymmetric stretch		s	2936–2915	5872–5831	8811–8749
		Symmetric stretch		s	2864–2843	5727–5685	8591–8532
	b.	Alkene, C_nH_{2n}	>C=C<, olefinic group				
		Monosubstituted	H >C=C<^H_H R	m m	3040–3012 3096–3077	6079–6024 6192–6154	9116–9033 9288–9231
		Cis, disubstituted	H >C=C<^H_R R	m	3040–3012	6079–6024	9116–9033
		Trans, disubstituted	H >C=C<^R_H R	m	3040–3012	6079–6024	9116–9033
		Gem, disubstituted	R >C=C<^H_H R	m	3096–3077	6192–6154	9285–9234
		Trisubstituted	R >C=C<^R_H R	m	3040–3012	6079–6024	9116–9033
	c.	Alkyne, C_nH_{2n-2}	$-C\equiv C-H$	s	3345–3295	6690–6595	10,050–9880
	d.	Aromatic, C_nH_n	Ar(C)-H	v	3030	6061	9900
II.	C-H bend						
	a.	Alkane, C_nH_{2n+2}	$CH_3-(CH_2)_N-CH_3$, normal paraffins				
		Methine	R-CHR'R"	w	1340	~2681	~4021
		Methylene	R-CH$_2$R'	m	1483–1445	2966–2889	4450–4335
		Methyl	R-CH$_3$	m s	1379–1370 1470–1429	2759–2740 2940–2858	4137–4110 4411–4286
		Gem, dimethyl	CH$_3$-CHR-CH$_3$	s s	1385–1379 1370–1364	2770–2759 2740–2729	4155–4144 4110–4093
		Tert-butyl	(CH$_3$)$_3$-CH	m s	1395–1385 ~1370	2789–2770 ~2740	4184–4155 ~4110
	b.	Alkene, C_nH_{2n}	>C=C<				
		Monosubstituted	H >C=C<^H_H R	s s s	952–985 915–905 1420–1410	1905–1970 1830–1810 2841–2821	2985–2956 2745–2715 4261–4232
		Cis, disubstituted	H >C=C<^H_R R	s	~690	~1379	~2069
		Trans, disubstituted	H >C=C<^R_H R	s m	970–960 1309–1295	1940–1919 2618–2591	2910–2879 3926–3887
		Gem, disubstituted	R >C=C<^H_H R	s s	895–885 1420–1410	1791–1770 2841–2821	2686–2655 4261–4232

Group		Molecular structure	Relative intensity	Fundamental	1st Overtone	2nd Overtone
	Trisubstituted	>C=C<^R_H R, R	s	840–790	1681–1580	2521–2370
c.	Alkyne, C_nH_{2n-2}	–C≡C–H	v, s	~680–605 ~629	1360–1210 ~1258	2040–1815 ~1887
d.	Aromatic, C_nH_n	Ar(C)-H				
	Monosubstituted (5 hydrogens)		v, s	~752 ~699	~1504 ~1399	~2256 ~2098
	Disubstituted (4 hydrogens)		v, s	~752	~1504	~2256
	Trisubstituted (3 hydrogens)		v, s	781	1563	2344
	Tetrasubstituted (2 hydrogens)		v, m	~833	~1667	~2500
	Pentasubstituted (1 hydrogen)		v, w	~885	~1770	~2655
III. C-C stretch						
a.	Alkene, C_nH_{2n} Symmetrical stretch	>C=C<	v	1681–1621	3361–3241	5043–4861
	Monosubstituted	As shown above	m	~1645	~3289	~4933
	Cis, disubstituted	As shown above	m	~1658	~3317	~4975
	Trans, disubstituted	As shown above	m	~1675	~3350	~5025
	Gem, disubstituted	As shown above	m	~1653	~3306	~6024
	Trisubstituted	As shown above	m	~1669	~3339	~5008
	Tetrasubstituted	As shown above	w	~1669	~3339	~5008
	Diene	>C=C< ··· >C=C<	w w	~1650 ~1600	~3300 ~3200	~4950 ~4801
b.	Alkyne, C_nH_{2n-2}					
	Acetylene	H-C≡C-H	none	—	—	—
	Monosubstituted	R-C≡C-H	m	2141–2101	4283–4202	6423–6301
	Disubstituted	R'-C≡C-R''	v,w	2262–2188	4525–4376	6789–6566
c.	Allene	>C=C=C<	m m	~1961 ~1064	~3922 ~2128	~5444 ~3192
d.	Aromatic	containing	v v m m	1640–1590 ~1580 1499–1450 ~1449	3280–3180 ~3160 ~2999 ~2899	4920–4770 ~4739 ~4498 ~4348
IV. Carbonyl		>C=O				
a.	Ketones	R >C=O R'				
	Saturated, acyclic	alkane (R) >C=O (R') alkane	s	1724–1704	3448–3407	5173–5110
	Saturated, cyclic					
	6- or more membered ring	 (or) AR >C=O R'	s	1724–1704	3448–3407	5173–5110

Group	Molecular structure	Relative intensity	Fundamental	1st Overtone	2nd Overtone
5-membered ring	O=C	s	1751–1770	3503–3540	5255–5216
4-membered ring	O=C	s	~1776	~3552	~5328
Alpha-, beta-unsaturated, acyclic		s	1684–1664	3367–3328	5051–4993
Alpha-, beta-unsaturated, cyclic					
6- or more membered ring		s	1684–1664	3367–3328	5051–4993
5-member ring		s	1724–1709	3448–3419	5173–5128
Alpha-, beta-, alpha'-, beta'-unsaturated, acyclic		s	1669–1664	3339–3328	5008–4993
Aryl		s	1701–1681	3401–3361	5102–5043
Diaryl		s	1669–1661	3339–3322	5008–4983
Alpha-diketones		s	1730–1709	3460–3419	5189–5128
Beta-diketones (enolic)		s	1639–1538	3279–3077	4919–4615
1,4-Quinones		s, m	1689–1661	3378–3322	5068–4983
Ketenes		s, m	~2151	~4301	~6667
b. Aldehydes, Carbonyl stretch	R >C=O H				
Saturated, aliphatic		s	1739–1721	3478–3442	5216–5163
Alpha-, beta-unsaturated, aliphatic		s	1704–1681	3407–3361	5110–5043
Alpha-, beta-, gamma-, delta-unsaturated, aliphatic		s	1681–1661	3361–3322	5043–4983
Aryl		s	1715–1695	3431–3390	5147–5084
C-H stretching, two bands		w w	2899–2817 2778–2703	5797–5634 5556–5405	8696–8453 8333–8110
c. Esters	R-C=O-OR'				
Saturated, acyclic		s	1751–1736	3503–3472	5255–5208
Saturated, cyclic: delta- and larger lactones		s	1751–1736	3503–3472	5255–5208
Gamma-lactones		s	1779–1761	3559–3521	5339–5283
Beta-lactones		s	~1818	~3636	~5456
Unsaturated, vinyl ester		s	1799–1770	3597–3540	5397–5311
Alpha-, beta-unsaturated and aryl		s	1730–1718	3460–3436	5189–5155
Alpha-, beta-unsaturated delta-lactone		s	1792–1718	3584–3436	5189–5155
Alpha-, beta-unsaturated gamma-lactone		s	1761–1739	3521–3478	5283–5216
Beta-, gamma-unsaturated gamma-lactone		s	~1799	~3597	~5397
Alpha-ketoesters		s	1754–1739	3509–3478	5263–5216
Beta-ketoesters (enolic)		s	~1650	~3300	~4950
Carbonates		s	1779–1739	3559–3478	5339–5216

Group	Molecular structure	Relative intensity	Fundamental	1st Overtone	2nd Overtone
V. Carboxylic acids, and ethers	–O-C=O–, and R-O-R				
Carbonyl stretching: saturated aliphatic		s	1724–1701	3448–3401	5173–5102
Alpha-, beta-unsaturated aliphatic aryl		s	1715–1689 2049–1681	3431–3378 4098–3361	5147–5068 5102–5043
Hydroxyl stretching (bonded), several bands		w	2703–2500	5405–5000	8110–7502
Carboxylate anion stretching		s	1610–1550	3221–3101	4831–4651
Ethers –C-O stretch	R'-O-R''	s	1205–1100	2410–2200	3615–3300
VI. Anhydrides	R'-C=O-O-C=O-R''				
Stretching, –C=O		s	1850–1750	3700–3500	5550–5250
Saturated, acyclic	–C=O stretching	s	1850–1750	3700–3500	5550–5250
	H_3CC-O stretching	s	1400–1350	2800–2700	4200–4050
	C-O-C stretching	s	1160–1100	2320–2200	3480–3300
Alpha-, beta-unsaturated and aryl, acyclic		s s	1828–1779 1770–1721	3656–3559 3540–3442	5485–5339 5311–5163
Saturated, 5-member ring		s s	1869–1821 1767–1751	3738–3643 3534–3503	5609–5464 5299–5255
Alpha-, beta-unsaturated, 5-member ring		s s	1848–1799 1828–1779	3697–3597 3656–3559	5546–5397 5485–5068
VII. Acyl halides					
Stretch: acyl fluorides		s	~1848	~3697	~5546
Acyl chlorides		s	~1795	~3591	~5385
Acyl bromides		s	~1808	~3617	~5426
alpha, beta-unsaturated and aryl		s m	1751–1748 1748–1718	3503–3497 3497–3436	5348–5244 5244–5155
COF_2		s	~1927	~3854	~5780
$COCl_2$		s	~1828	~3656	~5485
$COBr_2$		s	~1828	~3656	~5485
VIII. Amides					
a. Carbonyl stretch					
Primary, concentrated solutions and solids		s	~1650	~3300	~4950
Primary, dilute solutions		s	~1718	~3436	~5068
Secondary, concentrated solutions and solids		s	1681–1629	3361–3257	5043–4885
Secondary, dilute solutions		s	1701–1669	3401–3339	5102–5008
Tertiary, solutions and solids		s	1669–1629	3339–3257	5008–4885
Cyclic, delta-lactams, dilute solutions		s	~1681	~3361	~5043
Cyclic, gamma-lactams, dilute solutions		s	~1701	~3401	~5102
Cyclic, gamma-lactams fused to second ring, dilute solution		s	1751–1701	3503–3401	5255–5102
Cyclic, beta-lactams, dilute solution		s	1761–1730	3521–3460	5283–5189

Group	Molecular structure	Relative intensity	Fundamental	1st Overtone	2nd Overtone
Cyclic, beta-lactams fused to second ring, dilute solutions		s	1779–1770	3559–3540	5339–5311
Urea, acyclic		s	~1661	~3322	~4983
Urea, cyclic, 6-membered ring		s	~1639	~3279	~4919
Ureas, cyclic 5-member ring		s	~1721	~3442	~5163
Urethanes		s	1739–1689	3478–3378	5216–5068
Imides, acyclic		s	~1709	~3419	~5128
Imides, cyclic, 6-membered ring		s s	~1709 ~1701	~3419 ~3401	~5128 ~5102
Imides, cyclic, alpha, beta-unsaturated, 6-membered ring		s s	~1730 ~1669	~3460 ~3339	~5189 ~5008
Imides, cyclic, 5-membered ring		s s	~1770 ~1701	~3540 ~3401	~5311 ~5102
Imides, cyclic, alpha, beta-unsaturated, 5-membered ring		s s	~1792 ~1709	~3584 ~3419	~5368 ~5128
b. N-H stretch					
Primary, free; two bands		m m	~3497 ~3401	~6993 ~6803	~10493 ~10204
Primary, bonded, two bands		m m	~3344 ~3175	~6689 ~6349	~10030 ~9524
Secondary, free; one band		m	~3425	~6849	~10256
Secondary, bonded; one band		m	3333–3125	6667–6250	10,000–9372
c. N-H bending					
Primary amides, dilute solution		s	1621–1590	3241–3180	4861–4769
Secondary amides, dilute solution		s	1548–1511	3101–3021	4651–4531
IX. Alcohols and phenolics, O-H stretch					
Free O-H		v, sh	3650–3584	7299–7168	10953–10753
With hydrogen bonding, single bridge compounds	O–H···H–	v,sh	3546–3448	7092–6897	10638–10341
Polymeric O-H		s,b	3401–3195	6803–6390	10204–9588
Chelated compounds		w,b	3226–2500	6452–5000	9671–7496
Primary alcohols		s s	~1053 1351–1266	~2105 2703–2532	~3158 4054–3798
Secondary alcohols		s s	~1099 1351–1266	~2198 2703–2532	~3297 4054–3798
Tertiary alcohols		s s	~1149 1408–1316	~2299 2817–2632	~3448 4225–3948
Phenols		s s	~1198 1405–1316	~2395 2810–2632	~3614 4225–3948

Group		Molecular structure	Relative intensity	Fundamental	1st Overtone	2nd Overtone
X. Amines, N-H stretch						
	a. Primary, free; two bands		m	~3497	~6993	~10493
			m	~3401	~6803	~10204
	Secondary, free; one band		m	3497–3311	6993–6623	10493–9930
	Imines; one band	=N-H	m	3401–3300	6803–6601	10204–9901
	Amine salts		m	3125–3030	6250–6061	9372–9091
	b. N-H bend					
	Primary		s-m	1650–1590	3300–3180	4950–4769
	Secondary		w	1650–1550	3300–3101	4950–4651
	Amine salts		s	1600–1575	3200–3150	4801–4724
			s	~1499	~2999	~4498
	c. C-N stretch					
	Aromatic, primary		s	1340–1250	2681–2500	4021–3750
	Aromatic, secondary		s	1350–1280	2699–2561	4049–3842
	Aromatic, tertiary		s	1359–1309	2717–2618	4077–3926
	Aliphatic		w	1220–1020	2439–2041	3741–3061
			w	~1427	~2853	~4225
XI. Unsaturated N compounds, C=N stretch and C≡N stretch						
	a. Alkyl nitriles		m	2262–2205	4525–4410	6789–6615
	Alpha-, beta-unsaturated alkyl nitriles		m	2237–2217	4474–4435	6711–6653
	Aryl nitriles		m	2242–2222	4484–4444	6725–6667
	Isocyanates		m	2273–2232	4545–4464	6817–6725
	Isocyanides		m	2222–2070	4444–4141	6667–6211
	b. Imines & oximes					
	Alkyl compounds		v	1689–1649	3378–3297	5068–4919
	Alpha-, beta-unsaturated		v	1661–1629	3322–3257	4831–4885
	–N=N– stretch, azo- compounds		v	1629–1575	3257–3150	4885–4724
	–N=C=N–, diimides		s	2155–2128	4310–4255	6464–6423
	–N$_3$ stretch, azides		s	2160–2137	4320–4274	6481–6382
			w	1340–1179	2681–2358	4021–3537
	c. C-NO$_2$, nitro- compounds					
	Aromatic		s	1570–1499	3140–2999	4577–4498
			s	1370–1299	2740–2597	4110–3896
	Aliphatic		s	1570–1550	3140–3101	4710–4651
			s	1379–1370	2759–2740	4137–4110
	O-NO$_2$, nitrates		s	1650–1600	3300–3200	4950–4801
			s	1299–1250	2597–2500	3896–3750
	C-NO, nitroso		s	1600–1499	3200–2999	4801–4498
	O-NO, nitrites		s	1681–1600	3361–3200	5043–4801
			s	1626–1610	3252–3221	4878–4830
	d. Nitriles, –C≡N stretch		s, w	2280–2210	4560–4420	6840–6630

Group	Molecular structure	Relative intensity	Fundamental	1st Overtone	2nd Overtone
XII. Halogen, C-X stretch, and X-C=O stretch					
C-F		s	1408–1000	2817–2000	4227–3000
C-Cl		s	800–602	1600–1205	2400–1807
C-Br		s	602–500	1205–1000	1807–1500
C-I		s	~500	~1000	~1500
X-C=O stretch		v	1820–1790	3640–3580	5460–5370
XIII. Sulfur compounds					
a. S-H stretch		w	2597–2551	5195–5102	7794–7651
b. C=S stretch		s	1200–1050	2401–2101	3601–3152
c. C-S stretch		m, w	750–690	1500–1380	2250–2070
d. S=O stretch					
Sulfoxides		s	1070–1030	2139–2060	3150–3089
Sulfones		s	1160–1140	2320–2281	3481–3421
		s	1350–1300	2699–2601	4049–3902
Sulfites		s	1230–1149	2460–2299	3690–3448
		s	1429–1350	2857–2699	4286–4049
Sulfonyl chlorides		s	1185–1164	2370–2328	3568–3493
		s	1370–1340	2740–2681	4110–4021
Sulfonamides		s	1179–1140	2358–2281	3537–3421
		s	1350–1300	2699–2601	4049–3902
Sulfonic acids		s	1209–1149	2418–2299	3627–3448
		s	1060–1030	2121–2060	3182–3089
		s	~649	~1299	~1948
XIV. Other					
C-F stretch	C-F	s	~1200	~2400–2300	~3600
C-F bend		v, b	650–475	1300–950	1950–1425
Si-O stretch	Si-O	s	1100–1050	2200–2100	3300–3150
C-Si stretch	C-Si	s	1270–1255 (~1261)	2540–2510 (~2522)	3810–3765
CO_3 (carbonate)		s, b	1520–1380	3040–2760	4560–4140
S-CN		s	2050–2000	4100–4000	6150–6000

21.

INFRARED SPECTRAL CORRELATION CHARTS

Infrared band locations for C-H stretching and bending modes

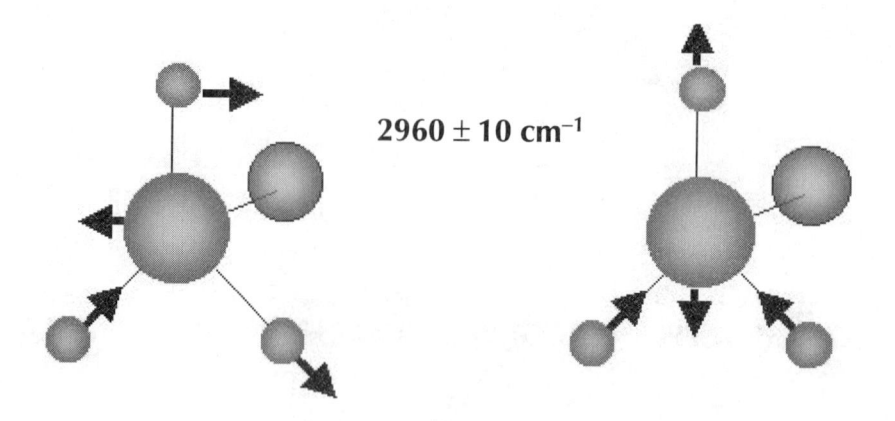

2960 ± 10 cm^{-1}

methyl C-H asymmetric stretch

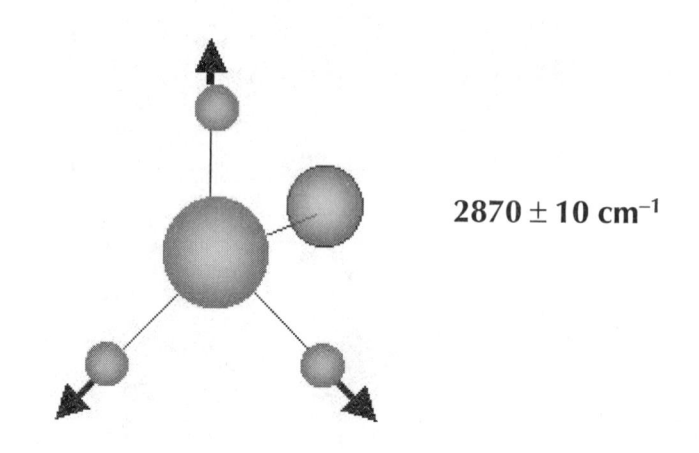

2870 ± 10 cm^{-1}

methyl C-H symmetric stretch

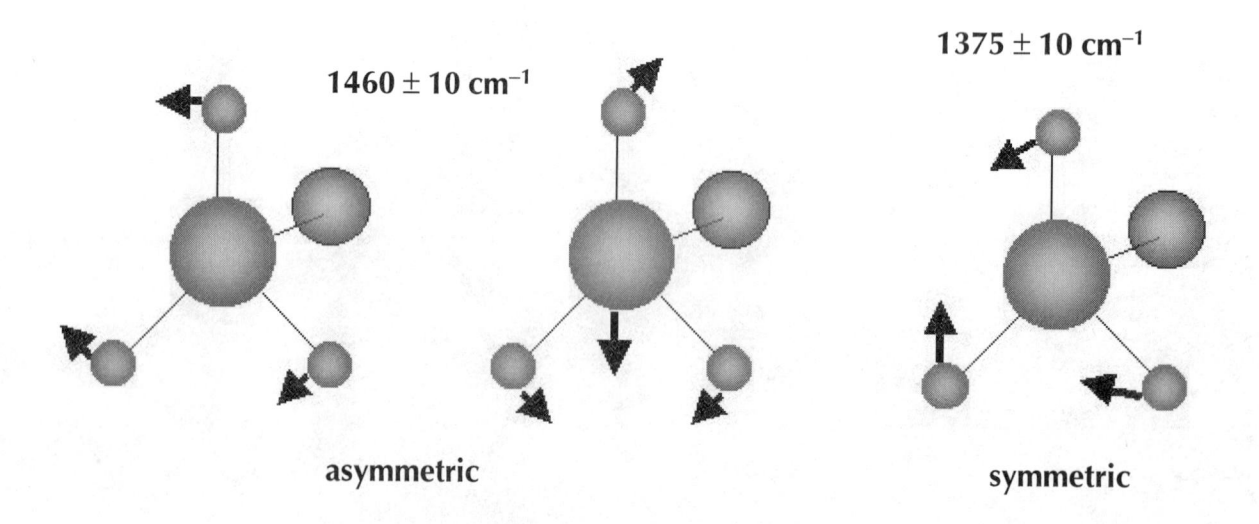

1460 ± 10 cm^{-1}

1375 ± 10 cm^{-1}

asymmetric

symmetric

methyl C-H bending mode

(See color plate 9.)

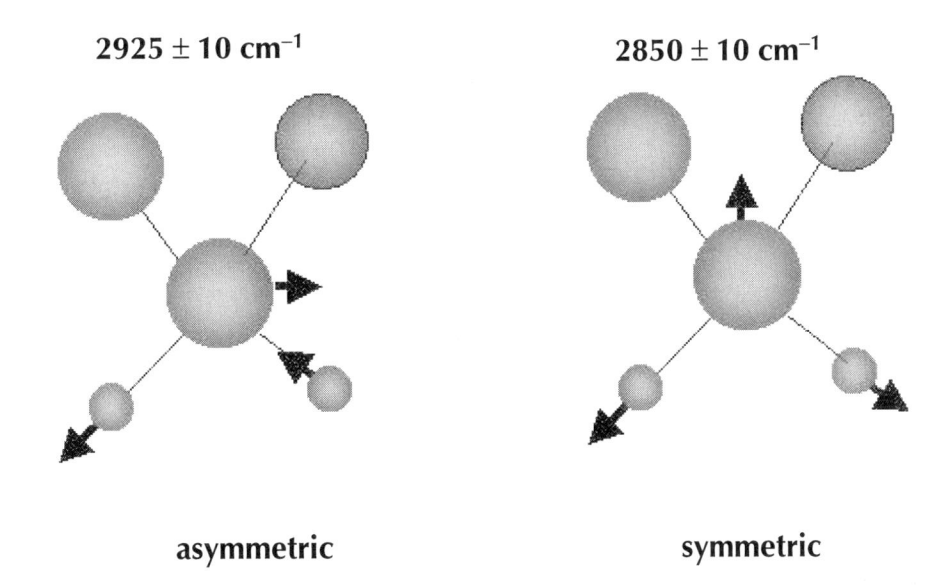

2925 ± 10 cm^{-1}　　　　　2850 ± 10 cm^{-1}

asymmetric　　　　　**symmetric**

methylene C-H stretching modes

(See color plate 10.)

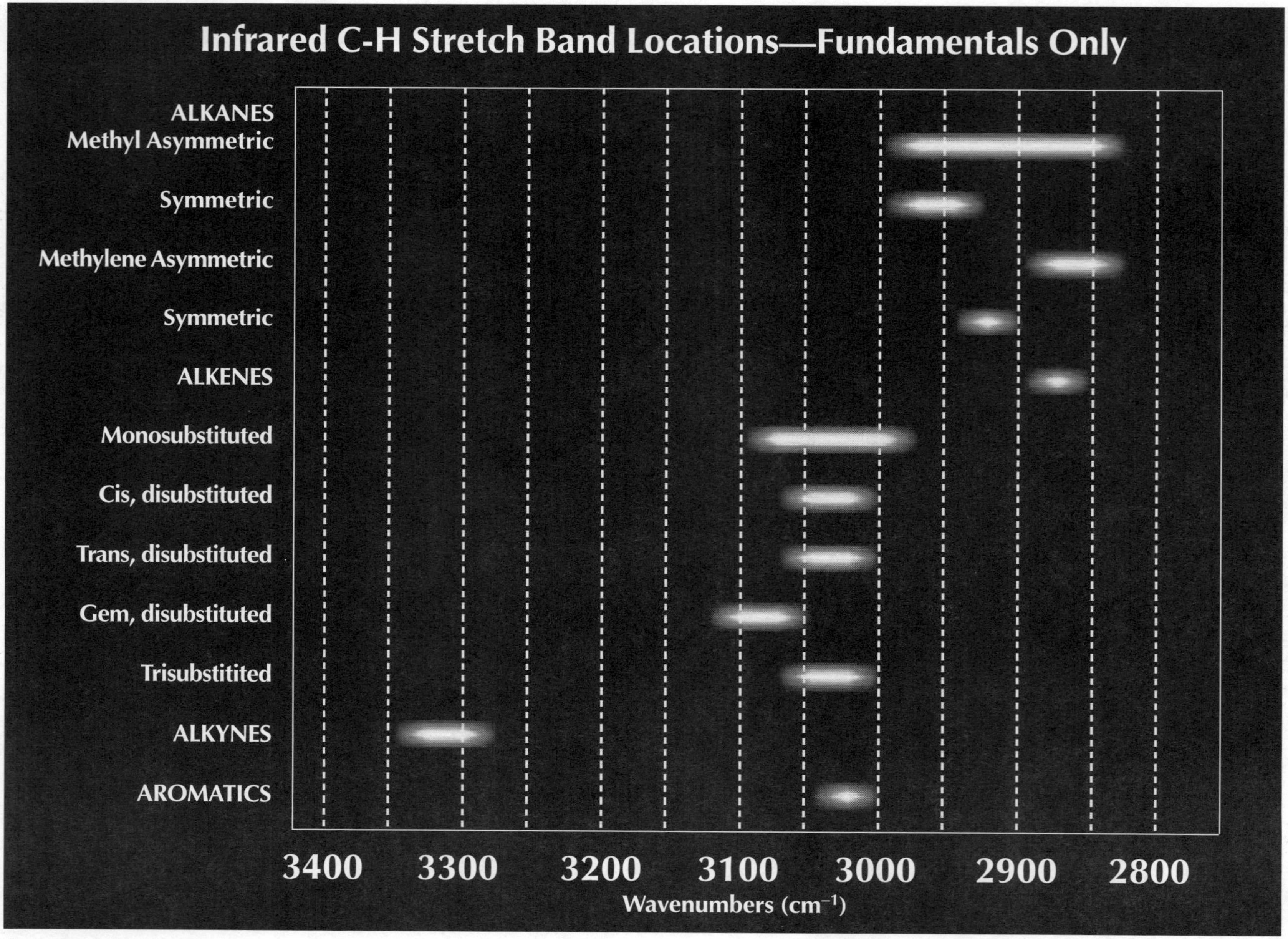

Infrared C-H Stretch Band Locations—Fundamentals Only

(See color plate 11.)

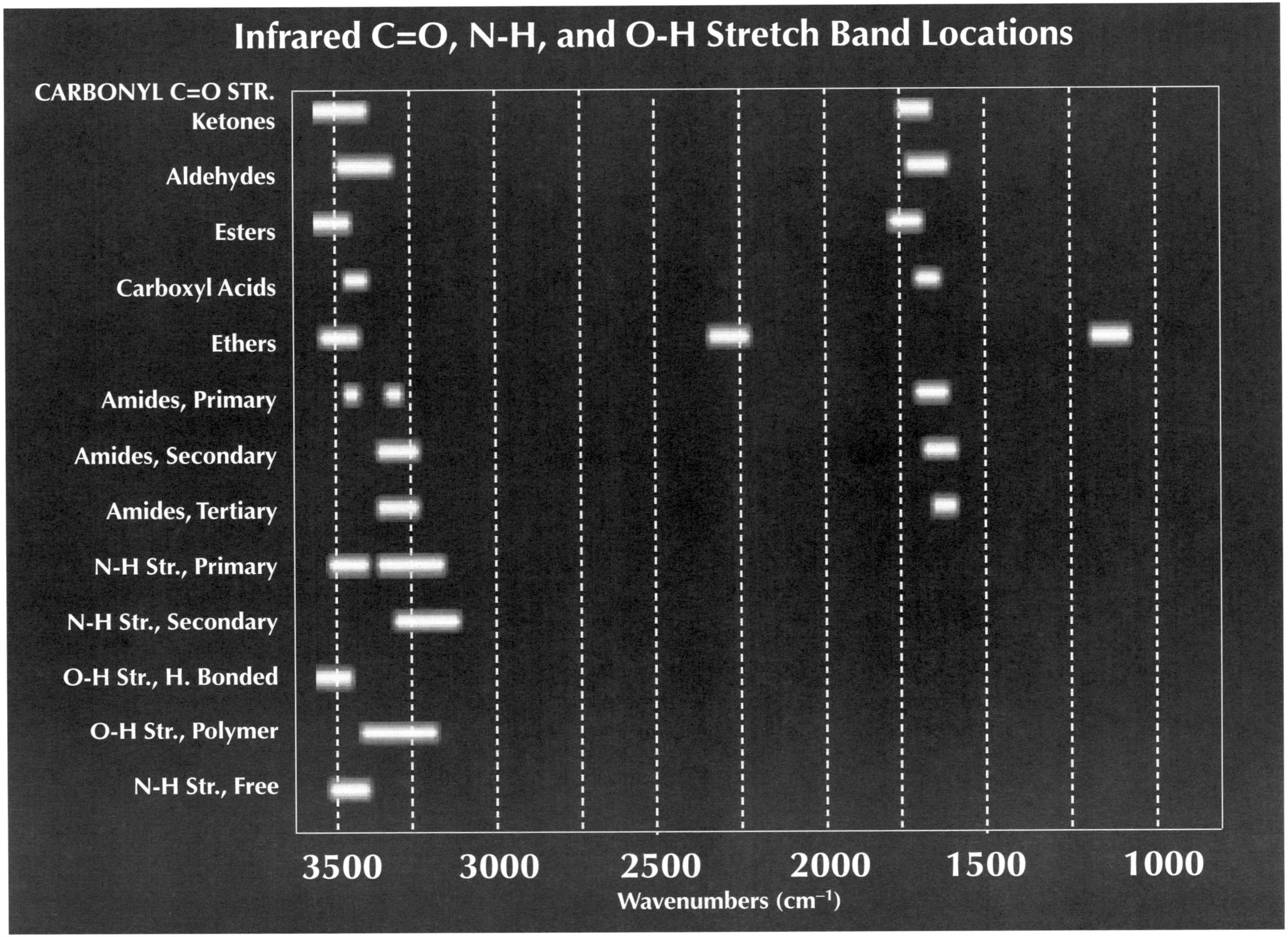

Infrared C=O, N-H, and O-H Stretch Band Locations

(See color plate 12.)

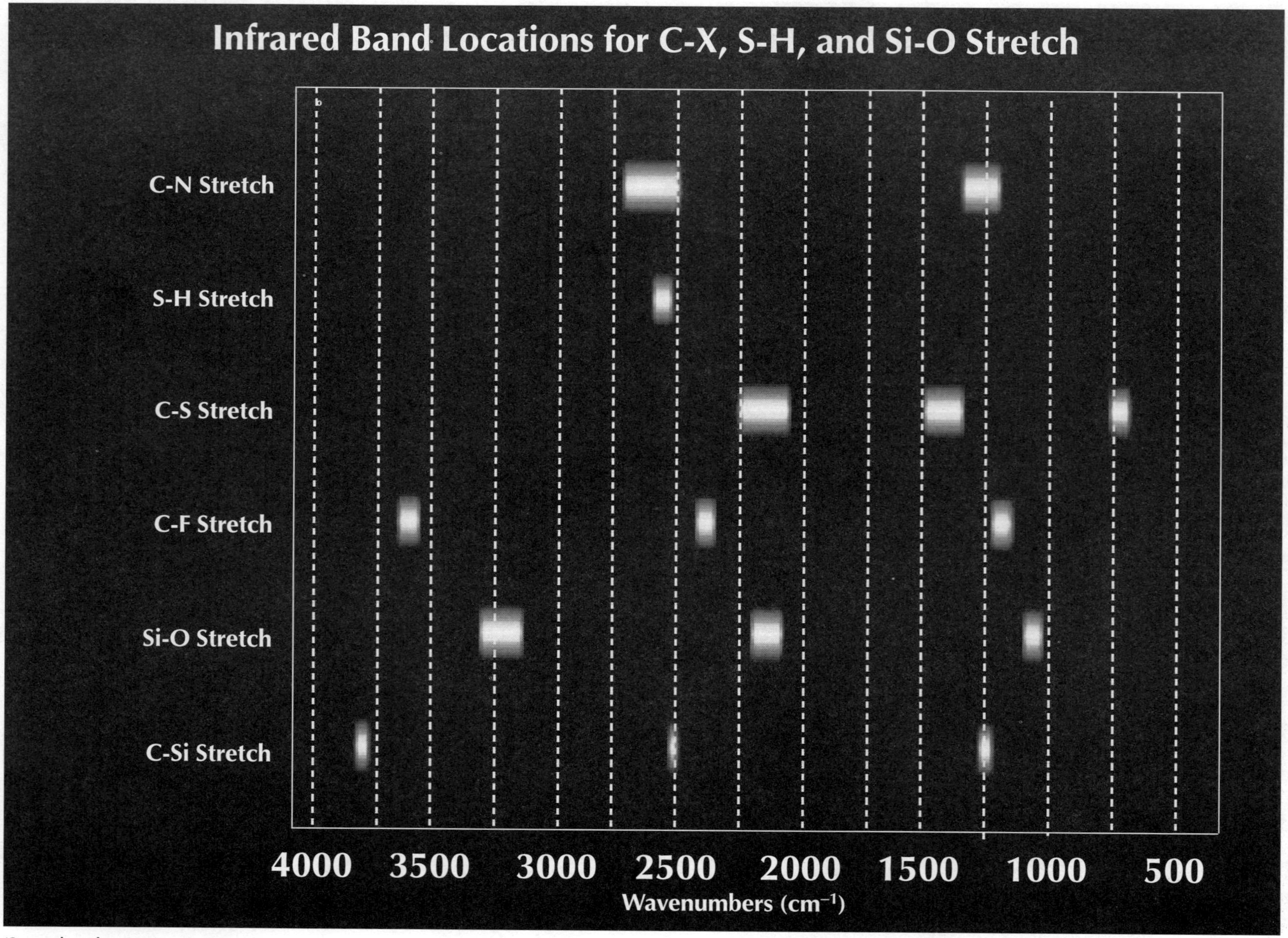

Infrared Band Locations for C-X, S-H, and Si-O Stretch

Wavenumbers (cm^{-1})

- C-N Stretch
- S-H Stretch
- C-S Stretch
- C-F Stretch
- Si-O Stretch
- C-Si Stretch

(See color plate 13.)

22.

FUNCTIONAL GROUPINGS AND CALCULATED LOCATIONS IN WAVENUMBERS (cm^{-1}) FOR RAMAN SPECTROSCOPY

Group	Molecular structure	Relative intensity	Fundamental
I. C-H stretch			
a. Alkane, C_nH_{2n+2}	$CH_3-(CH_2)_N-CH_3$, normal paraffins	m-s	
Methyl $-C-H_3$			2973–2959
Antisymmetric stretch		s	
Symmetric stretch		s	2882–2862
Methylene	$-C-H_2$		
Antisymmetric stretch		m	2936–2915
Symmetric stretch		m	2864–2843
b. Alkene, C_nH_{2n}	>C=C<, olefinic group		
Ethylene	H_2-C=C-H_2	s	~1340
Monosubstituted	H >C=C<^H_H R	w m	3040–3012 3096–3077
Cis, disubstituted	H >C=C<^H_R R	m	3040–3012
Trans, disubstituted	H >C=C<^R_H R	m	3040–3012
Gem, disubstituted	R >C=C<^H_H R	m	3096–3077
Trisubstituted	R >C=C<^R_H R	m	3040–3012
c. Alkyne, C_nH_{2n-2}	$-C\equiv C-H$	s	3345–3295
Acetylene	$H-C\equiv C-H$	w	3345–3295
d. Aromatic, C_nH_n	Ar(C)-H	v	3030
II. C-H bend			
a. Alkane, C_nH_{2n+2}	$CH_3-(CH_2)_N-CH_3$, normal paraffins		
Methine	R-CHR'R"	w	1340
Methylene	R-CH_2R'	w	1483–1445
Methyl	R-CH_3	w w	1379–1370 1470–1429
Gem, dimethyl	CH_3-CHR-CH_3	w w	1385–1379 1370–1364
Tert-butyl	$(CH_3)_3$-CH	w w	1395–1385 ~1370
b. Alkene, C_nH_{2n}	>C=C<		
Monosubstituted	H >C=C<^H_H R	s s s	952–985 915–905 1420–1410
Cis, disubstituted	H >C=C<^H_R R	s	~690
Trans, disubstituted	H >C=C<^R_H R	s m	970–960 1309–1295

Group		Molecular structure	Relative intensity	Fundamental
	Gem, disubstituted	R >C=C<^H_H R	s s	895–885 1420–1410
	Trisubstituted	R >C=C<^R_H R	s	840–790
c.	Alkyne, C_nH_{2n-2}	–C≡C–H	v, s	~680–605 ~629
	Acetylene	H-C≡C-H	m	680–605
d.	Aromatic, C_nH_n	Ar(C)-H		
	Monosubstituted (5 hydrogens)		v, s	~752 ~699
	Disubstituted (4 hydrogens)		v, s	~752
	Trisubstituted (3 hydrogens)		v, s	781
	Tetrasubstituted (2 hydrogens)		v, m	~833
	Pentasubstituted (1 hydrogen)		v, w	~885
III. C-C stretch				
a.	Alkene, C_nH_{2n} Symmetrical stretch	>C=C<	v	1681–1621
	Monosubstituted	As shown above	s	~1645
	Cis, disubstituted	As shown above	s	~1658
	Trans, disubstituted	As shown above	s	~1675
	Gem, disubstituted	As shown above	s	~1653
	Trisubstituted	As shown above	s	~1669
	Tetrasubstituted	As shown above	v	~1669
	Diene	>C=C< ··· >C=C<	v v	~1650 ~1600
b.	Alkyne, C_nH_{2n-2}			
	Acetylene	H-C≡C-H	s	2265–2020
	Monosubstituted	R- C≡C-H	m	2265–2020
	Disubstituted	R'-C≡C-R"	v,w	2262–2188
c.	Allene	>C=C=C<	m m	~1961 ~1064
d.	Aromatic	containing	v v m m	1640–1590 ~1580 1499–1450 ~1449
IV. Carbonyl		>C=O		
a.	Ketones	R >C=O R'		
	Saturated, acyclic	alkane (R) >C=O (R') alkane	m	1724–1704

Group		Molecular structure	Relative intensity	Fundamental
	Saturated, cyclic			
	6- or more membered ring	O=C⟨ (or) AR >C=O R'	m	1724–1704
	5-membered ring	O=C⟨	m	1751–1770
	4-membered ring	O=C⟨	m	~1776
	Alpha-, beta-unsaturated, acyclic		s	1684–1664
	Alpha-, beta-unsaturated, cyclic			
	6- or more membered ring		s	1684–1664
	5-membered ring		s	1724–1709
	Alpha-, beta-, alpha'-, beta'-unsaturated, acyclic		s	1669–1664
	Aryl		s	1701–1681
	Diaryl		s	1669–1661
	Alpha-diketones		s	1730–1709
	Beta-diketones (enolic)		s	1639–1538
	1,4-Quinones		s, m	1689–1661
	Ketenes		s, m	~2151
b.	Aldehydes, Carbonyl stretch	>C=O		
	Saturated, aliphatic		s	1739–1721
	Alpha-, beta-unsaturated, aliphatic		s	1704–1681
	Alpha-, beta-, gamma-, delta-unsaturated, aliphatic		s	1681–1661
	Aryl		s	1715–1695
	C-H stretching, two bands		w w	2899–2817 2778–2703
c.	Esters	R-C=O-OR'		
	Saturated, acyclic		m, s	1751–1736
	Saturated, cyclic: delta- and larger lactones		m, s	1751–1736
	Gamma-lactones		s	1779–1761
	Beta-lactones		s	~1818
	Unsaturated, vinyl ester		s	1799–1770
	Alpha-, beta-unsaturated and aryl		s	1730–1718
	Alpha-, beta-unsaturated delta-lactone		s	1792–1718
	Alpha-, beta-unsaturated gamma-lactone		s	1761–1739
	Beta-, gamma-unsaturated gamma-lactone		s	~1799
	Alpha-ketoesters		s	1754–1739
	Beta-ketoesters (enolic)		s	~1650
	Carbonates		s	1779–1739

Group	Molecular structure	Relative intensity	Fundamental
V. Carboxylic acids, and ethers	–O-C=O— and R-O-R		
Carbonyl stretching: saturated aliphatic		s	1724–1701
Alpha-, beta-unsaturated aliphatic aryl		s	1715–1689
			2049–1681
Hydroxyl stretching (bonded), several bands		w	2703–2500
Carboxylate anion stretching		s	1610–1550
Ethers –C-O str.	R'-O-R"	w	1205–1100
		m	925–700
VI. Anhydrides	R'-C=O-O-C=O-R"		
Stretching, –C=O		m	1850–1750
Saturated, acyclic	–C=O stretching	m	1850–1750
	H$_3$CC-O stretching	m	1400–1350
	C-O-C stretching	m	1160–1100
Alpha-, beta-unsaturated and aryl, acyclic		m	1828–1779
		m	1770–1721
Saturated, 5-member ring		m	1869–1821
		m	1767–1751
Alpha-, beta-unsaturated, 5-member ring		m	1848–1799
		m	1828–1779
VII. Acyl halides			
Stretch: acyl fluorides		s	~1848
Acyl chlorides		s	~1795
Acyl bromides		s	~1808
Alpha-, beta-unsaturated and aryl		s	1751–1748
		m	1748–1718
COF$_2$		s	~1927
COCl$_2$		s	~1828
COBr$_2$		s	~1828
VIII. Amides			
a. Carbonyl stretch			
Primary, concentrated solutions and solids		s	~1650
Primary, dilute solutions		s	~1718
Secondary, concentrated solutions and solids		m	~1650
Secondary, dilute solutions		s	1701–1669
Tertiary, solutions and solids		s	1669–1629
Cyclic, delta-lactams, dilute solutions		s	~1681
Cyclic, gamma-lactams, dilute solutions		s	~1701
Cyclic, gamma-lactams fused to second ring, dilute solution		s	1751–1701
Cyclic, beta-lactams, dilute solution		s	1761–1730
Cyclic, beta-lactams fused to second ring, dilute solutions		s	1779–1770
Urea, acyclic		s	~1661
Urea, cyclic, 6-membered ring		s	~1639
Ureas, cyclic 5-member ring		s	~1721

Group		Molecular structure	Relative intensity	Fundamental
	Urethanes		s	1739–1689
	Imides, acyclic		s	~1709
	Imides, cyclic, 6-membered ring		s	~1709
			s	~1701
	Imides, cyclic, alpha, beta-unsaturated, 6-membered ring		s	~1730
			s	~1669
	Imides, cyclic, 5-membered ring		s	~1770
			s	~1701
	Imides, cyclic, alpha, beta-unsaturated, 5-membered ring		s	~1792
			s	~1709
b.	N-H stretch			
	Primary, free, two bands		m	~3497
			m	~3401
	Primary, bonded, two bands		m	~3344
			m	~3175
	Secondary, free; one band		m	~3425
	Secondary, bonded; one band		m	3333–3125
c.	N-H Bending			
	Primary amides, dilute solution		s	1621–1590
	Secondary amides, dilute solution		s	1548–1511
IX. Alcohols & phenolics, O-H stretch				
	Free O-H		Very weak to none	3650–3584
	With hydrogen bonding, single bridge compounds	O-H $\cdots$ H–	Very weak to none	3546–3448
	Polymeric O-H		Very weak to none	3401–3195
	Chelated compounds		Very weak to none	3226–2500
	Primary alcohols		Very weak to none	~1053
				1351–1266
	Secondary alcohols		Very weak to none	~1099
				1351–1266
	Tertiary alcohols		Very weak to none	~1149
				1408–1316
	Phenols		Very weak to none	~1198
				1405–1316
X. Amines, N-H stretch				
a.	Primary, free; two bands		s	3350–3320
	Secondary, free; one band		m	3497–3311
	Imines; one band	=N-H	m	3401–3300
	Amine salts		m	3125–3030
b.	N-H bend			
	Primary		s-m	1650–1590
	Secondary		w	1650–1550
	Amine salts		s	1600–1575
			s	~1499
c.	C-N stretch			
	Aromatic, primary		s	1340–1250
	Aromatic, secondary		s	1350–1280

Group	Molecular structure	Relative intensity	Fundamental
	Aromatic, tertiary	s	1359–1309
	Aliphatic	w	1220–1020
		w	~1427

XI. Unsaturated N compounds, C=N stretch and C≡N stretch

a.	Alkyl nitriles	m	2262–2205
	Alpha-, beta-unsaturated alkyl nitriles	m	2237–2217
	Aryl nitriles	m	2242–2222
	Isocyanates	m	2273–2232
	Isocyanides	m	2222–2070
b.	Imines & oximes		
	Alkyl compounds	v	1689–1649
	Alpha-, beta-unsaturated	v	1661–1629
	–N=N– stretch, azo- compounds	v	1629–1575
	–N=C=N–, diimides	s	2155–2128
	–N$_3$ stretch, azides	s	2160–2137
		w	1340–1179
c.	C-NO$_2$, nitro- compounds		
	Aromatic	s	1570–1499
		s	1370–1299
	Aliphatic	s	1570–1550
		s	1379–1370
	O-NO$_2$, nitrates	s	1650–1600
		s	1299–1250
	C-NO, nitroso	s	1600–1499
	O-NO, nitrites	s	1681–1600
		s	1626–1610
d.	Nitriles, –C≡N stretch	s	2280–2210

XII. Halogen, C-X stretch and X-C=O stretch

C-F		m	1408–1000
C-Cl		m	800–602
C-Br		m	602–500
C-I		m	~500
Aromatic, 1,3,5 meta substitution		v	1000–990
X-C=O stretch		w	1820–1790

XIII. Sulfur compounds

a.	S-H stretch	s	2597–2551
b.	C=S stretch	s	1200–1050
c.	C-S stretch	s	700–650
d.	S=O stretch		
	Sulfoxides	s	1070–1030
	Sulfones	s	1160–1140
		s	1350–1300
	Sulfites	s	1230–1149
		s	1429–1350

Group	Molecular structure	Relative intensity	Fundamental
Sulfonyl chlorides		s	1185–1164
		s	1370–1340
Sulfonamides		s	1179–1140
		s	1350–1300
Sulfonic acids		s	1209–1149
		s	1060–1030
		s	~649
XIV. Other			
P=O stretch	P=O	s	1290
		s, b	1000–900

23.

RAMAN SPECTRAL CORRELATION CHARTS

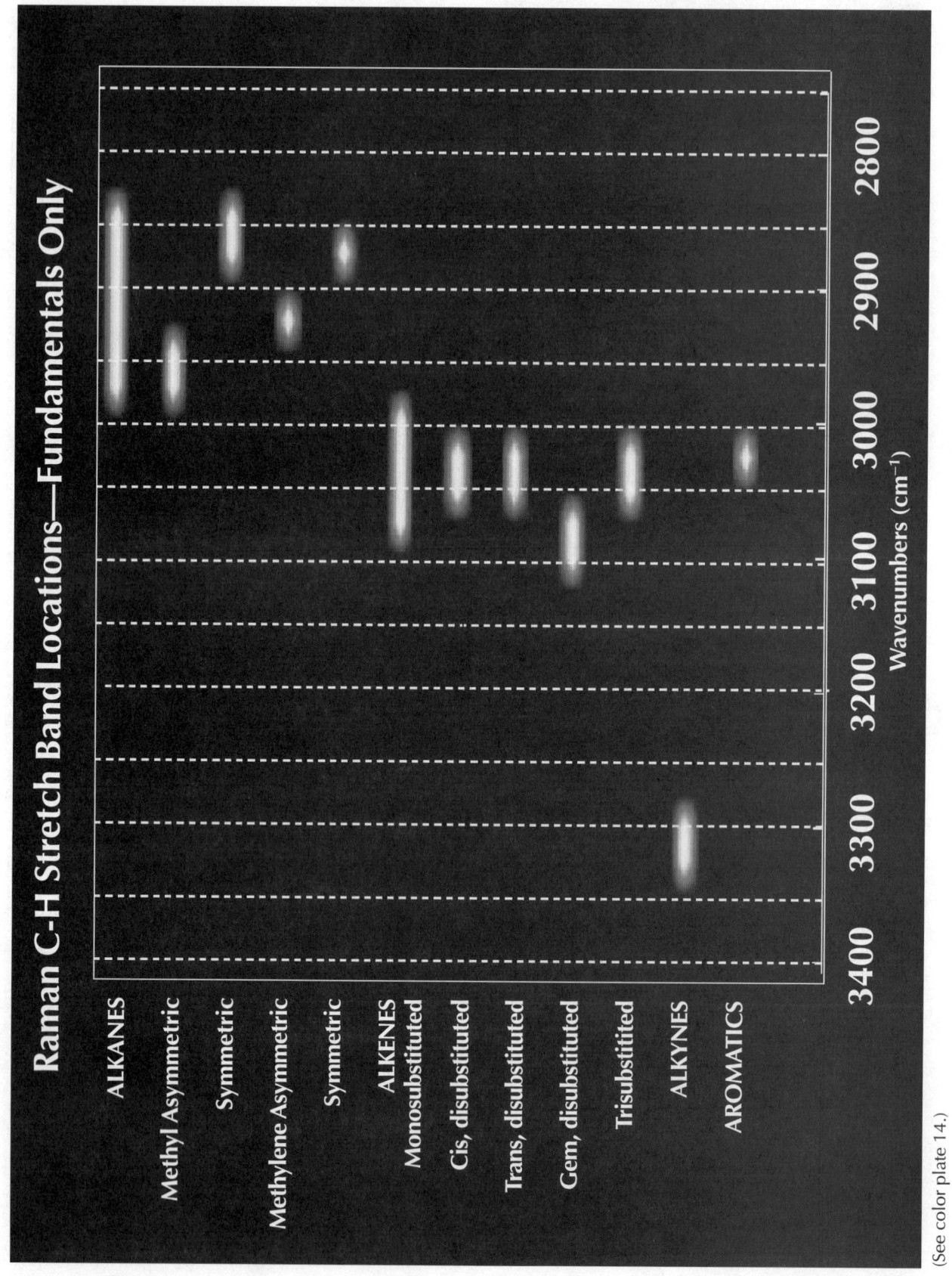

Raman C-H Stretch Band Locations—Fundamentals Only

ALKANES
Methyl Asymmetric
Symmetric
Methylene Asymmetric
Symmetric
ALKENES
Monosubstituted
Cis, disubstituted
Trans, disubstituted
Gem, disubstituted
Trisubstitited
ALKYNES
AROMATICS

Wavenumbers (cm^{-1})

(See color plate 14.)

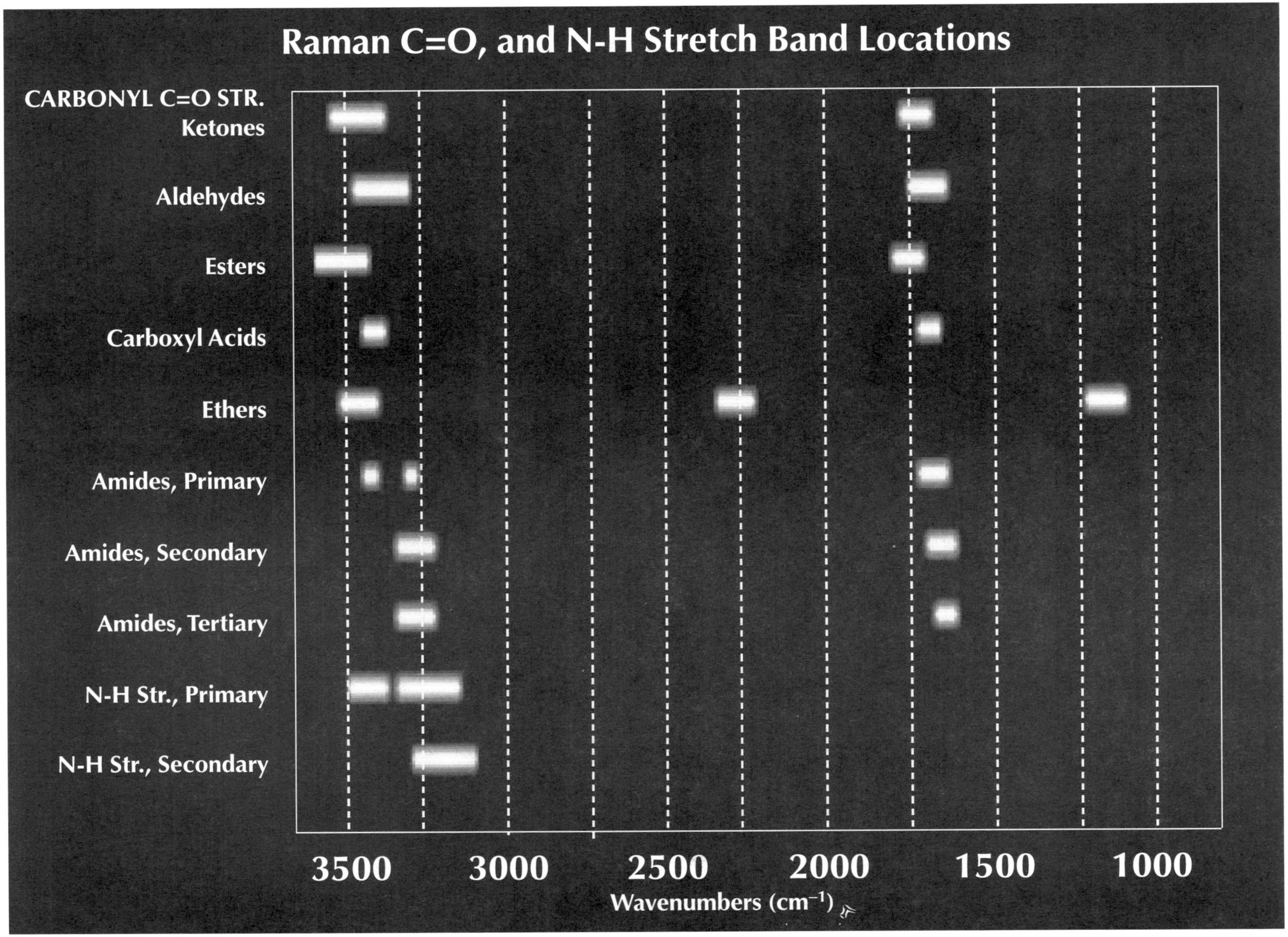

(See color plate 15.)

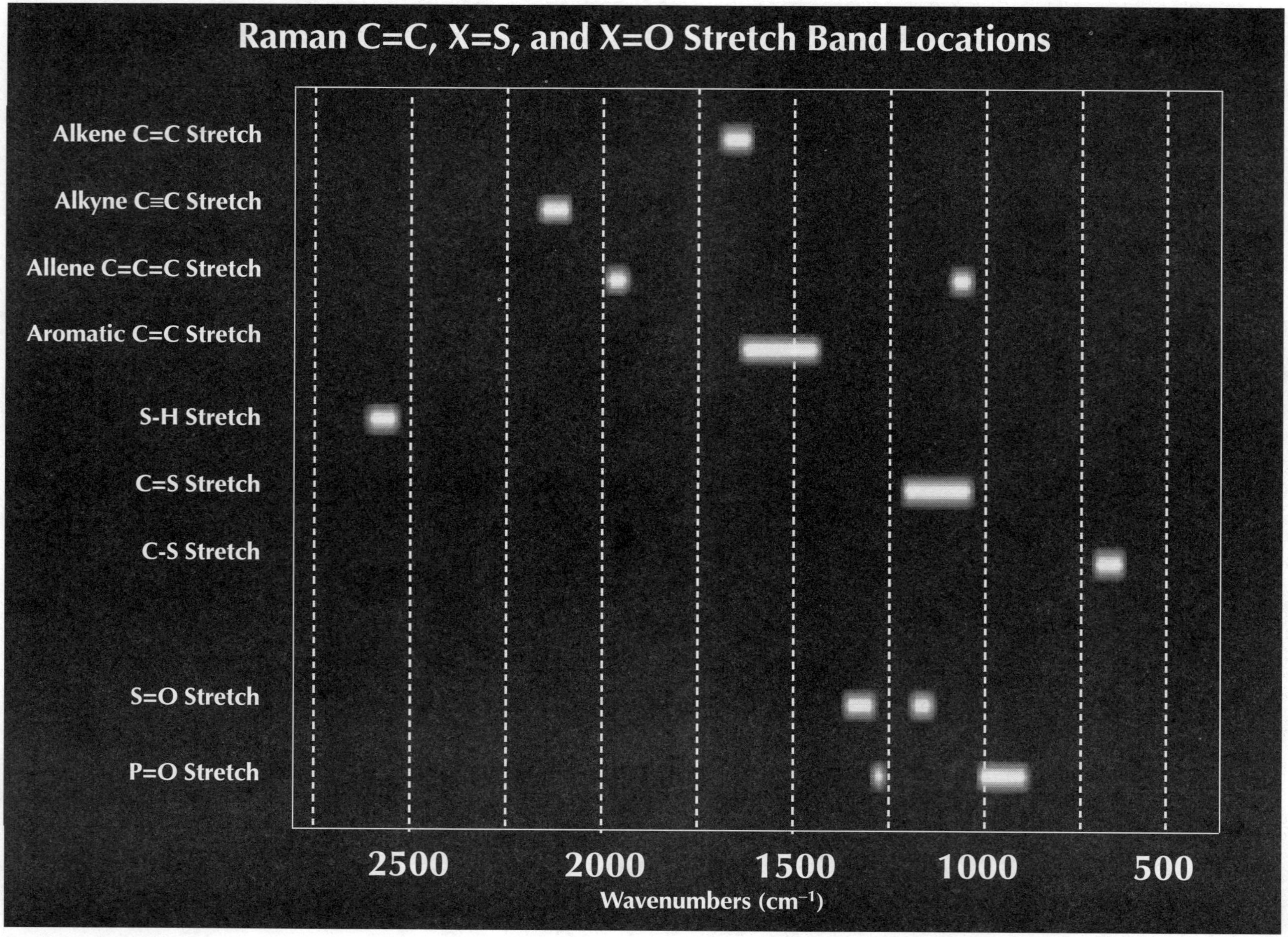

Raman C=C, X=S, and X=O Stretch Band Locations

24.

SURFACTANT AND POLYMER SPECTRA

SURFACTANT GROUPS

Surfactants (or surface-active agents) exist in a variety of forms and have often been referred to as *detergents*. General types of surfactants include *nonionics, anionics, cationics,* and *amphoterics,* among others. In designating and naming surfactant compounds, a number of abbreviations are used. For example, *P.O.E.* in a surfactant name indicates a polyethoxylated or polyoxyethylated surfactant. *HLB* in a surfactant name indicates that the compound has a "hydrophile-lipophile-balance" number. As a general rule, the higher the HLB, the more hydrophilic (water attracting) the surfactant. *EtO* indicates the presence of a specific molar concentration of ethylene oxide; when alcohols are treated with ethylene oxide in the presence of a base, an ethoxylate detergent results. *P.O.P.* indicates the presence of polyoxypropylene. The infrared spectra for these different classes of surfactants are included as spectra numbers 1327–1890. Examples of each surfactant group are listed under the appropriate category, and the appropriate spectra numbers are given in parentheses.

Nonionic Surfactants

Nonionics include the following types of compounds.

1. Esters of polyhydric alcohols
   a. Esters of glycerol and polyglycerol
      Glycerol monolaurate (1426)
      Glyceryl dilaurate (1425)
      Glycerol monostearate (1663)
      Glycerol distearate (1664)
      Glycerol trioleate (1367)
      Glycerol monoricinoleate (1665)
      Modified glyceryl phthalate resin (1755)
      Polyglycerol ester of oleic acid (1366)
   b. Esters of polyoxyethylated (P.O.E.) glycerides
      Polyoxyethylated (P.O.E.) castor oil, 20 moles EtO (1609)
   c. Esters of polyoxyethylated (P.O.E.) sorbitol, sorbitan, or sorbide
      P.O.E. Sorbitan monolaurate, HLB 13.3 (1586)
      P.O.E. Sorbitan monolaurate, HLB 16.7 (1666)
      P.O.E. Sorbitan monopalmitate, HLB 15.6 (1667)
      P.O.E. Sorbitan tristearate, 20 moles EtO, HLB 10.5 (1362)

P.O.E. Sorbitan monooleate, HLB 13.9 (1576)
P.O.E. Sorbitan trioleate, HLB 11.0 (1359)
  d. Esters of polyhydric alcohols
     Sorbitan monolaurate (1358)
     Sorbitan monopalmitate (1378)
     Sorbitan monostearate (1361)
     Sorbitan tristearate (1668)
     Sorbitan sesquioleate (1680)
     Sorbitan trioleate (1357, 1681)
     Sucrose monomyristate (1688)
     Sucrose monopalmitate (1360)
     Sucrose monostearate (1356)
     Sucrose dioleate (1355)
     Sucrose monotallowate (1354)
     Pentaerythritol monolaurate (1670)
     Pentaerythritol distearate (1330)
     Pentaerythritol tetrastearate (1671)
     Pentaerythritol dioleate (1830)
2. Alkoxylated amides
     Polyoxyethylated (P.O.E.) coco amide, 5 moles EtO (1350)
     P.O.E. Hydrogenated tallow amide, 50 moles EtO (1349)
     Oleic acid monoisopropanolamide (1578)
3. Esters of polyoxyalkylene glycols
  a. Laurates
     Laurate of diethylene glycol (1347)
     Laurate of polyethylene glycol 200 (1346)
     Laurate of polyethylene glycol 300 (1684)
     Laurate of polyethylene glycol 400 (1548)
     Laurate of polyethylene glycol 600 (1685)
     Laurate of polyethylene glycol 1540 (1433)
     P.O.E. Lauric acid, 9 moles EtO (1831)
     P.O.E. Lauric acid, 14 moles EtO (1377)
     Dilaurate of polyethylene glycol 200 (1342)
     Dilaurate of polyethylene glycol 300 (1686)
     Dilaureate of polyethylene glycol 400 (1428)
     Dilaureate of polyethylene glycol 600 (1687)
     Dilaureate of polyethylene glycol 1540 (1595)
  b. Stearates
     Stearate of polyethylene glycol 200 (1341)
     Stearate of polyethylene glycol 300 (1674)
     Stearate of polyethylene glycol 600 (1371)
     Stearate of polyethylene glycol 1540 (1589)
     Stearate of polyethylene glycol 4000 (1370)
     Stearate of polyethylene glycol 6000 (1675)
     Polyoxyethylated Stearic acid, 5 moles EtO (1676)
     P.O.E. Stearic acid, 9 moles EtO (1376)
     Polyoxyethylated Stearic acid, 10 moles EtO (1677)
     P.O.E. Stearic acid, 15 moles EtO (1375)
     Polyoxyethylated Stearic acid, 40 moles EtO (1678)
     Distearate of polyethylene glycol 300 (1340)
     Distearate of polyethylene glycol 400 (1703)
     Distearate of polyethylene glycol 600 (1588)
     Distearate of polyethylene glycol 1000 (1339)
     Distearate of polyethylene glycol 6000 (1369)

c. Oleates
   Oleate of ethylene glycol (1338)
   Oleate of diethylene glycol (1689)
   Oleate of polyethylene glycol 200 (1337)
   Oleate of polyethylene glycol 300 (1691)
   Oleate of polyethylene glycol 400 (1368)
   Oleate of polyethylene glycol 600 (1692)
   Oleate of polyethylene glycol 1000 (1592)
   P.O.E. oleic acid, 6 moles EtO (1374)
   Polyoxyethylated oleic acid, 9 moles EtO (1449)
   Dioleate of polyethylene glycol 200 (1336)
   Dioleate of polyethylene glycol 400 (1427)
   Dioleate of polyethylene glycol 600 (1335)
   Dioleate of polyethylene glycol 1000 (1705)
   Dioleate of polyethylene glycol 1540 (1334)
   Dioleate of polyethylene glycol 4000 (1570)
d. Ricinoleates
   Ricinoleate of ethylene glycol (1424)
   Ricinoleate of diethylene glycol (1693)
   Ricinoleate of polyethylene glycol 400 (1593)
   Ricinoleate of polyethylene glycol 600 (1442)
e. Coco fatty acids
   Polyoxyethylated Coco Fatty Acid, 5 moles EtO (1469)
f. Tall oils
   Polyoxyethylated Red oil, 10 moles EtO (1466)
   Polyoxyethylated Tall oil, 12 moles EtO (1611)
   P.O.E. Tall oil, 16 moles EtO (1372)
   Polyoxyethylated Rosin (1445)
g. Lanolins
   Polyoxyethylated Lanolin (1513)
h. Cocoates
   Methoxypolyethylene glycol 500 "cocoate" (1467)
i. Esters of polyoxypropylene glycols
   Stearate of propylene glycol (1842)
   Ricinoleate of propylene glycol (1841)
j. Esters of polyoxyethylene polyoxypropylene glycols
   Polyoxyethylated oxypropylated stearic acid (1836)
4. Ethers of polyoxyalkylene glycols
   a. Alkylphenyl ethers
      Polyoxyethylated tert-octylphenol, 3 moles EtO (1838)
      Polyoxyethylated tertiary octylphenol, 5 moles EtO (1577)
      Polyoxyethylated tert-octylphenol, 9–10 moles EtO (1837)
      Polyoxyethylated tert-octylphenol, 30 moles EtO (1839)
      Polyoxyethylated Nonylphenol, 1–2 moles EtO (1523)
      Polyoxyethylated Nonylphenol, 4 moles EtO (1587)
      Polyoxyethylated Nonylphenol, 6 moles EtO (1696)
      Polyoxyethylated Nonylphenol, 8 moles EtO (1834)
      Polyoxyethylated Nonylphenol, 9–10 moles EtO (1464)
      Polyoxyethylated Nonylphenol, 10–11 moles EtO (1833)
      Polyoxyethylated Nonylphenol, 15 moles EtO (1573)
      Polyoxyethylated Nonylphenol, 20 moles EtO (1483)
      Polyoxyethylated Nonylphenol, 30 moles EtO (1435)
   b. Alkyl ethers of polyethylene glycols
      P.O.E. Lauryl alcohol, 4 moles EtO (1460)

P.O.E. Lauryl alcohol, 23 moles EtO (1430)
Polyoxyethylated tridecyl alcohol, 3 moles EtO (1451)
Polyoxyethylated tridecyl alcohol, 9 moles EtO (1575, 1785, 1818)
Polyoxyethylated tridecyl alcohol, 12 moles EtO (1840)
Polyoxyethylated tridecyl alcohol, 15 moles EtO (1579)
Polyoxyethylated tetradecyl alcohol, 7 moles EtO (1704)
P.O.E. stearyl alcohol, 20 moles EtO (1446)
Polyoxyethylated cetyl alcohol, 20 moles EtO (1470)
Polyoxyethylated oleyl alcohol, 20 moles EtO (1471, 1835)
  c. P.O.E. and P.O.P. glycols and their ethers
Polyoxypropylene (M.W. 1200) and 40% EtO (1472)
Polyoxypropylene (M.W. 1750) and 10% EtO (1473)
Polyoxypropylene (M.W. 1750) and 20% EtO (1832)
Polyoxypropylene (M.W. 1750) and 40% EtO (1474)
Polyoxypropylene (M.W. 1750) and 80% EtO (1479)
Polyoxypropylene (M.W. 2100) and 50% EtO (1456)
  d. Thioethers of polyoxyethylene glycols
5. Tertiary acetylenic glycols
6. Polyoxyethylated alkyl phosphates

Anionic Surfactants

Anionics include the following.

1. Carboxylic acids and soaps
   a. Free acids
      Capric acid (1478)
      Lauric acid (1614)
      Myristic acid (1484)
      Palmitic acid (1485)
      Stearic acid (1562)
      Behenic acid (1706)
      Undecylenic acid (1486)
      Oleic acid (168–170, 627–628)
      Erucic acid (1493)
      Linoleic acid (117–119, 1487, 623–624)
      Naphthenic acid (1489)
      Coconut acids (1490)
      Tallow acids, distilled (1844)
      Tall oil fatty acids (1516)
   b. Alkali metal soaps
      Lithium stearate (1495)
      Sodium caprate (1699)
      Sodium laurate (1497)
      Sodium myristate (1707)
      Sodium palmitate (1501)
      Sodium undecylenate (1700)
      Sodium oleate (1491)
      Sodium linoleate (1498)
      Sodium ricinoleate (1492)
      Sodium naphthenate, 4% sodium (1420)
      Potassium laurate (1585)
      Potassium myristate (1500)

Potassium palmitate (1499)
Potassium stearate (1496)
Potassium undecylenate (1702)
Potassium linoleate (1601)
Potassium ricinoleate (1708)
Potassium naphthenate (1503)
Potassium abietate (1504)
- c. Ammonium and amine soaps
Ammonium caprate (1716)
Ammonium myristate (1722)
Ammonium palmitate (1327)
Ammonium stearate (1721)
Ammonium undecylenate (1820)
Ammonium oleate (1720)
Ammonium linoleate (1845)
Ammonium ricinoleate (1709)
Ammonium naphthenate ((1846)
Ammonium abietate (1710)
Morpholine laurate (1711)
Morpholine myristate (1712)
Morpholine palmitate (1505)
Morpholine stearate (1713)
Morpholine undecylenate (1590)
Morpholine oleate (1714)
Morpholine linoleate (1717)
Morpholine ricinoleate (1507)
Morpholine naphthenate (1715)
Morpholine abietate (1508)
Triethanolamine caprate (1718)
Triethanolamine laurate (1509)
Triethanolamine myristate (1511)
Triethanolamine palmitate (1512)
Triethanolamine stearate (1567)
Triethanolamine undecylenate (1719)
Triethanolamine oleate (1514)
Triethanolamine linoleate (1529)
Triethanolamine ricinoleate (1723)
Triethanolamine naphthenate (1530)
Triethanolamine abietate (1724)
- d. "Heavy metal soaps" (not water soluble)
Aluminum palmitate (1736)
Aluminum stearate (1515)
Aluminum oleate (797, 798, 1751)
Barium stearate (1737)
Barium naphthenate (1725)
Calcium stearate (1028, 1517)
Calcium oleate (1060, 1580)
Calcium linoleate (1726)
Calcium ricinoleate (1518)
Calcium naphthenate (1531)
Copper oleate (1550)
Copper naphthenate (1727)
Iron stearate (1572)

Iron naphthenate (1532)
Lead stearate (1519)
Lead naphthenate (1728)
Magnesium stearate (1566)
Manganese naphthenate (1729)
Nickel oleate (1730)
Zinc palmitate (1520)
Zinc stearate (1521)
Zinc oleate (1731)
Zinc linoleate (1565)
Zinc Resinate (1522)
2. Sulfated esters, salts
   a. Esters of simple alcohols
      Sulfated propyl oleate, Na salt (1533)
      Sulfated isopropyl oleate, Na salt (1732)
      Sulfated butyl oleate, Na salt (1607)
      Sulfated amyl oleate, Na salt (1534)
   b. Sulfated glycerides
      Sulfated glycerol trioleate, Na salt (1606)
      Sulfated castor oil, 2.0% organic sulfate (1600)
      Sulfated cod oil, Na salt (1535)
      Sulfated rice brand oil (1734)
      Sulfated soybean oil, 4.0% organic sulfate (1597)
      Sulfated tallow, Na salt (1594)
3. Sulfated amides, salts
   a. Sulfated alkloamine–carboxylic acid condensates
      Sulfated ethanolamine–lauric acid condensate, Na salt (1843)
      Sulfated ethanolamine–myristic acid condensate, Na salt (1524)
   b. Sulfated amides of unsaturated carboxylic acid
4. Sulfated alcohols, salts
   a. Alkyl sulfates
      Sodium *n*-octyl sulfate (1536)
      Sodium 2-ethylhexyl sulfate (1537)
      Sodium lauryl sulfate (1079)
      Sodium *sec*-tetradecyl sulfate (1538)
      Sodium cetyl sulfate (1528)
      Sodium oleyl-stearyl sulfate (1539)
      Magnesium Lauryl Sulfate (1540)
      Ammonium lauryl sulfate (1746)
      Diethanolammonium lauryl sulfate (1596)
      Triethanolammonium lauryl sulfate (1541)
      Miranol SM salt of lauryl sulfate (1738)
      Miranol C2M salt of lauryl sulfate (1739)
5. Sulfated ethers
   a. Sulfated alkylphenyl ethers of P.O.E. glycols
      Sulfated polyoxyethylated octylphenol, Na salt (1563)
      Sulfated polyoxyethylated nonylphenol, Na salt (1542)
6. Sulfated carboxylic acids, salts
   Sulfated castor oil fatty acid, Na salt (1544)
7. Sulfonated petroleum, salts
   Sodium petroleum sulfonate, M.W. 340–360 (1545)
   Sodium petroleum sulfonate, M.W. 513 (1740)
   Calcium petroleum sulfonate, ave. M.W. 900 (1556)

Barium petroleum sulfonate, ave. M.W. 1000 (1741)
Ammonium petroleum sulfonate, ave. M.W. 445 (1742)
Isopropylamine salt—sulfonated petroleum (1546)
Triethanolamine salt—sulfonated petroleum (1547)
8. Sulfonated aromatic hydrocarbons, salts
   a. Alkylnaphthalenes and polymeric alkylnaphthalenes
      Sodium diisopropylnaphthalene sulfonate (1551)
      Sodium dibutylnaphthalene sulfonate (1527)
      Sodium mono- and diamylnaphthalene sulfonates (1787)
      Sodium benzylnaphthalene sulfonate (1613)
      Sodium poly-alkylnaphthalene sulfonate (1436)
      Ethanolamine salt—dibutylnaphthalene sulfonate (1549)
   b. Alkyldiphenyl and polymeric alkylbenzenes
   c. Alkylbenzenes
      Sodium toluene sulfonate (1558)
      Sodium xylene sulfonate (1557)
      Sodium dodecylbenzene sulfonate (1525)
      Sodium kerylbenzene sulfonate (1553)
      Calcium dodecylbenzene sulfonate (1788)
9. Sulfonated aliphatic hydrocarbons, salts, or acids
10. Sulfonated esters, salts
    a. Esters of hydrooxysulfonic acids
    b. Esters of long-chain sulfocarboxylic acids
       Propyl ester of sulfooleic acid, sodium salt (1458)
    c. Esters of short-chain sulfocarboxylic acids
       Sodium lorol sulfoacetate (1561)
       Sodium diisobutyl sulfosuccinate (1454)
       Sodium diamyl sulfosuccinate (1618)
       Sodium dihexyl sulfosuccinate (1624)
       Sodium dioctyl sulfosuccinate (789–790, 1453)
       Sodium ditridecyl sulfosuccinate (1331)
11. Sulfonated amides, salts
    a. Amides of short sulfocarboxylic acids
       Sodium N-alkylsulfoacetamide (1452)
    b. N-Acylated short aminosulfonic acids
       Sodium N-methyl-N-palmitoyl taurate (1619)
       Sodium N-cyclohexyl-N-palmitoyl taurate (1747)
       Sodium N-methyl-N-oleyl taurate (1333)
       Sodium N-methyl-N-tallow-acid taurate (1620)
       Sodium N-methyl-N-tall-oil-acid taurate (1625)
12. Sulfonated amines, salts
       Ammonium alkylbenzimidazole sulfonate (1789)
13. Sulfonated ethers, salts
14. Sulfonated carboxylic acids, salts
       Sodium sulfooleate (1443)
15. Sulfonated phenols, salts
       Sodium monobutylphenylphenol monosulfonate (1439)
       Disodium dibutylphenylphenol disulfonate (1626)
       Potassium monoethylphenylphenol monosulfonate (1822)
       Guanidinium monoethylphenylphenol sulfonate (1627)
16. Sulfonated lignins, salts
       Sodium lignosulfonate, 5.4% SO$_3$Na (1621)
       Sodium lignosulfonate, 14.3% SO$_3$Na (1753)

Sodium lignosulfonate, 2 moles SO$_3$Na (1332)
Sodium lignosulfonate, 3 moles SO$_3$Na (1752)
17. Acylated amino acids, salts
n-Stearoyl-palmitoyl sarcosine (1623)
n-Oleoyl sarcosine (1461, 1824)
Sodium *N*-lauroyl sarcosinate (1628)
18. Acylated polypeptides, salts
19. Phosphates and Phosphatides
   a. Phosphated glycerides
      Soy phosphatides, lecithin (1608)
   b. Alkyl polyphosphates
      Sodium di(2-ethylhexyl) phosphate (1602)

Cationic Surfactants

These include the following.

1. Amines
   a. Primary
      Tertiary-C11–14H23–29 amine (1468)
      n-Octadecylamine (1748)
      Oleyl amine (1583)
      Coco amine (1642)
      Hydrogenated tallow amine (1632)
      Tallow amine (1582)
      Soya amine (1633)
   b. Secondary
      Dicoco amine (1581)
      Dihydrogenated tallow amine (1634)
   c. Tertiary
      Dimethyl hexadecyl amine (1643)
      Dimethyl octadecyl amine (1457)
      Dimethyl coco amine (1744)
      Dimethyl soya amine (1441)
   d. Diamines
      n-Coco-propylenediamine (1440)
      n-Soya-propylenediamine (1638)
      n-Tallow-propylenediamine (1437)
2. Amine salts
      Acetic acid salt of dodecylamine (1429)
      Acetic acid salt of hexadecylamine (1749)
      Acetic acid salt of octadecylamine (1636)
      Acetic acid salt of oleylamine (1574)
      Acetic acid salt of hydrogenated coco amine (1750)
      Acetic acid salt of hydrogenated tallow amine (1612)
      Acetic acid salt of tallow amine (1635)
      Acetic acid salt of soya amine (1615)
3. Trialkylamine oxides
      Cetyl dimethylamine oxide (1825)
   a. N-polyoxyethylated (P.O.E.) long-chain amines
      Polyoxyethylated tertiary amine, 5 moles EtO (1645)
      P.O.E. Tertiary amine—C12–14H25–29NH(C$_2$H$_4$O)$_5$H (1462)

P.O.E. Stearyl amine, 5 moles EtO (1434)
P.O.E. Stearyl amine, 15 moles EtO (1423)
P.O.E. Oleyl amine, 5 moles EtO (1416)
P.O.E. Tertiary amine—C18–24H37–49NH(C_2H_4O)$_{15}$H (1415)
P.O.E. Coco amine, 5 moles EtO (1414)
Polyoxyethylated coco amine, 10 moles EtO (1646)
Polyoxyethylated tallow amine, 2 moles EtO (1639)
P.O.E. Tallow amine, 5 moles EtO (1761)
P.O.E. Tallow amine, 15 moles EtO (1413)
P.O.E. Soya amine, 2 moles EtO (1760)
P.O.E. Soya amine, 5 moles EtO (1762)
P.O.E. Soya amine, 10 moles EtO (1412)
P.O.E. Soya amine, 15 moles EtO (1763)
P.O.E. Rosin amine, 5 moles EtO (1764)
n-b-Hydroxyethyl stearyl imidazoline (1410)
n-b-Hydroxyethyl coco imidazoline (1409)
n-b-Hydroxyethyl oleyl imidazoline (1637)

4. Quaternary ammonium salts
   a. Tetraalkylammonium salts
      Dodecyltrimethyl ammonium chloride (1408)
      Hexadecyltrimethyl ammonium chloride (1654)
      Octadecyltrimethyl ammonium chloride (1407, 1647)
      Cetyltrimethyl ammonium chloride (1658)
      Cetyldimethylethyl ammonium bromide (1765)
      Alkenyldimethylethyl ammonium bromide (1657)
      Soya trimethyl ammonium chloride (1406)
      Dilauryldimethyl ammonium bromide (1766)
   b. Trialkylarylammonium salts
      Lauryldimethylbenzyl ammonium chloride (1404)
      Stearyldimethylbenzyl ammonium chloride (1648)
      Alkyldimethyl-3,4-dichlorobenzyl ammonium chloride (1403)
   c. Polyethoxyethylammonium salts
5. Acylated polyamines, salts
      n-Oleoylethylenediamine, formate salt (1401)
      n-Stearoylethylenediamine, formate salt (1400)
   a. Quaternary salts
6. Heterocyclic amines, salts
   a. Quaternary imidazolinium salts
      b-Hydroxyethylbenzyl "coco" imidazolinium chloride (1655)
      Diethyl heptadecyl imidazolinium ethyl sulfate (1397)
   b. Pyridinium and quinolinium salts
      Laurylpyridinium chloride (1396)
      Cetylpyridinium bromide (1660)
      Laurylisoquinolinium bromide (1395, 1819)
7. Alkylolamine–fatty acid condensates (oxazolines)
      Substituted oxazoline, alkaterge C (1394)
      Substituted oxazoline, alkaterge A (1656)
      Substituted oxazoline, alkaterge E (1393)
8. Alkyl phosphonamides

Amphoterics

These are generally substituted imidazolinium salts.

> Sodium *n*-coco-*b*-aminopropionate (1390)
> Sodium *n*-lauryl-myristyl-*b*-aminopropionate (1661)
> Disodium *n*-lauryl-*b*-iminodipropionate (1389)

Perfluoro compounds

These are named as perfluoro surfactants, both anionic and cationic.

> Perfluoro surfactant–anionic (1388)
> Perfluoro surfactant–cationic (1387)

Sequestrants

These include compounds such as ethylendiamine tetraacetic acid, sodium (Na) salt.

> Ethylenediamine tetraacetic acid, 2 sodium salt (1386)
> Sodium dihydroxyethyl glycinate (1652)
> Trisodium nitrilotriacetate (1653)

Silicones

These include compounds such as silicone-based defoamers.

> Silicone (dimethicone-based) defoamer–water dispersible (1385)

Inorganics

Common inorganic surfactants are sodium borate, sodium carbonate, sodium phosphate (tribasic), and sodium silicate.

> Sodium borate, tetra (1384)
> Sodium carbonate (1017)
> Sodium phosphate, tribasic (1382)
> Sodium silicate (1381)

POLYMERS

Polymers are large molecules consisting of repeating units. The repeating units are referred to as *monomers*. Polymers exist as natural polymers, such as starches or polysaccharides, and synthetic polymers, commonly termed *plastics*. Polymers vary in molecular formula and molecular weight due to variation in the number of repeating units. Polymer backbone structures often have attached molecular groups. The molecular arrangement of these groups determines the stereochemical configuration for any given polymer. If all the attached groups are in the identical position along the polymer backbone chain, the polymer is in an *isotactic* configuration. If the attached groups alternate in their attached positions with a regular pattern, the *syndiotactic* configuration is ascribed. When attached groups are randomly attached to the polymer backbone, the polymer is said to have an *atactic* configuration. The isotactic configuration represents the most crystalline (rigid) of the configuration types.

Copolymers involve the use of two or more monomer types in a single backbone structure to achieve specific material performance properties. Structures are classified as *alternating* (-A-B-A-B-A-B-), *random* (-A-A-B-A-B-B-A-), *block* (-A-A-B-B-B-B-A-A-), or *graft* (-A-A-A-A-A-A-A-).

                                                                                         B   B
                                                                                         B   B

Polymerization involves the reaction of the monomer building blocks into polymers. The polymerization reaction types involve *addition* reactions and *condensation* reactions. Addition reactions typically involve the use of ethene to form polyethene and polyethylene. Condensation reactions typically involve different reaction products reacting to form a heteropolymer and a small molecular byproduct. The reaction of 1,6-diaminoethane and hexanedioic acid to form nylon and water is a classic example. Polymers made from one monomer type are termed *homopolymers;* those formed with two different monomers are referred to as *copolymers;* those formed using three monomer types are termed *terpolymers.* Polymer spectra are included throughout this handbook. Note in particular that spectra numbers 204–560 (UV-Vis, 4th-overtone NIR), 811–1006 (NIR), and 1227–1326 (IR) and several Raman spectra represent standard polymer material spectra.

VII.

DIELECTRIC SPECTROSCOPY

25.

MEASUREMENT OF DIELECTRIC CONSTANT, DIELECTRIC LOSS, AND LOSS TANGENT, AND RELATED CALCULATIONS FOR LIQUIDS AND FILM SAMPLES

INTRODUCTION

A method is described for measuring dielectric constant and dielectric loss for liquids and film samples using a network analyzer and coaxial reflectance probe. Calculations are given for loss tangent and half-power penetration depth for liquids. Examples are given for graphical display and comparison of these measurements for a set of polymer film samples in Figures 25.1–25.7; these figures are given only for illustrative purposes. Measurements of the dielectric constant, dielectric loss, and calculated loss tangent for material samples are given in the section following.

A network analyzer and coaxial reflectance probe provide the capability to measure the broad band dielectric relaxation spectra of various liquids, pastes, and polymer films over different temperatures. This configuration is capable of measuring ε' (dielectric constant) and ε'' (dielectric loss) for a variety of materials, such as liquids, suspensions, pastes, and films. From these basic measurements, calculations for a variety of parameters, including those listed later, can be made.

MEASUREMENT METHODS

The measurements are made using a network analyzer with a low-power external electric field (i.e., 0–5 dBm), typically over a frequency range of 300 kHz to 3 GHz, although network analyzers to 20 GHz are readily available, e.g., the HP 8720D [1]. Samples are measured by placing them in contact with a coaxial reflectance probe yielding low-loss resolution measurements for liquids, semisolids, and solid films. Solid samples must have a substantially flat surface to ensure solid contact between the flat probe surface and the test material. The coaxial reflectance method is not recommended for powdered or crystalline solids, due to irreproducible and poor contact between the probe surface and the test material. Solid materials with high dielectric loss are also not measured well using the reflectance probe method, due to precision error and sensitivity of the measurement to slight variations in the integrity of probe contact/pressure with the test material.

The instrument is calibrated for each set of measurements using ambient air, a short (circuit), and deionized water (25°C). Water is then remeasured to check the calibration; the resultant dielectric constant must measure between 79 and 80 across the range 600 kHz to 2.9

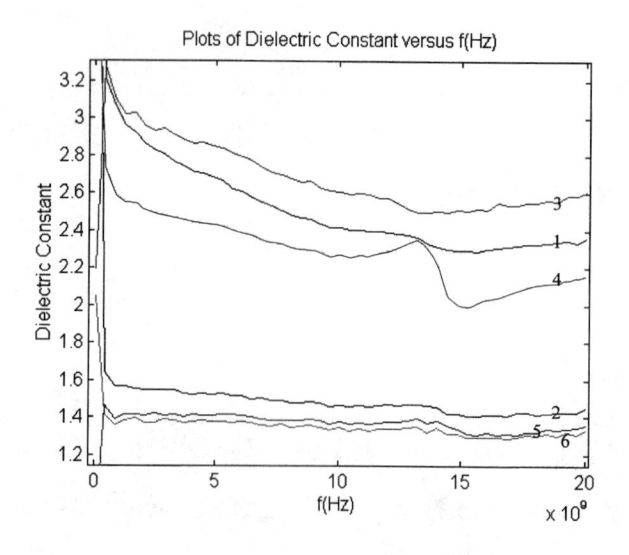

Fig. 25.1 Plot of dielectric constant (ε') as a function of frequency.

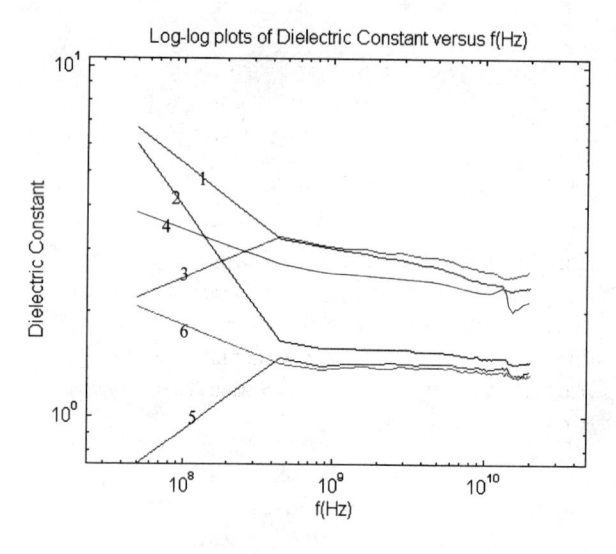

Fig. 25.2 Log-log plot of dielectric constant (ε') as a function of frequency.

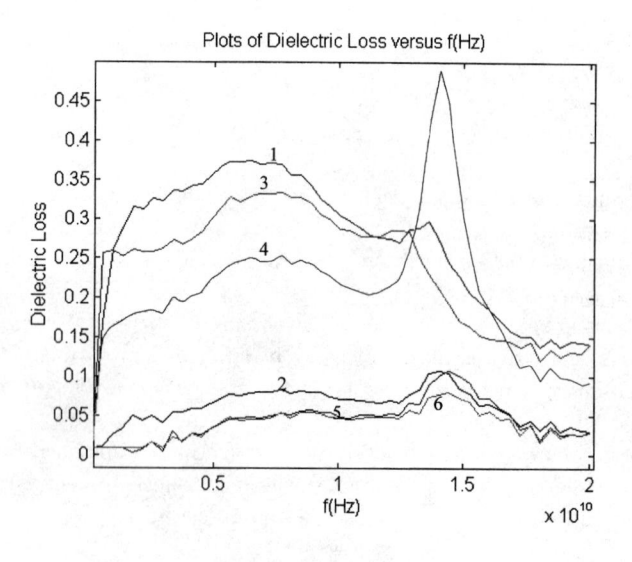

Fig. 25.3 Plot of dielectric loss (ε'') as a function of frequency.

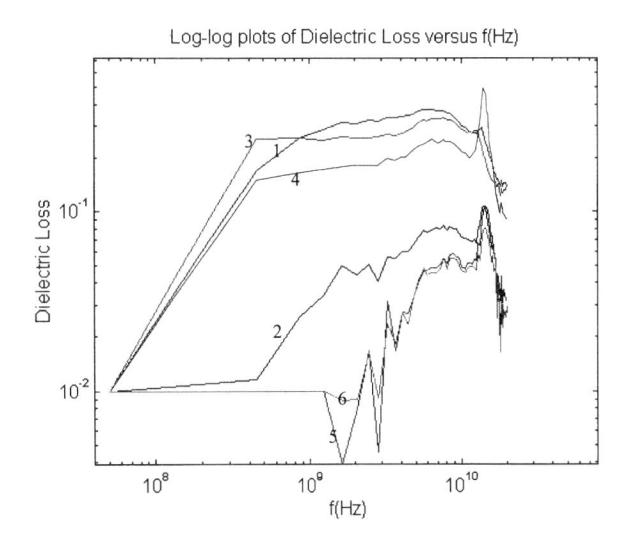

Fig. 25.4 Log-log plot of dielectric loss (ε'') as a function of frequency.

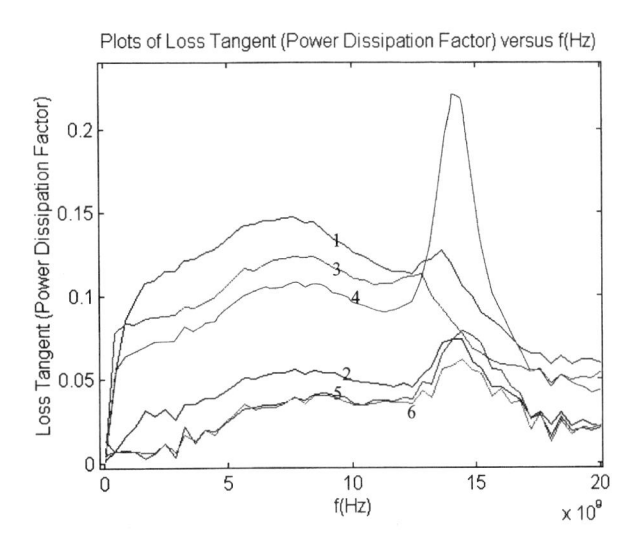

Fig. 25.5 Plot of loss tangent ($\varepsilon''/\varepsilon'$) as a function of frequency.

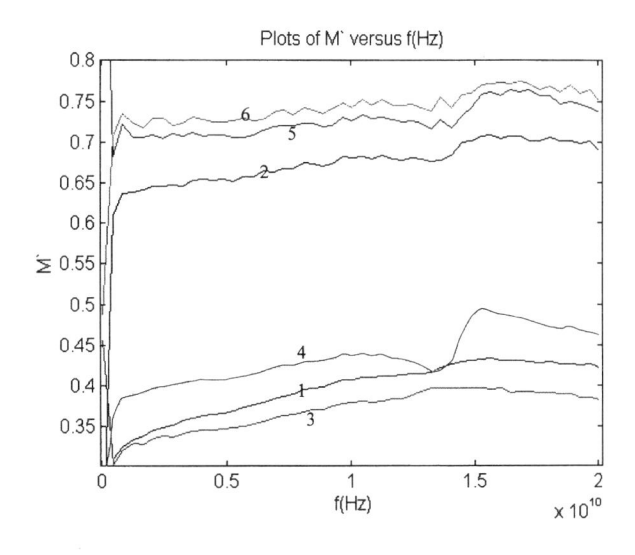

Fig. 25.6 Plot of imaginary part (M′) of the electric modulus as a function of frequency.

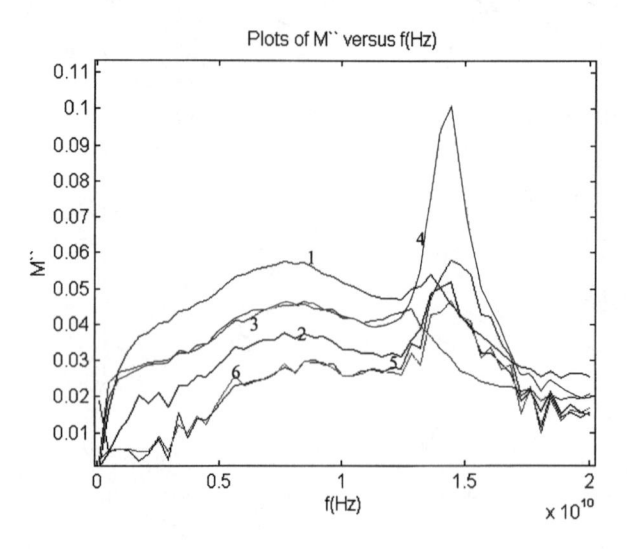

Fig. 25.7 Plot of imaginary part (M″) of the electric modulus as a function of frequency.

GHz. Specifically, HP 8752C (300 kHz to 3 GHz) and HP 8720D (50 MHz to 20 GHz) radio frequency (RF) network analyzers and an HP 85070B reflectance dielectric probe have been used for dielectric determinations [1]. Once calibrated, these instruments are used to measure directly dielectric constant and dielectric loss factor. From this information, calculations can be made for electric modulus, power dissipation factor (loss tangent), half-power penetration depth, and reciprocal half-power penetration depth. These parameters can be used to study the dielectric properties of materials.

The network analyzer/reflectance probe measurement determines the real (ε') and imaginary (ε'') parts of the complex relative permittivity (ε_r) of a sample. The network analyzer measures the reflection coefficient of the MUT (material under test), and an internal model in the microprocessor converts the reflection coefficient to the permittivity. The dielectric error sources include probe model accuracy (3% to 5%) and uncertainty due to the accuracy of the calibration method. Relative errors as large as 25% or more can occur at low frequency for samples with dielectric constant values less than 5. The complex relative permittivity describes the interaction of a material with an applied electric field. The dielectric constant (κ) is equivalent to the complex relative permittivity (ε_r):

$$\varepsilon_r = \kappa = \left(\frac{\varepsilon'}{\varepsilon_0}\right) - j\left(\frac{\varepsilon''}{\varepsilon_0}\right) \tag{1}$$

where ε_r is the complex relative permittivity and ε_0 is the permittivity in free space (i.e., 8.854×10^{-12} F/m). The real part of the complex permittivity (ε') as measured is proportional to the dielectric constant (κ) and is a measure of the energy stored in a material when an electric field is applied across that material. This value is greater than 1 for most solids and liquids. The imaginary part of the complex permittivity (ε'') is called the *loss factor*. It is a measure of how much energy is lost from a material when an electric field is applied across that material. This value is always greater than 0, for all materials have some loss under normal conditions. The loss factor includes the effects of conductivity of the material as well as its dielectric loss.

COMPUTATIONAL METHODS

As noted, the measured value of ε' is most often referred to as the *dielectric constant*, while the measurement of ε'' is denoted as the *dielectric loss factor*. These are measured directly, using the network analyzer, as the *real [Permittivity]* and *imaginary [Permittivity]* parts, respectively [1]. By definition ε'' is always positive; a value of less than zero is occasionally observed when ε'' is near zero, due to the measurement error of the analyzer.

The *loss tangent* or *power dissipation factor* is defined as the calculated ratio $\varepsilon''/\varepsilon'$. This loss tangent results as the vector sum of the orthogonal real (ε') and imaginary (ε'') parts of the complex relative permittivity (ε_r) of a sample. The vector sum of the real and imaginary vectors creates an angle (δ) where $\tan \delta$ is the analytical geometry equivalent to the ratio of $\varepsilon''/\varepsilon'$. Thus the following relationship holds:

$$\tan \delta = \frac{\varepsilon''}{\varepsilon'} = D = \frac{1}{Q} \tag{2}$$

where D is the power dissipation factor and Q is the power quality factor (often referred to in the literature). The *permeability* refers to the interaction of a material with a magnetic field and thus is not applicable to this discussion.

Multiple dielectric mechanisms can contribute to the complex relative permittivity (ε_r) of a sample. These include: dielectric constant, dielectric loss, conductivity, and various polarization effects. The sample thickness, temperature, density, homogeneity, and the like will also affect the measured dielectric properties of a material under test (MUT). An abrupt increase in dielectric loss for an MUT at a region over the measured frequency range indicates the occurrence of a dielectric transition. For example, in the microwave region near 1 GHz, the dipolar or rotational mechanism of molecules resonates with the applied field; in the infrared region near 10^3 GHz, the molecular vibrational mechanisms resonate with the applied field; and in the visible and ultraviolet regions near 10^6 GHz, the electronic transitions resonate with an applied field. The imaginary part of the complex permittivity (ε''), or the loss factor, exhibits an increase in value at each resonance transition. The real part of the complex permittivity (ε') is orthogonal to the imaginary part and thus exhibits a decrease in value at the resonance points. These transitions can be measured and recorded for each MUT. For measurements in the microwave region, an abrupt increase in the loss factor indicates a resonance frequency where rotational energy is maximized relative to the applied field. Thus the application of an energy field at this frequency would contribute to the maximum rotational energy and resultant conformational changes and heating.

Other calculations and models allow the prediction of the microwave affinity for a material and the role of the dielectric constant in the measured heating rate and volatilization of that material. One such model of relating the dielectric properties to the microwave affinity (absorption) is by the use of the concept of *penetration depth* [2]. The half-power penetration depth (HPPD) has been reported as an indication of the microwave penetration into organic liquids. A smaller HPPD indicates that the microwave power is more readily absorbed and dissipated into the material and that it has potential for conversion into heat or activation energy. A high HPPD indicates the material is nearly transparent to microwave energy. The overall HPPD is calculated using an equation from Ref. 2 modified as

$$\text{HPPD} = \left[0.9563 \varepsilon' \left(\sqrt{1 + \left(\frac{\varepsilon''}{\varepsilon'}\right)^2} - 1 \right) \right]^{-1/2} \tag{3}$$

The reciprocal of HPPD (1/HPPD) is directly proportional to the measured heating rate for organic compounds as liquids. Table 25.1 indicates the correlation between each measured or calculated microwave property of an organic compound (as liquid) and that material's heating rate in degrees Celsius per second. *Note:* As a simple measure, the dielectric loss factor is nearly as well correlated as 1/HPPD to the microwave absorption and/or heating rate.

Table 25.1 Correlations Between Microwave Properties of an Organic Compound (as Liquid) and Heating Rate

Material Property	Correlation (r) with heating rate (°C/s)
Dielectric constant (ε')	0.86
Dielectric loss factor (ε'')	0.98
Loss tangent ($\varepsilon''/\varepsilon'$)	0.85
HPPD	−0.76
1/HPPD	1.00

The electric modulus (M^*) and the imaginary parts of the equation are calculated as follows [3]:

$$M^* = \frac{\varepsilon'}{\varepsilon'^2 + \varepsilon''^2} + j\frac{\varepsilon''}{\varepsilon'^2 + \varepsilon''^2} \qquad (4)$$

where

$$M' = \frac{\varepsilon'}{\varepsilon'^2 + \varepsilon''^2} \qquad (5)$$

and

$$M'' = \frac{\varepsilon''}{\varepsilon'^2 + \varepsilon''^2} \qquad (6)$$

Plots showing the M' and M'' imaginary parts of the electric modulus are given in Figures 25.6 and 25.7 as a function of microwave frequency.

DIELECTRIC PROPERTIES OF MATERIALS

The following chapter (Chapter 26) lists the measured dielectric constant, dielectric loss, and calculated loss tangent for a comprehensive list of materials.

REFERENCES

1. Hewlett-Packard. HP 85070B Dielectric Probe Kit User's Manual. © Hewlett-Packard Company, 1997, pp. 8–7, 8–8; HP Application Note 1217–1, Basics of measuring dielectric properties of materials.
2. T. R. Lindstron, T. H. Parliment. Microwave volatilization of aroma compounds. Chapter 35, p. 406. In: T. H. Parliment, M. J. Morello, R.J. McGorrin, eds. *Thermally Generated Flavors.* Washington, D.C.: American Chemical Society, 1994.
3. A. Kyritsis, P. Pissis. Dielectric studies of polymer–water interactions and water organization in PEO/water systems. *J. Polym Sci B: Polym Phys 35:* 1545–1560, 1997.

ADDITIONAL SOURCES

Runt, James P., Fitzgerald, John J. *Dielectric Spectroscopy of Polymeric Materials.* Washington, D.C.: American Chemical Society, 1997.
Townes, Charles H., Schawlow, Arthur L. *Microwave Spectroscopy.* New York: Dover Publications, 1975.

26.

DIELECTRIC CONSTANTS OF MATERIALS (IN ALPHABETICAL ORDER)

Note: Some materials are included more than once at different temperatures.

1, 2-Dichloroethane (77°F) 10.7
1-Diethoxyethane (75°F) 3.8
1-Heptene (68°F) 2.1
1–Octanol (68°F) 10.3
2-Methyl-1-propanol (77°F) 17.7
3-Chloro-1-dihydroxyprone (68°F) 31.0
3-Dimethyl-2-butanone (293°F) 13.1
3 Dimethyl-2-butanone 13.1
ABS resin, lump 2.4–4.1
ABS resin, pellet 1.5–2.5
Acenaphthene (70°F) 3.0
Acetal (70°F) 3.6
Acetal bromide 16.5
Acetal doxime (68°F) 3.4
Acetaldehyde (41°F) 21.8
Acetamide (180°F) 59.0
Acetamide (68°F) 41
Acetanilide (71°F) 2.9
Acetic acid (36°F) 4.1
Acetic acid (68°F) 6.2
Acetic anhydride (66°F) 21.0
Acetone (127°F) 17.7
Acetone (32°F) 1.0159

Acetone (77°F) 20.7
Acetonitrile (70°F) 37.5
Acetophenone (75°F) 17.3
Acetoxime (24°F) 3
Acetyl acetone (68°F) 23.1
Acetyl bromide (68°F) 16.5
Acetyl chloride (68°F) 15.8
Acetylene (32°F) 1.0217
Acetylmethyl hexyl ketone (66°F) 27.9
Acrylic resin 2.7–4.5
Acteal 21.0–3.6
Air (dry) (68°F) 1.000536
Air 1
Alcohol, industrial 16–31
Alkyd resin 3.5–5
Allyl alcohol (58°F) 22.0
Allyl bromide (66°F) 7.0
Allyl chloride (68°F) 8.2
Allyl iodide (66°F) 6.1
Allyl isothiocyanate (64°F) 17.2
Allyl resin (cast) 3.6–4.5
Alumina 4.5
Alumina 9.3–11.5

DIELECTRIC CONSTANTS OF MATERIALS

Alumina china 3.1–3.9

Aluminum bromide (212°F) 3.4

Aluminum fluoride 2.2

Aluminum hydroxide 2.2

Aluminum oleate (68°F) 2.4

Aluminum phosphate 6.0

Aluminum powder 1.6–1.8

Amber 2.8–2.9

Aminoalkyd resin 3.9–4.2

Ammonia (–30°F) 22.0

Ammonia (–74°F) 25

Ammonia (40°F) 18.9

Ammonia (69°F) 16.5

Ammonia (gas?) (32°F) .0072

Ammonium bromide 7.2

Ammonium chloride 7.0

Amyl acetate (68°F) 5.0

Amyl alcohol (–180°F) 35.5

Amyl alcohol (140°F) 11.2

Amyl alcohol (68°F) 15.8

Amyl benzoate (68°F) 5.1

Amyl bromide (50°F) 6.3

Amyl chloride (52°F) 6.6

Amyl ether (60°F) 3.1

Amyl formate (66°F) 5.7

Amyl iodide (62°F) 6.9

Amyl nitrate (62°F) 9.1

Amyl thiocyanate (68°F) 17.4

Amylamine (72°F) 4.6

Amylene (70°F) 2.0

Amylene bromide (58°F) 5.6

Amylenetetrararboxylate (66°F) 4.4

Amylmercaptan (68°F) 4.7

Aniline (212°F) 5.5

Aniline (32°F) 7.8

Aniline (68°F) 7.3

Aniline formaldehyde resin 3.5–3.6

Aniline resin 3.4–3.8

Anisaldehyde (68°F) 15.8

Anisaldoxine (145°F) 9.2

Anisole (68°F) 4.3

Anitmony trichloride 5.3

Antimony pentachloride (68°F) 3.2

Antimony tribromide (212°F) 20.9

Antimony trichloride (166°F) 33.0

Antimony trichloride 33.2

Antimony triiodide (347°F) 13.9

Apatite 7.4

Argon (–376°F) 1.5

Argon (68°F) 1.000513

Arsenic tribromide (98°F) 9.0

Arsenic trichloride (150°F) 7.0

Arsenic trichloride (70°F) 12.4

Arsenic triiodide (302°F) 7.0

Arsine (–148°F) 2.5

Asbestos 3.0–4.8

Ash (fly) 1.7–2.0

Asphalt (75°F) 2.6

Asphalt, liquid 2.5–3.2

Azoxyanisole (122°F) 2.3

Azoxybenzene (104°F) 5.1

Azoxyphenitole (302°F) 6.8

Bakelite 3.5–5.0

Ballast 5.4–5.6

Ballmill feed (cement) 4.5

Balm, refuse 3.1

Barium chloride (2H$_2$O) 9.4

Barium chloride (anhyd.) 11.0

Barium chloride 9.4

Barium nitrate 5.8

Barium sulfate (60°F) 11.4

Barley flour 3.0–4.0

Barley powder 3.0–4.0

Beeswax 2.7–3.0

Benzal chloride (68°F) 6.9

Benzaldehyde (68°F) 17.8

Benzaldoxime (68°F) 3.8

Benzene (275°F) 2.1

Benzene (68°F) 2.3

Benzene (700°F) 1.0028

DIELECTRIC CONSTANTS OF MATERIALS

Benzil (202°F) 13.0

Benzonitrile (68°F) 26.0

Benzophenone (122°F) 11.4

Benzophenone (68°F) 13.0

Benzotrichloride (68°F) 7.4

Benzoyl chloride (32°F) 23.0

Benzoyl chloride (70°F) 22.1

Benzoylacetone (68°F) 29.0

Benzyl acetate (70°F) 5.0

Benzyl alcohol (68°F) 13.0

Benzyl benzoate (68°F) 4.8

Benzyl chloride (68°F) 6.4

Benzyl cyanide (155°F) 6.0

Benzyl cyanide (68°F) 18.3

Benzyl salicylate (68°F) 4.1

Benzylamine (68°F) 4.6

Benzylethylamine (68°F) 4.3

Benzylmethylamine (67°F) 4.4

Beryl 6.0

Biphenyl 20

Biwax 2.5

Bleaching powder 4.5

Bone black 5.0–6.0

Bornyl acetate (70°F) 4.6

Boron bromide (32°F) 2.6

Boronyl chloride (202°F) 5.2

Bromaceytal bromide 12.6

Bromal (70°F) 7.6

Bromine (32°F) 1.0128

Bromine (68°F) 3.1

Bromo-2-ethoxypentane (76°F) 6.5

Bromoacetyl bromide (68°F) 12.6

Bromoaniline (68°F) 13

Bromoanisole (86°F) 7.1

Bromobenzene (68°F) 5.4

Bromobutylene (68°F) 5.8

Bromobutyric acid (68°F) 7.2

Bromoctadecane 3.53

Bromodecane (76°F) 4.4

Bromodeodecane (76°F) 4.1

Bromodocosane (130°F) 3.1

Bromododecane (75°F) 4.07

Bromoform (68°F) 4.4

Bromoheptane (76°F) 5.3

Bromohexadecane (76°F) 3.7

Bromohexane (76°F) 5.8

Bromoisovaleric acid (68°F) 6.5

Bromomethane (32°F) 9.8

Bromonaphthalene (66°F) 5.1

Bromooctadecane (86°F) 3.5

Bromopentadecane (68°F) 3.9

Bromopropionic acid (68°F) 11.0

Bromotoluene (68°F) 5.1

Bromotridecane (50°F) 4.2

Bromoundecane (15°F) 4.7

Bronyl chloride (94°F) 5.21

Butane (30°F) 1.4

Butanol (1) (68°F) 17.8

Butanone (68°F) 18.5

Butycic anhydride (20°F) 12.0

Butyl chloral (64°F) 10.0

Butyl chloride (68°F) 9.6

Butyl oleate (77°F) 4.0

Butyl stearate (80°F) 3.1

Butylacetate (66°F) 5.1

Butylamine (70°F) 5.4

Butyraldehyde (79°F) 13.4

Butyric acid (68°F) 3.0

Butyric anhydride (68°F) 12.0

Butyronitrile (70°F) 20.7

Cable oil (80°F) 2.2

Calcite 8.0

Calcium 3.0

Calcium carbonate 6.1–9.1

Calcium fluoride 7.4

Calcium oxide, granule 11.8

Calcium sulfate (H_2O) 5.6

Calcium sulfate 5.6

Calcium superphosphate 14–15

Camphanedione (398°F) 16.0

DIELECTRIC CONSTANTS OF MATERIALS

Camphene (104°F) 2.3

Camphene (68°F) 2.7

Campher, crystal 10–11

Camphoric imide 4 (80°F) 5.5

Camphorpinacone (68°F) 3.6

Caprilic acid (18°F) 3.2

Caproic acid (160°F) 2.6

Caprolactam monomer 1.7–1.9

Caprylic acid (65°F) 3.2

Carbide 5.8–7.0

Carbide, powder 5.8–7.0

Carbon black 2.5–3.0

Carbon dioxide (32°F) 1.6

Carbon dioxide (68°F) 1.000921

Carbon dioxide, liquid 1.6

Carbon disulfide, liquid 2.6

Carbon disulphide (180°F) 2.2

Carbon disulphide (68°F) 2.6

Carbon tetrachloride (68°F) 2.2

Carnauba wax 2.9

Carvenone (68°F) 18.4

Carvol (64°F) 11.2

Carvone (71°F) 11.0

Casein 6.1–6.8

Casein resin 6.7

Cassiterite 23.4

Castor oil (60°F) 4.7

Castor oil (80°F) 2.6

Castor oil (hydrogenated) (80°F) 10.3

Cedrene (76°F) 3.2

Cellophane 3.2–6.4

Celluloid 3.3–11

Cellulose 3.2–7.5

Cellulose acetate (molding) 3.2–7.0

Cellulose acetate (sheet) 4.0–5.5

Cellulose acetate 3.2–7

Cellulose acetate butyrate 3.2–6.2

Cellulose nitrate (proxylin) 6.4

Cement (plain) 1.5–2.1

Cement 1.5–2.1

Cement, portland 2.5–2.6

Cement, powder 5–10

Cereals (dry) 3.0–5.0

Cerese wax 2.4

Cesium iodine 5.6

Cetyl iodide (68°F) 3.3

Charcoal 1.2–1.81

Chinaware, hard 4–7

Chloracetic acid (140°F) 12.3

Chloracetone 29.8

Chloral (68°F) 4.9

Chlorhexanone oxime 3

Chlorine (–50°F) 2.1

Chlorine (142°F) 1.5

Chlorine (32°F) 2.0

Chlorine, liquid 2

Chloroacetic acid (68°F) 21.0

Chloroacetone (68°F) 29.8

Chlorobenzene (100°F) 4.7

Chlorobenzene (230°F) 4.1

Chlorobenzene (77°F) 5.6

Chlorobenzine, liquid 5.5–6.3

Chlorocyclohexane (76°F) 7.6

Chloroform (212°F) 3.7

Chloroform (32°F) 5.5

Chloroform (68°F) 4.8

Chloroheptane (71°F) 5.5

Chlorohexanone oxime (192°F) 3.0

Chlorohydrate (68°F) 3.3

Chloromethane -4 12.6

Chloronaphthalene (76°F) 5.0

Chlorooctane (76°F) 5.1

Chlorophetane 5.4

Chlorotoluene (68°F) 4.7

Chlorotoluene, liquid 4–4.5

Cholesterin 2.86

Cholestral (80°F) 2.9

Chorine (170°F) 1.7

Chrome, ore 7.7–8.0

Chrome, pure 12

Chromite 4.0–4.2

Chromyl choride (68°F) 2.6

Cinnamaldehyde (75°F) 16.9

cis-3-Hexene (76°F) 2.1

Citraconic anhydride (68°F) 40.3

Citraconic nitrile 27

Clay 1.8–2.8

Clinker (cement) 2.7

CO_2 (32°F) 1.6

Coal tar 2.0–3.0

Coal, powder, fine 2–4

Cocaine (68°F) 3.1

Coffee refuse 2.4–2.6

Coke 1.1–2.2

Compound 3.6

Copper catalyst 6.0–6.2

Copper oleate (68°F) 2.8

Copper oxide 18.1

Corderite 2.5–5.4

Corn (dry granulars) 1.8

Corn 5–10

Corn, refuse 2.3–2.6

Corning glass 6.5

Cotton 1.3–1.4

Cotton seed oil 3.1

Creosol (63°F) 10.6

Creosol (75°F) 5.0

Creosol, liquid 9–11

Crotonic nitrice (68°F) 28.0

Crystale 3.5–4.7

Cumaldehyde (59°F) 11.0

Cumene (68°F) 2.4

Cumicaldehyde (58°F) 10.7

Cupric oleate 2.8

Cupric oxide (60°F) 18.1

Cupric sulfate ($5H_2O$) 7.8

Cupric sulfate (anhyd.) 10.3

Cupric sulfate 10.3

Cyanoacetic acid (40°F) 33.0

Cyanoethyl acetate (68°F) 19.3

Cyanogen (73°F) 2.6

Cyclohedane (20°F) 2.0

Cyclohenanone (68°F) 18.2

Cycloheptasiloxane (68°F) 2.7

Cyclohexane (68°F) 2.0

Cyclohexane, liquid 18.5

Cyclohexanecarboxylic acid (88°F) 2.6

Cyclohexanemethanol (140°F) 9.7

Cyclohexanol (77°F) 15.0

Cyclohexanone (68°F) 18.2

Cyclohexanone oxime (192°F) 3.0

Cyclohexene (68°F) 18.3

Cyclohexylamine-5 5.3

Cyclohexylphenol (130°F)

Cyclohexyltrifluoromethane-1 (68°F) 11.0

Cyclopentane (68°F) 2.0

Cymene 62 2.3

D-Cocaine 3.1

D.M.T. (Dacron powder) 1.33

D_2O 78.3

Decahydronaphtolene (68°F) 2.2

Decamethylcyclopentasiloxane (68°F) 2.5

Decamethyltetrasiloxane (68°F) 2.4

Decanal 8.1

Decane (68°F) 2.0

Decanol (68°F) 8.1

Decylene (62°F) 2.7

Decyne (68°F) 2.2

Deuterium (68°F) 1.3

Deuterium oxide (77°F) 78.3

Dextrin 2.2–2.4

Diacetoxybutane (76°F) 6.64

Diallyl sulfide (68°F) 4.9

Diamond 5.5–10.0

Diapalmitin 3.5

Diaphenylmethane 2.7

Dibenzofuran (212°F) 3.0

Dibenzyl sebacate (68°F) 4.6

Dibenzylamine (68°F) 3.6

Dibroheptane (24°F) 5.08

DIELECTRIC CONSTANTS OF MATERIALS

Dibromobenzene (68°F) 8.8

Dibromobutane (68°F) 5.7

Dibromoethylene (cis-1,2) (32°F) 7.7

Dibromoheptane (76°F) 5.1

Dibromohexane (76°F) 5.0

Dibromomethane (50°F) 7.8

Dibromopropane (68°F) 4.3

Dibromopropyl alcohol (70°F) 9.1

Dibutyl phthalate (86°F) 6.4

Dibutyl sebacate (86°F) 4.5

Dibutyl tartrate 109 9.4

Dichloracetic acid (20°F) 10.7

Dichloracetic acid (72°F) 8.2

Dichloracetone (68°F) 14.0

Dichlorobenzene (127°F) 2.8

Dichloroethane (1,2) (77°F) 10.3

Dichloroethane (68°F) 16.7

Dichloroethylene (62°F) 4.6

Dichloromethane (68°F) 9.1

Dichlorostyrene (76°F) 2.6

Dichlorotoluene (68°F) 6.9

Dictyl phthalate 5.1

Dicyclohexyl adipate (95°F) 4.8

Diebenzylamine (68°F) 3.6

Diethyl-dimalate 10.2

Diethyl 1-malate (68°F) 9.5

Diethyl benzalmalonate (32°F) 8.0

Diethyl disulfide (66°F) 15.9

Diethyl DL-malate (64°F) 10.2

Diethyl glutarate (86°F) 6.7

Diethyl ketone (58°F) 17.3

Diethyl L-malate (68°F) 9.5

Diethyl malonate (70°F) 7.9

Diethyl oxalate (70°F) 8.2

Diethyl oxaloacetate (66°F) 6.1

Diethyl racemate (68°F) 4.5

Diethyl sebacate (86°F) 5.0

Diethyl succinate (86°F) 6.6

Diethyl succinosuccinate (66°F) 2.5

Diethyl sulfide (68°F) 7.2

Diethyl sulfite (68°F) 15.9

Diethyl tartrate (68°F) 4.5

Diethyl zinc (68°F) 2.6

Diethylamine (68°F) 3.7

Diethylaniline (66°F) 5.5

Dihydrocaroone (66°F) 8.7

Dihydrocarvone (66°F) 8.5

Diimylamine (64°F) 2.5

Diisoamyl (62°F) 2.0

Diisoamylene (62°F) 2.4

Diisodoethylene 1 (80°F) 4.0

Diisodomethane (77°F) 5.3

Diisobutylamine (71°F) 2.7

Dimethoxybenzene (73°F) 4.5

Dimethyl-1-hydroxybenzene (62°F) 4.8

Dimethyl-2-hexane (68°F) 2.4

Dimethyl ethyl (68°F) 11.7

Dimethyl ethyl carbinol (68°F) 11.7

Dimethyl malonate (68°F) 10.4

Dimethyl oxalate (68°F) 3.0

Dimethyl pentane (20°F) 1.912

Dimethyl phthalate (75°F) 8.5

Dimethyl sulfate (68°F) 55.0

Dimethyl sulfide (68°F) 6.3

Dimethylamine (32°F) 6.3

Dimethylaniline (68°F) 4.4

Dimethylbromoethylene (68°F) 6.7

Dimethylheptane (68°F) 1.9

Dimethylpentane (68°F) 1.9

Dimethylquinoxaline (76°F) 2.3

Dimethyltouidine (68°F) 3.3

Dinitrogen oxide (32°F) 1.6

Dinitrogen tetroxide (58°F) 2.5

Dioctyl phthalate (76°F) 5.1

Dioxane 1,4 (77°F) 2.2

Dipalmitin (161°F) 3.5

Dipentene (68°F) 2.3

Diphenyl 1 (66°F) 2.5

Diphenyl ether (82°F) 3.9

Diphenylamine (124°F) 3.3

Diphenylamine (125°F) 3.3
Diphenylethane (110°F) 2.38
Diphenylethane (230°F) 2.4
Diphenylethane (62°F) 12.6
Diphenylmethane (62°F) 2.6
Dipropyl ketone (62°F) 12.6
Dipropylamine (70°F) 2.9
Distearin (172°F) 3.3
Docosane (122°F) 2.0
Dodecamethylcyclohexisloxane (68°F) 2.6
Dodecamethylpentasiloxane (68°F) 2.5
Dodecane (68°F) 2.0
Dodecanol (76°F) 6.5
Dodecyne (76°F) 2.2
Dolomite 6.8–8.0
Dowtherm (70°F) 3.4
Ebonite 2.5–2.9
Emery sand 16.5
Epichlorchydrin (68°F) 22.9
Epoxy resin (cast) 3.6
EPR 2.24
Ethanediamine (68°F) 14.2
Ethanethiol (58°F) 6.9
Ethanethiolic acid (68°F) 13.0
Ethanol (77°F) 24.3
Ethelene diamine (18°F) 16.0
Ethelene oxide-1 13.9
Ethoxy-3-methylbutane (68°F) 4.0
Ethoxybenzene (68°F) 4.2
Ethoxyethyl acetate (86°F) 7.6
Ethoxynaphthalone (66°F) 3.3
Ethoxypentane (73°F) 3.6
Ethoxytoluene (68°F) 3.9
Ethyl-1-brobutyrate (68°F) 8.0
Ethyl-2-iodopropionate (68°F) 8.8
Ethyl acetate (77°F) 6.0
Ethyl acetoacetate (71°F) 15.9
Ethyl acetoneoxalate (66°F) 16.1
Ethyl acetophenoneoxalate (66°F) 3.3
Ethyl alcohol (77°F) 24.3

Ethyl alcohol (*see* Ethanol)
Ethyl amyl ether (68°F) 4.0
Ethyl benzene (68°F) 2.5
Ethyl benzoate (68°F) 6.0
Ethyl benzoylacetate (68°F) 12.8
Ethyl benzoylacetoacetate (70°F) 8.6
Ethyl benzyl ether (68°F) 3.8
Ethyl bromide (64°F) 4.9
Ethyl bromoisobutyrate (68°F) 7.9
Ethyl bromopropionate (68°F) 9.4
Ethyl butyrate (66°F) 5.1
Ethyl carbonate (121°F) 14.2
Ethyl carbonate (68°F) 3.1
Ethyl cellulose 2.8–3.9
Ethyl chloracetate (68°F) 11.6
Ethyl chloroformate (68°F) 11.3
Ethyl chloropropionate (68°F) 10.1
Ethyl cinnamate (66°F) 5.3
Ethyl cyanoacetate (68°F) 27.0
Ethyl cyclobutane (68°F) 2.0
Ethyl dodecanoate (68°F) 3.4
Ethyl ether (−148°F) 8.1
Ethyl ether (−40°F) 5.7
Ethyl ether (68°F) 4.3
Ethyl ethoxybenzoate (70°F) 7.1
Ethyl formate (77°F) 7.1
Ethyl formylphenylacetate (68°F) 3.0
Ethyl fumarate (73°F) 6.5
Ethyl hydroxy-tetracarboxylate 5.9
Ethyl hydroxy-tetrocarboxylate 2.7
Ethyl hydroxymethylenephenylacetate 5.00
Ethyl hydroxymethylenomalonate 6.6
Ethyl iodide (68°F) 7.4
Ethyl isothiocyanate (68°F) 19.7
Ethyl levulinete (70°F) 12.1
Ethyl maleate (73°F) 8.5
Ethyl mercaptan (68°F) 8.0
Ethyl nitrate (68°F) 19.7
Ethyl oleate (80°F) 3.2
Ethyl palmitate (68°F) 3.2

DIELECTRIC CONSTANTS OF MATERIALS

Ethyl phenylacetate (70°F) 5.4
Ethyl propionate (68°F) 5.7
Ethyl salicylate (70°F) 8.6
Ethyl silicate (68°F) 4.1
Ethyl stearate (104°F) 3.0
Ethyl thiocyanate (68°F) 29.6
Ethyl trichloracetate (68°F) 7.8
Ethyl undecanoate (68°F) 3.6
Ethyl valerate (68°F) 4.7
Ethylamine (70°F) 6.3
Ethylaniline (68°F) 5.9
Ethylbenzene (76°F) 3.0
Ethylene chloride (68°F) 10.5
Ethylene chlorohydrin (77°F) 26.0
Ethylene cyanide (136°F) 58.3
Ethylene diamine (64°F) 16.0
Ethylene gylcol (68°F) 37.0
Ethylene iodide 3.4
Ethylene oxide 25 14.0
Ethylene tetraflouride 1.9–2.0
Ethylenechlorohydrin (75°F) 25.0
Ethylenediamine (64°F) 16.0
Ethylic resin 2.2–2.3
Ethylpentane (68°F) 1.9
Ethyltoluene (76°F) 2.2
Etibine (–58°F) 2.5
Eugenol (64°F) 6.1
Fab (from box, 8% moisture) 1.3
Fenchone (68°F) 12.0
FEP (cellular) 1.5
Fermanium tetrachloride (76°F) 2.4
Ferric oleate (68°F) 2.6
Ferrochromium 1.5–1.8
Ferromanganese 5.0–5.2
Ferrous oxide (60°F) 14.2
Ferrous sulfate (58°F) 14.2
Flour 2.5–3.0
Flourine (–332°F) 1.5
Flourspar 6.8
Fluorosulfonic acid 120

Fluorotoluene (86°F) 4.2
Fly ash 1.9–2.6
Formalin 23
Formamide (68°F) 84.0
Formic acid (60°F) 58.0
Forsterite 6.2
Freon 11 (70°F) 3.1
Freon 113 (70°F) 2.6
Freon 12 (70°F) 2.4
Fuller's earth 1.8–2.2
Furan (77°F) 3.0
Furfural (68°F) 42.0
Furfuraldehyde (68°F) 41.9
Gasoline (70°F) 2.0
Gerber oatmeal (in box) 1.5
Germanium tetrachloride (77°F) 2.4
Glass (silica) 3.8
Glass 3.7–10
Glass, bead 3.1
Glass, granule 6–7
Glass, raw material 2.0–2.5
Glucoheptitol (248°F) 27.0
Glycerin, liquid 47–68
Glycerol (32°F) 47.2
Glycerol (77°F) 42.5
Glycerol phthalate (cast alkyd) 3.7–4.0
Glyceryl triocetate (70°F) 6.0
Glycol (122°F) 35.6
Glycol (77°F) 37.0
Glycolic nitrile (68°F) 27.0
Grain 3–8
Graphite 12–15
Guaiacol 0 11.0
Gypsum 2.5–6.0
Hagemannie ester (68°F) 10.6
Halowax 4.5
HCN 116
Heavy oil 3
Heavy oil, C 2.6
Helium, liquid 1.05

Helium-3 (58°F) 1.055

Heptadecanone (140°F) 5.3

Heptane (68°F) 1.9

Heptane, liquid 1.9–2.0

Heptanoic acid 2.5

Heptanoic acid (71°F) 2.59

Heptanoic acid (160°F) 2.6

Heptanone (68°F) 11.9

Heptyl alcohol (70°F) 6.7

Hexamethyldisiloxane (68°F) 2.2

Hexane (−130°F) 2.0

Hexane, liquid 5.8–6.3

Hexanol (77°F) 13.3

Hexanone (59°F) 14.6

Hexdecamethylcycloheptasiloxane (68°F) 2.7

Hexyl iodide (68°F) 6.6

Hexylene (62°F) 2.0

HF 83.6

Hydrazine (68°F) 52.0

Hydrochloric acid 4.12

Hydrocyanic acid (32°F) 158.0

Hydrocyanic acid (70°F) 2.3

Hydrogen (212°F) 1.000284

Hydrogen (440°F) 1.23

Hydrogen bromide (−120°F) 7.0

Hydrogen bromide (24°F) 3.8

Hydrogen chloride (−188°F) 12.0

Hydrogen chloride (82°F) 4.6

Hydrogen cyanide (70°F) 95.4

Hydrogen fluoride (−100°F) 17

Hydrogen fluoride (32°F) 84.2

Hydrogen iodide (72°F) 2.9

Hydrogen peroxide (32°F) 84.2

Hydrogen sulfide (−84°F) 9.3

Hydrogen sulfide (48°F) 5.8

Hydroxy-4-methyl-2-pentanone (76°F) 18.2

Hydroxymethylene camphor (86°F) 5.2

Hydroxymethylenebenzyl cyanide (68°F) 6.0

Hydroxymethylenehydroxymethyleneacetoacetate 7.8

Hydrozine (68°F) 52.9

Ido-iodohexadecane (68°F) 3.5

Idoheptane (71°F) 4.9

Idohexane (68°F) 5.4

Idomethane (68°F) 7.0

Idopoctane (76°F) 4.6

Idotoluene (68°F) 6.1

Ilmenite 6.0–7.0

Inadol (140°F) 7.8

Indonol (60°F) 7.8

Iodine (107°F) 118.0

Iodine (250°F) 118.0

Iodine (granular) 4.0

Iodooctane (24°F) 4.62

Iodobenzene (68°F) 4.6

Iodoheptane (22°F) 4.92

Iodohexane (20°F) 5.37

Iodomethane (20°F) 7.0

Iodotoluene (20°F) 6.1

Iron oxide 14.2

Isoamyl acetate (68°F) 5.6

Isoamyl alcohol (74°F) 15.3

Isoamyl bromide (76°F) 6.1

Isoamyl butyrate (68°F) 3.9

Isoamyl chloracetate (68 f) 7.8

Isoamyl chloride (64°F) 6.4

Isoamyl chloroacetate 7.8

Isoamyl chloroformate (68°F) 7.8

Isoamyl iodide (65°F) 5.6

Isoamyl propionate 4.2

Isoamyl salicylate (68°F) 5.4

Isoamyl valerate (19°F) 3.6

Isoamyl valerate (66°F) 3.6

Isobutyl acetate (68°F) 5.6

Isobutyl alcohol (−112°F) 31.7

Isobutyl alcohol (32°F) 20.5

Isobutyl alcohol (68°F) 18.7

Isobutyl alcohol 18.7–31.7

Isobutyric acid 2.7

Isobutyl benzoate (68°F) 5.9

DIELECTRIC CONSTANTS OF MATERIALS

Isobutyl bromide (20°F) 4.0
Isobutyl bromide (68°F) 6.6
Isobutyl butyrate (68°F) 4.0
Isobutyl chloride (68°F) 7.1
Isobutyl chloroformate (68°F) 9.2
Isobutyl cyanide (74°F) 13.3
Isobutyl formate (66°F) 6.5
Isobutyl iodide (68°F) 5.8
Isobutyl nitrate (66°F) 11.9
Isobutyl resin 1.4–2.1
Isobutyl rininoleate (70°F) 4.7
Isobutyl valerate (66°F) 3.8
Isobutylamine (70°F) 4.5
Isobutylbenzene (62°F) 2.3
Isobutylene bromide (68°F) 4.0
Isobutyric acid (122°F) 2.7
Isobutyric acid (68°F) 2.6
Isobutyric acid (68°F) 2.7
Isobutyric anhydride (68°F) 13.9
Isobutyronitrile (77°F) 20.8
Isobutyronitrile (75°F) 20.8
Isobutyronitrile 23.9–20.8
Isocapronitrile (68°F) 15.7
Isoiodohexadecane 3.5
Isooctane 2.1–2.3
Isophthalic acid 1.4
Isoprene (77°F) 2.1
Isopropyl alcohol (68°F) 18.3
Isopropyl benzene (68°F) 2.4
Isopropyl nitrate (66°F) 11.5
Isopropylamine (68°F) 5.5
Isopropylether (77°F) 3.9
Isoquinoline (76°F) 10.7
Isosafrol (70°F) 3.4
Isovaleric acid (68°F) 2.7
Isovaleric acid (68°F) 2.6
Jet fuel (JP4) (70°F) 1.7
Jet fuel (military JP4) 1.7
Kent wax 6.5–7.5
Kerosene (70°F) 1.8

Kynar 2.0
Kynar® 6.4
Lactic acid (61°F) 22.0
Lactronitrile (68°F) 38.4
Lead acetate 2.5
Lead carbonate (60°F) 18.1
Lead chloride 4.2
Lead nitrate 37.7
Lead nomoxide (60°F) 25.9
Lead oleate (64°F) 3.2
Lead oxide 25.9
Lead sulfate 14.3
Lead sulfite 17.9
Lead tetrachloride (68°F) 2.8
Lime 2.2–2.5
Limonene (68°F) 2.3
Linde 5A molecular sieve, dry 1.8
Linoleic acid (32°F) 2.6–2.9
Linseed oil 3.2–3.5
Liquified air 1.5
Liquified hydrogen 1.2
Lithium chloride 11.1
Lonone (65°F) 10.0
LPG 1.6–1.9
m-Bromoaniline (66°F) 13.0
m-Bromotoluene (137°F) 5.4
m-Chloroanaline (66°F) 13.4
m-Chlorotoluene (68°F) 5.6
m-Creosol 5
m-Dichlorobenzene (77°F) 5.0
m-Dinitro benzene (68°F) 2.8
m-Nitrotoluene (68°F) 23.8
m-Sylene 2.4
m-Toluidine (64°F) 6.0
m-Xylene (68°F) 2.4
Maganese dioxide 5–5.2
Magnesium oxide 9.7
Magnesium sulfate 8.2
Malachite 7.2
Maleic anhydride (140°F) 51.0

DIELECTRIC CONSTANTS OF MATERIALS

Malolic anhydride 51

Malonic nitrile (97°F) 47.0

Mandelic nitrile (73°F) 18.1

Mandelitrile (73°F) 17.0

Mannitol (71°F) 3.0

Margarine, liquid 2.8–3.2

Melamine formaldehyde (see MF)

Melamine resin 4.7–10.9

Menthol (107°F) 4.0

Menthol (42°F) 3.95

Menthonol (110°F) 2.1

Menthonol (43°F) 2.1

Mercuric chloride 3.2

Mercurous chloride 9.4

Mercury (298°F) 1.00074

Mercury chloride 7–14

Mercury diethyl (68°F) 2.3

Mesityl oxide (68°F) 15.4

Methalamine

Mesitylene (68°F) 2.4

Mesitylene 3.4

Methalmine (77°F) 9.4

Methal cyanoacetate (69°F) 29.4

Methane (−280°F) 1.7

Methane, liquid 1.7

Methanol (77°F) 32.6

Methlene iodide 5.1

Methoxy-4-methylphenol (60°F) 11.0

Methoxybenzene (76°F) 4.3

Methoxyethyl stearate (140°F) 3.4

Methoxyphenol (82°F) 11.0

Methoxytoluene (68°F) 3.5

Methyl-1-cyclopentanol (35°F) 6.9

Methyl-2-pentanone (68°F) 13.1

Methyl-2,4-pentandeiol (86°F) 24.4

Methyl-5-ketocyclohexylene (68°F) 24.0

Methyl acetate (77°F) 6.7

Methyl acetophenoneoxalate (64°F) 2.8

Methyl alcohol (−112°F) 56.6

Methyl alcohol (32°F) 37.5

Methyl alcohol (68°F) 33.1

Methyl benzoate (68°F) 6.6

Methyl butane (68°F) 1.8

Methyl butyl ketone (62°F) 12.4

Methyl butyrate (68°F) 5.6

Methyl chloride (77°F) 12.9

Methyl chloroacetate (68°F) 12.9

Methyl ether (78°F) 5.0

Methyl ethyl ketone (72 °F) 18.4

Methyl ethyl ketoxime (68°F) 3.4

Methyl formate (68°F) 8.5

Methyl heptanol (68°F) 5.3

Methyl hexyl ketone (62°F) 10.7

Methyl iodide (68°F) 7.1

Methyl methacrylate (cast) 2.7–3.2

Methyl nitrobenzoate (80°F) 27.0

Methyl O-methoxybenzoate (70°F) 7.8

Methyl P-toluate (91°F) 4.3

Methyl propionate (66°F) 5.4

Methyl propyl ketone (58°F) 16.8

Methyl salicylate (68°F) 9.0

Methyl thiocyanate (68°F) 35.9

Methyl valerate (66°F) 4.3

Methylal (68°F) 2.7

Methylaniline (68°F) 6.0

Methylbenzylamine (65°F) 4.4

Methylcyclohexanol (68° c) 13.0

Methylcyclohexanone (192°F) 18.0

Methylcylopentane (68°F) 2.0

Methylene iodide (70°F) 5.1

Methyleneaceloacetate (70°F) 7.8

Methylenemalonate (72°F) 6.6

Methylenephenylacetate (68°F) 5.0

Methylether, liquid 5

Methylhexane (68°F) 1.9

Methylisocyanate (69°F) 29.4

Methyloctane (69°F) 30.0

Methylomine (21°F) 10.5

Methylphenyl hydrazine (66°F) 7.3

Methylpyridine (2) (68°F) 9.8

DIELECTRIC CONSTANTS OF MATERIALS

MF molding resin 5.5–6.0

MF with alpha cellulose filler 7.2–8.2

MF with asbestos filler 6.1–6.7

MF with cellulose filler 4.7–7.0

MF with flock filler 5.0–6.0

MF with macerated fabric filler 6.5–6.9

Mica (glass bonded) 6.9–9.2

Mica 2.6–3.2

Mica 7.0

Micanite 1.8–2.6

Mills (dry powder) 1.8

Mineral oil (80°F) 2.1

Monomyristin (158°F) 6.1

Monopalmitin (152°F) 5.3

Monostearin (170°F) 4.9

Morpholine (77°F) 7.3

n-Butyl alcohol (66°F) 7.8

n-Butyl bromide (68°F) 6.6

n-Butyl formate (–317°F) 2.4

n-Butyl iodide (77°F) 6.1

n-Butylacetate (19°F) 5.1

n-Butyricaid (68°F) 2.9

n-Hexane (68°F) 1.9

n-Methylaniline (68°F) 6.0

n-Pentane (68°F) 1.8

Naphthy ethyl ether (67°F) 3.2

Napthalene (185°F) 2.3

Napthalene (68°F) 2.5

Napthonitrile (70°F) 6.4

Napthyl ethyl ether (67°F) 3.2

Neon (68°F) 1.000127

Neoprene 6–9

Nitroanisole (68°F) 24.0

Nitrobenzal doxime (248°F) 48.1

Nitrobenzene (176°F) 26.3

Nitrobenzene (68°F) 35.7

Nitrobenzene (77°F) 34.8

Nitrobenzyl alcohol (68°F) 22.0

Nitrocellulose 6.2–7.5

Nitroethane (68°F) 19.7

Nitrogen (336°F) 1.454

Nitrogen (68°F) 1.000580

Nitroglycerin (68°F) 19.0

Nitromethane (68°F) 39.4

Nitromethane 22.7–39.4

Nitrosodimethylamine (68°F) 54.0

Nitrosyl bromide (4°F) 13.0

Nitrosyl chloride (10°F) 18.0

Nitrotoluene (68°F) 1.96

Nitrous oxide (32°F) 1.6

Nonane (68°F) 2.0

Nylon-6 3.7

Nylon-6/12 3.5

Nylon 4.0

Nylon 4.0–5.0

Nylon resin 3.0–5.0

o-Bromotoluene (137°F) 4.3

o-Chlorophenol (66°F) 8.2

o-Chlorotoluene (68°F) 4.5

o-Cresol (77°F) 11.5

o-Cresol (77°F) 11.5

o-Dichlorobenzene (77°F) 7.5

o-Nitroanaline (194°F) 34.5

o-Nitrotoluene (68°F) 27.4

o-Toluidine (64°F) 6.3

o-Xylene (68°F) 2.6

Octadecanol (136°F) 3.4

Octamethylcyclotetrasiloxane (68°F) 2.4

Octamethyltrisiloxane (68°F) 2.3

Octane (24°F) 1.061

Octane (68°F) 2.0

Octanone (68°F) 10.3

Octene (76°F) 2.1

Octyl alcohol (64°F) 3.4

Octyl iodide (68°F) 4.9

Octylene (65°F) 4.1

Oil, almond (68°F) 2.8

Oil, cotton seed (57°F) 3.1

Oil, grapeseed (61°F) 2.9

Oil, lemon (70°F) 2.3

Oil, linseed 3.4
Oil, olive (68°F) 3.1
Oil, paraffin (68°F) 2.2–4.7
Oil, peanut (52°F) 3.0
Oil, petroleum (68°F) 2.1
Oil, pyranol (68°F) 5.3
Oil, sesame (55°F) 3.0
Oil, sperm (68°F) 3.2
Oil, terpentine (68°F) 2.2
Oil, transformer (68°F) 2.2
Oleic acid (68°F) 2.5
Oleic acid 2.4–2.5
Opal wax 3.1
Organic cold-molding compound 6.0
Oxygen (–315°F) 1.51
Oxygen (68°F) 1.000494
p-Bromotoluene (137°F) 5.5
p-Chlorophenol (130°F) 9.5
p-Chlorotoluene (68°F) 6.1
p-Cresol (137°F) 9.9
p-Cresol (24°F) 5.0
p-Cresol (70°F) 5.6
p-Cymene (63°F) 2.3
p-Dibromobenzene (190°F) 4.5
p-Dichlorobenzine (120°F) 2.4
p-Dichlorobenzine (68°F) 2.86
p-Nitro analine (320°F) 56.3
p-Nitrotoluene (137°F) 22.2
p-Toludine 3.0
p-Toluidine (130°F) 5.0
p-Xylene (68°F) 2.3
Paint 5–8
Palmitic acid (160°F) 2.3
Paper (dry) 2.0
Paraffin 1.9–2.5
Paraffin wax 2.1–2.5
Paraldehyde (68°F) 14.5
Paraldehyde (77°F) 13.9
Parawax 2.3
Parrafin chloride 2.0–2.3

Penanthiene (68°F) 2.8
Pentachlorgethane (60°F) 3.7
Pentadiene 1,3 (77°F) 2.3
Pentane (68°F) 1.8
Pentanol (77°F) 13.9
Pentanone (2) (68°F) 15.4
Pentene (1) (68°F) 2.1
Pentochlorethane 3.7
Perlite 1.3–1.4
Petroleum 2.0–2.2
PFR with asbestos filler 5.0–7.0
PFR with glass fiber filler 6.6–7.0
PFR with mica filler 4.2–5.2
PFR with mineral filler (cast) 9.0–15.0
PFR with sisal fiber 3.0–5.0
PFR with wood flour filler 4.0–7.0
Phenanthrene (230°F) 2.7
Phenathiene (68°F) 2.8
Phenathrene (110°F) 2.72
Phenetole (70°F) 4.5
Phenol (104°F) 15.0
Phenol (118°F) 9.9
Phenol (50°F) 4.3
Phenol ether (85°F) 9.8
Phenol formaldehyde resin PFR 4.5–5.0
Phenol resin 4.9
Phenol resin, cumulated 4.6–5.5
Phenolic-based polymers 7.5
Phenoxyacetylene (76°F) 4.8
Phentidine (70°F) 7.3
Phenyl-L-propane (68°F) 2.7
Phenyl acetate (68°F) 6.9
Phenyl ether (86°F) 3.7
Phenyl isothiocyanate (68°F) 10.7
Phenyl isocyanate (68°F) 8.9
Phenyl urethane 2.7
Phenylacetaldehyde (68°F) 4.8
Phenylacetic (68°F) 3.0
Phenylacetonitrile (80°F) 18.0
Phenylethanol (68°F) 13.0

DIELECTRIC CONSTANTS OF MATERIALS

Phenylethyl acetate (58°F) 4.5

Phenylethylene (77°F) 2.4

Phenylhydrazine (72°F) 7.2

Phenylsalicylate (122°F) 6.3

Phosgene (32°F) 4.7

Phosphine (−76°F) 2.5

Phosphorus (93°F) 4.1

Phosphorus oxychloride (72°F) 14.0

Phosphorus pentachloride (320°F) 2.8

Phosphorus tribromide (68°F) 3.9

Phosphorus trichloride (77°F) 3.4

Phosphorus, red 4.1

Phosphorus, yellow 3.6

Phosphoryl chloride (70°F) 13.0

Phthalic acid 5.1–6.3

Phthalide (74°F) 36.0

Phthalide (166°F) 36.0

Pinacolin (62°F) 12.8

Pinacone (75°F) 7.4

Pine tree resin, powder 1.5–1.8

Pinene (68°F) 2.7

Piperidine (68°F) 5.9

Plaster 2.5–6.0

Plastic grain 65–75

Plastic pellets 1.1–3.2

Plastic sulphur, unground 1.5

Platinum catalyst 6.5–7.5

Poly(ethylene) amorphous 2.3

Poly(methylmethacrylate) 4.5

Poly(propylene) 1.5

Poly(propylene) amorphous 2.6

Poly(styrene) 3.8

Poly(vinyl acetate) 4.0

Poly(vinyl chloride) 5.0

Polyacetal 3.6–3.7

Polyacetol resin 2.6–3.7

Polyacrylic ester 3.5

Polyamide 2.5–2.6

Polybutylene 2.2–2.3

Polycaprolactam 2.0–2.5

Polycarbonate 2.9–3.0

Polycarbonate 2.96

Polycarbonate resin 2.9–3.0

Polyester 2.80

Polyester resin (flexible) 4.1–5.2

Polyester resin (glass fiber filled) 4.0–4.5

Polyester resin (rigid cast) 2.8–4.1

Polyester resin 2.8–4.5

Polyesters 4.0

Polyether chloride 2.9

Polyether resin 2.8–8.1

Polyether resin, unsaturated 2.8–5.2

Polyethylene (cellular) 1.50

Polyethylene (cross-linked) 2.30

Polyethylene (high density) 2.34

Polyethylene (low density) 2.28

Polyethylene 2.2–2.4

Polyethylene 2.26

Polyethylene, pellet 1.5

Polymide 2.8

Polymonochloro pifluoroethylene 2.5

Polypropylene 1.5

Polypropylene 2.24

Polypropylene powder 1.25

Polypropylene, pellet 1.5–1.8

Polystyrene resin 2.4–2.6

Polystyrol 2.0–2.6

Polysulfones 3.7

Polysulphonic acid 2.8

Polytetra fluoroethylene 2.0

Polyurethane 7.5

Polyvinyl alcohol 1.9–2.0

Polyvinyl chloride (irradiated) 2.7

Polyvinylchloride resin 5.8–6.8

Polyvinyl chloride (semirigid) 4.3

Polyvinyl chloride 2.7

Polyvinyl chloride 3.4

Porcelain 5.0–7.0

Porcelain with zircon 7.1–10.5

Potassium aluminum sulphate 3.8

DIELECTRIC CONSTANTS OF MATERIALS

Potassium carbonate (60°F) 5.6
Potassium chlorate 5.1
Potassium chloride 4.6
Potassium chloride 5.0
Potassium chloromate 7.3
Potassium chloronate 7.3
Potassium iodide 5.6
Potassium nitrate 5.0
Potassium sulfate 5.9
Propane (liquid) (32°F) 1.6
Propanediol (68°F) 32 .0
Propanol (177°F) 20.1
Propene (68°F) 1.9
Propionaldehyde (62°F) 18.9
Propionic acid (58°F) 3.1
Propionic anhydride (60°F) 18.0
Propionitrile (68°F) 27.7
Propyl acetate (68°F) 6.3
Propyl alcohol (68°F) 21.8
Propyl benzene (68°F) 2.4
Propyl bromide (68°F) 7.2
Propyl butyrate (68°F) 4.3
Propyl chloroformate (68°F) 11.2
Propyl ether (78°F) 3.4
Propyl formate (66°F) 7.9
Propyl nitrate (64°F) 14.2
Propyl propionate (68°F) 4.7
Propyl valerate (65°F) 4.0
Propylene liquid 11.9
Psuedocumene (60°F) 2.4
Pulegone (68°F) 9.5
Pulezone (66°F) 9.7
PVC, powder 1.4
Pyrex 4.8
Pyrex glass 4.3–5.0
Pyridine (68°F) 12.5
Pyroceram 3.5–4.5
Pyrrole (63°F) 7.5
Quartz 4.2
Quinoline (−292°F) 2.6

Quinoline (77°F) 9.0
Reburned lime 2.2
Refractory (cast) 6.7
Refractory (for casting) 1.8–2.1
Resorcinol 3.2
Rice (dry) 3.5
Rice bran 1.4–2.0
Rouge (jeweler's) 1.5–1.6
Rouge 1.5
Rubber (chlorinated) 3.0
Rubber (hard) 2.8
Rubber (isomerized) 2.4–3.7
Rubber 3.0
Rubber cement 2.7–2.9
Rubber chloride 2.1–2.7
Rubber, raw 2.1–2.7
Rubber, sulphurized 2.5–4.6
Ruby 11.3
Rutile 6.7
Safrol (70°F) 3.1
Salicylaldehyde (68°F) 13.9
Salt 3.0–15.0
Sand (dry) 5.0
Sand (silicon dioxide) 3–5.0
Santowax (70°F) 2.3
Selenium (482°F) 5.4
Selenium 11
Selenium (249°F) 5.4
Sesame 1.8–2.0
Shellac 2.0–3.8
Silica aluminate 2
Silica sand 2.5–3.5
Silicon 11.0–12.0
Silicon dioxide 4.5
Silicon tetrachloride (60°F) 2.4
Silicone molding compound (see SMC)
Silicone oil 2.2–2.9
Silicone resin, liquid 3.5–5.0
Silicone rubber 2.6
Silicone varnish 2.8–3.3

DIELECTRIC CONSTANTS OF MATERIALS

Silk 2.5–3.5

Silver bromide 12.2

Silver chloride 11.2

Silver cyanide 5.6

Slaked lime, powder 2.0–3.5

Slate 6.0–7.5

SMC (glass fiber filled) 3.7

Smithsonite 9.3

Soap powders 1.2–1.7

Sodium carbonate (10H$_2$O) 5.3

Sodium carbonate (anhyd.) 8.4

Sodium carbonate 5.3–8.4

Sodium chloride (salt) 6.1

Sodium chloride 5.9

Sodium cyanide 7.55

Sodium dichromate 2.9

Sodium nitrate 5.2

Sodium oleate (68°F) 2.7

Sodium perchlorate 5.4

Sodium phosphate 1.6–1.9

Sodium sulphide 5

Sorbitol (176°F) 33.5

Soy beans 2.8

Stannec chloride (72°F) 3.2

Starch 3–5

Starch, paste 1.7–1.8

Stearic acid (160°F) 2.3

Stearine 2.3

Steatite 5.5–7.5

Styrene (77°F) 2.4

Styrene (modified) 2.4–3.8

Styrene (phenylethane) (77°F) 2.4

Styrene resin 2.3–3.4

Succinamide (72°F) 2.9

Succinic acid (78°F) 2.4

Sucrose (mean) 3.3

Sucrose 3.3

Sugar 3.0

Sugar, granulated 1.5–2.2

Sulfur 1.6–1.7

Sulfur dioxide (–4°F) 17.6

Sulfur dioxide (32°F) 15.0

Sulfur monochloride (58°F) 4.8

Sulfur trioxide (64°F) 3.1

Sulfuric acid 100

Sulfurous oxychloride (72°F) 9.1

Sulfuryl chloride (72°F) 10.0

Sulfur (450°F) 3.5

Sulfur dioxide (32°F) 15.6

Sulfur trioxide (70°F) 3.6

Sulfur (244°F) 3.5

Sulfur, liquid 3.5

Sulfur, powder 3.6

Sulfuric acid (68°F) 84.0

Sulfuric oxychloride (72°F) 9.2

Syrup 50–80

Syrup wax 2.5–2.9

Tantalum oxide 11.6

Tartaric acid (14°F) 35.9

Tartaric acid (68°F) 6.0

Teflon® (4f) 2.0

Teflon® 2.0

Teflon®, FEP 2.1

Teflon®, PCTFE 2.3–2.8

Teflon®, PTFE 3.1

Teflon®, FEP 2.15

Teflon®, TFE 2.15

Tefzel 2.6

Tepineol 2.8

Terpinene (70°F) 2.7

Terpineol (72°F) 2.8

Tetrabromoethane (72°F) 7.0

Tetrachloroethylene (70°F) 2.5

Tetradecamethylhexosiloxane (68°F) 2.5

Tetradecamethyltetradecamethylcycloheptasiloxan 2.7

Tetradecanol (100°F) 4.7

Tetraethyl amylenetetracarboxylate 4.40

Tetraethyl hexane-1-phenyl tetracarboxylate (66°F) 5.9

Tetraethyl pentane diphenyl tetracarboxylate (68°F) 2.7

Tetraethyl propane tetracarboxylate (66°F) 5.2

Tetraethyl propylene tetracarboxylate (66°F) 6.0

Tetraethyl silicate (68°F) 4.1

Tetrafluoroethylene 2.0

Tetrahydro-*b*-naphthol (68°F) 11.0

Tetranitromethane (68°F) 2.2

Tetratriacontadiene (76°F) 2.8

TFE (cellular) 1.4

Thallium chloride 46.9

Thinner 3.7

Thioacetic acid (68°F) 13.0

Thionyl bromide (68°F) 9.1

Thionyl chloride (68°F) 9.3

Thiophene (60°F) 2.8

Thiophosphoryl chloride (70°F) 5.8

Thorium oxide 10.6

Thrichloroethylene (61°F) 3.4

Thujone (32°F) 10.0

Tide (loose from box) 1.6

Tin tetrachloride (68°F) 2.9

Titanium dioxide 110.00

Titanium oxide 40–50

Titanium tetrachloride (68°F) 2.8

Tobacco 1.6–1.7

Tobacco dust (6% moisture) 1.7

Toluene (68°F) 2.4

Toluene, liquid 2.0–2.4

Toluidine (68°F) 6.0

Tolunitrile (73°F) 18.8

Tolyl methyl ether (68°F) 3.5

Totane (111°F) 5.5

Tourmaline 6.3

trans-3-Hexene (76°F) 2.0

Transmission oil (80°F) 2.2

Tribromopropane (68°F) 6.4

Tributylphosphate (86°F) 8.0

Trichloroacetic acid (140°F) 4.6

Trichloroethane 7.5

Trichlorethylene (61°F) 3.4

Trichloropropane (76°F) 2.4

Trichlorotoluene (70°F) 6.9

Trichlorotoluene (69°F) 6.9

Trichlorotoluene 6.9

Tricosanone (176°F) 4.0

Tricresyl phosphate (104°F) 6.9

Triethyl aconitate (68°F) 6.4

Triethyl aluminum (68°F) 2.9

Triethyl ethanetricarboxylate (66°F) 6.5

Triethyl isoaconitate (68°F) 7.2

Triethylamine (21°F) 3.2

Triethylamine (77°F) 2.4

Trifluoroactic acid (68°F) 39.0

Trifluorotoluene (86°F) 9.2

Trimethyl-3-heptene (68°F) 2.2

Trimethyl borate (68°F) 8.2

Trimethylamine (77°F) 2.5

Trimethylbenzene (68°F) 2.3

Trimethylbutane (68°F) 1.9

Trimethylpentane (68°F) 2.9

Trimethylpentane 1.9

Trimethylsulfanilic acid (64°F) 89.0

Trinitrobenzene (68°F) 2.2

Trinitrotoluene (69°F) 22.0

Triolein (76°F) 3.2

Triphenylmethane (212°F) 2.3

Tripolmitin (140°F) 2.9

Tristearin (158°F) 2.8

Turpentine (wood) (68°F) 2.2

Undecane (68°F) 2.0

Undecanone (58°F) 8.4

Urea (71°F) 3.5

Urea 5–8

Urea formaldehyde (cellulose filler) 6.4–6.9

Urea resin 6.2–9.5

Urethane (121°F) 14.2

Urethane (74°F) 3.2

Urethane resin 6.5–7.1

Valeraldehyde (58°F) 11.8

DIELECTRIC CONSTANTS OF MATERIALS

Valeric acid (68°F) 2.6

Valeronitrile (70°F) 17.7

Vanadium oxybromide (78°F) 3.6

Vanadium oxychloride (78°F) 3.4

Vanadium sulfide 3.1

Vanadium tetrachloride (78°F) 3.0

Vaseline 2.2–2.9

Veratrol (73°F) 4.5

Vinyl alcohol resin 2.6–3.5

Vinyl butyral 3.3–3.9

Vinyl chloride (acetate) 3.0–3.1

Vinyl chloride (flexible) 3.5–4.5

Vinyl chloride (rigid) 2.8–3.0

Vinyl chloride resin, hard 5.8–6.4

Vinyl chloride resin, soft 2.8–4.0

Vinyl ether (68°F) 3.9

Vinyl formal 3.0

Vinyllidene chloride 3.0–4.0

Vycor glass 3.8

Water (212°F) 55.3

Water (32°F) 88.0

Water (390°F) 34.5

Water (4°F) 88

Water (68°F) 80.4

Water (80°F) 80.0

Water (steam) 1.00785

Wax 2.4–6.5

Wheat flour (dry powder) 1.6

Wheat flour 3.0–5.0

White mica 4.5–9.6

Wood, dry 2–6

Wood, pressed board 2.0–2.6

Wood, wet 10–30

Xylene (68°F) 2.4

Xylene, liquid 2.2–2.6

Xylenol (62°F) 3.9

Xylenol 17

Xylidine (68°F) 5.0

Zinc oxide 1.7–2.5

Zinc sulfide 8.2

Zircon 12.0

Zirconium oxide 12.5

Zirconium silicate 5.0

VIII.

CHEMOMETRICS AND DATA PROCESSING

27.

PRACTICES FOR DATA PREPROCESSING FOR OPTICAL SPECTROPHOTOMETRY

A variety of methods are used to condition spectra prior to manual observation, spectral searching, or multivariate analysis. This chapter is a basic primer in the assortment of data treatments available.

SCALING METHODS

Mean Centering

The mean spectrum is subtracted from all the spectra in the teaching set prior to calibration (Figure 27.1).

Autoscaling

The mean spectrum is subtracted from all the spectra in a teaching set (mean centering), with the additional step of dividing all the spectra in the teaching set by the standard deviation spectrum prior to calibration (Fig. 27.2).

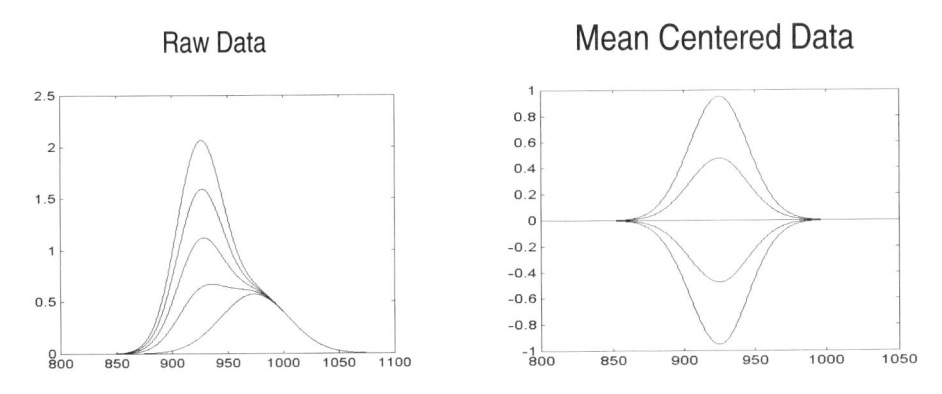

Fig. 27.1. Raw data vs. mean centered data.

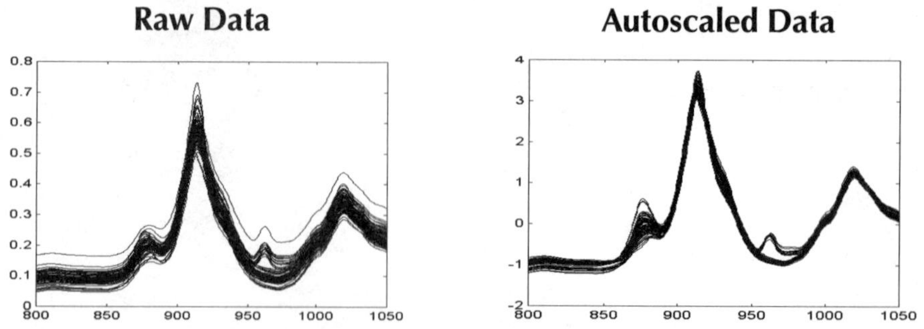

Fig. 27.2. Raw data vs. autoscaled data.

SMOOTHING

Boxcar Smoothing

Attempts are made to improve the signal-to-noise ratio in data by averaging successive data points to remove random variation. This process also broadens bands and removes some fine structure in the data. Oversmoothing can cause x-axis shifts in the signal data.

Fourier-Domain (FFT) Smoothing

Fourier-domain smoothing involves a Fourier transformation of the signal, the application of a filter function (with a set filter factor), and then performing a reverse Fourier transform to the data. A triangular filter function is generally applied. The FFT smoothing can remove high-frequency noise from the signal, but it can also add artifacts that appear as structure when overfiltering is applied (Fig. 27.3).

Savitsky–Golay Smoothing

This smoothing procedure performs a best-fit quadratic polynomial through successive data points. This smoothing technique determines the best-fit center point for the polynomial fit as constrained by the data point segment (or window). The signal-to-noise ratio is improved, bands are broadened, and fine structure can be lost. (Figure 27.3, from *Anal. Chem.* 36:1627, 1964).

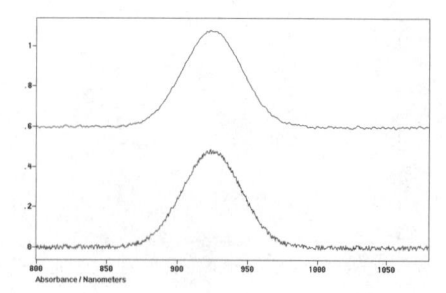

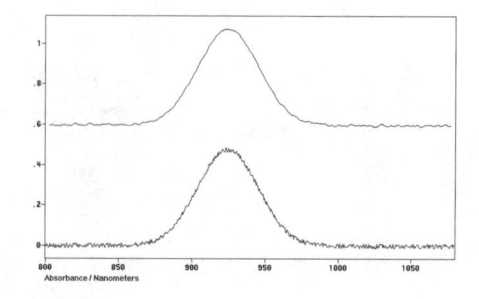

Fig. 27.3. FFT smoothing and Savitsky–Golay smoothing.

NORMALIZATION

Normalization by Means of a Pathlength Correction

Individual pathlength data is used as a scalar multiplier term for correction of an individual spectrum or groups of spectra.

Normalization by Area

All band areas are set to a single value for the purpose of signal comparison and/or correlation techniques (Figure 27.4).

Normalization by means of Reference Band

Divide the set of spectra for calibration by peak height or peak area of a reference band of consistent height or area, respectively. This ratio will "correct" the signal for quantitative analysis when the baseline is poorly resolved.

Normalization Using Multiplicative Signal Correction

This technique, also known as multiplicative scatter correction, was developed to correct for light scattering in reflectance measurements (or measurements containing a strong multiplicative component). Offset and scaling are performed for this algorithm. (Fig. 27.5, from *Multivariate Calibration,* H. Martens and T. Naes, Chichester, U.K.: Wiley, p. 345, 1989.)

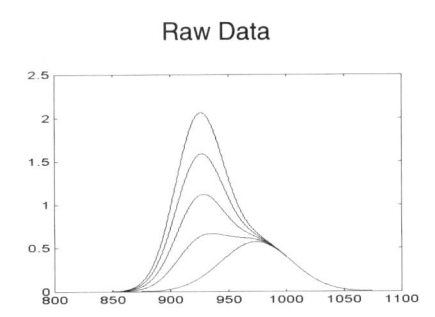

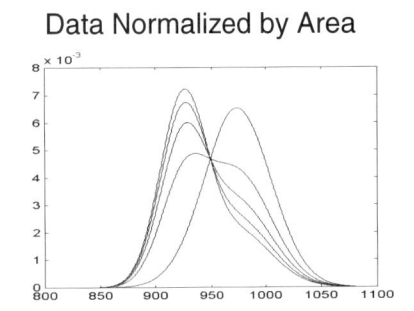

Fig. 27.4. Raw data vs. data normalized by area.

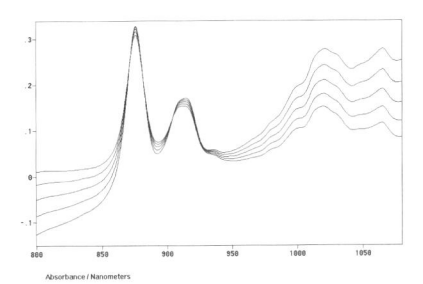

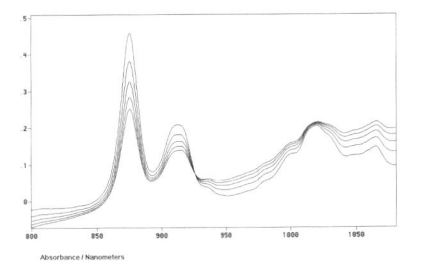

Fig. 27.5. (*left*) Using multiplicative signal correction, five equal increments of slope have been added to a single spectrum (with increasing multiplicative effect dependent upon the degree of slope). This effect is typical with reflectance measurements of crystalline materials. (*right*) MSC as applied to five equal increments of slope added to a single spectrum (with increasing multiplicative effect dependent upon the degree of slope). The multiplicative effect remains following the MSC treatment.

Normalization Using the Kubelka–Munk Transform

The K-M model is applied as a linearization function to signals with scattering and absorptive characteristics as often encountered in diffuse reflectance. This relationship (from V. P. Kubelka and F. Munk, *Z. Tech. Physik,* 12:593, 1931) is given as:

$$Signal_{K-M} = \frac{k}{s} = \frac{(1-R_{\lambda_i})^2}{2R_{\lambda_i}}$$

BASELINE CORRECTION

Baseline Offset Correction

Offset correction is performed by selecting a single point or multiple points on a spectrum and adding or subtracting a *y*-value (intensity value) from the point or points to correct the baseline offset (Figure 27.6). This preprocessing step is used to align the baseline of two or more spectra, causing them to overlap; or it is used to bring the minimum point to zero.

Derivatives

Derivatives can be used to remove offset and slope due to background differences (Figure 27.7).

Flat Baseline Correction

Two data points are selected on the signal, and a line connecting these points is subtracted from the signal (Figure 27.8). This procedure removes first-order uniform offset and slope characteristics.

KRAMERS–KRONIG CORRECTION

Specular reflectance of infrared energy from a sample surface typically interacts with approximately 5 microns of the sample surface when the angle of incidence for the radiation is 30° or more. Spectra measured using specular reflectance are useful for observations of surface

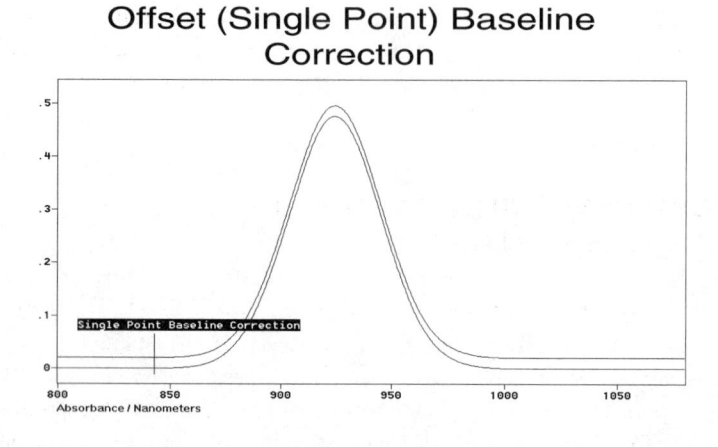

Fig. 27.6. Offset (single-point) baseline correction.

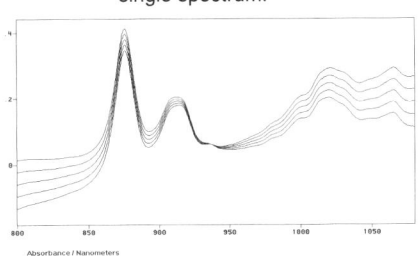

Fig. 27.7. (*left*) Five equal increments of slope have been added to a single spectrum. (*right*) Second-derivative (Savitsky–Golay, 10 convolution points) has been applied to five equal increments of slope added to a single spectrum. The derivative treatment removes slope for this case.

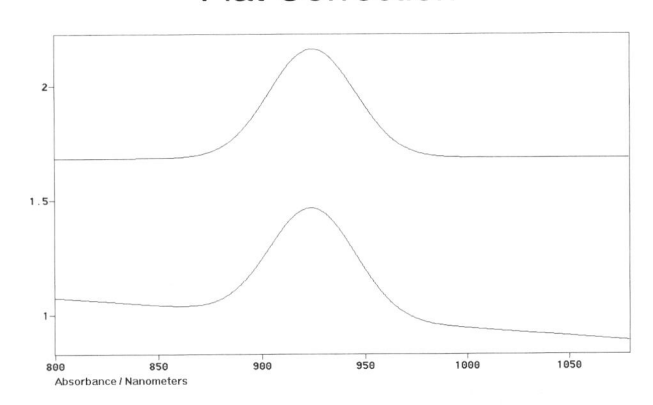

Fig. 27.8. Flat baseline correction.

degradation on coated surfaces. The specular-reflectance spectrum consists of two major spectral components convoluted into a single spectrum. These components are: (1) the molecular absorption spectrum, and (2) the refractive index spectrum. The unusual appearance of a specular-reflectance spectrum is due to the principle that some frequencies of light reflect more than others for any particular sample. The Kramers–Kronig correction is also termed the *dispersion correction.*

TRADITIONAL UV-VIS AND NIR SPECTROSCOPIC DATA-PREPROCESSING METHODS

The common linearization of reflectance (R) or transmission (T) data to absorbance is given by

$$\text{Absorbance:} \quad A = -\log R_\lambda \quad \text{and} \quad A = -\log T_\lambda$$

where A is absorbance expressed as reflectance or transmission when R or $T = I/I_o$.

An absorbance ratio has been used for NIR data by K. Norris and B. Hrushka (In: P. Williams and K. Norris, eds., *Near-Infrared Technology,* AACC, St. Paul, MN: 1987). This is the ratio of absorbance at one wavelength divided by the absorbance at a second wavelength. This ratio method is common to infrared spectroscopists, because it yields a term that compensates for baseline offset. The ratio of transmission data is also useful for normalizing or reducing baseline offset in spectroscopic measurements. A typical absorbance ratio is

$$\frac{A_{\lambda_1}}{A_{\lambda_2}}$$

The use of the moving-averaged-segment convolution (MASC) method for computing derivatives, brings about the following expression describing a first-derivative term:

$$A_{\lambda_2} - A_{\lambda_1} = A_{\lambda_{1+\delta}} - A_{\lambda_{1-\delta}}$$

where the derivative is given as the difference in absorbance values at two different wavelengths, with the position of each wavelength being determined as the gap (δ) distance + or − from a center wavelength (λ). Ratios of derivative terms have been applied to NIR data and have been shown empirically to produce lower prediction errors in special cases. The following expression is the first-derivative ratio where two different center wavelengths have been used:

First-derivative ratio (with two different center wavelengths): $\quad \dfrac{A_{\lambda_2} - A_{\lambda_1}}{A_{\lambda_4} - A_{\lambda_3}} = \dfrac{A_{\lambda_{1+\delta}} - A_{\lambda_{1-\delta}}}{A_{\lambda_{2+\delta}} - A_{\lambda_{2-\delta}}}$

The MASC form of the second derivative is as follows [from W. F. McClure, *NIR News* 4(6), 1993, p.12; and 5(1), 1994, pp. 12–13]. In this case, the second-derivative term is defined by the sum of absorbances at two wavelengths ($\lambda_{1+\delta}$ and $\lambda_{1-\delta}$), minus twice the absorbance at a center wavelength (λ_1). In this case the second-derivative gap size is designated as δ. The second-derivative preprocessing step is quite effective in removing slope and offset variations in spectral measurement baselines. It also "assists" the calibration mathematics in defining spectral regions where small response changes can be useful in calibration modeling. Without the use of derivatives, these regions would not be beneficial for use in calibration.

$$A_{\lambda_1} + A_{\lambda_3} - 2A_{\lambda_2} = A_{\lambda_{1-\delta}} + A_{\lambda_{1+\delta}} - 2A_{\lambda_1}$$

Second-derivative ratio preprocessing has also been demonstrated in the Norris regression method. Empirically this data-preprocessing technique has proved useful for some applications. The following relationship designates a second-derivative ratio with two different center wavelengths.

Second-derivative ratio (with two different center wavelengths): $\quad \dfrac{A_{\lambda_1} + A_{\lambda_3} - 2A_{\lambda_2}}{A_{\lambda_4} + A_{\lambda_6} - 2A_{\lambda_5}} = \dfrac{A_{\lambda_{1-\delta}} + A_{\lambda_{1+\delta}} - 2A_{\lambda_1}}{A_{\lambda_{2-\delta}} + A_{\lambda_{2+\delta}} - 2A_{\lambda_2}}$

The traditional Kubelka–Munk expression is shown next. This expression is used for samples exhibiting scatter, and it has proved beneficial for some measurements.

Kubelka–Munk expression: $\quad \dfrac{\left(1 - R_{\lambda_1}\right)^2}{2R_{\lambda_1}}$

The log[Kubelka–Munk] expression has been demonstrated and may prove empirically useful for some scattering sample data sets:

log Kubelka – Munk expression: $\quad \log\left[\dfrac{\left(1 - R_{\lambda_1}\right)^2}{2R_{\lambda_1}}\right]$

28.

REVIEW OF CHEMOMETRICS APPLIED TO SPECTROSCOPY: QUANTITATIVE AND QUALITATIVE ANALYSIS

INTRODUCTION

Chemometrics can generally be described as the application of mathematical and statistical methods to (1) improve chemical measurement processes and (2) extract more useful chemical information from chemical and physical measurement data. Recent advances in this discipline have led to a new breed of analytical tools—microprocessor-controlled "intelligent" instrumentation and data-analysis systems. Chemometrics represents, in the broadest sense, the penetration of serious mathematical science into the realm of the chemist. Prior to the introduction of a formal subdiscipline of chemistry, termed *chemometrics,* chemists applied as much mathematics as they had available for research and problem solving. But, in general, the use of statistical experimental design and data analysis were relegated to the engineer or the specialist statistician; chemists did chemistry, and life was simple. The current trend in scientific thinking across disciplines involves a multivariate approach. The world is not as simple as was once postulated. In fact, it is common for scientists to approach problems realizing that there are deeply hidden relationships between variables that can be wrestled from an experiment only by the use of newer data-analysis techniques. Thus chemometrics becomes a necessity, not the luxury it once might have been. The purpose of this three-part review is to summarize the use of chemometrics in spectroscopy over the past 10 years. Basic chemometric resources and techniques as applied to mass spectrometry, ultraviolet-visible spectrophotometry, and infrared spectroscopy (near-IR and mid-IR) will be reviewed. General techniques will be covered in the first part of this review, with a special emphasis on sample selection for calibration (teaching) sets and qualitative techniques. The second part of this review will cover quantitative analysis, with the third part covering the more advanced chemometric methods in greater detail.

This chapter represents a general overview of the major topics associated with the use of chemometrics in spectroscopy. Although most chemometric methods could be applied to spectroscopic data, the focus of this review is those techniques that *have been demonstrated* using spectroscopic data. The field is expanding rapidly and is well documented within its own specialist journals, mentioned later in this review.

Trends in Chemometrics

So what does the future look like for chemometric tutorials and education? First, most data analysis will have to move in the direction of multivariate approaches (rather than univariate), more than likely using nonlinear methods. Second, the quality and confidence limits for multivariate mathematical approaches will be evaluated at a basic, first-principles level. Third, since people can visualize only in three dimensions, this opens the opportunities for greater development of projection and graphical capabilities to allow users to perceive beyond three-dimensional space. Last, the trend in computer science is toward simulation of the brain and thought processes through neural networks (simulating the architecture and physiology of the brain), fuzzy logic (a metaphor for human approximation thinking), speech interactive networks, termed *ASR* (automatic speech recognition), and VR (virtual reality). Specialization is occurring at an alarming rate; e.g., there are already job advertisements for independent knowledge engineers, neural net consultants, virtual reality programmers, genetic algorithm experts, and multivariate financial consultants. There is a World Congress on Computational Intelligence (which includes three technologies in one meeting: neural networks, fuzzy systems, and evolutionary computation), sponsored by the IEEE Neural Networks Council, as well as an International Congress on Neural Networks, sponsored by the International Neural Network Society [1]. Chemometric tools are powerful enough to completely alter the science of chemistry as it is practiced today. As Albert Einstein stated, "The whole of science is nothing more than a refinement of everyday thinking" [2].

Tutorials

A multiplicity of tutorials are beginning to be published and made available for a more general audience. *Chemolab* and the *Journal of Chemometrics* generally, by their own admission, write for the "ivory tower" chemometrician. More advanced tutorials include Refs. 3–12. Readers have to decide which are the more useful tutorials for their own level of expertise. Recommended tutorials for the general chemist or spectroscopist are found listed later in the section on Journals. References 13–15 provide those stronger in chemometric knowledge with instructional material. Fundamental self-help texts are included in Refs. 16–18; these are more useful for providing a foundation prior to delving into the specialized chemometric tutorials and textbooks. In addition, high-quality short courses offered through the sponsorship of NAmICS [19], FACSS [20], and ACS [21] offer the practicing spectroscopist an opportunity to upgrade skills at the Pittsburgh or FACSS conferences. An audiovisual course describing principal components has been developed by Massart and Lewi [22].

General-Purpose Chemometric Software Packages

The software programs mentioned in this review are common to UV-Vis, NIR-IR users. The packages included here do not require any instrument purchase and are not necessarily designed for specific instrumental data.

>   *Grams 386 PLSPlus* from Galactic Industries [23]
>   MatLab-based (from The Mathworks) Packages [24]
>   >   *Chemometrics Expert Toolbox* from Perkin-Elmer [25]
>   >   *MatLab Application Toolbox: Chemometrics* [26]
>   >   *PLS Toolbox* from Eigenvector Technologies [27]
>   *Pirouette* from InfoMetrix [28]
>   *Unscrambler* from Camo [29]
>   *SIMCA* from Umetrics [30]

Internet Resources

The number of Internet resources is changing almost daily. In fact, the International Chemometrics Society conference is held over the network [refer to item a in the following list]. Listed next are a number of sites valuable for their chemistry contents. Internet sites a, c, g, h, i, and k are especially useful for chemometric information. A very recent reference for World Wide Web sites is found in Ref. 31. Here are some computer bulletin board examples:

a. *ICS-L (International Chemometrics Society) network:* Send message: SUBSCRIBE ICS-L FIRSTNAME LASTNAME to: LISTSERV@UMDD.UMD.EDU (an alternate address is LISTSERV@UMDD.BITNET).
b. *American Chemical Society Publications:* To access ACS publication information directly, gopher to INFX.INFOR.COM.4500; links to other chemistry gophers are through GOPHER ACSINFO.ACS.ORG.
c. *CHEM-COMP (Computational Chemistry):* Send message as: JOIN CHE-COMP FIRSTNAME LASTNAME to: AILBASE@MAILBASE.AC.UK.
d. *CHEMED-L (Chemistry Education Discussion List):* Send message as: SUBSCRIBE CHEMED-L FIRSTNAME LASTNAME to: LISTSERV@PSUVM.PSU.EDU.
e. *CHEM-MOD (Modeling Aspects of Computational Chemistry):* Send the message as: JOIN CHEM-MOD FIRSTNAME LASTNAME to: MAILBASE@MAILBASE.AC.UK.
f. *CHINF-L (Chemical Information Sources Discussion List):* Covers the broad subject of questions related to all aspects of chemistry. Send the message as: SUBSCRIBE CHMINF-L FIRSTNAME LASTNAME to: LISTSERVE@IUBVM (or alternate address LISTSERV@IUBVM.UCS.INDIANA.EDU.
g. *Computational Chemistry Newsletter:* Gopher to: BLANCA.TC.CORNELL.EDU and follow the path: Forefronts, HotTips, and Other News > Computational Chemistry Newsletter.
h. *Genetic Algorithms Users Group:* Send request to: GA-MOLECULE-REQUEST@TAMMY.HARVARD.EDU.
i. *Internet Resources for Mathematics and Science Education* (compiled by T.C. O'Haver): FTP to: INFORM.UMD.EDU and get the file MATH-SCIENCE-EDU in the following directory: inforM/Computing_Resources/NetInfo/ReadingRoom/InternetResources.
j. *Macintosh Chemistry Tutorials:* FTP to: ARCHIVE.UMICH.EDU and go to: /MAC/MISC/CHEMISTRY.
k. *MathSource software:* For instructions, send Internet message: HELP INTRO to: MATHSOURCE@WRI.COM.
l. *WWW (World Wide Web) sites:*
http://zeus.bris.ac.uk/~chph/brischem/news.html
http://www.aksess.no/camo/
http://www-chaos.umd.edu/
http://ac.dal.ca/~dtandrew/chemomet.html
http://odin.chemistry.uakron.edu/chemometrics/
http://www.elsevier.nl/catalogue/SA2/205/06040/06041/502682/
http://newton.foodsci.kvl.dk/foodtech.html
http://sass2.wlv.ac.uk:8080/
http://www.emsl.pnl.gov:2080/docs/incinc/homepage.html
http://www.halcyon.com/infometrix/welcome.htmlhttp://sunsite.nus.sg/wiley-text/che/che48.html
http://www-cac.sci.kun.nl/cac/
http://www.emsl.pnl.gov:2080/docs/cie/neural/gateway.html
http://www.chem.duke.edu./~reese/
http://stork.ukc.ac.uk/IMS/minitab/index.html
http://www.it-center.se/umetri/welcome.html

Books

The books most commonly used by practitioners of chemometrics, particularly "calibrationists," are still very few [32–37]. One of the best overall texts for matrix mathematics at beginning to advanced levels is found in Ref. 38.

Journals

Media useful to communicate basic concepts include:

> Series or sets of articles in journals and magazines, e.g., *Analytical Chemistry* A [39,40], tutorials in the *Journal of Chemometrics,* "Statistics in Spectroscopy" (*Spectroscopy*) [41], "Chemometrics in Spectroscopy" (*Spectroscopy*) [42], The Chemometric Space (*NIRnews*) [43].

> Basic and tutorial series in journals:
> *Chemometrics and Intelligent Laboratory Systems* [44]
> *Journal of Chemometrics*
> *Journal of Chemical Information and Computer Science*
> *Technometrics*
> *Analytica Chimica Acta*
> *Journal of the Royal Statistical Society*
> *Applied Statistics*
> *American Statistician*
> *Journal of the American Statistical Association*
> *Biometrika*

> Newsletters:
> *NAmICS (North American International Chemometrics Society) Newsletter* [19]
> *NIRNews (Near-Infrared)* [43]
> *U-POST* [45]
> *The Mathworks Newsletter* [46].

Bibliographies and literature searches related to specialized chemometric topics include *Window on Chemometrics* [47], which is extremely useful for keeping up to date. The references in the next section can start a beginner on the basic chronology of chemometrics, particularly as related to calibration. The first place to start looking for current research is in the fundamental reviews of chemometrics as presented biannually in *Analytical Chemistry* [48–55].

Nonspectroscopy Sources

History of Chemometrics

P. Geladi and K. Esbensen perform interviews with selected chemometricians on the origins, use, and future of chemometrics. The interviews consist of two parts: interviews [56] and discussion [57].

Reviews

I. Frank and coworkers presented a general description of the use of PLS for prediction of product quality in the *Journal of Chemical Information and Computer Science,* 1984 [58]. This paper describes the general use of PLS regression with spectroscopic data and compares the PLS regression method to multiple linear stepwise regression. The paper uses data from both infrared and sonic spectra. B. R. Kowalski and M. B. Seasholtz have written a review entitled "Recent Developments in Multivariate Calibration" as an invited paper to the *Journal of Chemometrics.* This paper [59], containing 74 key references, describes the advantages of multivariate techniques over univariate methods for many complex situations. The paper describes the methods and use of such linear regression techniques as principal component regression (PCR) and partial least squares (PLS). In addition, nonlinear PCR, nonlinear PLS, and locally weighted regression (LWR) are presented as calibration techniques suited to nonlinear data problems. The paper also briefly touches on second-order calibration methods.

PLS regression is described by mathematical and statistical presentation in a paper by A. Hoskuldsson. The operation and use of PLS regression is explained in minute detail [60]. W. F. McClure gives a general review of the basic history and application of NIR spectroscopy, with 43 references [61]. Kirsch and Drennen review the use of chemometric techniques in the pharmaceutical industry and include 58 references [62].

Official Methods/Practices Using Multivariate Calibration Chemometrics

A limited number of "official" analytical methods exist using chemometric practices. The following three are the major documents in existence. In all cases the approval process for these methods or practices was a difficult and time-consuming activity.

> The American Society for Testing and Materials (ASTM) Practice E1655–97, 1997 [63], is a method describing the standard practices for infrared, multivariate, quantitative analyses. Under the auspices of ASTM a prescribed set of practices has been compiled for basic-to-advanced quantitative calibration. This is a new standard contained in volume 03.06, 1999.
>
> The Association of Official Analytical Chemists (AOAC) Method 989.03, 1989 [64], presents a near-infrared method for determination of acid detergent fiber and crude protein in forage materials using multivariate calibration. This method is also referenced in *JAOAC* 71:1162, 1988. Sample preparation and calibration procedures are described.
>
> American Association of Cereal Chemists (AACC) Method 39–10 [65] describes a near-infrared method for protein determination for all classes of wheat based on near-infrared reflectance spectroscopy. This is a general method for protein calibration based on data from filter or full-wavelength instruments. The wheat samples must be ground to a powder passing a 1.0-mm screen using a prescribed cyclone grinding mill. The specifics related to calculating the calibration coefficients are not given.

QUANTITATIVE ANALYSIS

The use of multivariate quantitative analysis to describe the relationships between the concentration of an analyte and its instrument response (absorbance, transmittance, emission, retention time, peak height, peak area, etc.) has been described by multiple sources since 1960. The common practice of relating spectrophotometric response to concentration has been accomplished for a number of applications using Beer's law combined with C-matrix, K-matrix, multiple linear regression, principal component regression, and partial least squares regression. Each application, and each set of experimental conditions, will produce slight differences in the selection of calibration wavelengths as well as the weighting vectors (or coefficients) used for the calibration. Neither the methods themselves nor the wavelengths selected using these common tools are new. Each property is simply a set of common tools applied to natural phenomena under specific experimental conditions. The mathematics and application of these methods are addressed later in this chapter.

Calibration Literature in Chronological Order

Quantitative analysis as a chemometric application will be addressed in greater detail later in this review. Earlier, a series of references giving the historical development of chemometric methods in spectroscopy since 1974 was presented, as background information [66–88].

QUALITATIVE CALIBRATION TECHNIQUES

Sample Selection

The overall definition of sample selection techniques for spectroscopic analysis includes any selection process attempting to exclude redundancy in sample populations intended for calibration or validation. Ideally, the best sample selection techniques would reduce sample populations to the minimum number of samples needed to represent all meaningful spectral variation. Sample selection is useful in producing greater robustness (or resistance to overfitting of data during the calibration step) and in reducing the number of reference laboratory analyses required, because of the smaller number of samples need for calibration.

The mathematical criteria used to evaluate subset selection methods include no reduction in prediction errors for the calibration step [SEP (standard error of performance), SECV (standard error of cross-validation), and PRESS (prediction residual error sum of squares)] between the calibration developed using the entire set, and a calibration developed using only the sample subset.

The basic rationale in selecting sample sets generally follows the statistical determination of the spectral "alikeness" or "uniqueness" of any sample compared to an entire sample set, a small subset of a sample set, or a single reference spectrum. When determining how alike a sample is to another sample (or set of samples) one may use RMSD (root mean square deviation) between two spectra (test sample and reference sample) or between the test sample spectrum and the mean spectrum of a complete sample set. One may also calculate r (correlation) or R^2 (coefficient of determination) between a sample spectrum and a reference spectrum. A third approach involves spectral subtraction of the test spectrum and the reference spectrum, with a calculation of the magnitude of the residual spectrum (as total area or average residual) as a measure of difference.

Although two spectra may be nearly identical, there may be wide differences in chemical composition due to small, yet important, absorption bands hidden by a larger band (or bands). There may also be important chemical information within a sample that does not appear (or is very weak) in the spectral data. For example, in the near-infrared spectrum, materials containing peroxides, thiols, and inorganics or compounds with nearly identical spectra (aromatics, aliphatics, etc.) may be poorly differentiated. Many users of near-infrared spectroscopy have found that the optimum subsampling technique is to select sample subsets based upon the reference laboratory results rather than using a purely mathematical approach. A potential explanation for this phenomenon is found in the fact that a large proportion of near-infrared spectral information (at least for solid samples) is the result of particle size, or light-particle interaction, and the moisture spectrum of a sample. In addition to these two factors, the largest spectral deviations are sampling errors (differences between calibration samples and test set samples), repack errors (differences in particle orientation for repeated measurements), sample non-homogeneity (poorly mixed samples), sample cell orientation differences (due to imperfections in sample cell and detector geometry), temperature variations from measurement to measurement (affecting both the density and the refractive index of samples), instrument bandpass and band shape, and individual instrument noise characteristics.

In spectral-matching algorithms, methods of comparison vary in the quality of the matching "resolution" (or capacity and confidence to be placed on the matching index value) as well as in the computational time required to search large numbers of data files. If search areas are restricted to specific spectral regions in order to achieve better resolution of differences between very similar samples (e.g., enantiomers or stereoisomers), rather large deviations in other areas of the spectrum will be missed. Limiting the spectral region for spectral matching will give low confidence for identifying small differences between spectra, including adulteration, impurities, or tampering. Full spectral comparisons using principal component scores can be used for both sample selection and spectral matching.

Sample selection methods are important in the routine use of near-infrared spectroscopy, for the technique generally relies on large learning (or teaching) sets for calibration. The simpler techniques described within the literature for sample selection include: random selection, stratified selection, nearest centroid clustering, and spectral difference calculations. More complex approaches to sample selection include: correlation analysis between spectra in the wavelength domain, correlation analysis between principal component scores of spectral data, and discriminant analysis between raw or transformed (derivatized) spectra or between the principal component scores of spectra. As has been mentioned, a typical procedure is to manually select samples using reference laboratory analysis by eliminating redundant composition results for each parameter calibrated.

In theory, one would begin with some a priori information regarding the sample population used for calibration. Frequency distributions for most parameters in naturally occurring (and randomly selected) populations are assumed to be normal (or Gaussian). In reality, parameter distributions for a population may exist as highly skewed normal distributions, multimodal distributions, Poisson distributions, or log-normal distributions. Regardless of the actual distribution of the teaching or calibration sample set, the traditional goal in calibration has been to define a linear model where the confidence limits for predicted values are approximated by the F-distribution, with the error distribution normal and identical variance for each sample. The final result of calibration is a least squares calibration fit with confidence bands that are larger at extreme concentrations (i.e., both high and low values) than near the mean concentration value. The following sections represent a review of current practice in sample selection methods.

Random Selection

True randomized selection of a sample subset from a larger sample population produces a subset following the original distribution of the larger population. When randomly selected subsets produce calibrations outperforming other sample selection techniques, it is by chance alone [89]. The possibilities for this event can be computed for individual cases [90,91]. As an illustration, the Gaussian (normal) distribution can be used where approximately 68.27% of samples are within one standard deviation (1 sigma) of the arithmetic mean, 95.45% are within 2 sigma of the arithmetic mean, and 99.73% are within 3 sigma of the arithmetic mean. True randomized selection from an infinitely large Gaussian distribution would, over a long period of time, yield precisely these results. Distributions largely skewed or exhibiting asymmetry would retain these unusual features when random subset selection is employed. In statistical terms, randomly selected samples will have the same mean and standard deviation as the original population "over time." There is no advantage in random selection other than the low probability that a chance event may produce a superior calibration subset; statistically poorer calibration results will most often occur for random subsets when compared to calibration results obtained using the larger, complete sample set. This effect is due to the high probability of having decreased variance in the subset population as compared to the total population.*

Manual Subset Selection

Manual subset selection techniques are performed by choosing samples using a priori knowledge about the samples, combined with results from manual chemical analysis or visual

---

*Random selection is generally held to be well behaved (statistically speaking), as indicated by Chebyshev's theorem, where the probability that a random variable will have a value differing from the mean by more than K standard deviations is always less than $1/K^2$.

inspection of spectral overlay plots. These procedures can be a useful alternative for users lacking subset selection software algorithms. The method is accomplished using reference analytical data for all samples and selecting a uniformly distributed sample set by eliminating redundancy, or by visual inspection of the total sample population, whereby overlapping spectra are eliminated visually. These procedures are time consuming and extremely user dependent. Problems arising with these procedures for near-infrared spectroscopy can be found where the spectral response to analyte concentration is convoluted, with interfering bands causing wide variation in spectral profiles unrelated to analyte concentration. Here, the use of manual selection is inappropriate to resolve meaningful differences in spectra, differences that should be included within the calibration set.

Spectral Subtraction Methods for "Uniqueness" Tests

A procedure for subset selection utilizing spectral subtraction to describe sample spectra most different from the mean sample spectrum are found in Ref. 92. In this procedure, the assumption is made that large or unusual spectral absorbance signals are associated with large analyte concentrations or some other important, unique sample property. Small time-dependent shifts in wavelength accuracy within an instrument create derivative-like peaks in difference spectra. Particle-size differences, sample-temperature differences, and changes in sample presentation all create large nonconcentration-related spectral changes. Spectral difference techniques tend to select samples based upon these large nonconcentration-related effects, so when significant physical variation exists for a sample set, caution must be exercised for this subset selection technique. The technique as described in Ref. 92 is limited to the number of wavelengths recorded for each sample spectrum; thus a K-wavelength instrument would allow a maximum subset selection of $N \leq K$. The technique has been proved useful for subset selection in natural product samples [93].

Stratified Sample Selection

Stratified sample selection involves one of the following [94]:

1. Samples populations are divided into between 5 and 15 groups (strata) using an equidistant analyte concentration range. Samples are then selected from each stratum using random selection.
2. Samples are grouped using constituent ranged quintiles, with equal numbers of observations in each quintile; samples are selected from each quintile using random selection.

Each method also has limited utility, for it requires reference laboratory analysis for every sample for a minimum of one analyte. The procedure is also blind to interference bands and spectral variations important for computing robust calibration models.

Wavelength-Based Discriminant Analysis

Discriminant analysis has been used in a variety of ways for subset selection. A nearest-centroid clustering technique has been demonstrated in which sample sets are grouped using Mahalanobis distance scores into various sample clusters, where each cluster represents a unique spectral type. After initial clustering, samples are randomly selected from each cluster for use in calibration. This technique has not been reported as advantageous over simple random sample selection.

Discriminant analysis has been described for use as a qualitative analysis technique [95], to compare variation in grinding and sample preparation for solid samples [96], and to select small sample sets from large sample populations [97]. Populations of spectra can be discriminated into small groups using absorbance data for specific wavelengths. The size of the data spread out in two- or multidimensional space can be calculated as an RMS (root mean square) group size for each cluster. Mahalanobis distances or normalized distances can be used to separate data into distinct groups [98].

In discriminant analysis the assumption is made that samples closest together in wavelength space are very similar to one another. Conversely, samples far apart in wavelength space are thought to be part of separate spectral groups (or clusters). By using samples selected at specific distances in multidimensional space it is possible to select representative subsets. Wavelength selection–based discriminant analysis for subset selection includes the difficulties associated in searching for the optimum wavelength set. Once wavelengths for discrimination are selected, one can question the specificity of the discrimination based upon the analyte (or parameter) of interest. A rational, yet computationally intensive, procedure would include a row-reduction calibration to be performed for the constituent of interest. Once calibration wavelengths are selected yielding the optimum signal-to-noise ratio (for the parameter of interest), the same wavelength set can be used for discriminant analysis. Ideally, a separate sample subset would be determined using discriminant analysis for each constituent; this method would produce the maximum spectral variance in a sample subset related to each unique calibration.

Principal Component Score–Based Discriminant Analysis

Discriminant analysis using principal component scores rather than absorbance data at specified wavelengths is a useful technique both for grouping data sets (clustering) and for selecting sample subsets based on distance determination. Distance is determined between individual samples or mean clusters in "principal component scores" or "principal component space" [99,100]. Since principal component methods generally use large numbers of data points to represent each spectrum, this procedure has advantages over discriminant analysis using wavelength-only space. As an overall grouping technique, principal component scores used in combination with non-Euclidean distance measurements (e.g., Mahalanobis distance or the H statistic) is an extremely powerful tool for using full spectral response data to separate or cluster spectra into separate groups. This soft modeling technique prevents the loss of relevant information in separating samples as might occur with wavelength-only–based discriminant analysis.

Correlation Matching

The "alikeness" of one test spectrum (or series of spectra) to a reference spectrum can be determined by calculating a point-by-point correlation between absorbance data for each test and reference spectrum. The correlation matching can be accomplished for all data points available or for a preselected set only. The more alike the test and reference spectra are, the higher (closer to 1.00) are the r (correlation coefficient) and R^2 (coefficient of determination) values. A perfect match of the two spectra would produce an r or R^2 value of 1.00000. The sensitivity of the technique can be increased by pretreating the spectra as first- to higher-order derivatives and then calculating the correlation between test and reference spectra. Full spectral data can also be truncated (or reduced) to include only spectral regions of particular interest, a practice that will further improve matching sensitivity for a particular spectral feature of interest. Sample selection using this technique involves selecting samples most different from the mean population spectrum for the full sample set. Those samples with correlations of the lowest absolute values (including negative correlations) are selected first, and then samples of the second lowest correlation are selected (and so on), until the single sample of highest correlation is found. The spectra are assumed to follow a normal distribution about the mean, with a computable standard deviation. This assumption indicates that a uniformly distributed sample set can be selected based on the correlation between test spectra and the mean spectrum of a population of spectra.

Solving the Small-Training-Set Problem

A procedure for calculating the minimum number of training samples required for a calibration (or training) set using wavelength search calibration techniques is described by Honigs et al. [101]. By plotting the significance of the ratio of SEE (standard error of estimate) to R_A (the analytical range of the sample concentrations) as a function of the number of samples, the paper graphically illustrates the probability of obtaining a particular SEE as a function of the number of constituents (n_w) and the number of samples (n_s) within the calibration set. Another paper by Honigs et al. [102] describes an approach to near-infrared calibration when a limited sample set is available; the study involved 23 powdered geological samples. The number of wavelengths required for the calibration were determined using factor analysis, the MLR (multiple linear regression) was solved using a row-reduction algorithm [103]. The calibration was validated using cross validation and spectral reconstruction [104].

General Qualitative Practices

The ASTM (American Society for Testing and Materials) has published a "Standard Practice for General Techniques for Qualitative Analysis" (Method E 1252–88). The method [105] describes techniques useful for qualitative evaluation of liquids, solids, and gases using the spectral measurement region of 4000–50 cm^{-1} (above 2500 nm). A revised method is being drafted to encompass the newer chemometrics techniques in current research practice [106]. Calculation of Mahalanobis distance is described in the next section of this chapter. Primary references describing the use of this mathematical approach are defined in Refs. 95, 98, 107, and 108. A general discussion on the use of the Hat matrix (or H statistic) for the discrimination of outliers (or highly influential samples) is found in Refs. 109–111.

Mahalanobis Distance Measurements

The Mahalanobis distance statistic (or more correctly the square of the Mahalanobis Distance, D^2), is a scalar measure of where the spectral vector **a** lies within the multivariate parameter space used in a calibration model. The Mahalanobis distance is used for spectral matching, for detecting outliers during calibration or prediction, or for detecting extrapolation of the model during analyses [95,99]. Various commercial software packages may use D instead of D^2, may use other related statistics as an indication of high leverage outliers, or may call the Mahalanobis distance by another name. D^2 is preferred here because it is more easily related to the number of samples and variables. Model developers should attempt to verify exactly what is being calculated. Both mean-centered and nonmean-centered definitions for Mahalanobis distance exist, with the mean-centered approach being preferred. Regardless of whether mean-centering of data is performed, the statistic designated by D^2 has valid utility for qualitative calculations.

If **a** is a spectral vector (dimension f by 1) and **A** is the matrix of calibration spectra (of dimension n by f), then the Mahalanobis Distance is defined as

$$D^2 = \mathbf{a}^t(\mathbf{AA}^t)^+\mathbf{a} \tag{1a}$$

For a mean-centered calibration, **a** and **A** in equation (1a) are replaced by $\mathbf{a} - \bar{\mathbf{a}}$ and $A - \bar{A}$, respectively.

If a weighted regression is used, the expression for the Mahalanobis distance becomes:

$$D^2 = \mathbf{a}^t(\mathbf{ARA}^t)^+\mathbf{a} \tag{1b}$$

In MLR, if **m** is the vector (dimension k by 1) of the selected absorbance values obtained from a spectral vector **a**, and **M** is the matrix of selected absorbance values for the calibration samples, then the Mahalanobis distance is defined as

$$D^2 = \mathbf{m}^t(\mathbf{MM}^t)^{-1}\mathbf{m} \tag{2a}$$

If a weighted regression is used, the expression for the Mahalanobis distance becomes:

$$D^2 = \mathbf{m}^t(\mathbf{MRM}^t)^{-1}\mathbf{m} \tag{2b}$$

In PCR and PLS, the Mahalanobis distance for a sample with spectrum **a** is obtained by substituting the decomposition for PCR, or for PLS, into Eq. (1a). The statistic is expressed as

$$D^2 = \mathbf{s}^t \mathbf{s} \tag{3a}$$

If a weighted PCR or PLS regression is used, the expression for the Mahalanobis distance becomes

$$D^2 = \mathbf{s}^t \left(\mathbf{S}^t \mathbf{R} \mathbf{S}\right)^{-1} \mathbf{s} \tag{3b}$$

The Mahalanobis distance statistic provides a useful indication of the first type of extrapolation. For the calibration set, one sample will have a maximum Mahalanobis distance, D^2_{max}. This is the most extreme sample in the calibration set, in that it is the farthest from the center of the space defined by the spectral variables. If the Mahalanobis distance for an unknown sample is greater than D^2_{max}, then the estimate for the sample clearly represents an extrapolation of the model. Provided that outliers have been eliminated during the calibration, the distribution of Mahalanobis distances should be representative of the calibration model, and D^2_{max} can be used as an indication of extrapolation.

Spectral Matching

Techniques for matching sample spectra include the use of Mahananobis distance [95,98,107,108] and cross-correlation techniques similar to that described in the earlier subsection on Correlation Matching. The general method for comparing two spectra (test versus reference), where the reference is a known compound, or the mean spectrum of a set of known spectra is given as the M.I. (match index). The M.I. is computed by comparing the vector dot products between the test and reference spectra. The theoretical values for these dot products range from –1.0 to +1.0, where –1.0 is a perfect negative (inverse) correlation and +1.0 is a perfect match. Since for near-infrared spectroscopy only positive absorbance values are used to compute the dot products, the values for the match index must fall within the 0.0–1.0 range. The method, or variations thereof, is described in Refs. 112–116. The mathematics are straightforward and are demonstrated next.

The match index (M.I.) is equal to the cosine of the angle (designated as α) between two row vectors (the test and reference spectra) projected onto a two-dimensional plane and is equivalent to the correlation (*r*) between the two spectra (row vectors):

$$\text{M.I.} = \cos \alpha = \left(\frac{T \bullet R}{|T||R|} \right) \tag{4}$$

where *T* is the test spectrum row matrix and *R* is the reference spectrum row matrix. Note the following:

$$(T \bullet R) = \sum_{i=1}^{n} T_i R_i \tag{5a}$$

where T_i represents the individual data points for the test spectrum (designated as the absorbance values of spectrum *T* from wavelengths $i = 1-n$), and R_i represents the individual data points for the reference spectrum (designated as the absorbance values of spectrum *R* from wavelengths $i = 1-n$), and where

$$|T||R| = \left(\sum_{i=1}^{n} T_i^2 \right)^{1/2} \left(\sum_{i=1}^{n} R_i^2 \right)^{1/2} \tag{5b}$$

Note, the angle (α), in degrees, between two vectors can be determined from the match index using

$$\alpha = \cos^{-1} \text{M.I.} \tag{6}$$

Spectral Reconstruction Methods

Work by Honigs et al. [117] describes a method for training a computer to obtain individual pure component spectral profiles from complex mixtures. The mathematics results in a "reconstructed spectrum" representing the spectral profile of a specific chemical parameter within a mixture of parameters. The method uses a training (or calibration) set made up of sample spectra and reference (chemical analysis) values to produce reconstructed pure component spectra. The method utilizes the cross-correlation function to derive a reconstructed spectrum for each parameter; the mathematics for this is shown next.

The reconstructed spectrum (R.S.) for any parameter is given by the following relationship, computed for every wavelength (i) and is shown graphically by plotting the calculated R.S.$_i$ at each wavelength (as the ordinate) versus the wavelength (i) as the abscissa. Note that the R.S.$_i$ represents the absorbance profile for the reconstructed pure component at each wavelength, and thus the final graphic appears as a pure component spectrum. Kemeny and Wetzel demonstrated the practical use of this method [118].

To calculate the R.S.$_i$ value at each wavelength (i), the following relationship is used:

$$\text{R.S.}_{\cdot i} = \left[\frac{\sum_n \left[A_i(k) - \overline{A}_i \right]\left[C(k) - \overline{C} \right]}{\sum_n \left[C(k) - \overline{C} \right]^2} \right] \left(1 - \overline{C}\right) + \overline{A}_i \tag{7}$$

where

$A_i(k)$ represents the absorbance signal for the ith wavelength and the kth sample (where $k = 1-n$).

$\overline{A}_i$ is the average absorbance signal for all samples within the training set, for the ith wavelength.

$C(k)$ represents the concentration value (reference chemical value) as a weight fraction (relative to 1.0 as 100%), for the parameter of interest for the kth sample.

$\overline{C}$ is the average concentration (as a weight fraction) for all samples within the training set.

n represents all the samples within the teaching set.

APPLICATIONS OF CHEMOMETRICS TO SPECTROSCOPIC/ SPECTROMETRIC MEASUREMENT TECHNIQUES

MS/MS

Second-Order Calibration

The application of second-order calibration to MS/MS data is demonstrated by Y. Wang et al. [119]. This work describes mathematics for calibration using second-order data. The paper describes nonbilinear rank annihilation (NBRA), rank linear additivity (RLA), net analyte rank (NAR), and net analyte signal (NAS). The matrix computations associated with these methods are delineated, and a theoretical analysis of NBRA is given.

UV/VIS

General Use

U. Horchner and J. Kalivas [120] utilize simulated-annealing (SA)–based optimization for strategies to select wavelengths used in calibrating ultraviolet-visible spectra. A detail of search strategy is delineated and results displayed. The authors conclude that the SA approach is superior to other methods of wavelength selection, including forward selection, backward elimination, and genetic algorithms. C. Jares-Garcia and B. Medina [121] present a method for the rapid prediction of a variety of parameters in red wines (i.e., total acidity, volatile acidity, pH, free sulfur dioxide, tannins, and total anthocyans) from UV/Vis spectral data. Partial least squares calibration was used to generate a model relating the spectral data to each parameter of interest. Low errors of prediction were obtained in young red wines; heterogeneous red wine calibrations also demonstrated good prediction quality. The method was then applied to a mixture of all types of red wines, with the results acceptable for each parameter. The relative prediction errors (1 sigma standard error of prediction) as presented were: total acidity, 5.0%; volatile acidity, 9.9%; free sulfur dioxide, 29%; pH, 3.3%; tannins, 11%; and total anthocyans, 22%.

Evolving factor analysis was used with UV absorption spectra of the photooxidation of cinnamaldehyde in methanol by Ceppan and others [122]. Three component spectra were isolated. A PCR model was developed for the prediction of *trans-* and *cis*-cinnamaldehyde and the di-methyl acetal of cinnamaldehyde. Fourth-derivative UV spectroscopy is combined with partial least squares for the determination of the composition of protein mixtures, an example for bovine caseins is demonstrated by G. Arteaga and coworkers [123]. This technique yielded acceptable standard error of prediction and correlations for three proteins in 4 M guanidine solutions. The method is proposed as a replacement for gel electrophoresis.

S. Salman and coworkers [124] have used UV-visible spectra of naphthaldehyde Schiff bases. Regression of the principal components of a set of experimental solvent parameters against the UV extinction coefficients allowed prediction of expected UV intensities that were in good agreement with observed values. The model could be used to predict the UV spectrum of these bases in other solvents.

Near-Infrared

General Use

Aromatics have been measured in gasoline and kerosene using multivariate modeling. NIR is reported to have diminished the time to analysis by 90%. Seip [125] has reported this work in a 1994 published industrial report. Techniques for detecting and adjusting for nonlinearity in NIR data is demonstrated by S. D. Oman et al. [126] A new regression method for addressing the problems associated with nonlinear NIR data is suggested. The paper proposes a method referred to as *principal component shrinkage* (PCS). Incorporated into the method is an extension of the Stein estimate. This statistic is reported by the authors as first presented in 1960 (W. James and C. Stein, *Proc. Fourth Berkeley Symposium on Math. Stat. Prob.,* Vol. 1, pp. 361–379, Berkeley, CA: University of California Press, 1960).

P. Brown [127] describes a method for exploring the linearity between spectrophotometer response (as absorbance) at specific wavelengths and composition in multivariate systems. This technique is used to prescreen spectral regions where linearity in response is reasonably certain. Preselection of regions with linear response provides a valuable screening method for selection of data regions to use for partial least squares (PLS) calibration or other linear regression methods. J. Sun [128] presents the use of principal component regression (PCR) in a slightly altered form termed *correlation principal component regression* (CPCR). NIR data is used to test traditional PCR, CPCR, and PLS in terms of prediction ability and the number of factors used to model the variance within the calibration data. Using two example data sets, the three methods were shown to have similar predictive performance, but the CPCR is reported to require fewer principal components (or factors) to achieve optimal prediction ability.

E. Vigneau et al. [129] propose an application of procrustean methods to mid-IR as well as NIR data. The authors define Procrustean methods as those that "allow the fitting of a given matrix to another given matrix observed on the same objects." This is a spectroscopic application of the standard rotation in subspaces problem, also known as the *orthoganal Procrustes problem.* Two methods of computation are proposed and compared: (1) the traditional approach using orthoganal constraints on the computed transformation matrix, and (2) a method where no constraints are applied to the computed transformation matrix. The aim was to reconstruct mid-infrared spectra from the near-infrared spectra. The unconstrained Procrustean analysis proved to be the more efficient method for both calibration and validation (verification). The authors propose that analysis of the transformation matrix between the two infrared regions allows the determination of NIR wavelengths and IR frequency data that correspond to the same chemical functionalities.

P. Brown [130] details the method of wavelength selection for calibration modeling using near-infrared (NIR) data. This paper compares a calibration method using wavelength selection to full-wavelength calibration approaches. In the wavelength selection method described, only those wavelengths are selected that are highly selective for each component of interest. The authors propose that the wavelengths chosen using the method "are little affected by interaction effects and consequent nonlinearities." The data set tested was composed of 125 samples of mixtures of three sugars, each varying at five concentration levels. The teaching set was a full factorial 5^3 design. The selection method outperformed the full-wavelength method for prediction error. The leave-one-out cross-validation produced results for partial least squares that may encourage overfitting, according to the authors. The authors compare results for each of the three sugars—sucrose, glucose, and fructose—for each of the calibration methods.

G. Downey et al. [131] compare stepwise multiple linear regression (SMLR) and principal component regression (PCR) techniques for calibration development using NIR reflectance data from dried grass silage A filter NIR instrument having 19 discrete interference filters was used for the calibration testing. Predictions of crude protein (CP) and in vitro dry matter digestibility (IVDMD) were determined using both calibration methods, with the authors claiming no improvement in standard error of prediction for the PCR model over the SMLR model. The authors did not see an improvement in prediction performance for either method by applying a data transformation to reduce granularity effects. The authors state that the PCR model may be more robust.

J. Tackett [132] demonstrates the use a reference cell measurement to adjust the measurements taken from a sample cell. Multivariate calibration models were computed by making measurements of a teaching set of samples with known physical parameter measurements. Contiguously, a measurement is made of a reference compound contained within a reference cell. The sample measurement is adjusted by a measurement of the reference sample, and the adjusted measurements are used for computing the multivariate calibration model. The reference hydrocarbon compounds used by the author were *p*-cymene and 2-methylhexane. Using NIR, P. Brown [133] describes methods for selecting wavelengths for multivariate calibrations that are unaffected by interactions in the samples used in the measuring instrument. The author describes the philosophy of using pure components for calibration requiring a reliance on Beer's law for a proper application.

J. Lebouille and W. Drost [134] describe their use of NIR for barley and malt analysis relative to the 1986–91 crop season. The authors discuss details for their sample selection, calibration, and standardization schemes. D. Haaland and H. Jones [135] explore the use of NIR for body fluid or tissue analysis. The authors cite the advantageous depth of penetration for NIR energy as a key attribute for using NIR for this type of analysis. By combining partial least squares (PLS) regression and Latin hypercube statistical design, the authors report precise near-infrared analysis of urea, creatinine, and NaCl in dilute aqueous solutions. Coefficients of determination (*R*-squared) values for the three analytes were all greater than 0.997.

T. Almoey and E. Haugland [136] compare a suite of calibration methods commonly used in calibrating NIR instruments. Data splitting and simulation experiments were performed to evaluate the properties of a variety of calibration techniques. The paper compares restricted principal component regression (RPCR), partial least squares regression (PLSR), and two other types of PCR: PCR1, selected according to the size of the eigenvalues, and PCR2, selected according to a *t*-statistic. The authors state that RPCR can be considered a compromise between PCR and PLSR, since the first principal components of both methods are identical. J. Lin and C. Brown [137] utilized multivariate regression techniques and NIR spectroscopy for the determination of the physicochemical properties of saline solutions. Aqueous solutions of 0–5 M NaCl were measured using two wavelength regions: 680–1230 nm and 1100–2500 nm, at a temperature range of 23–28.5°C. Regression methods such as MLR and PCR were used for calibration. The authors report very good agreement between predicted results and expected values from the literature. They report that NIR spectroscopy can be used for the determination of water concentration, density, refractive index, freezing point depression, relative viscosity, activity coefficient of NaCl, electrical conductance, and osmolality. A fiber-optic probe consisting of silica fibers can be used for remote sensing in the 1100–1870 nm region.

B. Wise [138] demonstrates the use of continuum regression (CR) for calibration of NIR spectral data, and the use piecewise direct standardization (PDS) for calibration transfer from one NIR spectrophotometer to a second, lower-resolution NIR instrument. Additional chemometric tools, such as PCA, are demonstrated for detection of multiple sensor failures. K. Alam and J. Callis [139] report on the intermolecular interactions between alcohols and water using near-IR spectroscopy in the 1100–1800 nm region. Multivariate regression methods were used to approximate speciation of the alcohol/water complexes present in different concentrations of the alcohol and water mixtures. The stoichiometry and equilibrium constants of the pure components and their complexes were derived. S. Gurden and coworkers [140] report the use of factor analysis of mid-IR spectra and apply this technique to 15 mixtures of three pollutants potentially found in the workplace.

Special Techniques

J. Sun [141] reports the results from a number of authors related to the overall modeling capabilities provided by using principal component regression (PCR), correlated PCR (CPCR), and partial least squares (PLS). The use of CPCR reportedly requires fewer factors than normal PCR or PLS to obtain the optimized calibration solution.

Image Analysis

D. Wienke et al. [142] analyzed macroscopic samples of household waste using NIR image analysis for the detection of plastics among nonplastic waste. A series of images taken at four discrete wavelength regions was used to reconstruct a three-dimensional composite profile. The profile served as a fingerprint for each sample. The data were processed using multivariate image rank analysis (MIRA) into a scalar value. The MIRA results were found to be independent of the physical sample size and geometric sample position within the camera image. MIRA was reported to be insensitive against image errors such as shadows or sample presentation geometry. Additional early references for the use of image analysis are given in Refs. 143–145.

Cluster Analysis

A paper by T. Naes and T. Isaksson [146] presents the use of cluster analysis in splitting calibration teaching sets into smaller subgroups, with improved linearity in each group. The method is based on a weighted average of the Mahalanobis distance statistic and a squared regression residual. The final solution is based on what the author's refer to as fuzzy clustering. A.M.C. Davies [147] utilizes canonical variables and principal component analysis for qualitative near-infrared analysis. The method is applied to discriminate between samples of chlorobenzoic acid isomers and benzoic acid derivatives.

Neural Networks

D. Wang et al. [148] present a comparison between classical least squares (CLS) and digital finite impulse response filtering, adaptive sampling, and artificial neural networks (ANN) in the ability to model and estimate concentrations of individual parameters in complex multicomponent mixtures. The improved predicted results for the ANN are demonstrated over CLS in a controlled laboratory experiment and in a process stack plume experiment.

Infrared

General Use

Factor analysis was used to determine both the presence and the quantity of antioxidants in extended thermoplastic rubber (ETPR) using FT-IR spectroscopy. The antioxidants measured were A01[thiodiethylenebis(3,5-di-tert-butyl-4-hydroxyhydrocinnamate)] and A02 [tetrakis [methylene(3,5-di-tert-butyl-4-hydroxyhydrocinnamate)]methane]. Individual concentrations of both antioxidants can be determined using a partial least squares (PLS-1) calibration algorithm. This paper, by W. J. Thaiman et al. [149] utilizes contour plots and trend data to illustrate performance of the method. B. Alsberg et al. [150] present a method of data compression using B-splines and its application to second-order FT-IR spectra. The B-splines compression method was selected for its suitability to modeling smooth-curved data. The authors state that the primary goals of data compression are to reduce the overall file size and to reduce the computation required to analyze the compressed data. The data reduction for the method presented was from 16 bits per uncompressed data element to 0.16 bits per compressed data element. A comparison of a principal component model generated using uncompressed and compressed data demonstrated that the models are comparable, even with the 100-times reduction in data size.

N. Dupuy et al. [151] demonstrate the use of MLR, PCR, and PLS for quantitative analysis of latex in paper coatings using FT-IR Spectroscopy and an ATR sample element. The styrene-butadiene latex on the surface of coated papers was quantified. The range of coating material varied between 5 and 25 parts, where 1 part is equivalent to grams of coating per 100 grams of mineral pigments. A relative error of 0.15 parts was found in the latex at 4 cm^{-1} resolution. The optimum error was obtained using PLS, with the resultant model performance yielding 0.156 parts for styrene determination and 0.161 parts for butadiene determination.

M. Blanco et al. [152] present a method for the determination of vulcanized rubber additives using FT-IR and partial least squares (PLS) regression for calibration. The method was reported to produce satisfactory performance for prediction of the individual industrial-grade additives. Parameters measured include: stearic acid, N-phentl-N'-isopropyl-p-phenylenediamine; aliphatic and aromatic processing oils; 1,2-dihydro-2,2,4-trimethyl-quinoline polymer; hindered bisphenol (NKF); nickel dimethyldithiocarbamate; trimethylolpropane trimethacrylate; and N-(cyclohexylthio)phthalimide. The batches were prepared as standard solutions using carbon tetrachloride. T. Marshall et al. [153] investigated the effects of spectral resolution on the performance of classical least squares (CLS) calibration in open-air long-path FT-IR spectrometry. The paper represents a report on the progress of combining multivariate calibration using FT-IR and open-air long-path FT-IR spectrometry technology for the measurement of volatile organic compounds (VOCs) in the atmosphere.

Part I of a two-part series by V. Bellon-Maurel et al. [154] describes the use of FT-IR and attenuated total reflectance (ATR) combined with multivariate calibration applied to the quantification of individual sugar concentrations (glucose, maltose, maltotriose, and maltodexrin) during a starch hydrolysis process. The solutions studied contained dry-matter concentrations in the 25–30% range. Glucose and maltose were predicted within desired accuracy; maltotriose and maltodextrin were not predicted within required accuracy. Predictions for glucose and maltose were tested for repeatability and reproducibility and were found to be acceptable. FT-IR attenuated total reflectance (ATR) was used to monitor sugar production from the hydrolysis of starch by V. Bellon-Maurel et al. [155]. In Part II of this two-part series, matrix interferents (i.e., proteins and salts), as well as temperature and wavelength selection, were evaluated for their effects on the prediction of sugars. The presence of proteins and salts in low concentrations were reported to be of no influence on sugar prediction. The influence of temperature had a dramatic effect upon sugar prediction. The authors propose that calibrations should be computed using samples measured not at a single temperature but at a variety of temperatures, such as the range expected during routine analysis. The variation in wavelengths selected for calibration also significantly affected sugar prediction results.

W. Bucsi [156] reported that FT-IR spectrometry combined with multivariate analysis provided a useful technique for the quality control of cold-rolling additives, antioxidant DBPC, and polybutene contamination in aluminum cold-rolling oils, with high precision. A standard deviation (for prediction) was reported to be 0.008 for the alcohol additive, 0.004 for the ester additive, 0.004 for the DBPC antioxidant, and 0.017 for the polybutene contamination. H. Heise and A. Bittner [157] apply multivariate calibration to the analysis of physiological samples using mid-IR measurement data. Human plasma analysis is performed for protein, glucose, cholesterol, triglycerides, and urea. The measuring instrument utilizes mid-infrared spectral data with an ATR cell. Partial least squares (PLS) regression was used to compute the calibration models. This work was republished as reference [158] using the NIR spectral region. The use of short-wave NIR (SW-NIR) resulted in mean-squared prediction errors (RMSEP) of 1.4% for total protein, 6.5% for total cholesterol, and 13.8% for triglycerides.

H. J. Luinge et al. [159] compare partial least squares (PLS) and artificial neural networks (ANN) as methods for multivariate calibrations using infrared data. The prediction results from both methods were equivalent; however, the computational time required for the PLS modeling approach is considerably shorter. J. Bak and A. Larsen [160] report the measurement of CO spectra in the range of 2.38–5100 mg/L (ppm) range. The spectra were measured at a resolution of 4 cm^{-1} in the mid-IR (2186–2001 cm^{-1}) region. A FT-IR instrument was used in combination with the partial least squares (PLS1) calibration to compute a CO model from the teaching set. The instrument response (as absorbance) versus concentration of CO was found to be highly nonlinear. A linearization routine was applied to the spectroscopic data prior to computing the calibration model. It was found that the integrated absorbance areas, rather than the absorbance values, were more useful for modeling. A fifth-order polynomial was used for the final model to calculate concentration from absorbance areas. The result was an improved calibration model with increased concentration prediction accuracy over the wide concentration range of CO.

A. Michell [161] used mid-IR spectra to rapidly assess pulpwood quality using samples of wood taken from Tasmanian *Eucalyptus globulus.* Measures of pulp wood quality included: yield, soda charge at kappa No. 15, total lignin, and hot water and alkali solubles. A PLS model was developed based on IR response information at 34 frequencies. The authors report success in this method of analysis (A40). D. Qin and P. Griffiths [162] have utilized PLS with vapor-phase IR spectra of mixtures of *o-, m-,* and *p*-xylene to perform quantitative multicomponent analysis. The errors in prediction for mixtures of the xylenes was 1% for each component over a temperature range of 35–65°C. When calibration spectra were all measured at a single temperature (35°C), rather than over the expected temperature range for the unknown samples, the error was increased by a factor of 5.

A modified least squares procedure was used in combination with FTIR spectra of coal samples by C. Alciaturi and M. Escobar [163] to derive improved practices for smoothing, band resolution, peak finding, and data compression for this application. A Ph.D. dissertation by P. Bhandare [164] describes the use of IR absorption spectroscopy and chemometrics for the determination of a variety of constituents in blood tissue. A technique termed *self-modeling multivariate mixture analysis* was presented by J. Guilment and coworkers [165] to extract pure-component information from complex sample mixtures, when spectra for the pure components within the mixture are unavailable. In the first experiment, differences between adjacent nonidentical monolayers of polymer laminate were distinguished. Spatial resolution was achieved for distinguishing between a 2–3-micron-thick inner layer from a 4-layer polymer laminate. A second experiment produced pure-component spectra from a KBr pellet made using a mixture of three compounds.

T. Visser and coworkers [166] use ANN and PLS combined with IR spectra to classify spectra of pesticides as either organophosphorus or nonorganophosphorus types. Classification by using these chemometric techniques resulted in a success rate approximately equal to results obtained from interpretation by spectroscopic experts. The difference between the methods tested lies primarily in the computational time required for optimizing the calibration model.

Multivariate calibration methods were compared for the determination of multiple constituents in blood serum using mid-IR spectra. P. Bhandare and coworkers [167] applied PLS-ANN for the determination of total proteins, glucose, and urea in human blood serum samples. The PLS-ANN calibration method did not yield significantly better prediction performance as compared with PLS and PCR.

IR Special

Window factor analysis (WFA) is tested as a general method for chromatographic and spectroscopic data. R. Brereton and coworkers [168] describe three approaches to WFA and compare these approaches as general data-analysis techniques. The authors report that WFA with mean-centered sequential scores is sensitive to peak shapes; mean-centered sequential loadings WFA is a method of choice when low variability exists in the nonsequential direction. WFA with no mean-centering (as is) is a poor performer when heteroscedastic noise is present.

IR Expert Calibration Systems

R. Aries and coworkers [169] report the development of an expert system for obtaining multivariate calibrations using PCR or PLS. The authors have created a rule-based system that produces calibrations automatically for FT-IR spectroscopic data and have provided a commercial software package. An expert calibration system has been proposed by Workman et al. that includes a three-phase development strategy [170]. This expert system would allow the use of MLR, PCA, PCR, PLS, LWR, and NNs in an automated fashion. The Phase I portion of this project is available as a commercial software package [25].

Vapor Phase

D. Qin [171] has produced a doctoral dissertation on the details of applying multivariate regression methods to multicomponent vapor-phase infrared spectra.

Cluster/Discriminant Analysis

L.-P. Choo et al. [172] describe a method for classifying human white and gray matter brain tissue using IR spectroscopy combined with a variety of multivariate techniques. The authors report that brain tissues identified by standard pathological methods as heavily, moderately, or minimally affected by Alzheimer's disease (AD) can be successfully classified at better than 90% accuracy using the IR-based classification method. Control tissues were distinguishable from known AD tissues with a success rate of 100% using the method. C. Gilbert and S. Kokot [173] used diffuse-reflectance infrared Fourier-transform spectroscopy (DRIFTS) to differentiate a series of cellulosic fabric samples, including dyed, undyed, and cotton based. Band assignments are discussed relative to previous work, and visual interpretation of spectra are explored. All spectra were qualitatively and quantitatively measured, with good differentiating results. The chemometric methods applied included principal component analysis (PCA) and soft independent modeling of class analogies (SIMCA). Discrimination of sample types was successful based on dye, fabric type, and textile-processing techniques.

H. Schultz et al. [174] reported the use of the IR region (700–3600 cm^{-1}) to examine a method for determining the information content of subsections of this spectral region. The method demonstrated the redundancy within selected spectral regions and provided a way to determine information content and compare the merits of prespecified wavenumber regions for their overall effective information content. M. Defernez and R. Wilson [175] utilized DRIFTS to determine the type of fruit present in samples of jam. The experiment had as its purpose the differentiation between strawberry-containing jams and nonstrawberry-containing jams. A group mean distance classification method was used following PCA or PLS decomposition of the data. Success in classification was near 100%. The use of ATR sample presentation resulted in a 91% classification success rate. The method was strongly influenced by the total sugar content of the jams tested, with the largest effect caused by the "normal" and "reduced" sugar content jams.

E. Bye [176] classified mixtures of crystalline and amorphous silica using IR spectroscopy with multivariate analysis. The PLS multivariate approach was used, and the author reports the method can provide a low-cost, rapid, and robust method for quantitative analysis of complex silica dust mixtures. Samples of quartz crystals were analyzed using infrared spectroscopy principal component analysis (PCA) and quantitative multivariate analysis by P. Zecchini et al. [177]. The calibration models were used to classify and quantify the sample population. The absorption spectra were measured from 3300 to 3600 cm^{-1}, with reference chemical analysis performed using inductively coupled argon plasma spectrometry (ICP).

FT-IR spectroscopy and chemometrics were used by M. Defernez and coworkers [178] to perform discriminant analysis to classify fruit purees into three predetermined group classifications: apple, raspberry, and strawberry. The groups were classified with 100% success. A separate discriminant analysis was performed to determine whether fresh or freeze-thawed fruit samples could be distinguished. This experiment met with a 98.3% success rate for strawberry samples and 75% success for raspberry samples. A third experiment attempted to discriminate the fruit from each group into two levels of ripeness. Results for level-of-ripeness classification resulted in 92.5% correct group for raspberry, but strawberry samples were not distinguishable. Experiment four was a test as to the classification of apple samples according to added (or none) sulfur dioxide, producing 90% correct classification. A set of apple samples was tested for a fifth experiment as to variety classification; 86% of samples tested were correctly identified as Bramley or not.

T. Visser and coworkers [179] have studied ANN and PLS regression for computerized studies of interpretive IR spectra. Classification by skilled interpreters was superior in all cases and experiments performed using the particular chemometric method proposed by the authors. ANN and PLS were nearly equivalent, with the exception that ANNs exhibit a higher score for individual band recognition. FT-IR and factor analysis (FA) were used to evaluate the potential for measuring the hydration states of four different natural and synthetic nucleic acids by W. Pohle and coworkers [180]. In a set of classification experiments, H. Hobert and coworkers [181] used PCA with IR spectral data to classify Saale River sediments of German origin.

Spectral Searching Algorithms

The authors introduce a new spectral retrieval algorithm termed the *adequate peak search* (APS) method. W. G. De Ruig and J. M. Weseman [182] present a definition for the APS algorithm as used for interpretation of infrared spectra. The authors evaluate the method for selectivity, trueness, ruggedness, and false results. The APS method is compared to "classical" interpretation of infrared spectra; the concepts of limit of detection, optimization, and automation are discussed relative to APS. The detection limit for APS was reported to be four to seven times lower than that attained by a number of experts.

SPECIALIZED CHEMOMETRIC TECHNIQUES

Calibration Transfer

Calibration transfer methods involve the computation of calibration models using data from a primary scanning or interference filter spectrometer, and transferring the calibration with a set of samples measured on a second (or secondary) instrument [183]. With this calibration transfer method, bias and/or slope adjustment is often required to minimize the residuals from the second instrument and to adjust the model for the transfer procedure. This practice increases overall error of analysis on the second instrument, especially when a broad range of samples are measured. Calibrations are generally developed using 50–500 samples that define the dimensions of the vectors for the analytical line in a calibration model; often a statistically *insignificant* set of 10–20 samples is used to adjust the bias and slope. Based on statistical criteria [63], slope adjustments should never be made, and bias adjustments are appropriate only after considerable testing. However, once a bias and or slope is made for any model, a separate test or validation set must be used to determine the prediction accuracy.

A second approach to calibration transfer requires sets of specially prepared reference samples (or a single special sample) measured on a primary, or reference, instrument (on which the calibration model was developed) and a secondary, or transfer, instrument. The spectra of the same samples from each instrument are used to adjust the spectra of the second instrument to "look" virtually the same as those obtained from the original or primary instrument [184,185]. The problem with this method (the *golden instrument* approach) is that the primary or reference instrument as well as the reference samples will change, so common samples measured with both instruments over time will not be related to the original model produced on the primary instrument. Likewise, if the primary instrument is repaired, it may only approximate its original spectral performance. Thus over time libraries of sample spectra will have no direct relationship to any existing physical instrument or prepared samples. Bouveresse et al. have described a method for instrument standardization that is a modification of the algorithms previously demonstrated but that follows the reference instrument approach [186].

A third approach to calibration transfer uses chemometric techniques to reduce the impact, within a model, of normal instrument-to-instrument variation on analytical precision [187–189]. Calibration methods that include interinstrument variations have the quality of producing models less sensitive to differences between spectrophotometers; these models make the requirements of instrument standardization potentially less important if equal performance can be achieved by using these techniques. Such modeling techniques decompose full spectral information into smaller spectral regions that are less subject to between-instrument variations than other regions of the spectrum. The disadvantage of these techniques is that areas of the spectrum most subject to intra- and interinstrument variation often contain valuable sample-concentration information. Such information should be included within a calibration for optimum analytical performance and signal-to-noise ratio; thus the method cannot be relied upon as a generalized approach to resolving the transfer problem.

A fourth series of approaches to the calibration transfer problem involves the use of internal standards such as polystyrene to adjust dispersive instruments for wavelength accuracy or precision. The limitation to this technique is that it does not address issues of photometric throughput (accuracy), variability in bandpass, differences in stray light characteristics, and detector nonlinearity. In addition, standards such as polystyrene provide wavelength or frequency calibration over limited spectral regions, defined by a few absorption bands. This approach does not ensure that the entire range of the instrument is calibrated or has linear dispersion.

Instrument Standardization

A more global approach to standardize spectrophotometers (make instruments more alike) was developed. If instruments are alike and remain stable enough, calibration transfer no longer becomes an analytical performance issue. Ganz et al. [190] developed an approach to account for both the major and the minor manufacturing and/or time-based variations between instruments. Although individual instrument noise characteristics are known to cause difficulty in calibration model transfer, it is generally considered that the major variable in transferring calibrations between instruments is the difference in wavelength calibration (or wavelength accuracy) between spectrophotometers [191–193]. This is considered to be true for both filter and continuous scanning instruments. Other problems requiring resolution are the stability of internal standards, linearity in dispersion, and band shape in dispersive instruments.

Workman and Coates [194] present a comprehensive approach to instrument characterization, leading to instrument standardization and calibration transferability involving five steps, a key part being the standardization of the instrument. This last factor is addressed to ensure that spectra produced from different instruments of the same design are essentially identical. The proposed five-step approach leading to transferability includes: (1) characterization and standardization of the instrument function, (2) stray-light characterization and correction, (3) sample cell photometric corrections, (4) continuous background referencing for lamp, fiber, and detector effects, and (5) chemometric tools for optimizing model performance.

A method for standardizing second-order instruments is presented by Wang and Kowalski [195] using GC/MS and LC/UV data. This paper proposes the use of the Gauss–Newton method and a common standard sample to reduce variation between sample measurements using second-order methods. A critical ingredient to the success of this method is the design of the common standard sample. More recently, a method presented by Wang and coworkers [196] involves a correction factor that can be calculated to accommodate for nonlinear instrument responses in calibration transfer, most often due to stray light. The correction factor is termed the *additive background* term and is applied with the method delineated in Ref. 189.

Continuum Regression (CR)

Three excellent references describing the background for CR include the following: M. Stone and R. J. Brooks [197] discuss the concept of continuum-type regression as early as 1990 in the *Journal of the Royal Statistical Society*. A. Lorber, L. E. Wangen, and B. R. Kowalski [198] describe this technique in the *Journal of Chemometrics*. B. M. Wise and N. L. Ricker [199] demonstrate the use of continuum regression "for the identification of finite impulse response (FIR) dynamic models." In CR, a model is chosen for a particular application using a minimum value selected over a response surface comprised of the abscissa (*x*-axis) as a power of the singular value decomposition (SVD) values for PCR, PLS, and MLR; the ordinate (*y*-axis) as the number of factors (or latent variables) in the model; and the *z*-axis as the prediction error sum of squares (PRESS) values corresponding to the *x,y* surface. In this way an optimum model can be selected based not only on the number of factors but also on the type of modeling algorithm.

ACKNOWLEDGEMENT

We would like to thank Debra Kaufman and Karen Lanigan of the Perkin-Elmer Library Service for performing the excellent quality literature searches for us.

REFERENCES

1. INNS (International Neural Networks Society), 1250 24th Street, NW, Suite 300, Washington, DC 20037; tel. 202-466-4667.
2. A. Einstein. Physics and Reality. *Journal of the Franklin Institute,* Philadelphia, pp. 313–382, March 1936.
3. S. Sekulic, B. R. Kowalski. MARS: A tutorial. *J. Chemom.* 6:199–216, 1992.
4. P. K. Hopke. Target transformation factor analysis. *Chemom. Intell. Lab. Syst.* 6:7–19, 1989.
5. P. J. Rousseeuw. Tutorial to robust statistics. *J. Chemom.* 5:1–20, 1991.
6. E. V. Thomas. A primer on multivariate calibration. *Anal. Chem.* 66:795A–803A, 1994.
7. R. P. Lippman. An introduction to computing with neural nets. *IEEE ASSP Magazine* April:4–22, 1987.
8. K. R. Beebe, B. R. Kowalski. An introduction to multivariate calibration and analysis. *Anal. Chem.* 57:1007A–1017A, 1987.
9. P. J. Gemperline. Mixture analysis using factor analysis I: Calibration and quantitation. *J. Chemom.* 3:549–568, 1989.
10. J. C. Hamilton, P. J. Gemperline. Mixture analysis using factor analysis I: Self-modeling curve resolution. *J. Chemom.* 5:1–13, 1990.

11. R. Browning. Multi-spectral imaging in materials microanalysis. *Surface Interface anal.* 20:495–502, 1993.
12. H. N. J. Poulisse, R. T. P. Jansen. Reduction of analytical variance by using a discrete-time data-weighting filter to estimate abrupt changes in batch-type processes, Parts 1 and 2. *Anal. Chim. Acta.* 151:433–446, 1983.
13. D. Haaland, E. Thomas. Partial least-squares methods for spectral analysis.1. Relation to other quantitative calibration methods and the extraction of qualitative information. *Anal. Chem.* 60:1193–1202, 1988.
14. D. Haaland, E. Thomas. Partial least-squares methods for spectral analysis 2. Application to simulated and glass spectral data. *Anal. Chem.* 60: 1202–1208, 1988.
15. A. Lorber, L. E. Wangen, B. R. Kowalski. A Theoretical Foundation for the PLS Algorithm. *J. Chemometrics.* 1:19–31, 1987.
16. M. R. Spiegel. *Theory and Problems of Statistics, Schaum's Outline Series.* New York: McGraw-Hill, 1961, 359 pp.
17. M. Fogiel, ed. *The Statistics Problem Solver.* New York: Research and Education Assoc., 1982, 1043 pp.
18. N. R. Draper, H. Smith. *Applied Regression Analysis.* New York: Wiley, 1966.
19. NAmICS (The North American Chapter of the International Chemometrics Society), contact David Lee Duewer, NAmICS Membership Secy., Bldg 222/Chemistry, Room B-156, NIST, Gaithersburg, MD 20899.
20. FACSS (Federation of Analytical Chemistry and Spectroscopy Societies) offers annual courses in statistics and chemometrics, contact FACSS Office, 198 Thomas Johnson Dr., Suite S-2, Frederick, MD 21702–0405; tel. 301–846–4797.
21. ACS (American Chemical Society), Department of Continuing Education, 1155 Sixteenth St., N.W., Washington, DC 20036; tel. 202-872-4600.
22. D. L. Massart, P. J. Lewi. *Principal Components, An Audiovisual Course.* New York: Elsevier Science, 1995.
23. Galactic Industries Corp., 395 Main Street, Salen, NH 03079.
24. The MathWorks, Inc., Cochituate Place, 24 Prime Parkway, Natick, MA 01760.
25. The Perkin-Elmer Corp., Real-Time Systems Division, 761 Main Avenue, Norwalk, CT 06859–0201.
26. MatLab, from The MathWorks, Inc., Cochituate Place, 24 Prime Parkway, Natick, MA 01760.
27. Eigenvector Technologies, P.O. Box 483, 196 Hyacinth, Manson, WA 98831.
28. Infometrix, Inc., 2200 Sixth, Suite 833, Seattle, WA.
29. A. S. Camo, Olav Tryggvasonsgt. 24, N-7011 Trondhei, Norway.
30. A. B. Umetri, 48, Box 7960, S-907 19 Umea, Sweden; and Umetrics Inc., 371 Highland Ave., Winchester, MA1890.
31. J. H. Krieger. Chemistry Sites Proliferate on the Internet's World Wide Web. *C&EN*, Nov. 13:35–46, 1995.
32. J. C. Miller, J. N. Miller, *Statistics for Analytical Chemistry.* New York: Wiley, 1984, 202 pp.
33. M. A. Sharaf, D. L. Illman, B. R. Kowalski. *Chemometrics.* New York: Wiley, 1986, 332 pp.
34. D. L. Massart, B. G. Vandeginste, S. N. Deming, Y. Michotte, L. Kaufman. *Chemometrics: A Textbook.* New York: Elsevier, 1988, 488 pp.
35. H. Martens, T. Naes. *Multivariate Calibration.* Chichester, U.K.: Wiley, 1989, 419 pp.
36. H. Mark, J. Workman. *Statistics in Spectroscopy.* Boston: Academic Press, 1991, 313 pp.
37. E. R. Malinowski, D. G. Howery. *Factor Analysis in Chemistry.* New York: Wiley-Interscience, 1980, 251 pp.
38. G. H. Golub, C. F. Van Loan. *Matrix Computations.* Baltimore: Johns Hopkins University Press, 1989, 642 pp.
39. K. R. Beebe, B. R. Kowalski. An introduction to multivariate calibration and analysis. *Anal. Chem.* 59:1007A–1017A, 1987.
40. Sonja Sekulic et al. Non-linear multivariate calibration methods in analytical chemistry. *Anal. Chem.* 65:835A–845A, 1993.
41. H. Mark, J. Workman. Statistics in spectroscopy. *Spectroscopy* 1986–1993 (series).
42. J. Workman, H. Mark. Chemometrics in spectroscopy. *Spectroscopy* 1993–1994 (series).
43. T. Naes, T. Isaksson. The chemometric space. *NIRnews,* 6 Charlton Mill, Charlton, Chichester, West Sussex PO18 0HY, UK.
44. *Chemometrics and Intelligent Laboratory Systems,* Elsevier Science Publishers, Amsterdam, The Netherlands.
45. *U-POST, The Unscrambler User's Club Newsletter,* Camo, Olav Tryggvasonsgt. 24 N-7011 Trondheim, Norway; tel (47) 73514966.
46. *The Mathworks Newsletter,* The Mathworks, Inc., 24 Prime Parkway, Natick, MA 01760-1500; tel. (508) 653-1415.
47. *Window on Chemometrics,* contact Judith Barnsby, The Royal Society of Chemistry, Thomas Graham House, Science Park, Milton Rd., Cambridge CB4 4WF, UK; tel. 44 (0) 223 420066. (Toll-free in U.S.: 1-800-473-9234).

REFERENCES

48. B. R. Kowalski. Chemometrics. *Anal. Chem.* 52:112R–122R, 1980.
49. I. E. Frank, B. R. Kowalski. *Anal. Chem.* 54:232R–243R, 1982.
50. M. F. Delaney. *Anal. Chem.* 56:261R–277R, 1984.
51. S. Ramos, K. R. Beebe, W. P. Carey, E. Sanchez, B. C. Erickson, B. E. Wilson, L. E. Wangen, B. R. Kowalski. *Anal. Chem.* 58:294R–315R, 1986.
52. S. D. Brown, T. Q. Barker, R. J. Larivee, S. L. Monfre, H. R. Wilk. *Anal. Chem.* 60:252R–273R, 1988.
53. S. D. Brown. *Anal. Chem.* 62:84R–101R, 1990.
54. S. D. Brown, R. S. Bear, Jr., T. B. Blank. *Anal. Chem.* 64:22R–49R.
55. S. D. Brown, T. B. Blank, S. T. Sum, L. G. Weyer. *Anal. Chem.* 66:315R–359R, 1994.
56. P. Geladi, K. Esbensen. The start and early history of chemometrics: Selected interviews. Part 1. *J. Chemometrics.* 4:337–354, 1990.
57. K. Esbensen, P. Geladi. The start and early history of chemometrics: Selected interviews. Part 2. *J. Chemometrics.* 4:389–412, 1990.
58. I. E. Frank, J. Feikema, N. Constantine, B. R. Kowaslki. Prediction of product quality from spectral data using the partial least-squares method. *J. Chem. Inf. Comput. Sci.* 24:20–24, 1984.
59. B. R. Kowalski, M. B. Seasholtz. Recent developments in multivariate calibration. *J. Chemometrics* 5:129–145, 1991.
60. A. Hoskuldsson. PLS Regression Methods. *J. Chemometrics* 2:211–228, 1988.
61. W. F. McClure. *Spectrosc. Tech. Food Anal.* New York: VCH, 1994, pp. 13–57.
62. J. D. Kirsch, J. K. Drennen. Near-infrared spectroscopy: applications in the analysis of tablets and solid pharmaceutical dosage forms. *Appl. Spectrosc. Rev.* 30:139–174, 1995.
63. The American Society for Testing and Materials (ASTM) Practice E1655–94. *ASTM Annual Book of Standards.* West Conshohocken, PA 19428–2959, Vol. 03.06, 1995.
64. AOAC. *Official Methods of Analysis.* 15th ed. Arlington, VA 22201–3301, 1990, pp. 74–76.
65. P. Williams, K. Norris, eds. *Near-Infrared Technology in the Agricultural and Food Industry.* St. Paul, MN: American Association of Cereal Chemists, 1987, p. 291.
66. J. Sustek. Method for the choice of optimal analytical positions in spectrophotometric analysis of multicomponent systems. *Anal. Chem.* 46:1676–1679, October 1974.
67. R. N. Cochran, F. H. Horne. Statistically weighted principal component analysis of rapid scanning wavelength kinetics experiments. *Anal. Chem.* 49:846–853, May 1977.
68. D. Metzler, C. M. Harris, R. L. Reeves, W. H. Lawton, M. S. Maggio. Digital analysis of electronic absorption spectra. *Anal. Chem.* 49:864A–874A, September 1977.
69. M. K. Antoon, J. H. Koenig, J. L. Koenig. Least squares curve-fitting of Fourier transform infrared spectra with applications to polymer systems. *Appl. Spectrosc.* 31:518–524, 1977.
70. M. K. Antoon, L. D'Esposito, J. L. Koenig. Factor analysis applied to Fourier transform infrared spectra. *Appl. Spectrosc.* 33:351–357, 1979.
71. W. R. Hruschka, K. H. Norris. Least-squares curve fitting of near infrared spectra predicts protein and moisture content of ground wheat. *Appl. Spectrosc.* 36:261–265, 1982.
72. D. M. Haaland, R. G. Easterling. Application of new least-squares methods for the quantitative infrared analysis of multicomponent samples. *Appl. Spectrosc.* 36:665–672, 1982.
73. H. J. Kisner, C. W. Brown. Kavarnos simultaneous determination of triglycerides, phospholipids, and cholesterol esters by infrared spectrometry. *Anal. Chem.* 54:1479–1485, August 1982.
74. W. Lindberg, J.-A. Persson, S. Wold. Partial least-squares method for spectrofluorimetric analysis of mixtures of humic acid and ligninsulfonate. *Anal. Chem.* 55:643–648, April 1983.
75. I. E. Frank, J. H. Kalivas, B. R. Kowalski. Partial least-squares solutions for multicomponent analysis. *Anal. Chem.* 55:1800–1804, September 1983.
76. M. A. Maris, C. W. Brown, Lavery. Nonlinear multicomponent analysis by infrared spectrophotometry. *Anal. Chem.* 55:1694–1703, September 1983.
77. H. J. Kisner, C. W. Brown, G. J. Kavarnos. Multiple Analytical frequencies and standards for the least-squares spectrometric analysis of serum lipids. *Anal. Chem.* 55:1703–1707, September 1983.
78. D. E. Honigs, J. M. Freelin, G. M. Hieftje. Hirschfeld near-infrared reflectance analysis by Gauss-Jordan linear algebra. *Appl. Spectrosc.* 37:491–497, 1983.
79. M. Otto, W. Wegscheider. Spectrophotometric multicomponent analysis applied to trace metal determinations. *Anal. Chem.* 57:63–69, January 1985.
80. S. D. Frans, J. M. Harris. Selection of analytical wavelengths for multicomponent spectrophotometric determinations. *Anal. Chem.* 57:2680–2684, November 1985.
81. I. A. Cowe, J. W. McNicol. The use of principal components in the analysis of near-infrared spectra. *Appl. Spectrosc.* 39:257–266, Mar./April 1985.
82. S. Kawata, K. Komeda, K. Sasaki, S. Minami. Advanced algorithm for determining component spectra based on principal component analysis. *Appl. Spectrosc.* 39:610–614, July/Aug. 1985.
83. P. Geladi, B. Kowalski. An example of 2-block predictive PL-SR with simulated data. *Analytica Chimica Acta* 185:19–32, 1986.
84. H. Mark. Comparative study of calibration methods for near-infrared reflectance analysis using a nested experimental design. *Anal. Chem.* 58:2814–2819, November 1986.

85. S. Wold, P. Geladi, K. Esbensen, J. J. Ohman. Multi-way PCS- and PLS-analysis. *Chemometrics* 1:41–56, January 1987.
86. D. Haaland, E. Thomas. Partial least-squares methods for spectral analysis.1. Relation to other quantitative calibration methods and the extraction of qualitative information. *Anal. Chem.* 60:1193–1202, June 1988.
87. D. Haaland, E. Thomas. Partial least-squares methods for spectral analysis. 2. Application to simulated and glass spectral data. *Anal. Chem.* 60:1202–1208, June 1988.
88. A. Lorber, L. E. Wangen, B. R. Kowalski. A theoretical foundation for the PLS algorithm. *J. Chemometrics* 1:19–31, January 1987.
89. H. Mark, J. Workman, Jr. *Spectrosc.* 2(1):58–59, 1987a.
90. H. Mark, J. Workman, Jr. *Spectrosc.* 2(2):60–64, 1987b.
91. H. Mark, J. Workman, Jr. *Spectrosc.* 2(3):47–49, 1987c.
92. D. E. Honigs, G. M. Hieftje, H. L. Mark, T. B. Hirschfeld. Unique sample selection via near-infrared spectral subtraction. *Anal. Chem.* 57:2299–2303, 1985.
93. J. Workman. Near-infrared spectroscopy: A comparison of random versus spectrally selected calibration sets for use in forage quality evaluation. In *Proc. American Forage and Grassland Conference*, Athens, GA, 1986, 132–136.
94. S. M. Abrams, J. S. Shenk, M. O. Westerhaus, F. E. Barton II. Determination of forage quality by near-infrared reflectance spectroscopy: Efficacy of broad-based calibration equations. *J. Dairy Sci.* 70:806, 1987.
95. H. L. Mark, D. Tunnell. Qualitative near-infrared reflectance analysis using Mahalanobis distances. *Anal. Chem.* 57:1449–1456, 1985.
96. H. Mark. Use of Mahalanobis distances to evaluate sample preparation methods for near-infrared reflectance analysis. *Anal. Chem.* 59:790–795, 1987.
97. W. R. Windham, D. R. Mertens, F. E. Barton II. Protocol for NIRS calibration: Sample selection and equation development and validation. In *Agric. Handbook No. 643*. Washington, DC: Agric. Res. Serv., USDA, 96–103, 1989.
98. H. Mark. Normalized distances for qualitative near-infrared reflectance analysis. *Anal. Chem.* 58:379–384, 1986.
99. P. J. Gemperline, L. D. Webber, F. O. Cox. Raw materials testing using soft-independent modeling of class analogy analysis of near-infrared reflectance spectra. *Anal. Chem.* 61:138–144, 1989.
100. J. S. Shenk, M. O. Westerhaus. Population definition, sample selection, and calibration procedures for near infrared reflectance spectroscopy. *Crop Sci.* 31:469–474, 1991.
101. D. E. Honigs, G. M. Hieftje, T. Hirschfeld. Number of Samples and wavelengths required for the training set in near-infrared reflectance spectroscopy. *Appl. Spectrosc.* 38:844–847, 1984.
102. D. E. Honigs, T. B. Hirschfeld, G. M. Hieftje. Near-infrared spectrophotometric methods development with a limited number of samples: Application to carbonate in geological Samples. *Appl. Spectrosc.* 39:1062–1064, 1985.
103. D. E. Honigs, J. M. Freelin, G. M. Hieftje, T. B. Hirschfeld. *Appl. Spectrosc.* 37:491, 1983.
104. D. E. Honigs, G. M. Hieftje, T. B. Hirschfeld. *Appl. Spectrosc.* 38:317, 1984.
105. ASTM. American Society for Testing and Materials, 1988. Method E 1252. *Standard Practice for General Techniques for Qualitative Infrared Analysis*. ASTM Committee E 13, 1916 Race Street, Philadelphia, PA 19103, 1998.
106. ASTM Committee E13.11. *Practice for Near Infrared Qualitative Analysis*. ASTM Committee E 13, 1916 Race Street, Philadelphia, PA 19103.
107. R. G. Whitfield, M. E. Gerber, R. L. Sharp. *Appl. Spectrosc.* 41:1204–1213, 1987.
108. P. C. Mahalanobis. *Natl. Inst. Science Proc.* 2:49, 1936.
109. D. C. Hoaglin, R. E. Welsch. The Hat Matrix in Regression and ANOVA. *Am. Statistician* 32(1):17–22, 1978.
110. N. R. Draper, J. A. John. Influential observations and outliers in regression. *Technometrics.* 23(1):21–26, 1981.
111. R. D. Cook, S. Weisberg. Characterizations of an empirical influence function for detecting influential cases in regression. *Technometrics.* 22(4):495–508, 1980.
112. J. C. Reid, E. C. Wong. Data-reduction and -search system for digital absorbance spectra. *Appl. Spectrosc.* 20:320–325, 1966.
113. P. M. Owens, T. L. Isenhour. Infrared spectral compression procedure for resolution independent search systems. *Anal. Chem.* 55:1548–1553, 1983.
114. K. Tanabe, S. Saeki. Computer retrieval of infrared spectra by a correlation coefficient method. *Anal. Chem.* 47:118–122, 1975.
115. L. V. Azarraga, R. R. Williams, J. A. de Haseth. Fourier encoded data searching of infrared spectra (FEDS/IRS). *Appl. Spectrosc.* 35:466–469, 1981.
116. J. A. de Haseth, L. V. Azarraga. Interferogram-based infrared search system. *Anal. Chem.* 53:2292–2296, 1981.

REFERENCES

117. D. E. Honigs, G. M. Hieftje, T. Hirschfeld. A new method for obtaining individual component spectra from those of complex mixtures. *Appl. Spectrosc.* 38:317–322, 1984.
118. G. J. Kemeny, D. L. Wetzel. On the usage of near-infrared spectral reconstruction. *Appl. Spectrosc.* 41:161–163, 1987.
119. Y. Wang, O. S. Borgen, B. R. Kowalski, M. Gu, F. Turecek. *J. Chemometrics* 7(2):117–130, 1993.
120. U. Horchner, J. H. Kalivas. Simulated-annealing-based optimization algorithms: Fundamentals and wavelength selection applications. *J. Chemometrics* 9:283–308, 1995.
121. C. Garcia-Jares, B. Medina. *Analyst* 120:1891, 1995.
122. M. Ceppan, R. Fiala, V. Brezova, J. Panak, V. Motlikova. *Chem. Pap.* 48:25, 1994.
123. G. E. Arteaga, Y. Horimoto, E. Li-Chan, S. Nakai. *J. Agric. Food Chem.* 42:1938, 1994.
124. S. R. Salman, A. G. Petros, B. C. Sweatman, J. C. Lindon. *Can. J. Appl. Spectrosc.* 39:1, 1994.
125. K. L. Seip. *Report NTIS, Order PB94-132123,* 1994, 41 pp.
126. S. D. Oman, T. Naes, A. Zube. *J. Chemometrics* 7(3):195–212, 1993.
127. P. J. Brown. *J. Chemometrics* 7:255–265, 1993.
128. J. Sun. A correlation principal component regression analysis of NIR data. *J. Chemometrics* 9:21–29, 1995.
129. E. Vigneau, M. F. Devaux, M. Safar. Application of procrustean methods to mid- and near-infrared spectral data. *J. Chemometrics* 9:125–135, 1995.
130. P. J. Brown. Wavelength selection in multicomponent near-infrared calibration. *J. Chemometrics* 6:151–161, 1992.
131. G. Downey, P. Robert, D. Bertrand, M-F Devaux. Dried grass silage analysis by NIR reflectance spectroscopy—A comparison of stepwise multiple linear regression and principal component techniques for calibration development on raw and transformed spectral data. *J. Chemometrics* 3:397–407, 1989.
132. J. E. Tackett. U.S. Patent no. 92–971886, 1992, 14 pp.
133. P. J. Brown. *Chemom. Intell. Lab. Syst.* 28:139, 1995.
134. J. L. M. Lebouille, W. C. Drost. Monograph for *Proceedings of the European Brew. Conv. Symposium on Instrumentation and Measurement,* 1992, pp. 14–22.
135. D. M. Haaland, H. D. T. Jones. *Proceedings of the 9th International Society for Optical Engineering (SPIE) Meeting.* International Conference on Fourier Transform Spectroscopy, 1993, pp. 448–449.
136. T. Almoey, E. Haugland. *Appl. Spectrosc.* 48:327, 1994.
137. J. Lin, C. W. Brown. *J. Near Infrared Spectrosc.* 1:109, 1993.
138. B. M. Wise. *Process Control Qual.* 5:73, 1993.
139. M. Alam, J. B. Callis. *Anal. Chem.* 66:2293, 1994.
140. S. P. Gurden, R. G. Brereton, J. A. Groves. *Chemometrics Intell. Lab. Syst.* 23:123, 1994.
141. J. Sun. *J. Chemometrics* 9:21, 1995.
142. D. Wienke, W. van den Broek, L. Buydens. *Anal. Chem.* 67:3760, 1995.
143. P. Geladi. Multivariate image analysis. *Chemometrics Intell. Lab. Syst.* 14:375–390, 1992.
144. K. H. Esbensen, P. L. Geladi, H. F. Grahn. Multivariate image regression. *Chemometrics Intell. Lab. Syst.* 14(1992):357–374, 1992.
145. H. F. Grahn, J. Saaf. Multivariate image regression and analysis for clinical magnetic resonance images. *Chemometrics Intell. Lab. Syst.* 14(1992), 391–396, 1992.
146. T. Naes, T. Isaksson. Splitting of calibration data by cluster analysis. *J. Chemometrics* 5(1):49–65, 1991.
147. A. M. C. Davies. *Spectrosc. Eur.* 5:40, 1993.
148. C. D. Wang, W. T. Waler, R. H. Kagann. *Proceedings of the International Society for Optical Engineering (SPIE)* 2366:251, 1995.
149. J. Thaiman, A. Debska, A. Eckard, J. A. Weaver. *Proceedings of the 42nd International Wire Cable Sym.* 1993, pp. 547–551.
150. B. K. Alsberg, E. Nodland, O. M. Kvalheim. Compression of nth-order data arrays by B-splines. Part 2: Application to second-order FT-IR spectra. *J. Chemometrics* 8:127–145, 1994.
151. N. Dupuy, L. Duponchel, B. Amram, J. P. Huvenne, P. Legrans. Quantitative analysis of latex in paper coating by ATR-FTIR spectroscopy. *J. Chemometrics* 8:333–347, 1994.
152. M. Blanco, J. Coello, H. Iturriaga, S. Maspoch, E. Bertran. *Appl. Spectrosc.* 49:747, 1995.
153. T. L. Marshall, C. T. Chaffin, V. D. Makepeace, R. M. Hoffman, R. M. Hammaker, W. G. Fateley, P. Saarinen, J. Kauppinen. *J. Mol. Struct.* 324(1–2):19, 1994.
154. V. Bellon-Maurel, C. Vallat, D. Goffinet. *Appl. Spectrosc.* 49:556, 1995.
155. V. Bellon-Maurel, C. Vallat, D. Goffinet. *Appl. Spectrosc.* 49:563, 1995.
156. W. G. Busci. *Lubr. Eng.* 51:131, 1995.
157. H. M. Heise, A. Bittner. *J. Mol. Struct.* 348:127, 1995.
158. A. Bittner, R. Marbach, H. M. Heise. *J. Mol. Struct.* 349:341, 1995.
159. H. J. Luinge, J. H. van der Maas, T. Visser. *Chemometrics Intell. Lab. Syst.* 28:129, 1995.
160. J. Bak, A. Larsen. *Appl. Spectrosc.* 49:437, 1995.

161. A. J. Michell. *Appita J.* 47:29, 1994.
162. D. Qin, P. Griffiths. *J. Quant. Spectrosc. Radiat. Transfer.* 52:51, 1994.
163. C. E. Alciaturi, M. E. Escobar. *Rev. Tec. Ing. Iniv. Zulia* 17:1, 1994.
164. P. S. Bhandare. Doctoral dissertation, Worcester Polytech. Inst. Worcester, MA, 327 pp., 1993.
165. J. Guilment, S. Markel, W. Windig. *Appl. Spectrosc.* 48:320, 1994.
166. T. Visser, H-J Luinge, J. van der Maas. *Proceedings of the 9th International Society for Optical Engineering (SPIE) Meeting, International Conference on Fourier Transform Spectroscopy,* 1993, pp. 232–3.
167. P. Bhandare, Y. Medelson, E. Stohr, R. A. Peura. *Appl. Spectrosc.* 48:271–3, 1994.
168. Brereton, Richard G., S. P. Gurden, J. A. Groves. *Chemometrics Intell. Lab. Syst.* 27:73, 1995.
169. R. E. Aries, D. P. Lidiard, R. A. Spragg. *Proceedings of the 9th International Society for Optical Engineering (SPIE) Meeting, International Conference on Fourier Transform Spectroscopy,* 1993, pp. 338–339.
170. J. Workman, B. Kowalski, P. Mobley. The design of an expert calibration system for spectroscopic based process analytical chemistry. *Proc. Instrument Society of America (ISA) Conference,* Toronto, Canada, 1995, pp. 97–106.
171. D. Qin. Doctoral dissertation, University of Idaho, 1994, 201 pp.
172. L.-P. Choo, J. R. Mansfield, N. Pizzi, R. L. Somorjai, M. W. Jackson, C. Halliday, H. H. Mantsch. *Biospectroscopy* 1:141, 1995.
173. Gilbert, C., S. Kokot. *Vib. Spectrosc.* 9:161, 1995.
174. H. Schultz, K. Tetzaff, L. Vogel. *J. Chemometrics* 9:143, 1995.
175. M. Defernez, R. H. Wilson. *J. Sci. Food Agric.* 67:461, 1995.
176. E. Bye. *Proc. 7th Int. Symp. on Inhaled Part. VII.* 1994, pp. 519–525.
177. P. Zecchini, K. Yamni, B. Viard, D. Dothee. *Proc. 48th IEEE Freq. Control Symp.,* 1994, pp. 91–98.
178. M. Defernez, E. K. Kemsley, R. H. Wilson. *J. Agric. Food Chem.* 43:109, 1995.
179. T. Visser, H. J. Luinge, J. H. van der Maas. *Anal. Chim. Acta.* 296:141, 1994.
180. W. Pohle, E. B. Starikov, W. Scheiding, A. Rupprecht. *Proc. 5th Int. Conf. Spectrosc. Biol. Mol.,* 1993, pp. 47–50.
181. H. Hobert, U. Schulz, J. Einax. *Acta Hydrochim. Hydrobiol.* 22:76, 1994.
182. W. G. De Ruig, J. M. Weseman. A new approach to confirmation by infrared spectrometry. *J. Chemometrics* 4:61–77, 1990.
183. J. Coates, T. Davidson, L. McDermott. *Spectroscopy* 7(9):40–49, 1992.
184. J. S. Shenk, M. O. Westerhaus. *Crop Sci.* 31:1694–1696, 1991.
185. J. S. Shenk, M. O. Westerhaus, W. C. Templeton. *Crop Sci.* 25:159–161, 1985.
186. E. Bouveresse, D. L. Massart, P. Dardenne. Modified algorithm for standardization of near-infrared spectrometric instruments. *Anal. Chem.* 67:1381–1389, 1995.
187. H. Mark, J. Workman. *Spectroscopy* 3(11):28–36, 1988.
188. K. G. Nordqvist. U.S. Patent no. 4,944,589, 1990.
189. Y. Wang, D. J. Veltkamp, B. Kowalski. *Anal. Chem.* 63:2750–2756, 1991.
190. A. Ganz, R. Hoult, D. Tracy. Standardizing and calibrating a spectrometric instrument. U.S. Patent No. 5,303,165, April 12, 1994.
191. J. S. Shenk. *Near-Infrared Reflectance Spectroscopy (NIRS): Analysis of Forage Quality.* Agriculture Handbook #643. Washington, D.C.: U.S. Department of Agriculture, pp. 41–44, 1985.
192. T. Hirschfeld, E. Stark. *Analysis of Foods and Beverages.* San Diego, CA: Academic Press, p. 510, 1984.
193. H. Mark, J. Workman, Jr. *Statistics in Spectroscopy.* Boston, MA: Academic Press, p. 284, 1991.
194. J. Workman, J. Coates. Multivariate calibration transfer. The importance of standardizing instrumentation. *Spectroscopy* 8(9):36–42, 1993.
195. Y. Wang, M. Stone, B. Kowalski. Standardization of second-order instruments. *Anal. Chem.* 65, 1993.
196. Z. Wang, T. Dean, B. Kowalski. *Anal. Chem.* 67:2379, 1995.
197. R. J. Brooks. *J. Royal Statistical Soc. B* 52:237–269, 1990.
198. A. Lorber, L. E. Wangen, B. R. Kowalski. *J. Chemometrics* 1:19, 1987.
199. B. M. Wise, N. L. Ricker. *J. Chemometrics* 7(1):1, 1993.

29.

REVIEW OF CHEMOMETRICS APPLIED TO SPECTROSCOPY: DATA PREPROCESSING

INTRODUCTION

Before beginning the next section, a caveat noted by Deming et al. is brought to the reader's attention [1]:

> Press et al. [2] have emphasized that data "consist of numbers, of course. But these numbers are fed into the computer, not produced by it. These are numbers to be treated with considerable respect, never to be tampered with, nor subjected to a numerical process whose character you do not completely understand. You are well advised to acquire reverence for data that is rather different from the 'sporty' attitude which is sometimes allowable, or even commendable, in other numerical tasks." Yet by and large within chemometrics, preprocessing often seems to be carried out with little understanding of its fundamental effect on the structure of the data.

The intent of this section of the review is to identify articles published in the recent literature pertaining to data-preprocessing methods and data transformations. In general data transformations are applied to each sample, and each sample would be transformed similarly whether a single sample or group of samples is transformed. Preprocessing methods alter each sample according to the characteristics of the entire data set, i.e., mean centering or autoscaling. Adding or subtracting a single sample from the data set would affect the preprocessing of each sample. In this review the preceding definitions will be followed. In addition, *pretreatment methods* may refer either to preprocessing methods or data transformations. Articles gathered in this search broke down into two different areas: data pretreatment used (1) prior to exploratory and classification methods that use only spectral or perhaps class information and (2) prior to calibration that involves predicting some quantity of a future unknown sample based on a measured spectral value. Data pretreatment as applied to exploratory and classification chemometric techniques will be considered first.

EXPLORATORY AND CLASSIFICATION: APPLICATION-DRIVEN DATA-PREPROCESSING RESEARCH

Methods That Use Noise Estimates

Of particular interest to spectroscopists is the fact that many chemometric methods assume homoscedastic noise (a constant noise level across the spectrum) when many spectroscopic methods produce data that contain heteroscedastic noise (errors vary in magnitude across the spectrum.) Toft and Kvalheim [3] clearly described the cause of heteroscedastic noise in infrared (IR) absorbance data as well as several potential problems that result. In addition a local windowing method was presented that estimates *pseudorank* (here loosely defined as the number of factors required to model the physically or chemically significant part of the data). In a later paper, Kvalheim et al. [4] proposed a modified Box–Cox power transform to convert heteroscedastic errors into homoscedastic errors. The authors described other benefits of the method: the enhancement of information contained in small peaks compared to large ones, and the preservation of strong linear correlations between signals. Also recognizing the problems raised by heteroscedastic noise, Paatero and Tapper [5] proposed "balanced scaling," which takes a data matrix, X, and scales in a three-step algorithm using a matrix of error estimates, s, of the same size as X. The authors suggested using balanced scaling prior to performing PCA and also proposed a new method of factorization that makes use of noise estimates to calculate factors that are more chemically relevant and meaningful.

The development of chemometric methods that use estimates of measurement errors means that spectroscopists and analytical chemists will need to do more work to obtain the required noise estimates. Hayashi and Matsuda [6] developed a method to estimate the relative standard deviation that would result in repeated measurements by high-performance liquid chromatography (HPLC.) The method for estimating measurement errors was compared to actual results obtained using HPLC and was found to estimate noise extremely well. The authors noted optimization of operating parameters as one of the most promising areas of future research.

Spectral Enhancement and Denoising

Research has also been conducted in the areas of spectral enhancement (narrowing the bandwidth of spectral bands) and noise reduction using numerical methods. Gorshkov and Kouzos [7] proposed a method called the data-reflection algorithm (DRA) that was used in the spectral enhancement of Fourier-transform ion cyclotron resonance (FT-ICR) mass spectroscopy and nuclear magnetic resonance (NMR) spectroscopy data. The method involved augmenting the time-reflected data with the original data to form a centerburst spectrum. The authors asserted that the method is analogous to the well-understood Michelson interferometer. In Michelson interferometry the centerburst interferogram is made up of the signal detected as a function of pathlength differences between light beam components. With the DRA method, the centerburst time-domain spectrum contains the phase information of the original data. The authors reported higher-resolving power compared to the original magnitude-mode FT spectrum and a higher signal-to-noise (S/N) ratio. Iwata and Koshouba [8, 9] published two papers involving a method of minimizing noise in a spectrum using the Fourier transformation, circulent matrices, and singular-value decomposition (SVD). The first paper described the algorithm and noted that little distortion of the signal waveform occurred during noise reduction. The second paper addressed difficulties in applying the method to continuum spectra and alleviated the problem of extensive computation time. Greek et al. [10] proposed a two-point maximum-entropy method to be used in S/N enhancement and/or deconvolution of Raman spectra. The method does not require a priori information and performed well in comparison to other commonly used methods.

Takeuchi et al. [11] examined the problem of spike noise and proposed robust summation as a solution. This article spurred on further comment by Hill [12], who suggested missing-point fitting as the optimal method for dealing with spike noise.

Comparison Papers

Several data-preprocessing comparison papers were written that involved applying many preprocessing methods on a single or small number of data sets to determine which performed best. Sanchez et al. [13] applied eight different preprocessing methods to several data sets collected using high-performance liquid chromatography coupled with diode array detection (HPLC-DAD.) Methods were compared in their ability to enhance detection of pure and impure zones in the chromatograms and spectra obtained by HPLC-DAD. In another study, Carlsson and Janne [14] studied several preprocessing methods in a classification study conducted to reduce or eliminate the need for experimenting on animals. Methods used first and second derivatives, multiplicative scatter correction (MSC), and piecewise multiplicative scatter correction (PMSC). The methods were applied to NIR data in the quality-control study of a pharmaceutical product. Deming et al. [1] explored the effects of closure, normalization, and ratioing on principal component analysis (PCA) of a simulated data set. It was found that in some cases the rank of the data increased following preprocessing. In some data sets the authors noted, "the rank of the . . . data set [was] increased by the constant length and maximum value transformations, even though a degree of freedom [was] removed from the description of each object in the data set" [2]. This problem is especially significant because one of the main goals of preprocessing is minimization of noise and instrumental effects so that the analyst can get at the chemical information. In pursuing this, preprocessing would ideally lower the rank of the data.

Derivatives

Derivatives are a commonly applied preprocessing technique in spectroscopy. The general rule of thumb is that first derivatives are useful in removing baseline offsets and that the second derivative corrects for baseline offsets and sloping baselines. From a calculation standpoint, Savitsky and Golay [15] delivered a landmark paper in 1962 that described what has become a standard method for smoothing spectra and calculating derivatives. Some years later, Gorry [16] extended this work to overcome the problem of truncating ends of the spectrum and also simplified computation for all derivatives. Holler et al. [17] used second derivatives in fitting curves. The advantage of the method was elimination of unnecessary parameters, which reduces the likelihood of overfitting the data and improves the extraction of chemical information from the raw data. The authors noted that the negative trade-off for the method is the potential for blurring profiles of small, overlapped peaks in the smoothing step. Windig [18] found similar advantages and disadvantages in a paper that took the second derivative of spectra as a step in pure-variable-based self-modeling mixture analysis. Windig noted that selection of window size was crucial in optimizing smoothing and preventing blurring of peaks.

Data Transformations

The use of data transformations is becoming increasingly important to the field of spectroscopy. A short list of these include the Hadamard transform, wavelet transform, and the now-commonplace spectroscopist's tool the Fourier transform (FT). Chapados et al. [19] used the Fourier deconvolution first proposed by Kaupinnen et al. [20] using a Cauchy–Gauss production function. The authors found that the Cauchy–Gauss product line shape more closely matches that of an experimental IR spectrum than functions proposed in earlier papers and is therefore more effective in enhancing spectra. Brereton [21] wrote a tutorial that described the mathematics behind the Fourier transformation as well as the instrumental applications of FT. The author discussed the use of pulsed FT methods in NMR and microwave research and the use of interferometric FT methods in IR and ultraviolet (UV) spectroscopy. In the applications section of this review, several papers will be discussed in which FT was used to reduce the size of the data.

Treado and Morris [22] described the Hadamard transform and explored possible uses of the transform in spectroscopy and analytical chemistry. In order to clearly describe the transform, the authors made comparisons to the Fourier transform, which is more generally known to spectroscopists and analytical chemists.

The wavelet transformation seems to be an area ripe with research possibilities for spectroscopists [25,26]. This is witnessed by the rise in the number of wavelet-related articles. As with any new research field, the aspiring newbie often faces an arduous task finding easily understood papers on the subject. Vidakovic and Muller [23] wrote *Wavelets for Kids: A Tutorial Introduction,* which fulfills the foregoing need (in addition to being kid tested and mother wavelet approved). A chapter by Press et al. [24] in *Numerical Recipes* also gives a useful overview of wavelets and provides PASCAL code so that the user may immediately begin to use the wavelet transformation. Source code for wavelet transforms in other computer languages is also available at several world wide web (WWW) sites [27–29]. In addition, *Wavelet Digest* [30] is an electronic journal with articles, conference announcements, questions, software, etc. *Mathsoft* also maintains a very useful resource page [31] that has many wavelet-related papers available for download via ftp (file transfer protocol).

Two papers were written on signal enhancement of flow injection analysis (FIA) data using digital filters. Lee et al. [32] used the Fourier transform generalized to also include the Gram and Meixner polynomials as basis functions. The goal of the paper was to find the basis function that most closely resembled the signal peak so that the differences between frequencies spanned by the signal were most sharply different from those of the noise. This increased the prospect of recovering the signal. Bos and Hoogendam [33] had similar goals in a paper that used the wavelet transform to minimize the effects of noise and baseline drift in determining peak intensities of FIA data. Research involving baseline drift showed that selection of the optimal frequency of the wavelet is dependent on the slope of the peak of interest.

Devaux et al. [34] recognized that a problem with some chemometric methods is the need to store prohibitively large amounts of data. The authors proposed use of the fast Fourier transform (FFT) followed by PCA to reduce the size of an NIR reflectance data set used in classification of wheat flours. The size of the data set was reduced considerably and classification remained accurate.

CALIBRATION-PREPROCESSING TECHNIQUES

Traditional Spectroscopic Data-Preprocessing Methods

The common linearization of reflectance (R) or transmission data (T) to absorbance is given by

$$A = -\log R_1 \quad \text{and} \quad A = -\log T_1 \tag{1}$$

where A is absorbance expressed as reflectance or transmission when R or $T = I/I_o$.

The absorbance ratio has been used for NIR data by K. Norris and Williams [35] and is the ratio of absorbance at one wavelength divided by the absorbance at a second wavelength. This ratio method is common to infrared spectroscopists because using it results in a term that compensates for baseline offset. The ratio of transmission data is also useful for normalizing or reducing baseline offset in spectroscopic measurements. A typical absorbance ratio is given as

$$\frac{A_{\lambda_1}}{A_{\lambda_2}} \tag{2}$$

The use of the moving-averaged-segment convolution (MASC) method for computing derivatives yields the following expression describing a first-derivative term:

$$A_{\lambda_2} - A_{\lambda_1} = A_{\lambda_{1+\delta}} - A_{\lambda_{1-\delta}} \tag{3}$$

where the derivative is given as the difference in absorbance values at two different wavelengths and the position of each wavelength is determined as the gap (d) distance + or – from a center wavelength (l_1). Ratios of derivative terms have been applied to NIR data and have empirically shown to produce lower prediction errors in special cases. The expression shown is the first-derivative ratio where two different center wavelengths have been used:

$$\frac{A_{\lambda_2} - A_{\lambda_1}}{A_{\lambda_4} - A_{\lambda_3}} = \frac{A_{\lambda_{1+\delta}} - A_{\lambda_{1-\delta}}}{A_{\lambda_{2+\delta}} - A_{\lambda_{2-\delta}}} \tag{4}$$

The MASC form of second derivative is shown in Eq. (5) [36,37]. In this case, the second-derivative term is defined by the sum of absorbances at two wavelengths ($l_{1+\delta}$ and $l_{1-\delta}$), minus twice the absorbance at a center wavelength (λ_1). In this case the second-derivative gap size is designated as δ (delta). The second-derivative preprocessing step is quite effective in removing slope and offset variations in spectral measurement baselines. It also "assists" the calibration mathematics in defining spectral regions where small response changes can be useful in calibration modeling. Without the use of derivatives, these regions would not be beneficial for use in calibration.

$$A_{\lambda_1} + A_{\lambda_3} - 2A_{\lambda_2} = A_{\lambda_{1-\delta}} + A_{\lambda_{1+\delta}} - 2A_{\lambda_1} \tag{5}$$

Second-derivative ratio preprocessing has also been demonstrated in the Norris regression method. Empirically this data-preprocessing technique has proved useful for some applications. The following relationship designates a second-derivative ratio with two different center wavelengths:

$$\frac{A_{\lambda_1} + A_{\lambda_3} - 2A_{\lambda_2}}{A_{\lambda_4} + A_{\lambda_6} - 2A_{\lambda_5}} = \frac{A_{\lambda_{1-\delta}} + A_{\lambda_{1+\delta}} - 2A_{\lambda_1}}{A_{\lambda_{2-\delta}} + A_{\lambda_{2+\delta}} - 2A_{\lambda_2}} \tag{6}$$

The traditional Kubelka–Munk expression [Eq. (7)] is used for samples exhibiting scatter and has proven beneficial for some measurements:

$$\frac{(1-R_{\lambda_1})^2}{2R_{\lambda_1}} \tag{7}$$

The log [Kubelka–Munk] expression [Eq. (8)] has been demonstrated and may empirically prove useful for some scattering sample data sets:

$$\log\left[\frac{(1-R_{\lambda_1})^2}{2R_{\lambda_1}}\right] \tag{8}$$

Noise Compensation in Modeling

Spectral interferences stemming from different spectral responses between instruments are complex and result from the convolution of the measured sample optical properties (e.g. refractive index, temperature, composition) and the total instrument function. The sample in effect becomes an active optical element of the instrument. Noise compensation [38–40] has been used by measuring calibration samples on different instruments while varying the measuring parameters (wavelength, temperature, sample presentation, etc.) for each instrument. If the type and magnitude of interferences were well understood, one could manually add these effects into a data set prior to modeling. Adding different types and magnitudes of instrument interferences (than those that exist in reality) reduces the quantity and quality of response data useful for calibration (thus reducing the model performance). The more exact the characterization of interference noise added during the modeling step, the more optimal the calibration performance becomes across instruments.

The generalized concept has existed for some time. U.S. Patent No. 4,944,589 [38] broadly describes:

> A method for reducing the susceptibility to interference of the measured value from a measuring instrument. . . . During the calibration procedure, a variation of one or several properties of the measuring instrument, *or of representations thereof in the processing of the measured values* [italics added], is introduced intentionally, said variation having the same order of magnitude as or being substantially greater than expected variations, when one or several test objects are being measured, without such information being supplied to the instrument.

Claim 4 states:

> The method as claimed in claim 2, the spectrometer consisting of a grating monochromator, wherein the wavelength is shifted by changing the angular position of the grating

of said monochromator or by changing the angular position of the grating of said monochromator *or by changing correspondingly said numerical representation of the angular position when calculating the wavelength, relative to the positions or values, for the main numbers of the measurements of the calibration test objects* [italics added].

In the body of this patent text it states:

Practical tests have shown that the interference thus intentionally introduced has implied a sharply increased insusceptibility to the much smaller variations in the measuring instrument parameter or parameters, which are normally expected during subsequent measurements of test objects of the same type as are used in the calibration procedure.

Extensions of this method to include errors associated with variation in operators, sample presentation, sample nonhomogeneity, and temperature (of sample and instrument) are discussed (and taught) in Refs. 39–41. Reference 41 adapts these techniques to octane measurement in gasoline (with the explicit addition of the terms *transmittance shifts, wavelength shifts, absorbance-baseline shifts, and absorbance-baseline tilts*).

One of the most popular NIR commercial software packages [42] contains another form of noise compensation referred to as "using a repeatability file in calibration." This technique generates a scaled noise vector, which when added into the calibration process will "minimize systematic differences among instruments . . . and make the calibration equation insensitive to changes in [sample] temperature *and instruments,* but not reduce the accuracy of the calibration." Although such tools are in common practice for food, pharmaceutical, and agricultural applications, their use has not been extended to petroleum products.

Reference 43 provides a basis for the statement that it is important to use regions in the teaching set spectral data (for a calibration) where the relationship applies as

$$\beta_i \frac{\delta_{s_i}}{\delta_{p_i}} = 0$$

where β_i is the regression coefficient at wavelength i, s_i is the "true" spectrum response at wavelength i, and p_i is the interference property associated with wavelength variation and photometric variation between (and within) instruments at wavelength i. It is implicit within the statement $\delta_{\beta_i} / \delta_{\lambda_i} = 0$ from Ref. 43 that

$$\beta_i \frac{\delta_{s_i}}{\delta_{\lambda_i}} = 0$$

is the desired condition for spectral data with respect to *the robustness condition* in calibration equations, relative to wavelength changes between (and within) instruments. Reference 41 extends this explanation to other interference properties related to octane analysis and other multivariate algorithms, although the details of explanation are not important for implementation of the method.

Trends in Data Pretreatment

In calibration, the expected benefits of data pretreatment are somewhat different from the goals of pretreatment prior to exploratory or classification methods. For example, with PCA the goal tends toward removing instrumental artifacts so that the chemical information in the spectra becomes more apparent and is maximized. In calibration, reference values as well as spectra may need to be pretreated, and the desire for predictive ability means that the data may be altered so that the spectral features in the usual sense are lost but predictive ability is improved.

Thomas [44] wrote a tutorial, "A Primer on Multivariate Calibrations," that explores calibration in general but has a good-sized discussion of data pretreatment. Several comparison papers have been written in which several methods were applied to a data set to determine the relative subsequent effect on predictive ability in calibration.

Davidian and Haaland [45] wrote a tutorial for dealing with data sets exhibiting heteroscedastic noise (nonconstant variance across the spectrum). For certain cases the authors recommended the use of generalized least squares (GLS) and the variance function estimation (VFE) method. This tutorial provides a foundation for dealing with situations in which the use of data preprocessing or transformations does not provide enough flexibility.

de Noord [46] studied the effect of data-preprocessing methods on the robustness and parsimony of multivariate calibration models. The point was made that proper data preprocessing eliminates nonrelevant or nonchemically related sources of variance and nonlinearities such that more parsimonious calibration models are constructed. In examining data consisting of NIR measurements of heavy oil products, the author studied the relationship between prediction errors and model complexity. With proper preprocessing, the bias decreased more sharply as parameters were added. In addition, once the optimal number of parameters was exceeded, the error associated with measurement noise increased more sharply. The author showed that reduction in bias through proper data preprocessing results in a less complex model. Although predictive ability may not necessarily be enhanced in the short term, a more parsimonious calibration model is expected to be more robust over time.

In a departure from the usual preprocessing literature, Jancar and Wegscheider [47] turn from primarily considering preprocessing of spectra and instead consider the importance of scaling the concentration values in partial least squares 2 (PLS2). PLS2 differs from PLS1 in that it performs calibration of multiple analytes simultaneously. The authors found that prediction of minor components is improved and prediction is more robust with proper scaling of the reference values.

Digital Filtering

As with the exploration and classification literature, much work has been done in calibration using digital filters as a preprocessing technique. Erickson et al. [48] discussed the similarities and differences between multivariate calibration methods and digital filters used in quantitative analysis. In particular, the authors examined the mathematics and statistics of the Kalman innovation filter as compared to calibration methods such as principal component regression (PCR) and classical least squares (CLS.)

Small et al. [49] employed Gaussian-shaped bandpass digital filters by using Fourier filter techniques to remove the absorbance associated with the background. The data set consisted of 300 calibration and 69 prediction samples of bovine plasma in which glucose was the analyte of interest. The samples were measured using an FT-NIR spectrometer. In order to deal with the complexity of the biological sample matrix, the authors used digital filtering coupled with PLS regression. Digital filtering was found to improve predictive ability, minimize the importance of selecting a spectral range, and reduce the number of factors required to build the optimal PLS model.

In another application of FT, Donahue et al. [50] used a limited number of Fourier coefficients to reduce data size and enhance the S/N ratio of UV measurements of deoxyribonucleotides. Multiple linear regression (MLR), PLS, and PCR were carried out using data in both the spectral and Fourier domains. Optimal results were obtained using PCR and PLS in the Fourier domain, although adequate predictive ability was found using simple MLR.

Mattu and Small [51] were able to isolate the spectral information related to a single analyte by applying a narrow-bandpass digital filter to a shortened FT-IR interferogram. The authors noted that since these filtered interferogram segments are analyte specific, univariate analysis may be used for quantitative analysis. The proposed method was found to perform similarly in predictive ability compared to conventional spectral calibration techniques, but was more efficient in terms of data storage and computation time.

In an application using a wavelet transformation, Stark et al. [52] performed quantitative analysis of 14 minerals contained in rock samples using FT-IR spectroscopy. The wavelet transform was used to roughly separate mineralogical information contained in the spectrum from measurement noise. The authors elected to ignore wavelet coefficients that varied too much in replicately measured samples, on the assumption those coefficients contained measurement errors. Samples that varied too little were also ignored, for they were believed to hold little discriminative or quantitative information. Using the remaining wavelet coefficients as the data set, the authors constructed an affine estimator for quantitation, and the method outperformed nonnegative least squares in many cases.

Two papers were published dealing with signal enhancement of electrochemical techniques by digital processing methods. Although the analytical methods veer away from spectroscopy, the problems posed by the data are similar to those encountered by spectroscopists. Economou and Fielden [53] compared the effectiveness of infinite impulse response filters and finite impulse response filters in denoising data collected by stripping voltammetry in flowing solutions. Chow et al. [54] used exponential low-pass filtering, moving-average smoothing, polynomial smoothing, and rectangular low-pass filtering to enhance the S/N ratio of potentiometric stripping analysis (PSA). A method involving Fourier transformation was proposed that resulted in a 2.3-times S/N ratio improvement while retaining a linear relation between peak area and concentration.

Mean Centering

Seasholtz and Kowalski [55] examined the implications of mean centering, one of the most commonly used preprocessing methods. The authors looked to error propagation formulas in order to build a theory concerning the effect of mean centering and then supported the work with a simulation study. The authors found that mean centering results in larger prediction errors when the data (1) varies linearly with concentration (2) has no baseline and (3) has no closure in the concentrations.

Diffuse Reflectance and Scatter Correction

Hellend et al. [56] examined several methods of correcting for light-scattering effects due to different sizes and shapes of particles. The authors included a theoretical discussion of the various types of multiplicative scatter correction (MSC) and the relationship to the standard normal variate (SNV) transformation. Eight different NIR reflectance and transmittance data sets were then used to investigate the effect on predictive ability using first derivatives, second derivatives, SNV, and three variations of MSC as pretreatment methods.

In dealing with NIR diffuse-reflectance data, Barnes et al. [57] noted problems with particle size, scattering, and multicollinearity. The authors proposed two mathematical transformations for dealing with this type of data. The standard normal variate (SNV) transformation is applied at each wavelength and serves to remove the multiplicative interferences of scatter and particle size. Detrending (DT) deals with curvilinearity of the baseline and baseline shifts. Several NIR diffuse-reflectance data sets were used to illustrate typical problems encountered, to show how the two transformations deal with those problems, and to provide a calibration set for testing the pretreatment methods' effect on predictive ability. Results showed that SNV and DT lowered prediction errors compared to using raw, first-derivative, or second-derivative data. Several years later the authors wrote a paper [58] that studied the effect of several scaling methods on NIR reflectance difference spectra. In particular, the authors indicated the relationship of SNV to multiplicative scatter correction (MSC), another popular preprocessing method commonly applied to NIR reflectance data. The authors also studied the best order in which to apply SNV and DT.

MSC was introduced into the spectroscopy literature by Isaksson and Naes [59]. In the original paper, the authors described the method and gave mention to the Kubelka–Munk transform [60,61], which is used to ensure good linear modeling. MSC was applied to several data sets and the relative predictive ability compared to the results using the raw data. Improvement in predictive ability was seen in all cases, but the biggest enhancements in predictive abililty were seen in data sets that had large variation in the constituents. The authors discussed the relation of MSC to other work and proposed several reasons for the improvement shown using MSC.

Piecewise multiplicative scatter correction (PMSC), a nonlinear version of MSC, was later proposed by Isaksson and Kowalski [62]. PMSC is similar to MSC except that a small moving window of wavelength channels is used to make linear regression fits to local regions of the spectra. In three NIR diffuse-transmittance data sets of meat, the predictive ability based on the root mean square error of prediction (RMSEP) was improved by as much as 31% using PMSC as opposed to MSC.

Other Preprocessing Techniques

In an approach different from the usual paradigm of preprocessing data prior to calibration, Karstang and Manne [63] proposed optimized scaling (OS-1 and OS-2), which combines the preprocessing step with the calibration step. Two variations of the method were presented. OS-1 was devised for the situation where all the constituent concentrations are known. OS-2 was developed for the more common case where the reference values for only one constituent is used in calibration. Both OS-1 and OS-2 were applied to several data sets and the results were compared to those obtained using the raw data, mean-centered data, and data pretreated using MSC. In a later study Isaksson et al. [64] compared OS-2 to MSC and several other preprocessing methods in a study involving eight data sets. The authors found OS-2 outperformed MSC in cases where the major constituents had large variations. In data sets where minor constituents were being predicted, MSC provided better predictive ability than OS-2.

Several other paper were written that blur the lines between preprocessing and calibration. These involve methods for performing first-order calibration in the presence of interferents not included in the calibration set. A typical approach in such a situation might be to progress to second-order methods, such as LC-UV, that have the second-order advantage of being capable of quantitation of an analyte in the presence of uncalibrated for interferents. Another approach might be to continue with first-order calibration but to collect more calibration samples that contain the new interferent. The following papers somehow attempt to eliminate or minimize the effect of uncalibrated interferents.

Karstang and Kvalheim [65] introduced a method involving local curve fitting combined with derivative spectroscopy. The authors reported good predictive ability of an analyte in the presence of an uncalibrated interferent. UV and X-ray diffraction (XRD) data sets were used in the study. In another paper, Karstang and Kvalheim [66] compared three methods of background correction that begins using a principal component model that describes the calibration space. This model was then combined with each of the three functions to determine which of the methods produced the smallest prediction errors. Methods used in the study were curve fitting, iterative target transformation analysis, and local curve fitting. IR, UV, and XRD data sets were used in the comparison study. Salit et al. [67] addressed the problem of background estimation and subtraction in inductively coupled plasma (ICP) emission spectroscopy and evaluated both a heuristic method and a statistical method for dealing with the problem.

Agbodjan and Rutan [68] proposed and demonstrated an information-based optimization of an adaptive Kalman filter for quantifying analytes in the presence of uncharacterized interferents. Kalman filtering has been used previously in similar applications, but it requires that the overlap-free region be known beforehand. The proposed method determines the overlap-free region based on statistics using the residuals.

REFERENCES

1. S. N. Deming, J. A. Palasota, J. M. Nocerino. The geometry of multivariate object preprocessing. *J. Chemo.* 7:393, 1993.

2. W. H. Press, B. P. Flannery, S. A. Teukolsky, W. T. Vetterling. *Numerical Recipes in Pascal: The Art of Scientific Computing*. New York: Cambridge University Press, 1989.
3. J. Toft, O. M. Kvalheim. Eigenstructure tracking analysis for revealing noise pattern and local rank in instrumental profiles: Application to transmittance and absorbance IR spectroscopy. *Chemo. Lab.* 19:65, 1993.
4. O. M. Kvalheim, F. Brakstad, Y. Liang. Preprocessing of analytical profiles in the presence of homoscedastic or heteroscedastic noise. *Anal. Chem.* 66:43, 1994.
5. P. Paatero, U. Tapper. Analysis of different modes of factor analysis as least squares fit problems. *Chemo. Lab.* 18:183, 1993.
6. Y. Hayashi, R. Matsuda. Deductive prediction of measurement precision from signal and noise in liquid chromatrography. *Anal. Chem.* 66:2874, 1994.
7. M. V. Gorshkov, R. T. Kouzes. Data reflection algorithm for spectral enhancement in Fourier transform ICR and NMR spectroscopies. *Anal. Chem.* 67:3412, 1995.
8. T. Iwata, J. Koshoubu. A new method to eliminate the background noise from a line spectrum. *Appl. Spec.* 48:1453, 1994.
9. T. Iwata, J. Koshoubu. Minimization of noise in spectral data. *Appl. Spec.* 50:747, 1996.
10. L. S. Greek, H. G. Schulze, M. W. Blades, A. V. Bree, B. B. Gorzalka, R. F. B. Turner. SNR enhancement and deconvolution of Raman spectra using a two-point entropy regularization method. *Appl. Spec.* 49:425, 1995.
11. H. Takeuchi, S. Hashimoto, I. Harada. Simple and efficient method to eliminate spike noise from spectra recorder on charge-coupled device detectors. *Appl. Spec.* 47:129, 1993.
12. W. Hill. Comment on a "Simple and efficient method to eliminate spike noise from spectra recorded on charge-coupled device detectors." *Appl. Spec.* 47:2171, 1993.
13. F. C. Sanchez, P. J. Lewi, D. L. Massart. Effect of different preprocessing methods for principal component analysis applied to the composition of mixtures: Detection of impurities in HPLC-DAD. *Chemo. Lab.* 25:157, 1994.
14. A. E. Carlsson, K. L. R. Janne. Near-infrared spectroscopy as an alternative to biological testing for quality control of hyaluronan: Comparison of data preprocessing methods for classification. *Appl. Spec.* 49:1037, 1995.
15. A. Savitzky, M. J. E. Golay. Smoothing and differentiation of data by simplified least squares procedures. *Anal. Chem.* 36:1627, 1964.
16. P. A. Gorry. General least-squares smoothing and differentiation by the convolution (Savitszky–Golay) method. *Anal. Chem.* 62:570, 1990.
17. F. Holler, D. H. Burns, J. B. Callis. Direct use of second derivatives in curve-fitting procedures. *Appl. Spec.* 43:877, 1989.
18. W. Windig. The use of second-derivative spectra for pure-variable based self-modeling mixture analysis techniques. *Chemo. Lab.* 23:71, 1994.
19. C. Chapados, M. Trudel, J. Miletic. Enhancing infrared spectra by Fourier deconvolution using the Cauchy–Gauss product function. *Chemo. Lab.* 22:209, 1994.
20. J. K. Kauppinen, D. J. Moffatt, H. H. Mantsch, D. G. Cameron. Fourier self-deconvolutions: A method for resolving intrinsically overlapped bands. *Appl. Spec.* 35:271, 1981.
21. R. G. Brereton. Fourier transforms: Use, theory and applications to spectroscopic and related data. *Chemo. Lab.* 1:17, 1986.
22. P. J. Treado, M. D. Morris. A thousand points of light: The Hadamard transform in chemical analysis and instrumentation. *Anal. Chem.* 61:723A, 1989.
23. B. Vidakovic, P. Muller. Wavelets for kids. Web pages http://www.cwi.nl/~pauldzwvl/start.html (2000), and http://www.uni-koeln.de/themen/statistik/methods.e.htm (2000).
24. W. H. Press, B. P. Flannery, S. A. Teukolsky, W. T. Vetterling. *Numerical Recipes in Fortran: The Art of Scientific Computing*. New York: Cambridge University Press, 1992.
25. M. Bos, J. A. M. Vrielink. The wavelet transform for pre-processing IR spectra in the identification of mono- and di-substituted benzenes. *Chemo. Lab.* 23:115, 1994.
26. B. Walczak, B. van den Bogaert, D. L. Massart. Application of wavelet packet transform in pattern recognition of near-IR data. *Anal. Chem.* 68:1742, 1996.
27. WaveLab .701, a toolbox developed at Stanford to be used with MATLAB, http://playfair.stanford.edu/~wavelab/.
28. Mathworks wavelet toolbox, http://www.mathworks.com/wavelet.html.
29. S+WAVELETS, a toolkit to be used with S-PLUS analyis software: http://www.mathsoft.com/splsprod/wavelets.html.
30. *Wavelet Digest*, electronic newsletter, http://www.wavelet.org/cm/ms/what/wavelet/index.html.
31. *Mathsoft Wavelet Resources*, http://mathsoft.com/wavelets.html.
32. O. Lee, A. P. Wade, G. A. Dumont. Generalized Fourier smoothing of flow injection analysis data. *Anal. Chem.* 66:4507, 1994.
33. M. Bos, E. Hoogendam. Wavlet transform for the evaluation of peak intensities in flow-injection analysis. *Anal. Chim. Acta* 267:73, 1992.

REFERENCES

34. M. Devaux, D. Bertrand, P. Robert, J. Morat. Extraction of near infra-red spectral information by fast Fourier transform and principal component analysis. Application to the discrimination of baking quality of wheat flours. *J. Chemo.* 1:103, 1987.
35. P. Williams, K. Norris, eds. *Near-Infrared Technology.* St. Paul, MN: Amer. Assoc. Cereal Chemists, 1987, pp. 44, 112, 113.
36. W. F. McClure. *NIRnews* 4(6):12, 1993.
37. W. F. McClure. *NIRnews* 5(1):12–13, 1994.
38. K. G. Nordquist. U.S. Patent No. 4,944,589; Filed Sept. 22, 1988; Awarded July 31, 1990.
39. J. Workman. In *Proc. 1991 Georgia Nutrition Conference,* 37–47, Nov. 1991.
40. J. Workman, H. Andren. *American Laboratory.* p. 33, Dec. 1991.
41. R. DiFoggio, M. Sadhukhan. U.S. Patent No. 5,397,899; Filed July 21, 1992; Awarded Mar. 14, 1995.
42. J. S. Shenk, M. Westerhaus. Infrasoft International Software (available from Perstorp Analytical, Silver Spring, MD).
43. H. Mark, J. Workman. *Spectrosc.* 3(11):28–36, 1988.
44. E. V. Thomas. A primer on multivariate calibration. *Anal. Chem.* 66:795A, 1994.
45. M. Davidian, P. D. Haaland. Regression and calibration with nonconstant error variance. *Chemo. Lab.* 9:231, 1990.
46. O. E. de Noord. The influence of data preprocessing on the robustness and parsimony of multivariate calibration models. *Chemo. Lab.* 23:65, 1994.
47. L. Jancar, W. Wegscheider. Effect of scaling regimes on the prediction of analytical results from multivariate calibration. *Anal. Chim. Acta* 248:459, 1991.
48. C. L. Erickson, M. L. Lysaght, J. B. Callis. Relationship between digital filtering and multivariate regression in quantitative analyis. *Anal. Chem.* 64:1155A, 1992.
49. G. W. Small, M. A. Arnold, L. A. Marquardt. Strategies for coupling digital filtering with partial least-squares regression: Application to the determination of glucose in plasma by Fourier transform near-infrared spectroscopy. *Anal. Chem.* 65:3279, 1993.
50. S. M. Donahue, C. W. Brown, M. J. Scott. Analysis of deoxyribonucleotides with principal component and partial least-squares regression of UV spectra after Fourier preprocessing. *Appl. Spec.* 44:407, 1990.
51. M. J. Mattu, G. W. Small. Quantitative analysis of bandpass-filtered Fourier transform infrared interferograms. *Anal. Chem.* 67:2269, 1995.
52. P. B. Stark, M. M. Herron, A. Matteson. Empirically minimax affine mineralogy estimates from Fourier transform infrared spectrometry using a decimated wavelet basis. *Appl. Spec.* 47:1820, 1993.
53. A. Economou, P. R. Fielden. Digital filtering in stripping analysis. *Anal. Chim. Acta* 305:165, 1995.
54. C. W. K. Chow, D. E. Davey, D. E. Mulcahy, T. C. W. Yeow. Signal enhancement of potentiometric stripping analysis using digital signal processing. *Anal. Chim. Acta* 307:15, 1995.
55. M. B. Seasholtz, B. R. Kowalski. The effect of mean centering on prediction in multivariate calibration. *J. Chemo.* 6:103, 1992.
56. I. S. Helland, T. Naes, T. Isaksson. Related versions of the multiplicative scatter correction method for preprocessing spectroscopic data. *Chemo. Lab.* 29:233, 1995.
57. R. J. Barnes, M. S. Dhanoa, S. J. Lister. Standard normal variate transformation and detrending of near-infrared diffuse reflectance spectra. *Appl. Spec.* 45:772, 1989.
58. M. S. Dhanoa, S. J. Lister, R. J. Barnes. On the scales associated with near-infrared reflectance difference spectra. *Appl. Spec.* 49:765, 1995.
59. T. Isaksson, T. Naes. The effect of multiplicative scatter correction (MSC) and linearity improvement in NIR spectroscopy. *Appl. Spec.* 42:1273, 1988.
60. P. Kubelka, F. Munk. *Z Tech Phys,* 12:593, 1931.
61. J. L. Sanderson. *J. Opt. Soc. Am.* 32:727, 1942.
62. T. Isaksson, B. R. Kowalski. Piece-wise multiplicative scatter correction applied to near-infrared diffuse transmittance data from meat products. *Appl. Spec.* 47:702, 1993.
63. T. V. Karstang, R. Manne. Optimized scaling: A novel approach to linear calibration with closed data sets. *Chemo. Lab.* 14:165, 1992.
64. T. Isaksson, Z. Wang, B. R. Kowalski. Optimised scaling (OS-2) regression applied to near infrared diffuse spectroscopy data from food products. *J. Near Infrared Spec.* 1:85, 1993.
65. T. V. Karstang, O. M. Kvalheim. Multivariate prediction and background correction using local modeling and derivative spectroscopy. *Anal. Chem.* 63:767, 1991.
66. T. V. Karstang, O. M. Kvalheim. Comparison between three techniques for background correction in quantitative analysis. *Chemo. Lab.* 12:147, 1991.
67. M. L. Salit, J. B. Collins, D. A. Yates. Heuristic and statistical algorithms for automated emission spectral background intensity estimation. *Appl. Spec.* 48:915, 1994.
68. A. A. Agbodjan, S. C. Rutan. Optimization of an adaptive Kalman filter based on information theory. *Chemo. Lab.* 24:137, 1994.

30.

REVIEW OF CHEMOMETRICS APPLIED TO SPECTROSCOPY: MULTIWAY ANALYSIS

INTRODUCTION

In this chapter's review on chemometrics in spectroscopy we will describe a recent methodology that has attracted increasing interest in spectroscopy, namely, multiway analysis. The application of multiway analysis in spectroscopy is still relatively new, hence many methodological improvements are being investigated currently. Part of this review will also be used to describe the algorithmic improvements made in the last decade.

There are three main of areas of interest in multiway analysis: curve resolution, exploratory analysis, and calibration. We will first describe the most common models and then review applications in these three areas. Note that we will mainly discuss three-way analysis. But keep in mind that the step from three-way to four-way analysis is relatively simple, and the methods are less complex than going from two- to three-way analysis.

Multiway Data

In standard multivariate data analysis, data is arranged in a two-way structure: a table or a matrix. A typical example is a table where each row corresponds to a sample and each column to the absorbance at a particular wavelength. The two-way structure explicitly implies that for every sample the absorbance is determined at every wavelength, and vice versa; thus the data can be indexed with two indices, one defining the sample number and one defining the wavelength number. This arrangement is intimately connected to the techniques subsequently used for analysis of the data (principal component analysis (PCA) etc.). However, for a wide variety of data types a more appropriate structure would be a three-way table or array. An example could be a situation where for every sample the fluorescence emission is determined for several wavelengths for several different excitation wavelengths. In this case every data element can be logically indexed by three indices, one identifying the sample number, one the excitation wavelength, and one the emission wavelength. Fluorescence and hyphenated methods [1] like chromatographic data are prime examples of data types that have been successfully exploited using multiway analysis. However, consider a situation where spectral data is acquired on a sample at different chemical or physical circumstances, e.g., an NIR spectrum measured at several different temperatures (or pH or additive concentration or other experimental conditions that affect the analytes in different relative proportions) on the same sample. Such data could also be arranged in a three-way structure, indexed by sample, temperature, and wavenumber. Clearly, three-way data occurs frequently, but is often not recognized as such due to lack of awareness. In Refs. 2–4 some interesting and innovative applications are shown using data that is not normally considered multiway.

Important Terminology

Some authors advocate using array notation [5], others prefer notation adapted from tensor algebra [6,7], and others use modified matrix algebra notation [8]. It is unfortunate that there is yet no consensus on the notation, but this reflects that multiway analysis is still a young discipline inspired by many different fields. The terms *mode, way,* and *order* are used more or less interchangeably, though a distinction is sometimes made between the geometrical dimension of the array/cube (the number of ways) and the number of independent ways (which is the order/mode) [9–11]. An ordinary two-way covariance matrix is thus a one-mode two-way array, because the variables are identical in the two ways.

In Ref. 12 a very intuitive notation is used to describe different data structures, depending on the number of modes referring to objects and the number of modes referring to variables in some sense. A three-way array consisting of a set of samples (objects) each being measured spectrofluorometrically at several excitation and emission wavelengths will be an OV^2 three-way array, because one mode refers to samples and two modes, excitation and emission wavelengths, refer to variables. A chromatographic run using fluorescence excitation-emission, on the other hand, will give a V^3 three-way array if only one sample is analyzed, because all three modes are variable modes. If several samples are analyzed, the resulting data will be an OV^3 four-way array. An RGB image can logically be arranged as an O^2V three-way, since two modes (height and width) refer to the object, and one mode (RGB) refers to variables. Using this notation it is quite obvious that most three-way analysis is performed on arrays with *at least* two variable modes. Using, e.g., a trilinear model on an RGB image has not been reported. This does not imply that such an analysis would be fruitless. In greytone image analysis, singular value decomposition is sometimes used for feature extraction with success, and hence the same should most likely be possible using multivariate images and multiway decomposition. However, the essence is that multiway models are used mostly in situations where latent variables in some form can be presumed to exist in at least two modes.

Software

A large number of multiway programs are available aimed at problems in the more social sciences (see, e.g., Ref. 13), some of which are also implemented in statistical packages such as SPSS and SAS. However, only programs relevant to spectral analysts will be mentioned in this section.

Harshman and Lundy offer a very extensive set of tools for analyzing three-way data by PARAFAC (parallel factor analysis) [14]. The program is described in Ref. 15.

Bro and Andersson offer a set of MATLAB M-files for estimating the PARAFAC, the N-PLS, and the *N*-mode principal component analysis models [16]. The algorithms are either true *N*-way or at least "highway." The models can be estimated under a variety of constraints (unimodality, nonnegativity, orthogonality, etc.), and procedures are also given for preprocessing.

Ross offers FORTRAN code for estimating the PARAFAC model and the Tucker2 and 3 models. The models handle two-, three-, and four-way data and optional nonnegativity constraints [17].

Paatero offers a stand-alone PC program for estimating bilinear and trilinear (PARAFAC) models [18]. He denotes the models PMF2 and PMF3, after his algorithm, which is characterized by being very fast compared to standard PARAFAC algorithms and supposedly more robust. The models can be estimated under nonnegativity constraints and using weights for each individual data element.

Kroonenberg offers a program for three-mode principal component analysis and PARAFAC [19] called *3waypack*. The Tucker2 and Tucker3 model can be estimated using the TUCKALS2 and TUCKALS3 methods [20]. The program also offers additional procedures for handling missing data, preprocessing, and core rotation. The programs have been described in Ref. 21.

Books, Reviews, and Tutorials

Three-mode Principal Component Analysis by P. M. Kroonenberg [22] is an extensive description of virtually all aspects of the Tucker models and related models such as PARAFAC. The book focuses on applications in the social sciences, but it is very valuable for descriptions, references, and thoroughly worked-through examples. The book *Research Methods for Multimode Data Analysis,* edited by Law et al. [23], contains a large number of articles on multiway analysis within psychometrics. Coppi and Bolasco edited a book of contributions presented at the International Meeting on the Analysis of Multiway Data Matrices in Rome, March 1988 [24]. It contains important papers on rank, preprocessing, degeneracy, multidimensional scaling, etc.

Geladi [10] gave an overview of multiway methods in psychometrics and chemometrics, and Smilde [25] wrote an overview of the "problems and prospects" of three-way analysis in chemometrics. He organizes the methods according to how many blocks of data are involved; e.g., exploratory analysis in the PCA sense is a one-block problem, while calibration/regression is normally a two-block problem. He outlines and describes the methods available for handling typical problems and also describes problems that need to be considered in future research. He also suggests several areas of potential application for multiway methods. P. M. Kroonenberg has written two reviews on three-way models, the first [26] concerned mainly with three-mode factor analysis, and the second [27] also concerning PARAFAC and related methods. The first review is interesting in that it claims to contain references to all known applications of three-way methods until 1983. Harshman and Lundy [15] have written a review on the PARAFAC model and the program they offer to estimate the model. The principle of PARAFAC and many aspects are described in detail, including analysis of longitudinal data, preprocessing, uniqueness, and analysis of covariance matrices. Tu and Burdick [28] provide a review on the application of PARAFAC in spectroscopy and compare what is essentially GRAM and PARAFAC. The results obtained are similar, though PARAFAC fits the data better than GRAM. Sanchez and Kowalski [7] give an overview of second-order calibration in a tensor perspective and with emphasis on GRAM.

Only a few tutorials on multiway analysis have been written. Henrion wrote a tutorial on *N*-mode principal component analysis, including descriptions of how to interpret and rotate the so-called core matrix [29], while Bro has written a tutorial on PARAFAC [30].

MULTIWAY MODELS AND ALGORITHMS

Most multiway models have their origin in the late 1960s psychometrics society, and the interest in chemometrics was initiated much later. In many chemometrics research groups there is now a fruitful cooperation between psycho- and chemometricians. This cooperation resulted in the conferences TRIC '93 (the first conference on Three-way Methods in Chemistry) on August 22–25, 1993, Epe, The Netherlands, and TRICAP '97 (Three-way Methods in Chemistry and Psychology) on May 2–5, 1997, Lake Chelan, WA. In this review some references to psychometric literature will be given, because many important methodological results come from that area. The workhorses in two-way chemometrics analysis are PCA and PLS (partial least squares regression), and these methods have also been successfully applied to multiway data by rearranging (unfolding) the data into two-way data. Since this approach is justified mostly because two-way methods are simply more accessible, and since two-way methods have already been reviewed in detail, we will discuss only methods that specifically use the multiway structure of the data in the modeling.

Kiers [11] discusses the relationship among the most common three-way models and orders the models according to their complexity. He shows that the unfold PCA/Tucker1 model is the most general model, in the sense that other models can be seen as constrained versions of unfold PCA. The most restricted and simple model is shown to be the trilinear model, for it uses the fewest parameters.

PARAFAC/GRAM

PCA is a bilinear model, which can easily be recognized by stating the mathematical model:

$$x_{ij} = \sum_{f=1}^{F} a_{if} b_{jf}$$

where i designates the ith row and j the jth column of the data matrix **X**. A natural extension to three-way data immediately follows:

$$x_{ijk} = \sum_{f=1}^{F} a_{if} b_{jf} c_{kf}$$

This trilinear model forms the basis of PARAFAC (parallel factor analysis), CANDECOMP (canonical decomposition), and GRAM (generalized rank annihilation method). The PARAFAC [31] and the CANDECOMP [32] models are identical, but were developed in different contexts, both in 1970. The term CANDECOMP is not often used in chemometrics, and we will thus use only the term PARAFAC. It is quite interesting to note that Appellof and Davidson [33] independently reinvented PARAFAC in 1981 for analyzing fluorescence data. They give very insightful ideas on how to speed up the algorithm by initially compressing the array using estimated bases in the different modes (see also Refs. 34–36). They base the algorithm on the alternating least squares principle and speed up the algorithm by extrapolation steps, very much in line with other approaches for estimating the PARAFAC model. For some reason this work was mostly forgotten for several years while most multiway research in chemometrics focused on methods based on rank annihilation. The PARAFAC model and algorithm seeks to decompose the three-way array into the parameters a_{if}, b_{jf}, and c_{kf}, so that the trilinear model is a (possibly weighted) least squares model of the data. Unlike the PCA model, no extra constraints, such as orthogonality, are needed to identify this model [37–39]. The implication is that the structural model is sufficient to uniquely estimate the parameters of the model. If the model approximately reflects an underlying latent structure in the data, then this structure can be directly estimated. This voids the necessity of using postrotations of the model to be able to interpret the model in line with the real world. For many hyphenated techniques, e.g., LC-VIS, one can therefore directly estimate the pure spectra and profiles up to a scaling by simply estimating the trilinear model. The uniqueness property was the prime reason for developing the PARAFAC model and was originally proposed as a principle by Cattell in 1944 [40].

The practical use of PARAFAC has been, and still is to some extent, hampered by practical problems in devising sound algorithms for estimating the model. It is *the* multiway model that can be most difficult to estimate, and requires the most experience by the user. Progress has been made in both psycho- and chemometrics in describing and remedying the possible problems, most notably swamps and (two-factor) degeneracies. Degeneracy is a situation where the algorithm either fails or has difficulties estimating the model when the parameters of two components are equal or almost equal in shape and have a high negative correlation. The reason this happens is only partly known [41–43]. It is most often related not to the reality of the data but to a numerical phenomenon; e.g., too many components are used, the trilinear model is inappropriate, or the data is not properly preprocessed. Degeneracy can be monitored by computing the angle between the components. Usually the *multiple cosine* or *uncorrected correlation coefficient* [39,44] is used for this purpose. The multiple cosine shows for each possible pair of components how related these are in terms of the angle between the components. Hence if any multiple cosine approaches –1 it is a sign that the model is at least temporarily degenerate.

It is now generally agreed that there are two types of degeneracies: the hard (true) and the soft (pseudo) degeneracy. A soft degeneracy—a swamp—occurs when the estimation path leads through situations in which the convergence slows down dramatically. By continuing the iterations further, the swamp will cease, and convergence to a sensible solution will be achieved. A true degeneracy occurs when further iterations do not cause the degeneracy to disappear. Mitchell and Burdick [45] investigated situations in which soft degeneracies occur. They suggest that the analysis be stopped and redone if the multiple cosine is too low. Rayens and Mitchell suggest using regularization for preventing swamps [46]. In Ref. 47 a different approach is suggested when degenerate solutions are consistently observed. Based on the observation that degeneracy causes components to correlate, the authors propose using a model that is essentially a PARAFAC model with orthogonality constraints in one mode, and then a subsequent Tucker3 core estimation. The orthogonality constraint helps the model circumvent the degeneracy, while the subsequently estimated core is helpful in understanding if and how the data deviates from the PARAFAC model. This ad hoc approach is shown to be beneficial in a study of rating of television shows. Because the procedure uses orthogonality constraints it is most likely not very helpful in spectral analysis, but it may be useful when trying to explore which model is the most appropriate in a given situation. Another, more suitable, approach for spectral analysis is suggested by Krijnen and ten Berge [48]. They simply propose using a nonnegativity constrained PARAFAC model. The parameters being forced to be nonnegative effectively prevents any components from being negatively correlated. Furthermore, the nonnegativity is often justified from a theoretical point of view, in that negative parameters (e.g., absorbance, concentration, pH) do not make sense. They show that the nonnegativity effect can remove problems with degeneracy and also can be helpful if the uniqueness of the model is not satisfactory.

The PARAFAC model, with its uniqueness property and its simple structure, is expected to achieve increased attention in the future in many different areas. But in order to make the use of PARAFAC possible for a broader audience it is of utmost importance to develop faster and more robust algorithms. In recent literature some efforts have been made to make better algorithms for estimating the PARAFAC model [49–53], and further progress is likely to be expected in that area because it is well known that the alternating least squares principle underlying most PARAFAC algorithms is not the most efficient one.

Note that in two-way PCA, the so-called scores of a spectral decomposition can be interpreted as concentrations or amounts of latent variables, the latent variable in this sense being defined by the loading vector. The same interpretation can be assigned to the PARAFAC model. But because the PARAFAC model is unique, the scores will be not pseudo-concentrations of abstract latent variables but estimates of the concentrations of real analytes. The units of these concentration estimates are not defined by the model, due to the scaling indeterminacy. But if the reference concentrations of the analytes are known for one or more samples, the scale will be defined and the concentration given directly as the scores. Hence the PARAFAC model is suitable for quantitative as well as qualitative analysis, i.e., curve resolution as well as calibration.

Another trilinear model that has received much attention in chemometrics has several acronyms: RAFA (rank annihilation factor analysis), GRAFA (generalized rank annihilation factor analysis), GRAM (generalized rank annihilation method), TLD (trilinear decomposition), and DTD (direct trilinear decomposition). GRAFA is identical to GRAM. TLD is identical to DTD, and the very broad term TLD should generally be avoided, for PARAFAC is also a trilinear decomposition. Ho and coworkers [54–56] developed an algorithm called rank annihilation factor analysis (RAFA) for estimating the concentration of an analyte in an unknown matrix using solely the unknown sample and a pure standard! This amazing property has later been coined the *second-order advantage,* because this property is obtained by using the second-order, or two-way, structure of the individual sample instead of merely unfolding the matrix to a first-order structure. The second-order advantage in essence is identical to the uniqueness of the trilinear structure, and the model underlying the GRAM/RAFA/DTLD algorithm is the same as in the PARAFAC model. The idea behind RAFA was based on reducing the rank of the calibration sample by subtracting the contribution from the analyte of interest. This was intuitively appealing, but the method itself was somewhat unsatisfactory and slow. In 1984 Lorber [57,58] saw that the problem of finding the optimal prediction could be automated by realizing that the sought reduction in rank could be expressed as a generalized eigenvalue problem. Sanchez and Kowalski [59] generalized the method into the *generalized rank annihilation method* (GRAM), in which several components could be present/absent in both calibration and standard samples. Wilson et al. [60] further improved GRAM by formulating it as a generalized eigenproblem, ostensibly giving increased stability of the algorithm. Faber et al. [61], in part I of a series on GRAM, compare the different formulations of GRAM and suggest formulating the algorithm as a standard eigenvalue problem, much in line with the eigenvalue solution originally developed by Lorber, in this way giving more insight to when and how the method works well. They also discuss the problems of complex and degenerate solutions. In the second part of the series [62], expressions for estimating bias and variance for GRAM/RAFA predictions are developed. It is shown that the bias and variance of predicted concentrations can vary considerably between the different implementations of GRAM/RAFA. In the third part [63], Faber et al. discuss in detail aspects of practical implementation with respect to possible numerical problems, as well as problems arising from deviations of the data from the theoretical model being used. These deviations include matrix effects, interaction effects, nonlinearities, retention time shifts, and heteroscedastic noise. They also propose remedies for the problems caused by such phenomena. Faber et al. [64] describe the standard errors in the eigenvalues of cross-product matrices by error propagation and use it to define error estimates for predictions from GRAM.

Wilson et al. [65] extended GRAM to setups where the contribution of a single analyte to the signal does not correspond to a rank-one signal as presupposed in the trilinear model underlying GRAM. This method, called nonbilinear rank annihilation (NBRA), is numerically equivalent to GRAM, but special attention is needed for correctly estimating the concentrations. Wang et al. [66] further elaborated on NBRA and showed that quantification of an analyte is possible if the presence of the analyte in a sample leads to an increase in the rank of the corresponding measured second-order data structure.

Sanchez and Kowalski [67] made an extension of GRAM called direct trilinear decomposition (DTD) for situations where more than two samples are available. The GRAM method is based on the trilinear model and seeks to estimate this model using a (generalized) eigenvalue problem. The method is restricted by one mode having maximally dimension 2 (i.e., two samples). However, estimating the parameters of the model using two samples will give the fixed spectral and chromatographic profiles in case of a chromatography-spectroscopy analysis. Because these parameters are fixed for new samples as well, the concentrations of analytes for new samples can be found by simple regression on the fixed parameters. Used in this way the method is called DTD. Since GRAM is intrinsically based on using two samples, the extension is based on defining a set of two synthetic samples involving linear combinations of the original samples. Sanchez and Kowalski advocate using a Tucker1 model for that purpose, while Sands and Young [68] propose a somewhat different scheme for an algorithm that can be considered equivalent to DTD.

A problem in GRAM and DTD arises because eigenvectors and values involved in estimating the model can be complex, for different reasons. This is undesirable because the pure spectra, profiles, etc. can henceforth not be estimated. Li et al. [69], Li and Gemperline [70], and Faber et al. [61] discuss different approaches to remedy this mathematical artifact of the method by similarity transformations. Booksh et al. [71] suggest that complex solutions arising from analytes having the same concentration ratio between samples can be remedied by carefully using the most appropriate expression for calculating the true profiles from the eigenvectors used to decompose the data. They also suggest using curve-fitting between scores (predicted relative concentrations) and true concentrations when the relationship between the concentration and the signal intensity is nonlinear. They show the benefit of their proposals on a data set obtained from a fiber-optic heavy-metal sensor.

As noted by Booksh and Kowalski [72], "it is not surprising that occasionally new [calibration] algorithms are actually reformulations of older accepted methods." The basic principle of GRAM/DTD has been invented several times. As early as 1972, Schönemann [73] developed a similar algorithm, essentially and interestingly based on the same idea of Cattell [40] (through Ref. 74) that caused Harshman to develop the PARAFAC model and algorithm. Schönemann's GRAM-like method has later been refined in a more stable and sensible way by, e.g., de Leeuw and Pruzansky [75]. In Refs. 68 and 76–78, several reinventions of GRAM are investigated and compared with PARAFAC for its use in curve resolution. They find GRAM inferior to PARAFAC but suggest using GRAM for initialization of the PARAFAC algorithm. Sanchez and Kowalski [67] suggest the same. As described by Booksh and Kowalski [72], the DATAN algorithm developed by Scarminio and Kubista [79] is yet another example of the reinvention of GRAM. These reinventions of similar methods are actually quite interesting. Much can be learned by scrutinizing different authors' views on similar methods and problems.

Since the DTD and PARAFAC models are structurally identical, one may ask what the difference is, and which model should be used. The mathematical difference between the two is that the PARAFAC model is a least squares model, whereas the DTD model has a less well-defined optimization criterion. For noise-free data it gives the correct solution, but for noisy data there is no provision for the quality of the model. The practical importance of this fact has not yet been systematically investigated, but it is known that both methods fail to give sound results in certain situations [80]. The advantage of DTD is the speed of the algorithm, whereas the advantage of the PARAFAC model is the possibility of modifying the algorithm according to external knowledge, such as using weighted regression when uncertainties are available or using nonnegativity when negative parameters are known not to conform to reality. Another advantage of the PARAFAC model is the ease with which it can be extended to higher-order data. No systematic investigations have been performed to compare DTD and PARAFAC.

Kiers and Smilde compared different versions of GRAM and PARAFAC [81]. They proved under which conditions these methods can be expected to be able to predict the concentration of analytes in an unknown sample, possibly with unknown interferences, using only one standard. They found that all but the original GRAM method developed by Sanchez and Kowalski will fail to predict in cases where some analyte profiles overlap completely; i.e., spectra, chromatographic profiles, etc. are identical in shape for some analytes.

Gerritsen et al. [82] compare GRAM with iterative target transformation factor analysis and residual bilinearization on simulated data. They find that GRAM and residual bilinearization in general perform better but are somewhat more sensitive to problems with reproducibility of the data.

N-Mode Principal Component Analysis

During the 1960s L. Tucker developed a series of three-way models now known as the Tucker1, Tucker2, and Tucker3 models, also collectively called three-mode principal component analysis or, originally, three-mode factor analysis, though the models would be conceived by most people as component models [83,84]. These models are not as unique as the trilinear models, but they have been extensively used primarily for exploratory analysis. The Tucker1 model is essentially the well-known unfolding technique, where the data array is rearranged to a matrix, which is then subjected to ordinary two-way PCA. The Tucker2 model is of intermediate complexity compared with the Tucker1 and Tucker3 models. It will not be discussed much here, for only few applications in chemistry have yet been seen. The Tucker3 model is a quadrilinear model in its three-way version, since the model consists of four sets of parameters, with each set being conditionally linear. In 1980, Kroonenberg and de Leeuw devised an alternating least squares algorithm called TUCKALS3 (or TUCKALS2 for the Tucker2 model) for estimating the model in a least squares sense with orthonormal loadings [20]. In Refs. 85 and 86, additional results on estimating the tucker models in a least squares sense are described, and in Refs. 52, 87, and 88, certain suggestions are given to speed up the algorithm when modeling very large arrays.

The Tucker3 model has rotational freedom. As with two-way analyses this has prompted the need for rotations of solutions in order to increase interpretability [22, 89–91].

Due primarily to the nonuniqueness, the Tucker models have not received much attention in spectral analysis but have been used primarily for exploratory analysis in, e.g., environmental chemistry [92–94]. However, the Tucker3 model forms the basis for what is termed *constrained* or *restricted* Tucker models, initially proposed by Kiers [89]. In restricted Tucker models chemical (or other) knowledge is used to restrict especially the core elements, forcing individual elements to attain specific values. In this way it is possible in some situations to define models that uniquely estimate sought properties like spectra, concentrations, etc. The use of restricted Tucker models has been promoted in chemometrics by Smilde [95,96]. It has been proposed as primarily a method for solving problems where the second-order signal from the analyte of interest has a medium rank as opposed to idealized fluorescence spectroscopy or chromatography/spectroscopy, where each analyte is supposed to contribute with a rank-one signal. The restricted Tucker models can be seen essentially as structural models accommodated to a specific chemical problem.

Other Models

Öhman et al. [97,98] developed a method called *residual bilinearization* for predicting samples with unknown interferences, i.e., the same scope as for GRAM/DTD. The method is based on unfolding the three-way array to a matrix and subsequently estimating a calibration model on this unfolded array. The residuals from this analysis are then subsequently estimated by a bilinear model, and the calibration and residual models are then iteratively estimated from each other's residuals until convergence. Wang et al. [99] compared residual bilinearization with nonbilinear rank annihilation and found them to be almost identical, though with different noise propagation properties.

In Ref. 100 a general scheme for multiway curve resolution is outlined, with a focus on establishing the presence of selective channels, i.e., variables where some analytes are present/absent. The method also deals with data that cannot be arranged into a multiway array. Grung and Kvalheim [101] described another method for resolving three-way arrays called *evolving projections by optimized search*.

There are several approaches to calibration when using multiway methods. The standard multivariate two-way approach in chemometrics involves the use of PCA or PLS, where the array of independent variables is decomposed into a set of scores that are then subsequently or simultaneously regressed on the dependent variable(s). The same approach is of course also applicable to three-way analysis. The analog to principal component regression (PCR) is then to use either PARAFAC or Tucker to decompose the array and subsequently to regress the scores on the dependent variable. For PARAFAC and Tucker it holds that a model with given number of components, e.g., a three-component PARAFAC model, is not obtainable as a subset of a four-component model. That is, in general one cannot expect that any components from a model of given rank will equal any components from a model of the same data but with a different dimensionality. This is referred to as the models being *nonnested,* as opposed to many two-way models (PCA, PLS). It is therefore necessary to reestimate the model for every number of components sought. Validation techniques based on resampling of any sort may therefore take very long time to compute, which can be a problem, especially for the PARAFAC model.

In 1989 Ståhle [102] extended the PLS model to three-way data by extending the two-way algorithm in a straightforward manner. The optimality of the proposed algorithm, however, was not substantiated. Later, Bro developed a general multiway PLS that was shown to be optimal according to the underlying theory of PLS [103]. In the three-way version, the three-way array is decomposed into a trilinear model, such as the PARAFAC model, only for *N*-PLS, the model is not a least squares model of the independent variables, but seeks in accordance with PLS to describe simultaneously the variation in the dependent and the independent variables.

Since the trilinear three-way model is unique it follows directly that scores obtained from a decomposition of spectral data will be the concentrations of the analytes. If the concentrations are known for at least one sample, one can directly compute the estimated concentrations from the scores. This is the basis for the second-order calibration performed normally with GRAM/DTD and PARAFAC. The same principle can be used for other models tailored for specific chemical situations, as shown by Refs. 89, 95, and 96.

Preprocessing

Preprocessing of three-way arrays is more complicated than in the two-way case, and much work has been done in psychometrics to elucidate the characteristics of scaling and centering. *Single-centering* is done by subtracting the mean of all elements corresponding to a specific combination of variables *across* the mode being centered. If centering is to be performed across more than one mode, one has to do this by first centering one mode and then centering the outcome of this centering. If two centerings are performed in this way, it is referred to as *double-centering. Triple-centering* means centering across all three modes, one at a time. In Refs. 41, 104, and 105, the effect of both scaling and centering on the trilinear behavior of the data is described. It turns out that centering one mode at a time is the only appropriate way of centering, with respect to the assumptions of the multilinear model. Centering one mode at a time essentially removes any constant levels in that particular mode. The same holds for other kinds of centering. Scaling in multiway analysis also has to be done, taking the trilinear model into account. One should not, as with centering, scale column-wise, but rather whole "slabs" of the array should be scaled. If variable *j* of the second mode is to be scaled (compared to the rest of the variables in the second mode), it is necessary to scale all columns where variable *j* occurs. This means that one has to scale whole matrices instead of columns. Scaling the first mode is referred to as scaling *within* the first mode. When scaling within several modes is desired, the situation is a bit complicated, because scaling one mode affects the scale of the other modes. If scaling is desired within several modes, this has to be done iteratively. Another complicating issue is the interdependence of centering and scaling. In general, scaling within one mode disturbs prior centering across the same mode but not across other modes. Centering across one mode disturbs scaling within all modes. Hence, only centering across arbitrary modes or scaling within one mode is straightforward, and not all combinations of iterative scaling and centering will converge. Note that an alternative to scaling and the complications arising from scaling is of course to use weighted regression for estimating the models.

APPLICATIONS OF MULTIWAY METHODS

Mass Spectrometry

Quantitative Analysis

Wilson et al. [106] used nonbilinear rank annihilation (NBRA) on MS/MS spectra of samples containing warfarin, different hydroxywarfarins, and phenylbutazone. NBRA was also compared to and outperformed different ordinary curve resolution and regression techniques.

UV/Visible Spectroscopy

Quantitative Analysis

Li et al. [69] showed how GRAM combined with similarity transformations of complex eigenvectors and values can be used to estimate low concentrations of hydrocortisone in urine, using LC-UV. They noted that using the original RAFA algorithm of Lorber [57] very inaccurate results were obtained, presumably because too much background signal is being attributed to the analyte.

Nørgaard and Ridder [107] described an interesting application of RAFA to data arising from FIA coupled with UV/VS detection. Samples containing three different though similar acids were injected into a FIA system with a pH profile induced over the sample plug. The differences in pK_a gave rise to conditions under which a second-order method should be able to quantify the analytes. However, due to problems arising from the fact that there is no dispersion of the sample constituents, the predictions using only one sample for calibration were not satisfactory.

In Ref. 108 a liquid chromatographic system with UV/VS detection was made for generating data applicable for RAFA. The system was tested on mixtures containing ethylbenzene, *o*-xylene, and *p*-xylene.

Lin et al. [109] described a heavy-metal sensor specifically designed for generating second-order data. Using time-dependent dialysis for separating different metal ions, a second-order signal is obtained by measuring VS spectra several times during the dialysis in the presence of a chelating chromophore. Through the use of GRAM and especially DTD it is possible to predict Pb^{2+} and Cd^{2+} in the presence of unknown interferences. Some problems, however, occur due to, e.g., nonlinearities in the intensity-concentration dependence. Some of these problems are investigated further in Ref. 71.

By the use of kinetic spectroscopic techniques Henshaw et al. [110] showed the benefit of using second-order calibration, viz. GRAM and DTD, as opposed to traditional univariate and multivariate techniques, for determining chlorinated hydrocarbons using the Fukiwara reagent. The goal was to predict the amount of trichloroethylene, chloroform, and 1,1, 1-trichloroethane, respectively, in samples containing the remaining ("unknown") analytes. GRAM did not succeed very well, at least not as well as hoped for. This was attributed to interaction effects between the analytes. When using DTD instead, the results were substantially improved, most likely due to the increased robustness of using several instead of two samples for decomposition. Tauler et al. [111] investigated similar data and found that by using multivariate curve resolution it was possible to both predict concentrations and estimate spectra of analytes in mixtures. In Ref. 96, Smilde et al. suggested yet another approach for quantifying and resolving the results from the Fukiwara reagent analysis. They proposed to use a restricted Tucker model for this purpose, and obtained very good results using only one standard and one unknown sample for each prediction.

In Öhman et al. [112] the residual bilinearization was compared to GRAM for estimating concentrations of diphenylamine and benzophenone in the presence of ("unknown") benzil and succinic acid dibenzyl ester. Data was generated by HPLC/UV/VS, and both methods succeeded reasonably, even though the spectral and chromatographic profiles overlapped severely.

Bro and Heimdal [113] used multilinear PLS to predict enzymatic activity of polyphenol oxidase from kinetic UV/VS spectra and also compared the calibration model with models obtained by more traditional analyses and other multiway methods.

Xie et al. [114] compared the merits of GRAM and DTD used for quantifying binary mixtures of *p*-, *o*, and *m*-amino benzoic acids and orciprenaline reacting with diazotized sulfanilamide in a kinetic UV/VS study.

Qualitative Analysis

Durell et al. [115] used PARAFAC for analyzing UV absorption spectra of plastocyanin obtained in different oxidation states and pH. Treating the pH and oxidation state as independent modes, a four-way array was obtained (sample × wavelength × oxidation state × pH) that was then successfully decomposed using a four-way PARAFAC model. Trilinear and bilinear models were also estimated by unfolding the four-way array accordingly. It was shown that unique two-way decomposition was only possible using external knowledge about target spectra, while for the three- and four-way analysis no such knowledge was required. The unique results enabled detailed exploring of different plastocyanins, verifying earlier-proposed hypotheses and elucidating new aspects. Gianelli et al. [116] use a single-beam imaging spectrophotometer in conjunction with thin-layer chromatography to obtain second-order data, subsequently analyzed with RAFA.

Fluorescence

Fluorescence is *the* spectroscopic technique for multiway analysis, due to the intrinsic trilinearity of fluorescence (excitation, emission, lifetime). Ross and Leurgans [51] discuss in detail the theoretical rationale for using PARAFAC and the Tucker2 model in fluorescence spectroscopy.

Quantitative Analysis

Ho et al. [54] exemplified their RAFA method with simple one- and two-component mixtures of perylene and anthracene and showed that they were able to determine the concentration of one analyte in the presence of the other using only one pure standard and using fluorescence excitation emission landscapes. In their following papers [55,56] they used samples of a six-component polynuclear aromatic hydrocarbon to show the same in a more complex matrix. In Ref. 117 DTD was used for quantifying initial concentrations of two components (glycine and glutamine) based on the kinetic development of fluorescence in a thin-layer chromatographic system. Poe and Rutan [118] compared the use of GRAM, peak height, peak area, and adaptive Kalman filter for quantifying analytes in a separation of polycyclic aromatic hydrocarbons using reversed-phase liquid chromatography coupled with fluorescence detection. GRAM outperformed the other methods but was more sensitive to retention-time shifts.

Qualitative Analysis

Keun et al. [119] used fluorescence excitation emission data of samples containing different amounts of proton acceptors and at different pH values to investigate the steady-state fluorescence of *N*-acetyl-L-tyrosinamide under different conditions in model systems. By using PARAFAC with nonnegativity constraints they were able to resolve the measured excitation emission matrices and to substantiate and develop earlier hypotheses regarding the basic configuration and characteristics of tyrosine. Lee et al. [120] used a similar setup to separate the fluorescence spectra of rhodamine 6G, rhodamine B, and sulforhodamine 101 from measured mixtures. Ross et al. [121] also used a similar setup to resolve spectra of pigment complexes in pea thylakoids using PARAFAC with nonnegativity constraints. Millican and McGown [122] used the DTD-like algorithm developed by Burdick et al. [123] to resolve single-sample two-component mixtures of different amounts of anthracene, 9,10-diphenylanthracene, and 1,3,5,8-tetraphenylpyrene. By introducing fluorescence lifetime besides excitation and emission, resolution of minor components was significantly improved. Karukstis et al. [124] characterized the steady-state room-temperature fluorescence of 2-(*p*-toluidino)naphtalene-6-sulfonate in various solvents using the PARAFAC model. Russell and Gouterman [36] used fluorescence excitation-emission to resolve systems containing platinum, palladium, and rhodium porphyrins even in the presence of quite severe spectral overlapping. In Ref. 125, DTD was used for resolving fluorescence excitation and emission spectra of porphyrins in several animal dental calculus deposits.

Other Applications

Quantitative analysis

Wilson et al. [65,106] showed that their nonbilinear rank annihilation method could satisfactorily predict concentrations of different sugars in mixtures dissolved in D_2O and measured by 2D NMR.

Qualitative Analysis

Herrmann et al. [126] used the Tucker3 model (called COMSTAT [85] in the paper) to explore EEG power spectral data from 65 healthy subjects.

REFERENCES

1. T. Hirschfeld. *Anal. Chem.* 52:297A–312A, 1980.
2. C. L. de Ligny, M. C. Spanjer, J. C. van Houwelingen, H. M. Weesie. *J. Chromatog* 301:311–324, 1984.
3. A. K. Smilde, P. H. van der Graff, D. A. Doornbos, T. Steerneman, A. Sleurink. *Anal. Chim. Acta.* 235:41–51, 1990.
4. A. K. Smilde, D. A. Doornbos. *J. Chemom.* 5:345–360, 1991.
5. S. Leurgans, R. T. Ross. *Stat. Sci.* 7:289–319, 1992.
6. D. S. Burdick. *Chemom. Intell. Lab. Syst.* 28:229–237, 1995.
7. E. Sanchez, B. R. Kowalski. *J. Chemom.* 2:265–280, 1988.
8. W. Wold, P. Geladi, K. H. Esbensen, J. Öhman. *J. Chemom.* 1:41–56, 1987.
9. J. D. Carroll, P. Arabie. *Ann. Rev. Psychol.* 31:607–649, 1980.
10. P. Geladi. *Chemom. Intell. Lab. Syst.* 7:1101–30, 1989.
11. H. A. L. Kiers. *Psychometrika* 56:449–470, 1991.
12. K. H. Esbensen, S. Wold, P. Geladi. *J. Chemom.* 3:33–48, 1988.
13. R. Coppi, Softstat '95. In: F. Faulbaum, W. Bandilla, eds. *Advances in Statistical Software 5.* Stuttgart: Lucius & Lucius, 1996, pp. 37–46.
14. R. A. Harshman. Department of Psychology, University of Western Ontario, London, Ontario, Canada, N6A 5C2.
15. R. A. Harshman, M. E. Lundy. *Comp. Stat. Data Anal.* 18:39–72, 1994.
16. http://newton.foodsci.kvl.dk/foodtech.html.
17. R. T. Ross Department of Biochemistry, The Ohio State University, 484 West 12th Ave., Columbus, OH 43210. http://www.biosci.ohio-state.edu/~rtr/multilin/muldoc.html.
18. P. Paatero. Department of Physics, University of Helsinki, P.O. Box 9, FIN-00014 Helsinki, Finland.
19. P. M. Kroonenberg. Department of Education, Leiden University, Wassenaarseweg 52, 2333 AK Leiden, The Netherlands.
20. P. M. Kroonenberg, J. de Leeuw. *Psychometrika* 45:69–97, 1980.
21. P. M. Kroonenberg. *Comp. Stat. Data Anal.* 18:73–96, 1994.
22. P. M. Kroonenberg. *Three-mode Principal Component Analysis. Theory and Applications.* Leiden: DSWO Press, 1989.
23. H. G. Law, C. W. Snyder, Jr., J. A. Hattie, R. P. McDonald, eds. *Research Methods for Multimode Data Analysis.* New York: Praeger, 1984.
24. R. Coppi, S. Bolasco, eds. *Multiway Data Analysis.* New York: Elsevier, 1989.
25. A. K. Smilde. *Chemom. Intell. Lab. Syst.* 15:143–157, 1992.
26. P. M. Kroonenberg. *British J. Math. Stat. Psych.* 36:81–113, 1983.
27. P. M. Kroonenberg. *Stat. Appl.* 4:619–633, 1992.
28. X. M. Tu, D. S. Burdick. *Stat. Sinica.* 2:577–593, 1992.
29. R. Henrion. *Chemom. Intell. Lab. Syst.* 25:1–23, 1994.
30. R. Bro. An interactive introduction to the PARAFAC and N-PLS in chemometrics at http://www.wiley.co.uk/wileychi/chemometrics/literature.html (2000).
31. R. Harshman. *UCLA Working Papers in Phonetics* 16:1–84, 1970.
32. J. D. Carroll, J. Chang. *Psychometrika.* 35:283–319, 1970.
33. C. J. Appellof, E. R. Davidson. *Anal. Chem.* 53:2053–2056, 1981.
34. M. D. Russell, M. Gouterman. *Spectrochim. Acta.* 1988, 44A, 857–861.
35. M. D. Russell, M. Gouterman. *Spectrochim. Acta.* 1988, 44A, 863–872.
36. M. D. Russell, M. Gouterman. *Spectrochim. Acta.* 1988, 44A, 873–882.
37. R. A. Harshman. *UCLA Working Papers in Phonetics* 22:111–117, 1972.
38. J. B. Kruskal. *Psychometrika* 41:281–293, 1976.
39. J. B. Kruskal. In R. Coppi, S. Bolasco, eds. *Multiway Data Analysis.* New York: Elsevier, 1989, pp. 8–18.
40. R. B. Cattell. *Psychometrika* 9:267–283, 1944.
41. R. A. Harshman, M. E. Lundy. In: H. G. Law, C. W. Snyder, J. A. Hattie, R. P. McDonald, eds. *Research Methods for Multimode Data Analysis.* New York: Praeger, 1984, pp. 216–284.
42. W. P. Krijnen. Ph.D. dissertation, University of Gronningen, 1993.
43. J. B. Kruskal. In: H. G. Law, C. W. Snyder, J. A. Hattie, R. P. McDonald, eds. *Research Methods for Multimode Data Analysis.* New York: Praeger, 1984, 36–62.
44. J. B. Kruskal, R. A. Harshman, M. E. Lundy. In: R. Coppi, S. Bolasco, eds. *Multiway Data Analysis.* New York: Elsevier, 1989, 115–122.
45. B. C. Mitchell, D. S. Burdick. *J. Chemom.* 8:155–168, 1994.
46. W. S. Rayens, B. C. Mitchell. *Chemom. Intell. Lab. Syst.* (accepted for publication), 1997.
47. M. E. Lundy, R. A. Harshman, J. B. Kruskal. In: R. Coppi, S. Bolasco, eds. *Multiway Data Analysis.* New York: Elsevier, 1989, pp. 123–130.
48. W. P. Krijnen, J. M. F. ten Berge. *Appl. Psychological Measurement* 16:295–305, 1992.
49. C. Hayashi, F. Hayashi. *Behaviormetrika* 11:49–60, 1982.

REFERENCES

50. H. A. L. Kiers, W. P. Krijnen. *Psychometrika* 56:147–152, 1991.
51. R. T. Ross, S. E. Leurgans. *Meth. Enzymology* 246:679–700, 1995.
52. H. A. L. Kiers, R. A. Harshman. *Chemom. Intell. Lab. Syst.* 36:31–40, 1997.
53. P. Paatero. A weighted non-negative least squares algorithm for three-way-PARAFAC-factor analysis. *Chemom. Intell. Lab. Syst.* 36:223–242, 1997.
54. C. Ho, G. D. Christian, E. R. Davidson. *Anal. Chem.* 50:1108–1113, 1978.
55. C. Ho, G. D. Christian, E. R. Davidson. *Anal. Chem.* 52:1071–1079, 1980.
56. C. Ho, G. D. Christian, E. R. Davidson. *Anal. Chem.* 53:92–98, 1981.
57. A. Lorber. *Anal. Chim. Acta.* 164:293–297, 1984.
58. A. Lorber. *Anal. Chem.* 57:2395–2397, 1985.
59. E. Sanchez, B. R. Kowalski. *Anal. Chem.* 58:496–499, 1986.
60. B. E. Wilson, E. Sanchez, B. R. Kowalski. *J. Chemom.* 3:493–498, 1989.
61. N. M. Faber, L. M. C. Buydens, G. Kateman. *J. Chemom.* 8:147–154, 1994.
62. N. M. Faber, L. M. C. Buydens, G. Kateman. *J. Chemom.* 8:181–203, 1994.
63. N. M. Faber, L. M. C. Buydens, G. Kateman. *J. Chemom.* 8:273–285, 1994.
64. N. M. Faber, L. M. C. Buydens, G. Kateman. *J. Chemom.* 7:495–526, 1993.
65. B. E. Wilson, W. Lindberg, B. R. Kowalski. *J. Am. Chem. Soc.* 111:3797–3804, 1989.
66. Y. Wang, O. S. Borgen, B. R. Kowalski, M. Gu, F. Turecek. *J. Chemom.* 1993, 7, 117–130.
67. E. Sanchez, B. R. Kowalski. *J. Chemom.* 4:29–45, 1990.
68. R. Sands, F. W. Young. *Psychometrika.* 45:39–67, 1980.
69. S. Li, J. C. Hamilton, P. J. Gemperline. *Anal. Chem.* 64:599–607, 1992.
70. S. Li, P. J. Gemperline. *J. Chemom.* 7:77–88, 1993.
71. K. S. Booksh, Z. Lin, Z. Wang, B. R. Kowalski. *Anal. Chem.* 66:2561–2569, 1994.
72. K. S. Booksh, B. R. Kowalski. *J. Chemom.* 8:287–292, 1994.
73. P. H. Schönemann. *Psychometrika* 37:441–451, 1972.
74. W. Meredith. *Psychometrika* 29:177–185, 1964.
75. J. de Leeuw, S. Pruzansky. *Psychometrika* 43:479–490, 1978.
76. B. C. Mitchell, D. S. Burdick. *Chemom. Intell. Lab. Syst.* 20:149–161, 1993.
77. D. S. Burdick, X. M. Tu, L. B. McGown, D. W. Millican. *J. Chemom.* 4:15–28, 1990.
78. S. E. Leurgans, R. T. Ross, R. B. Abel. *SIAM J. Matrix Anal. Appl.* 14:1064–1083, 1993.
79. L. Scarminio, M. Kubista. *Anal. Chem.* 65:409–416, 1993.
80. H. A. L. Kiers, A. K. Smilde. *J. Chemom.* 9:179–195, 1995.
81. H. A. L. Kiers, A. K. Smilde. *J. Chemom.* 9:179–195, 1995.
82. M. J. P. Gerritsen, H. Tanis, B. G. M. Vandeginste, G. Kateman. *Anal. Chem.* 64:2042–2056, 1992.
83. L. R. Tucker. In: C. W. Harris, ed. *Problems in Measuring Change.* Madison: University of Wisconsin Press, 1963, pp. 122–137.
84. L. R. Tucker. *Psychometrika* 31:279–311, 1966.
85. J. Röhmel, B. Streitberg, W. M. Herrmann. *Neuropsychobiology* 10:157–163, 1983.
86. J. M. F. ten Berge, J. de Leeuw, P. M. Kroonenberg. *Psychometrika* 52:183–191, 1987.
87. B. K. Alsberg, O. M. Kvalheim. *Chemom. Intell. Lab. Syst.* 24:31–42, 1994.
88. B. K. Alsberg, O. M. Kvalheim. *Chemom. Intell. Lab. Syst.* 24:43–54, 1994.
89. H. A. L. Kiers. *Stat. Appl.* 4:659–667, 1992.
90. P. Brouwer, P. M. Kroonenberg. *J. Classification* 8:93–98, 1991.
91. R. Henrion. *J. Chemom.* 7:477–494, 1993.
92. R. Henrion, G. Henrion, G. C. Onuoha. *Chemom. Intell. Lab. Syst.* 16:87–94, 1992.
93. P. J. Gemperline, K. H. Miller, T. L. West, J. E. Weinstein, J. C. Hamilton, J. T. Bray. *Anal. Chem.* 64:523A–532A, 1992.
94. G. Henrion, D. Nass, G. Michael, R. Henrion. *Fres. J. Anal. Chem.* 352:431–436, 1995.
95. A. K. Smilde, Y. Wang, B. R. Kowalski. *J. Chemom.* 8:21–36, 1994.
96. A. K. Smilde, R. Tauler, J. M. Henshaw, L. W. Burgess, B. R. Kowalski. *Anal. Chem.* 66:3345–3351, 1994.
97. J. Öhman, P. Geladi, S. Wold. *J. Chemom.* 4:79–90, 1990.
98. J. Öhman, P. Geladi, S. Wold. *J. Chemom.* 4:135–146, 1990.
99. Y. Wang, O. Borgen, B. R. Kowalski. *J. Chemom.* 7:439–445, 1993.
100. R. Tauler, A. Smilde, B. R. Kowalski. *J. Chemom.* 9:31–58, 1995.
101. B. Grung, O. M. Kvalheim. *Chemom. Intell. Lab. Syst.* 29:213–221, 1995.
102. L. Ståhle. *Chemom. Intell. Lab. Syst.* 7:95–100, 1989.
103. R. Bro. *J. Chemom.* 10:47–62, 1996.
104. J. M. F. ten Berge. In: R. Coppi, S. Bolasco, eds. *Multiway Data Analysis.* New York: Elsevier, 1989, pp. 53–63.
105. J. B. Kruskal. *Proc. Symp. Appl. Math.* 28:75–104, 1983.
106. B. E. Wilson, B. R. Kowalski. *Anal. Chem.* 61:2277–2284, 1989.
107. L. Nørgaard, C. Ridder. *Chemom. Intell. Lab. Syst.* 23:107–114, 1994.
108. M. McCue, E. R. Malinowski. *J. Chromat. Sci.* 21:229–234, 1983.

109. Z. Lin, K. S. Booksh, L. W. Burgess, B. R. Kowalski. *Anal. Chem.* 66:2552–2560, 1994.
110. J. M. Henshaw, L. W. Burgess, K. S. Booksh, B. R. Kowalski. *Anal. Chem.* 66:3328–3336, 1994.
111. R. Tauler, A. K. Smilde, J. M. Henshaw, L. W. Burgess, B. R. Kowalski. *Anal. Chem.* 6:3337–3344, 1994.
112. J. Öhman, P. Geladi, S. Wold. *J. Chemom.* 4:135–146, 1990.
113. R. Bro, H. Heimdal. *Chemom. Intell. Lab. Syst.* 34:85–102, 1996.
114. Y. Xie, J. J. Baeza-Baeza, G. Ramis-Ramos. *Chemom. Intell. Lab. Syst.* 32:215–232, 1996.
115. S. R. Durell, C. Lee, R. T. Ross, E. L. Gross. *Arch. Biocehm. Biophys.* 278:148–160, 1990.
116. M. L. Gianelli, D. H. Burns, J. B. Callis, G. D. Christian, N. H. Andersen. *Anal. Chem.* 55:1858–1862, 1983.
117. M. Gui, S. Rutan, A. Agbodjan. *Anal. Chem.* 67:3293–3299, 1995.
118. R. B. Poe, S. C. Rutan. *Anal. Chim. Acta.* 283:845–853, 1993.
119. J. Keun, R. T. Ross, S. Thampi, S. Leurgans. *J. Phys. Chem.* 96:9159–9162, 1992.
120. C. Lee, K. Kim, R. T. Ross. *Korean Biochem. J.* 24:374–379, 1991.
121. R. T. Ross, C. Lee, C. M. Davis, B. M. Ezzedine, E. A. Fayyad, S. Leurgans. *Biochim. Biophys. Acta.* 1056:317–320, 1991.
122. D. W. Millican, L. B. McGown. *Anal. Chem.* 62:2242–2247, 1990.
123. D. S. Burdick, X. M. Tu, L. B. McGown, D. W. Millican. *J. Chemom.* 4:15–28, 1990.
124. K. K. Karukstis, D. A. Krekel, D. A. Weinberger, R. A. Bittker, N. R. Naito, S. H. Bloch. *J. Phys. Chem.* 99:449–453, 1995.
125. M. M. C. Ferreira, M. L. Brandes, I. M. C. Ferreira, K. S. Booksh, W. C. Dolowy, M. Gouterman, B. R. Kowalski. *Appl. Spectrosc.*, 49:1317–1325, 1995.
126. W. M. Herrmann, J. Röhmel, B. Streitberg, J. Willmann. *Neuropsycho-biology* 10:164–172, 1983.

NUMERICAL INDEX TO SPECTRA 1–2130

SPECTRA NUMBERS 1–560

UV-Vis (200–900 nm) and SW-NIR (650–850 nm): Organic compounds and Polymers

Spectra Numbers	Compound	Spectra Numbers	Compound
1–2	Quartz cuvet	63–65	Ethanol
3–5	Acetone	66–68	2-Ethyl-1-butanol
6–8	Chloroform	69–71	2-Methyl-1-butyn-2-ol
9–11	Water, deionized	72–74	Formic acid
12–14	Isopropanol	75–77	3-Methylcyclohexanol
15–17	Methylal	78–80	4-Methylcyclohexanol
18–20	Acetic acid	81–83	2-Octanol
21–23	Dichloroacetic acid	84–86	Octyl alcohol
24–26	Gluconic acid	87–89	Propyl alcohol
27–29	Methoxacetic acid	90–92	2-Propyn-1-ol
30–32	Butyric acid	93–95	Heptanoic acid
33–35	iso-Butyric acid	96–98	Diethylene glycol monobutyl ether
36–38	Hexanoic acid	99–101	Diethylene glycol monoethyl rther
39–41	2-Ethylbutyric acid	102–104	Ethylene glycol monobutyl ether
42–44	Acetic anhydride	105–107	3-Methoxy-1-butanol
45–47	Butyric anhydride	108–110	1-Methoxy-2-propanol
48–50	Propionic anhydride	111–113	Diacetone alcohol
51–53	tert-Amyl alcohol	114–116	1,3-Butanediol
54–56	Butyl alcohol	117–119	Linoleic acid
57–59	tert-Butyl alcohol	120–122	2-Ethyl-1,3-hexanediol
60–62	2-Ethylhexanoic acid	123–125	Propylene glycol

NUMERICAL INDEX TO SPECTRA 1–2130

Spectra Numbers	Compound
126–128	Diethylene glycol
129–131	Dipropylene glycol
132–134	Nonanoic acid
135–137	Trethylene glycol
138–140	Benzyl alcohol
141–143	DL-a-Methylbenzyl alcohol
144–146	2-Phenylethyl alcohol
147–149	3-Phenyl-1-propanol
150–152	2-Phenoxyethanol
153–155	o-Hydroxyacetophenone
156–158	n-Butyraldehyde
159–161	Citronellal
162–164	Crotonaldehyde
165–167	Formaldehyde
168–170	Oleic acid
171–173	Glyoxal
174–176	Propionaldehyde
177–179	Benzaldehyde
180–182	trans-Cinnamaldehyde
183–185	Salicylaldehyde
186–188	Propionic acid
189–191	N,N-Dimethylacetamide
192–194	N,N-Dimethylformamide
195–197	10-Undecenoic acid
198–200	Valeric acid
201–203	iso-Valeric acid
204–207	Acrylonitrile/butadiene/styrene resin
208–211	Alginic acid, sodium salt
212–215	Butyl methacrylate/isobutyl methacrylate copolymer
216–219	Cellulose acetate
220–223	Cellulose acetate butyrate
224–227	Cellulose propionate
228–230	Cellulose triacetate
231–233	Ethyl cellulose
234–236	Ethylene/acrylic acid copolymer
237–240	Ethylene/ethyl acrylate, 82/18 copolymer
241–244	Ethylene/propylene, 60/40 copolymer
245–248	Ethylene/vinyl acetate, 86/14 copolymer
249–252	Ethylene/vinyl acetate, 82/18 copolymer
253–256	Ethylene/vinyl acetate, 75/25 copolymer
257–260	Ethylene/vinyl acetate, 72/28 copolymer
261–264	Ethylene/vinyl acetate, 67/33 copolymer
265–268	Ethylene/vinyl acetate, 60/40 copolymer
269–271	Hydroxybutyl methyl cellulose, 8% hydroxybutyl, 20% methoxyl
272–274	Hydroxypropyl cellulose
275–277	Hydroxypropyl methyl cellulose, 10% hydroxypropyl, 30% methoxyl
278–280	Methyl cellulose
281–283	Methyl vinyl ether/maleic acid, 50/50 copolymer
284–286	Methyl vinyl ether/maleic anhydride, 50/50 copolymer
287–290	Nylon 6 (polycaprolactam)
291–295	Nylon 6/6 (polyhexamethylene adipamide)
296–301	Nylon 6/9 (polyhexamethylene nonanediamide)
302–305	Nylon 6/10 (polyhexamethylene sebacamide)
306–309	Nylon 6/12 (polyhexamethylene dodecanediamide)
310–312	Nylon 6/T (polytrimethyl hexamethylene terephthalamide)
313–316	Nylon 11 (polyundecanoamide)
317–320	Nylon 12 (polylaurylactam)
321–324	Phenoxy resin
325–328	Polyacetal
329–331	Polyacrylamide
332–334	Polyacrylamide, carboxyl modified (low content)
335–337	Polyacrylamide, carboxyl modified (high content)
338–340	Poly(acrylic acid)
341–343	Polyamide resin

NUMERICAL INDEX TO SPECTRA 1–2130

Spectra Numbers	Compound
344–347	1,2-Polybutadiene
348–351	Poly(1-butene), isotactic
352–354	Poly(n-butyl methacrylate)
355–358	Polycaprolactone
359–362	Polycarbonate resin
363–365	Poly(diallyl isophthalate)
366–368	Poly(diallyl phthalate)
369–371	Poly(2,6-dimethyl-p-phenylene oxide)
372–374	Poly(4,4-dipropoxy-2,2-diphenyl propane fumarate)
375–378	Poly(ethyl methacrylate)
379–382	Polyethylene, high density
383–386	Polyethylene, chlorinated (25% Cl)
387–389	Polyethylene, chlorinated (36% Cl)
390–393	Polyethylene, chlorinated (42% Cl)
394–396	Polyethylene, chlorinated (48% Cl)
397–399	Polyethylene, chlorosulfonated
400–403	Poly(ethylene oxide)
404–407	Polyethylene, oxidized
408–410	Poly(ethylene terephthalate)
411–413	Poly(2-hydroxyethyl methacrylate)
414–417	Poly(isobutyl methacrylate)
418–420	Polyisoprene, chlorinated
421–424	Poly(methyl methacrylate)
425–428	Poly(4-methyl-1-pentene)
429–432	Poly(alpha-methylstyrene)
433–435	Poly(p-phenylene ether-sulphone)
436–438	Poly(phenylene sulfide)
439–441	Polypropylene, isotactic, chlorinated
442–445	Polypropylene, isotactic
446–449	Polystyrene
450–452	Polysulfone resin
453–455	Poly(tetrafluoroethylene)
456–458	Poly(2,4,6-tribromostyrene)
459–461	Poly(vinyl acetate)
462–465	Poly(vinyl alcohol), 100% hydrolyzed
466–469	Poly(vinyl alcohol), 98% hydrolyzed
470–473	Poly(vinyl butyral)

Spectra Numbers	Compound
474–477	Poly(vinyl chloride)
478–481	Poly(vinyl chloride), carboxylated
482–484	Poly(vinyl formal)
485–487	Poly(vinyl pyrrolidone)
488–491	Poly(vinyl stearate)
492–494	Poly(vinylidene fluoride)
495–498	Styrene/acrylonitrile, 75/25 copolymer
499–501	Styrene/acrylonitrile, 70/30 copolymer
502–505	Styrene/allyl alcohol copolymer
506–509	Styrene/butadiene, ABA block copolymer
510–513	Styrene/butyl methacrylate copolymer
514–517	Styrene/ethylene/butylene, ABA block copolymer
518–520	Styrene/isoprene, ABA block copolymer
521–524	Styrene/maleic anhydride, 50/50 copolymer
525–528	Vinyl alcohol/vinyl butyral copolymer (80% vinyl butyral)
529–531	Vinyl chloride/vinyl acetate copolymer (81% vinyl chloride)
532–535	Vinyl chloride/vinyl acetate copolymer (88% vinyl chloride)
536–538	Vinyl chloride/vinyl acetate copolymer (90% vinyl chloride)
539–541	Vinyl chloride/vinyl acetate copolymer carboxylated (86% vinyl chloride)
542–545	Vinyl chloride/vinyl acetate/hydroxypropyl acrylate terpolymer (80% vinyl chloride)
546–548	Vinyl chloride/vinyl acetate/vinyl alcohol terpolymer (91% vinyl chloride)
549–551	Vinylidene chloride/acrylonitrile copolymer (20% acrylonitrile)
552–554	Vinylidene chloride/vinyl chloride copolymer (5% vinylidene chloride)
555–557	N-Vinyl pyrrolidone/vinyl acetate copolymer
558–560	Zein, purified

SPECTRA NUMBERS 561–592

SW-NIR (800–1100 nm): Organic Compounds and Mixtures

Spectra Numbers	Compound	Spectra Numbers	Compound
561–562	Acetone	577–578	n-Decane
563–564	Cyclohexane	579–580	n-Heptane
565–566	Ethylbenzene	581–582	Pentane
567–568	Gasoline (High ethanol content)	583–584	p-Xylene
569–570	Gasoline (High aromatics content)	585–586	tert-Butanol
571–572	Gasoline (Low aromatics content)	587–588	Toluene
573–574	Isopropanol	589–590	Trimethyl pentane
575–576	tert-Butyl methyl ether	591–592	Water, deionized

SPECTRA NUMBERS 593–1006

LW-NIR (1000–2600 nm): Organic Compounds and Polymers

Spectra Numbers	Compound	Spectra Numbers	Compound
593–594	Acetone	633–634	Valeric acid
595–596	Polyester	635–636	iso-Valeric acid
597–598	Chloroform	637–638	Dichloroacetic acid
599–600	Water, deionized (0.2 cm)	639–640	Gluconic acid
601–602	Isopropanol	641–642	Lactic acid
603–604	Acetic acid	643–644	Methoxyacetic acid
605–606	Methylal	645–646	Butyric anhydride
607–608	Butyric acid	647–648	Propionic anhydride
609–610	iso-Butyric acid	649–650	tert-Amyl alcohol
611–612	2-Ethybutyric acid	651–652	Butyl alcohol
613–614	Hexanoic acid	653–654	tert-Butyl alcohol
615–616	2-Ethylhexanoic acid	655–656	Ethyl alcohol
617–618	Formic acid	657–658	2-Ethyl-1-butanol
619–620	Heptanoic acid	659–660	2-Methyl-3-butyn-2-ol
621–622	Nonanoic acid	661–662	3-Methylcyclohexanol
623–624	Linoleic acid	663–664	4-Methylcyclohexanol
625–626	Octanoic acid	665–666	2-Octanol
627–628	Oleic acid	667–668	Octyl alcohol
629–630	Propionic acid	669–670	Propyl alcohol
631–632	10-Undecenoic acid	671–672	2-Propyn-1-ol

NUMERICAL INDEX TO SPECTRA 1–2130

Spectra Numbers	Compound
673–674	Diethylene glycol monobutyl ether
675–676	Diethylene glycol monoethyl ether
677–678	Ethylene glycol monobutyl ether
679–680	3-Methoxy-1-butanol
681–682	1-Methoxy-2-propanol
683–684	Diacetone alcohol
685–686	1,3-Butanediol
687–688	2-Ethyl-1,3-hexanediol
689–690	Propylene glycol
691–692	Diethylene glycol
693–694	Dipropylene glycol
695–696	Triethylene glycol
697–698	Benzyl alcohol
699–700	DL-a-Methylbenzyl alcohol
701–702	2-Phenylethyl alcohol
703–704	3-Phenyl-1-propanol
705–706	2-Phenoxyethanol
707–708	o-Hydroxyacetophenone
709–710	n-Butyraldehyde
711–712	Citronellal
713–714	Crotonaldehyde
715–716	Formaldehyde
717–718	Glyoxal
719–720	Propionaldehyde
721–722	Benzaldehyde
723–724	trans-Cinnamaldehyde
725–726	Anisaldehyde
727–728	Salicylaldehyde
729–730	N,N-Dimethylacetamide
731–732	N,N-Dimethylformamide
733–734	Acetic acid
735–736	Styrene:butadiene:styrene ABA film
737–738	Styrene:isoprene:styrene
739–740	Polystyrene:polybutadiene/polystyrene:polyisoprene
741–742	Poly-alpha-olefins, amorphous
743–744	Styrene:ethylene butylene:styrene copolymer
745–746	Polystyrene:polyisoprene/polycyclopentadiene resin

Spectra Numbers	Compound
747–748	Polypropylene, atactic
749–750	Ethylene vinyl acetate with C9 and C5 hydrocarbon resins
751–752	Styrene:ethylene butylene:styrene copolymer II
753–754	n-Decane
755–756	Isooctane
757–758	Dimethicone (Silicone)
759–760	Starch
761–762	Menthol
763–764	Polypropylene (66%) and polyester (34%)
765–766	Cellulose (63%) and polypropylene (37%)
767–768	Polypropylene and polyethylene
769–770	Rayon and polyester
771–772	Polypropylene/polyethylene (60%) and polyester (40%)
773–774	Camphor
775–776	Silicone fluid (Dow, 350 cs)
777–778	Silicone fluid (SWS, 350 cs)
779–780	Prolpylene glycol
781–782	Methanol
783–784	Toluene
785–786	Polypropylene, crystalline
787–788	Rayon
789–790	Sodium di-octyl sulfosuccinate
791–792	Alcohol ethoxylate
793–794	Phospholipid
795–796	Gyceryl phthalate
797–798	Aluminum oleate
799–800	Silicone fluid (Dow, 1000 cs)
801–802	Silicone fluid (Dow 2–1922)
803–804	Triethanolamine
805–806	Poly(acrylic acid)
807–808	Polystyrene (32 cm^{-1} resolution)
809–810	Polystyrene (4 cm^{-1} resolution)
811–812	Acrylonitrile/butadiene/styrene resin
813–814	Alginic acid, sodium salt
815–816	Butyl methacrylate/isobutyl methacrylate copolymer

NUMERICAL INDEX TO SPECTRA 1–2130

Spectra Numbers	Compound
817–818	Cellulose acetate
819–820	Cellulose acetate butyrate
821–822	Cellulose propionate
823–824	Cellulose triacetate
825–826	Ethyl cellulose
827–828	Ethylene/acrylic acid copolymer
829–830	Ethylene/ethyl acrylate, 82/18 copolymer
831–832	Ethylene/propylene, 60/40 copolymer
833–834	Ethylene/vinyl acetate, 86/14 copolymer
835–836	Ethylene/vinyl acetate, 82/18 copolymer
837–838	Ethylene/vinyl acetate, 75/25 copolymer
839–840	Ethylene/vinyl acetate, 72/28 copolymer
841–842	Ethylene/vinyl acetate, 60/40 copolymer
843–844	Hydroxybutyl methyl cellulose, 8% hydroxybutyl, 20% methoxyl
845–846	Hydroxypropyl cellulose
847–848	Hydroxypropyl methyl cellulose, 10% hydroxypropyl, 30% methoxyl
849–850	Methyl cellulose
851–852	Methyl vinyl ether/maleic acid, 50/50 copolymer
853–854	Nylon 6 (polycaprolactam)
855–856	Nylon 6/6 (Polyhexamethylene adipamide)
857–858	Nylon 6/9 (polyhexamethylene nonanediamide)
859–860	Nylon 6/10 (polyhexamethylene sebacamide)
861–862	Nylon 6/12 (polyhexamethylene dodecanediamide)
863–864	Nylon 6/T (polytrimethyl hexamethylene terephthalamide)
865–866	Nylon 11 (polyundecanoamide)
867–868	Nylon 12 (polylaurylactam)
869–870	Phenoxy resin
871–872	Polyacetal
873–874	Polyacrylamide

Spectra Numbers	Compound
875–876	Polyacrylamide, carboxyl modified (low content)
877–878	Polyacrylamide, carboxyl modified (high content)
879–880	Poly(acrylic acid)
881–882	Polyamide resin
883–884	1,2-Polybutadiene
885–886	Poly(1-butene), isotactic
887–888	Poly(n-butyl methacrylate)
889–890	Polycaprolactone
891–892	Polycarbonate resin
893–894	Poly(diallyl isophthalate)
895–896	Poly(diallyl phthalate)
897–898	Poly(2,6-dimethyl-p-phenylene oxide)
899–900	Poly(4,4-dipropoxy-2,2-diphenyl propane fumarate)
901–902	Poly(ethyl methacrylate)
903–904	Polyethylene, high density
905–906	Polyethylene, chlorinated (25% Cl)
907–908	Polyethylene, chlorinated (36% Cl)
909–910	Polyethylene, chlorinated (42% Cl)
911–912	Polyethylene, chlorinated (48% Cl)
913–914	Polyethylene, chlorosulfonated
915–916	Poly(ethylene oxide)
917–918	Polyethylene, oxidized
919–920	Poly(ethylene terephthalate)
921–922	Poly(2-hydroxyethyl methacrylate)
923–924	Poly(isobutyl methacrylate)
925–926	Polyisoprene, chlorinated
927–928	Poly(methyl methacrylate)
929–930	Poly(4-methyl-1-pentene)
931–932	Poly(alpha-methylstyrene)
933–934	Poly(p-phenylene ether-sulphone)
935–936	Poly(phenylene sulfide)
937–938	Polypropylene, isotactic, chlorinated
939–940	Polypropylene, isotactic
941–942	Polystyrene
943–944	Polysulfone resin
945–946	Poly(tetrafluoroethylene)

Spectra Numbers	Compound	Spectra Numbers	Compound
947–948	Poly(2,4,6-tribromostyrene)	983–984	Styrene/maleic anhydride, 50/50 copolymer
949–950	Poly(vinyl acetate)	985–986	Vinyl alcohol/vinyl butyral copolymer (80% vinyl butyral)
951–952	Poly(vinyl alcohol), 100% hydrolyzed	987–988	Vinyl chloride/vinyl acetate copolymer (81% vinyl chloride)
953–954	Poly(vinyl alcohol), 98% hydrolyzed	989–990	Vinyl chloride/vinyl acetate copolymer (88% vinyl chloride)
955–956	Poly(vinyl butyral)	991–992	Vinyl chloride/vinyl acetate copolymer (90% vinyl chloride)
957–958	Poly(vinyl chloride)	993–994	Vinyl chloride/vinyl acetate copolymer carboxylated (86% vinyl chloride)
959–960	Poly(vinyl chloride), carboxylated		
961–962	Poly(vinyl formal)		
963–964	Poly(vinyl pyrrolidone)		
965–966	Poly(vinyl stearate)	995–996	Vinyl chloride/vinyl acetate/hydroxypropyl acrylate terpolymer (80% vinyl chloride)
967–968	Poly(vinylidene fluoride)		
969–970	Styrene/acrylonitrile, 75/25 copolymer	997–998	Vinyl chloride/vinyl acetate/vinyl alcohol terpolymer (91% vinyl chloride)
971–972	Styrene/acrylonitrile, 70/30 copolymer		
973–974	Styrene/allyl alcohol copolymer	999–1000	Vinylidene chloride/acrylonitrile copolymer (20% acrylonitrile)
975–976	Styrene/butadiene, ABA block copolymer	1001–1002	Vinylidene chloride/vinyl chloride copolymer (5% vinylidene chloride)
977–978	Styrene/butyl methacrylate copolymer	1003–1004	N-Vinyl pyrrolidone/vinyl acetate copolymer
979–980	Styrene/ethylene/butylene, ABA block copolymer	1005–1006	Zein, purified
981–982	Styrene/isoprene, ABA block copolymer		

SPECTRA NUMBERS 1007–2000

Infrared (4000–500 cm^{-1}): Organic Compounds, Polymers, Surfactants, and HATR

Spectra Numbers	Compound	Spectra Numbers	Compound
1007	Kymene	1013	Cellulose, 100% soft wood bleached kraft
1008	Polyvinyl alcohol, standard		
1009	Polyacrylamide	1014	Polyethyleneterphthalate fiber, melted film
1010	Cellulose 35 by 70 micron single fiber	1015	2,6-Di-*tert*-butyl-*p*-cresol (BHT)
1011	Starch	1016	Sodium sulfate (in KBr)
1012	1,2-Propanediol, capillary film	1017	Sodium carbonate

NUMERICAL INDEX TO SPECTRA 1–2130

Spectra Numbers	Compound	Spectra Numbers	Compound
1018	Solium silicate, soluble	1052	Diatomaceous earth (in KBr)
1019	Cellophane	1053	Zinc sulfate (in KBr)
1020	Carboxymethyl cellulose (CMC) absorbent polymer	1054	Sesame seed oil, between KCl windows
1021	Wood pulp, thermomechanical (in KBr)	1055	Machine oil, capillary film between KCl windows
1022	Polyamide—epichlorohydrin resin	1056	Water (in KBr)
1023	Paraffin wax, food grade	1057	Brown kraft bag, fiber (in KBr)
1024	Protein, human hair (IR microspectroscopy)	1058	Grease, heavy lubricating, thin film
1025	Polyacrylamide, modified dried film	1059	Grease, light lubricating, thin film
1026	Triton X-102	1060	Calcium oleate
1027	Polyvinyltoluene, mixed isomers	1061	Silicone, Y-12226,
1028	Calcium stearate (in KBr)	1062	Sodium dihydrogen phosphite (in KBr)
1029	Polyethylene film	1063	Starch, cast film on AgBr
1030	Bis(2-ethylhexyl) phthalate, capillary film	1064	Polyvinyl alcohol, 78000 M.W.
1031	D-Sorbitol, 99% (in KBr)	1065	Methylparabenzoic acid (in KBr)
1032	Glycerol monostearate, cast film on KCl	1066	Ethylparabenzoic acid (in KBr)
1033	Glycerol distearate, cast film on KCl	1067	Propylparabenzoic acid (in KBr)
1034	Calcium carbonate (in KBr)	1068	Butylparabenzoic acid (in KBr)
1035	Magnesium silicate (talc) (in KBr)	1069	Citric acid, cast film on AgBr
1036	Nujol oil, between KCl windows	1070	Sorbic acid (in KBr)
1037	Cola, Coke Classic, cast film on AgBr	1071	Dimethicone, 350 cs, film cast on KCl
1038	Corn syrup, dark, cast film on AgBr	1072	Benzoic acid, 99% + (in KBr)
1039	Corn syrup, light, cast film on AgBr	1073	Ammonium bicarbonate (in KBr)
1040	Molasses, cast film on AgBr	1074	Sodium bicarbonate (in KBr)
1041	Dextrose, cast film on AgBr	1075	Siloxane, wetting agent
1042	Lemon-lime soda, cast film on AgBr	1076	Malic acid (in KBr)
1043	Erucamide	1077	Fluorocarbon surfactant, cast film
1044	Cotton seed oil, between KCl windows	1078	Silicone (FTS-226), between KCl windows
1045	Polyvinyl alcohol/polyvinyl acetate	1079	Sodium lauryl sulfate (in KBr)
1046	Coffee, on AgBr window	1080	Ethyl cellulose, chloroform cast film
1047	Coffee, instant, on AgBr window	1081	Protein, human blood, cast film on AgBr
1048	Tea, water extract, cast film	1082	Menthol
1049	Urine, synthetic, cast film on AgBr	1083	Ceresin wax
1050	Sugar, brown, cast film on AgBr	1084	Polypropylene (66%) and polyester (34%)
1051	Glycerol oleate, liquid film		

NUMERICAL INDEX TO SPECTRA 1–2130

Spectra Numbers	Compound
1085	Wood pulp (63%), polyethylene (21%), polypropylene (16%)
1086	Wood pulp (68%), polypropylene (32%)
1087	Polypropylene/polyethylene
1088	Paraffin (bees) wax (on KCl)
1089	Sodium periodate (in KBr)
1090	Sodium iodate (in KBr)
1091	Potassium iodate (in KBr)
1092	Calcium oxide (in KBr)
1093	Methylacrylate, neat liquid
1094	Poly(propylene glycol), M.W. approximately 425
1095	Benzoyl peroxide, 97% (in KBr)
1096	Aloe vera, capillary film between KCl
1097	Aluminum ammonium sulfate (in KBr)
1098	Aluminum potassium sulfate (in KBr)
1099	Aluminum sodium sulfate (in KBr)
1100	Sucrose
1101	Hexanol
1102	Glucose
1103	Fructose
1104	Dextrose, cast film on AgBr
1105	Cola, generic
1106	Heptane
1107	Starch, extract cast film on AgBr
1108	Aniline, liquid
1109	Polycarbonate
1110	Decane, liquid
1111	FD&C Blue no. 1 dye (in KBr)
1112	Toluene
1113	Stearyl alcohol
1114	Methylene (3,5-di-tert-butyl-4-hydroxyhydrocinnamate)
1115	Polyamide-epichlorohydrin resin
1116	Bromophenol blue
1117	Poneac S red, dye
1118	Sodium bisulfite

Spectra Numbers	Compound
1119	Silica, flint glass
1120	Cellulose (natural cotton)
1121	Silk
1122	Wool
1123	Menthol (in KBr)
1124	Boric Acid (in KBr)
1125	Camphor (in KBr)
1126	Triethanolamine
1127	Silicone fluid, 1000cs
1128	Silicone fluid, 350cs
1129	Silicone fluid, Dow 2–1922
1130	Malic acid (in KBr)
1131	Benzethonium chloride (in KBr)
1132	Polypropylene: polyethylene (60%) and polyester (40%)
1133	Lycra, elastic thread
1134	Glyoxolated cationic polyamide, cast film
1135	Imidazoline-based debonder, cast film on KCl
1136	tert-Amyl alcohol (ATR)
1137	Butyl alcohol (ATR)
1138	tert-Butyl alcohol (ATR)
1139	Ethyl alcohol (ATR)
1140	2-Ethyl-1-butanol (ATR)
1141	2-Methyl-3-butyn-2-ol (ATR)
1142	3-Methylcyclohexanol (ATR)
1143	4-Methylcyclohexanol (ATR)
1144	2-Octanol (ATR)
1145	Octyl alcohol (ATR)
1146	Propyl alcohol (ATR)
1147	2-Propyn-1-ol (ATR)
1148	Diethylene glycol monobutyl ether (ATR)
1149	Diethylene glycol monoethyl ether (ATR)
1150	Ethylene glycol monobutyl ether (ATR)
1151	3-Methoxy-1-butanol (ATR)
1152	1-Methoxy-2-propanol (ATR)

NUMERICAL INDEX TO SPECTRA 1–2130

Spectra Numbers	Compound
1153	1,3-Dichloro-2-propanol (ATR)
1154	2,3-Dichloro-1-propanol (ATR)
1155	Acetoin (3-hydroxy-2-butanone) (ATR)
1156	Diacetone alcohol (ATR)
1157	1,3-Butanediol (ATR)
1158	2-Ethyl-1,3-hexanediol (ATR)
1159	1,2,6-Hexanetriol (ATR)
1160	1,3-Propanediol (ATR)
1161	Propylene glycol (ATR)
1162	Diethylene glycol (ATR)
1163	Dipropylene glycol (ATR)
1164	Benzyl alcohol (ATR)
1165	DL-a-Methylbenzyl alcohol (ATR)
1166	2-Phenylethyl alcohol (ATR)
1167	3-Phenyl-1-propanol (ATR)
1168	2-Phenoxyethanol
1169	o-Hydroxyacetophenone (ATR)
1170	N-Butyraldehyde (ATR)
1171	Citral (ATR)
1172	Citronellal (ATR)
1173	Crotonaldehyde (ATR)
1174	Formaldehyde (ATR)
1175	Glyoxal (ATR)
1176	Propionaldehyde (ATR)
1177	Aldol
1178	Benzaldehyde (ATR)
1179	trans-Cinnamaldehyde (ATR)
1180	p-Tolualdehyde (ATR)
1181	Anisaldehyde (ATR)
1182	Salicylaldehyde (ATR)
1183	N,N-Dimethylacetamide (ATR)
1184	N,N-Dimethylformamide (ATR)
1185	Acetone (ATR)
1186	C35-C60 Hydrocarbon wax (ATR)
1187	Chloroform (ATR)
1188	Dimethicone, 10,000 cs (ATR)
1189	Glycerol (ATR)

Spectra Numbers	Compound
1190	Isopropanol (ATR)
1191	Methanol (ATR)
1192	Petrolatum (ATR)
1193	Silicon wax (ATR)
1194	Soy sterol (ATR)
1195	Sunflower seed, oil (ATR)
1196	Water, deionized (ATR)
1197	Ethyl cellulose, chloroform extract cast film
1198	Glycerol, cast film on KCl
1199	Polypropylene standard
1200	Polyethylene (high density) and polypropylene
1201	Polyethylene (high density)
1202	Polyvinyl alcohol
1203	Polyethylene/polyvinyl acetate standard
1204	Polylactic acid (30%) and polyvinyl alcohol (70%) blend
1205	Polyvinyl pyrrolidone
1206	Poly(butyl acrylate)
1207	Poly(glycolide-co-lactide), approximately 18% oxylactoyl units
1208	Poly(hydroxy ethyl methacrylate) melt
1209	Poly(hydroxy ethyl methacrylate)
1210	Poly(lactic acid)
1211	Poly(lactic acid-g-AA)
1212	Poly(lactic acid-g-methyl methacrylate) (30%) and polyvinyl alcohol (70%)
1213	Poly(maleic acid)
1214	Poly(methyl methacrylate)
1215	Poly(vinyloctadecylether-co-maleic anhydride-co-maleic acid)
1216	Poly(lactic acid), reflectance
1217	Polyvinyl methyl ether/isobutyl vinyl ether (12.0)
1218	Polyvinyl methyl ether/isobutyl vinyl ether (4.0)
1219	Polyvinyl methyl ether/isobutyl vinyl ether (8.3)
1220	Poly(acrylic acid)

NUMERICAL INDEX TO SPECTRA 1–2130

Spectra Numbers	Compound
1221	Polystyrene (17%) and polyisoprene (83%)
1222	Polystyrene:polybutadiene copolymer
1223	Styrene:ethylene/butylene (86%) and polystyrene (14%)
1224	Polystyrene (Dow)
1225	Polypropylene with trace Polyethylene
1226	Polyvinyl alcohol/polyvinyl acetate
1227	Acrylonitrile/butadiene/styrene resin
1228	Alginic acid, sodium salt
1229	Butyl methacrylate/isobutyl methacrylate copolymer
1230	Cellulose acetate
1231	Cellulose acetate butyrate
1232	Cellulose propionate
1233	Cellulose triacetate
1234	Ethyl cellulose
1235	Ethylene/acrylic acid copolymer
1236	Ethylene/ethyl acrylate, 82/18 copolymer
1237	Ethylene/propylene, 60/40 copolymer
1238	Ethylene/vinyl acetate, 86/14 copolymer
1239	Ethylene/vinyl acetate, 82/18 copolymer
1240	Ethylene/vinyl acetate, 75/25 copolymer
1241	Ethylene/vinyl acetate, 72/28 copolymer
1242	Ethylene/vinyl acetate, 67/33 copolymer
1243	Ethylene/vinyl acetate, 60/40 copolymer
1244	Hydroxybutyl methyl cellulose, 8% hydroxybutyl, 20% methoxyl
1245	Hydroxypropyl cellulose
1246	Hydroxypropyl methyl cellulose, 10% hydroxypropyl, 30% methoxyl
1247	Methyl cellulose
1248	Methyl vinyl ether/maleic acid, 50/50 copolymer

Spectra Numbers	Compound
1249	Methyl vinyl ether/maleic anhydride, 50/50 copolymer
1250	Nylon 6 (polycaprolactam)
1251	Nylon 6/6 (polyhexamethylene adipamide)
1252	Nylon 6/9 (polyhexamethylene nonanediamide)
1253	Nylon 6/10 (polyhexamethylene sebacamide)
1254	Nylon 6/12 (polyhexamethylene dodecanediamide)
1255	Nylon 6/T (polytrimethyl hexamethylene terephthalamide)
1256	Nylon 11 (polyundecanoamide)
1257	Nylon 12 (polylaurylactam)
1258	Phenoxy resin
1259	Polyacetal
1260	Polyacrylamide
1261	Polyacrylamide, carboxyl modified (low content)
1262	Polyacrylamide, carboxyl modified (high content)
1263	Poly(acrylic acid)
1264	Polyamide resin
1265	1,2-Polybutadiene
1266	Poly(1-butene), isotactic
1267	Poly(n-butyl methacrylate)
1268	Polycaprolactone
1269	Polycarbonate resin
1270	Poly(diallyl isophthalate)
1271	Poly(diallyl phthalate)
1272	Poly(2,6-dimethyl-p-phenylene oxide)
1273	Poly(4,4-dipropoxy-2,2-diphenyl propane fumarate)
1274	Poly(ethyl methacrylate)
1275	Polyethylene, high density
1276	Polyethylene, chlorinated, (25% Cl)
1277	Polyethylene, chlorinated (36% Cl)
1278	Polyethylene, chlorinated (42% Cl)
1279	Polyethylene, chlorinated (48% Cl)
1280	Polyethylene, chlorosulfonated

NUMERICAL INDEX TO SPECTRA 1–2130

Spectra Numbers	Compound
1281	Poly(ethylene oxide)
1282	Polyethylene, oxidized
1283	Poly(ethylene terephthalate)
1284	Poly(2-hydroxyethyl methacrylate)
1285	Poly(isobutyl methacrylate)
1286	Polyisoprene, chlorinated
1287	Poly(methyl methacrylate)
1288	Poly(4-methyl-1-pentene)
1289	Poly(alpha-methylstyrene)
1290	Poly(p-phenylene ether-sulphone)
1291	Poly(phenylene sulfide)
1292	Polypropylene, isotactic, chlorinated
1293	Polypropylene, isotactic
1294	Polystyrene
1295	Polysulfone resin
1296	Poly(tetrafluoroethylene)
1297	Poly(2,4,6-tribromostyrene)
1298	Poly(vinyl acetate)
1299	Poly(vinyl alcohol), 100% hydrolyzed
1300	Poly(vinyl alcohol), 98% hydrolyzed
1301	Poly(vinyl butyral)
1302	Poly(vinyl chloride)
1303	Poly(vinyl chloride), carboxylated
1304	Poly(vinyl formal)
1305	Polyvinyl pyrrolidone
1306	Poly(vinyl stearate)
1307	Poly(vinylidene fluoride)
1308	Styrene/acrylonitrile, 75/25 copolymer
1309	Styrene/acrylonitrile, 70/30 copolymer
1310	Styrene/allyl alcohol copolymer
1311	Styrene/butadiene, ABA block copolymer
1312	Styrene/butyl methacrylate copolymer
1313	Styrene/ethylene/butylene, ABA block copolymer
1314	Styrene/isoprene, ABA block copolymer
1315	Styrene/maleic anhydride, 50/50 copolymer
1316	Vinyl alcohol/vinyl butyral copolymer (80% vinyl butyral)
1317	Vinyl chloride/vinyl acetate copolymer (81% vinyl chloride)
1318	Vinyl chloride/vinyl acetate copolymer (88% vinyl chloride)
1319	Vinyl chloride/vinyl acetate copolymer (90% vinyl chloride)
1320	Vinyl chloride/vinyl acetate copolymer carboxylated (86% vinyl chloride)
1321	Vinyl chloride/vinyl acetate/hydroxypropyl acrylate terpolymer (80% vinyl chloride)
1322	Vinyl chloride/vinyl acetate/vinyl alcohol terpolymer (91% vinyl chloride)
1323	Vinylidene chloride/acrylonitrile copolymer (20% acrylonitrile)
1324	Vinylidene chloride/vinyl chloride copolymer (5% vinylidene chloride)
1325	N-Vinyl pyrrolidone/vinyl acetate copolymer
1326	Zein, purified
1327	Ammonium palmitate
1328	SC10–008 Cetiol 1414E surfactant
1329	Diethanolamine–oleic acid condensate
1330	Pentaerythritol distearate
1331	Sodium ditridecyl sulfosuccinate
1332	Sodium lignosulfonate, 2 moles sodium/lignin unit
1333	Sodium N-methyl-N-oleyl taurate
1334	Dioleate of polyethylene glycol 1540
1335	Dioleate of polyethylene glycol 600
1336	Dioleate of polyethylene glycol 200
1337	Oleate of polyethylene glycol 200
1338	Oleate of ethylene glycol
1339	Distearate of polyethylene glycol 1000
1340	Distearate of polyethylene glycol 300

NUMERICAL INDEX TO SPECTRA 1–2130

Spectra Numbers	Compound	Spectra Numbers	Compound
1341	Stearate of polyethylene glycol 200	1374	P.O.E. Oleic acid (6 moles EtO)
1342	Dilaurate of polyethylene glycol 200	1375	P.O.E. Stearic acid (15 moles EtO)
1343	Known aerosol OT WS6651	1376	P.O.E. Stearic acid (9 moles EtO)
1344	P.O.E. sorbitan monooleate (20 moles EtO)	1377	P.O.E. Lauric acid (14 moles EtO)
		1378	Sorbitan monopalmitate
1345	Known Vinol 165 polyvinyl alcohol	1379	P.O.E. Sorbitan monostearate (20 moles EtO)
1346	Laurate of polyethylene glycol 200		
1347	Laurate of diethylene glycol	1380	Known myristic acid—95% purity
1348	Monoisopropanolamide–lauric acid	1381	Sodium silicate—soluble
1349	P.O.E. Hydrogenated tallow amide	1382	Sodium phosphate, tribasic (sodium phosphate·12HOH)
1350	Polyoxyethylated coco amide (5 moles EtO)	1383	Sodium carbonate
1351	Polyethoxylated oleamide (5 moles EtO)	1384	Sodium borate, tetra (sodium borate·10HOH)
1352	Diethanolamine-coconut fatty acid condensate (90%)	1385	Silicone defoamer—water dispersible
1353	Monoethanolamide lauric acid	1386	ethylenediamine tetraacetic acid, 2 sodium salt
1354	Sucrose monotallowate	1387	Perflouro surfactant—cationic
1355	Sucrose dioleate	1388	Perflouro surfactant—anionic + B76 ionic
1356	Sucrose monostearate		
1357	Sorbitan trioleate	1389	Disodium *n*-lauryl-b-iminodipropionate
1358	Sorbitan monolaurate		
1359	P.O.E. Sorbitan trioleate (20 moles EtO)	1390	Sodium *n*-coco-b-aminopropionate
		1391	Substituted imidazolinium salt
1360	Sucrose monopalmitate	1392	Substituted imidazolinium salt, example 2
1361	Sorbitan monostearate		
1362	P.O.E. Sorbitan tristearate (20 moles EtO)	1393	Substituted oxazoline (Alkaterge E)
		1394	Substituted oxazoline (Alkaterge C)
1363	P.O.E. Sorbitan monostearate	1395	Laurylisoquinolinium bromide
1364	P.O.E. Sorbitan monolaurate	1396	Laurylpyridinium chloride
1365	P.O.E. Castor oil (40 moles EtO)	1397	Diethyl heptadecyl imidazolinium ethyl sulfate
1366	Polyglycerol ester of oleic acid		
1367	Glyceryl trioleate	1398	Quaternary imidazolinium salt—stearic acid
1368	Oleate of polyethylene glycol 400		
1369	Distearate of polyethylene glycol 6000	1399	Stearamido propyldimethyl-b-hydroxethyl
1370	Stearate of polyethylene glycol 4000	1400	*n*-Stearoylethylenediamine, formate salt
1371	Stearate of polyethylene glycol 600	1401	*n*-Oleoylethylenediamine, formate salt
1372	P.O.E. Tall oil (16 moles EtO)		
1373	P.O.E. Coco fatty acids (15 moles EtO)	1402	Diiso-C4–0-*o*-Et-*o*-C2 dimethyl-0 ammonium chloride

NUMERICAL INDEX TO SPECTRA 1–2130

Spectra Numbers	Compound
1403	Alkyldimethyl-3,4-dichlorobenzyl ammonium chloride
1404	Lauryldimethylbenzyl ammonium chloride
1405	Di"coco" dimethyl ammonium chloride
1406	Soya trimethyl ammonium chloride
1407	Octadecyltrimethyl ammonium chloride
1408	Dodecyltrimethyl ammonium chloride
1409	n-b-Hydroxyethyl coco imidazoline
1410	n-b-Hydroxyethyl stearyl imidazoline
1411	P.O.E. Duomeen T (3 moles EtO)
1412	P.O.E. Soya amine (10 moles EtO)
1413	P.O.E. Tallow amine (15 moles EtO)
1414	P.O.E. Coco amine (5 moles EtO)
1415	P.O.E. Tertiaryamine C18–24H37–49NH(C$_2$H$_4$O)15H
1416	P.O.E. Oleyl amine (5 moles EtO)
1417	Oleoyl polypeptide–sodium salt
1418	Known zinc stearate in KBr
1419	Pure Triton X-102
1420	Sodium naphthenate (4% sodium)
1421	Known sodium stearate
1422	Sodium poly-alkylbenzene sulfonate
1423	P.O.E. Stearyl amine (15 moles EtO)
1424	Monorincinoleate of ethylene glycol
1425	Glyceryl dilaurate
1426	Glyceryl monolaurate
1427	Dioleate of polyethylene glycol 400
1428	Dilaurate of polyethylene glycol 400
1429	Acetic acid salt of dodecylamine
1430	P.O.E. Lauryl alcohol (23 moles EtO)
1431	Sulfated laural ether of tetraethylene glycol, sodium salt
1432	Known kristalex 3100 resin melted film
1433	Laurate of polyethylene glycol 1540
1434	P.O.E. stearyl amine (5 moles EtO)
1435	Polyoxyethylated nonylphenol (30 moles EtO)
1436	Sodium poly-alkylnaphthalene sulfonate
1437	n-Tallow-propylenediamine—80% diamine
1438	Known calcium stearate in KBr
1439	Sodium monobutylphenylphenol monosulfonate
1440	n-Coco-propylenediamine—84% diamine
1441	Dimethyl soya amine—92% tertiary
1442	Ricinoleate of polyethylene glycol 600
1443	Sodium sulfooleate
1444	Known sorbitan monooleate
1445	Polyoxyethylated rosin
1446	P.O.E. Stearyl alcohol (20 moles EtO)
1447	Alkylarylpolyether sulfonate–sodium salt
1448	Polyoxyethylated nonylphenol
1449	Polyoxyethylated oleic acid (9 moles EtO)
1450	Ammonium + alkylbenzimidazole sulfonate
1451	Polyoxyethylated tridecyl alcohol (3 moles EtO)
1452	Sodium n-alkylsulfoacetamide
1453	Sodium dioctyl sulfosuccinate
1454	Sodium diisobutyl sulfosuccinate
1455	Polyoxyethylated T. octylphenol (9–10 Moles EtO)
1456	Polyoxypropylene & 50% EtO (M.W. ~2100)
1457	Dimethyl octadecyl amine (92% tertiary)
1458	Propyl ester of sulfooleic acid–sodium salt
1459	Coconut f.a. ester-2-hydroxyethane sulfamide
1460	P.O.E. Lauryl alcohol (4 moles EtO)
1461	n-Oleoyl sarcosine

NUMERICAL INDEX TO SPECTRA 1–2130

Spectra Numbers	Compound
1462	P.O.E. Tertiary amine C12–14H25–29NH(C$_2$H$_4$O)5H
1463	Cetyldimethylamine oxide
1464	Polyoxyethylated nonylphenol (9–10 Moles EtO)
1465	Known sodium lauryl sulfate
1466	Polyoxyethylated red oil (10 moles EtO)
1467	Methoxypolyethylene glycol 500 "cocoate"
1468	Tertiary-C11–14 H23–29 amine
1469	Polyoxyethylated coco fatty acid (5 moles EtO)
1470	Polyoxyethylated cetyl alcohol (20 moles EtO)
1471	Polyoxyethylated oleyl alcohol (20 moles EtO)
1472	Polyoxypropylene & 40% EtO (M.W. ~1200)
1473	Polyoxypropylene & 10% EtO (M.W. ~1750)
1474	Polyoxypropylene & 40% EtO (M.W. ~1750)
1475	2-Tertiary-dodecylmercaptoethanol
1476	P.O.E. tertiary-Dodecyl mercaptan (12 moles EtO)
1477	Polyoxyethylated octyl phosphate
1478	Capric acid
1479	Polyoxypropylene & 80% EtO (M.W. ~1750)
1480	3,6-Dimethyl-4-octyne-3,6-diol
1481	2,4,7,9-Tetramethyl-5-decyne-4,7-diol
1482	Ammonium monoethylphenylphenol monosulfonate
1483	Polyoxyethylated nonylphenol (20 moles EtO)
1484	Myristic acid
1485	Palmitic acid—KBr disk
1486	Undecylenic acid
1487	Linoleic acid
1488	Alkyl-NH$_4$ dodecylbenzene sulfonate

Spectra Numbers	Compound
1489	Naphthenic acids
1490	Coconut acids
1491	Sodium oleate
1492	Sodium ricinoleate
1493	Erucic acid
1494	Tallow acids (distilled)
1495	Lithium stearate in KBr
1496	Potassium stearate in KBr
1497	Sodium laurate in KBr
1498	Sodium linoleate
1499	Potassium palmitate
1500	Potassium myristate
1501	Sodium palmitate
1502	Sodium sulfate in KBr
1503	Potassium naphthenate
1504	Potassium abietate
1505	Morpholine palmitate
1506	Known oleic acid
1507	Morpholine ricinoleate
1508	Morpholine abietate
1509	Triethanolamine laurate
1510	Known glyceryl monooleate
1511	Triethanolamine myristate
1512	Triethanolamine palmitate
1513	Polyoxyethylated lanolin
1514	Triethanolamine oleate
1515	Aluminum stearate in KBr
1516	Tall oil fatty acids
1517	Calcium stearate in KBr
1518	Calcium ricinoleate in KBr
1519	Lead stearate in KBr
1520	Zinc palmitate in KBr
1521	Zinc stearate in KBr
1522	Zinc resinate in KBr
1523	Polyoxyethylated nonylphenol (1–2 moles EtO)
1524	Sulfated ethanolamine-myristic acid, sodium salt
1525	Sodium dodecylbenzene sulfonate

NUMERICAL INDEX TO SPECTRA 1–2130

Spectra Numbers	Compound
1526	Known ricinoleic acid
1527	Sodium dibutylnaphthalene sulfonate
1528	Sodium cetyl sulfate
1529	Triethanolamine linoleate
1530	Triethanolamine naphthenate
1531	Calcium naphthenate, calcium liquid 4%
1532	Iron naphthenate, 6% Fe
1533	Sulfated propyl oleate, sodium salt
1534	Sulfated amyl oleate, sodium salt
1535	Sulfated cod oil, sodium salt
1536	Sodium *n*-octyl sulfate
1537	Sodium 2-ethylhexyl sulfate
1538	Sodium *sec*-tetradecyl sulfate
1539	Sodium oleyl-stearyl sulfate
1540	Magnesium lauryl sulfate
1541	Triethanolammonium lauryl sulfate
1542	Sulfated polyoxyethylated nonylphenol, sodium salt
1543	Sulfated 9-phenyl ether-4-ethylene glycol, ammonium salt
1544	Sulfated castor oil fatty acid, sodium salt
1545	Sodium petroleum sulfonate (M.W. 340–360)
1546	Isopropylamine salt–sulfonated petroleum
1547	Triethanolamine salt–sulfonated petroleum
1548	Laurate of polyethylene glycol 400
1549	Ethanolamine salt—dibutylnaphthalene, sulfonate
1550	Copper oleate in KBr
1551	Sodium diisopropylnaphthalene sulfonate
1552	K Polymerized alkylnaphthalene sulfonate
1553	Sodium kerylbenzene sulfonate
1554	Sodium monobutyldiphenyl sulfonate
1555	Ca Polymerized alkylbenzene sulfonate
1556	Calcium petroleum sulfonate
1557	Sodium xylene sulfonate
1558	Sodium toluene sulfonate
1559	Triethanol-ammonium dodecylbenzene sulfonate
1560	Oleate 2-hydroxy-ethane sulfonic acid, sodium salt
1561	Sodium lorol sulfoacetate
1562	Stearic acid
1563	Sulfated polyoxyethylated octylphenol-sodium
1564	Sodium petroleum sulfonate
1565	Zinc linoleate KBr
1566	Magnesium stearate KBr
1567	Triethanolamine stearate
1568	P.O.E. Hydrogenated tallow amide (5 moles EtO)
1569	Known abietic acid
1570	Dioleate of polyethylene glycol 4000
1571	Di"hydrogenated tallow" dimethyl ammonium chloride
1572	Iron stearate in KBr
1573	Polyoxyethylated nonylphenol (15 moles EtO)
1574	Acetic acid salt of oleylamine
1575	P.O.E. Tridecyl alcohol (9 moles EtO)
1576	P.O.E. Sorbitan monooleate HLB 13.9
1577	Polyoxyethylated tertiary octylphenol (5 moles EtO)
1578	Oleic acid monoisopropanolamide
1579	Polyoxyethylated tridecyl alcohol (15 moles EtO)
1580	Calcium oleate in KBr
1581	Dicoco amine—85% secondary
1582	Tallow amine—95% primary
1583	Oleylamine—95% primary
1584	*n*-Hexadecylamine—95% primary
1585	Potassium laurate
1586	P.O.E. Sorbitan monolaurate HLB 13.3

NUMERICAL INDEX TO SPECTRA 1–2130

Spectra Numbers	Compound
1587	Polyoxyethylated nonylphenol (4 moles EtO)
1588	Distearate of polyethylene glycol 600
1589	Stearate of polyethylene glycol 1540
1590	Morpholine undecylenate
1591	Ethylene glycol monostearate
1592	Oleate of polyethyene gylcol 1000
1593	Ricinoleate of polyethylene glycol 400
1594	Sulfated tallow, sodium salt
1595	Dilaurate of polyethylene glycol 1540
1596	Diethanolammonium lauryl sulfate
1597	Sulfated soybean oil, sodium salt (org. SO_3—4%)
1598	Polyoxyethylated octylphenol (30 moles EtO)
1599	Polyoxyethylated octylphenol (12–13 moles EtO)
1600	Sulfated castor oil, sodium salt (org. SO_3—2%)
1601	Potassium linoleate
1602	Sodium di(2-ethylhexyl) phosphate
1603	Known sulfated oleic acid—sodium salt
1604	P.O.E. Oleyl alcohol (20 moles EtO)
1605	Sodium-sulfated ethanolamine-lauric acid con.
1606	Sulfated glyceryl trioleate, sodium salt
1607	Sulfated butyl oleate, sodium salt
1608	Soy phosphotides (95%) (lecithin)
1609	Polyoxyethylated castor oil (20 moles EtO)
1610	Polyoxyethylated octylphenol (3 moles EtO)
1611	Polyoxyethylated tall oil (12 moles EtO)
1612	Acetic acid salt of hydrogenated tallow amine
1613	Sodium benzylnaphthalene sulfonate
1614	Lauric acid

Spectra Numbers	Compound
1615	Acetic acid salt of soya amine
1616	P.O.E. Lauric amide (5 moles EtO)
1617	Ca Polymerized alkylbenzene sulfonate
1618	Sodium diamyl sulfosuccinate
1619	Sodium *n*-methyl-*n*-palmitoyl taurate
1620	Sodium *n*-methyl-*n*-tallow acid taurate
1621	Sodium lignosulfonate (5.4% sodium sulfate)
1622	Ca lignosulfonate (12.2% clcium sulfate groups)
1623	*n*-Stearoyl-palmitoyl sarcosine
1624	Sodium dihexyl sulfosuccinate
1625	Sodium *n*-methyl-*n*-tall-oil-acid taurate
1626	Disodium dibutylphenylphenol disulfonate
1627	Guanidinium monoethylphenylphenol sulfonate
1628	Sodium *n*-lauroyl sarcosinate
1629	Alkyl polyphosphate $Na_5R_5(P_3O_{10})_2$ R
1630	Alkyl polyphosphate $Na_5R_5(P_3O_{10})_2$ R—second sample
1631	*n*-Tetradecylamine
1632	Hydrogenated tallow amine 95% primary
1633	Soya amine 95% primary
1634	Dihydrogenated tallow amine (85% sec.)
1635	Acetic acid salt of tallow amine
1636	Acetic acid salt of octadecylamine
1637	*n*-b-Hydroxyethyl oleyl imidazoline
1638	*n*-soya-propylenediamine (80% diamine)
1639	Polyoxyethylated tallow amine (2 moles EtO)
1640	Polyethoxylated tertiary amine (15 moles EtO)
1641	Tertiary-C18–24 H37–49 amine
1642	Coco amine (95% primary)

NUMERICAL INDEX TO SPECTRA 1–2130

Spectra Numbers	Compound
1643	Dimethyl hexadecyl amine (92% tertiary)
1644	Polyoxyethylated oleyl amine (2 moles EtO)
1645	Polyoxyethylated tertiaryamine (5 moles EtO)
1646	Polyoxyethylated coco amine (10 moles EtO)
1647	Octadecyltrimethyl ammonium chloride
1648	Stearyldimethylbenzyl ammonium chloride
1649	Substituted oxazoline (Alkaterge T)
1650	Alkyl phosphonamide RNHP(O)(OR')ONH$_3$R
1651	P.O.E. Sorbitan monooleate HLB = 10.0
1652	Sodium dihydroxyethyl glycinate
1653	Trisodium nitrilotriacetate
1654	Hexadecyltrimethyl ammonium chloride
1655	b-Hydroxyethylo"coco" imidazolinium chloride
1656	Substituted oxazoline (Alkaterge A)
1657	Alkenyldimethylethyl ammonium bromide
1658	Cetyltrimethyl ammonium chloride in KBr
1659	Diisobutylcresoxy-EtO-Et)dimethyl-0 ammonium chloride
1660	Cetylpyridinium bromide in KBr
1661	Sodium n-lauryl-myristyl-b-aminopropionate
1662	Disodium n-tallow-b-iminodipropionate in KBr
1663	Glyceryl monostearate
1664	Glyceryl distearate
1665	Glyceryl monoricinoleate
1666	P.O.E. Sorbitan monolaurate HLB = 13.3
1667	P.O.E. Sorbitan monopalmitate HLB = 15.6
1668	Sorbitan tristearate
1669	Sucrose monooleate
1670	Pentaerythritol monolaurate
1671	Pentaerythritol tetrastearate

Spectra Numbers	Compound
1672	Diethanolamine stearic acid condensate
1673	Diethylene glycol monostearate
1674	Stearate of polyethylene glycol 300
1675	Stearate of polyethylene glycol 6000
1676	Polyoxyethylated stearic acid (5 moles EtO)
1677	Polyoxyethylated stearic acid (10 moles EtO)
1678	Polyoxyethylated stearic acid (40 moles EtO)
1679	Diethanolamine myristic acid condensate, 86%
1680	Sorbitan sesquioleate
1681	Sorbitan trioleate
1682	Diethanolamine lauric acid condensate, 90%
1683	Diethanolamine oleic acid condensate
1684	Laurate of polyethylene glycol 300
1685	Laurate of polyethylene glycol 600
1686	Dilaurate of polyethylene glycol 300
1687	Dilaurate of polyethylene glycol 600
1688	Sucrose monomyristate
1689	Oleate of diethylene glycol
1690	P.O.E. Tridecyl alcohol (6 moles EtO)
1691	Oleate of polyethylene glycol 300
1692	Oleate of polyethylene glycol 600
1693	Ricinoleate of diethylene glycol
1694	P.O.E. tertiary-octylphenol (7–8 moles EtO)
1695	P.O.E. tertiary-octylphenol (16 moles EtO)
1696	Polyoxyethylated nonylphenol (6 moles EtO)
1697	P.O.E. trimethylnonyl alcohol (8 moles EtO)
1698	P.O.E. tridecyl alcohol (12 moles EtO)
1699	Sodium caprate
1700	Sodium undecylenate
1701	Sodium resinate (abietate)
1702	Potassium undecylenate

NUMERICAL INDEX TO SPECTRA 1–2130

Spectra Numbers	Compound
1703	Distearate of polyethylene glycol 400
1704	Polyoxyethylated tetradecyl alcohol (7 moles EtO)
1705	Dioleate of polyethylene glycol 1000
1706	Behenic acid
1707	Sodium myristate
1708	Potassium ricinoleate, cast film on AgBr
1709	Ammonium ricinoleate, liquid film on AgBr
1710	Ammonium abietate, cast film on AgBr
1711	Morpholine laurate, liquid film on AgBr
1712	Morpholine myristate, liquid film on AgBr
1713	Morpholine stearate, cast film on AgBr
1714	Morpholine oleate, liquid film on AgBr
1715	Morpholine naphthenate, liquid on AgBr
1716	Ammonium caprate, liquid film on AgBr
1717	Morpholine linoleate, liquid film on AgBr
1718	Triethanolamine caprate, cast film on AgBr
1719	Triethanolamine undecylenate, film on AgBr
1720	Ammonium oleate, cast film on AgBr
1721	Ammonium stearate, cast film on AgBr
1722	Ammonium myristate, cast film on AgBr
1723	Triethanolamine ricinoleate, liquid film
1724	Triethanolamine abietate, liquid on AgBr
1725	Barium naphthenate, liquid on AgBr
1726	Calcium linoleate, liquid on AgBr
1727	Copper naphthenate, liquid on AgBr
1728	Lead naphthenate, liquid on AgBr

Spectra Numbers	Compound
1729	Manganese naphthenate, liquid on AgBr
1730	Nickel oleate, liquid on AgBr
1731	Zinc oleate, liquid on AgBr
1732	Sulfated isopropyl oleate—sodium salt, cap. film
1733	Sodium sulfonated neatsfoot oil, cap. film
1734	Sulfated rice brand oil—sodium salt, cap. film
1735	Sodium sulfonated sperm oil, liquid on AgBr
1736	Aluminum palmitate on KBr disk
1737	Barium stearate on KBr disk
1738	Miranol SM salt of lauryl sulfate
1739	Miranol C2M salt of lauryl sulfate
1740	Sodium petroleum sulfonate (M.W. 513)
1741	Barium petroleum sulfonate (ave. M.W. 1000)
1742	Ammonium petroleum sulfonate (M.W. 445)
1743	K monoethyl phenylphenol monosulfonate
1744	Dimethyl coco amine—liq. film on AgBr
1745	Polyethoxylated stearyl amine (10 moles EtO)
1746	Ammonium lauryl sulfate—film on AgBr
1747	Sodium n-cyclohexyl-n-palmitoyl taurate
1748	n-Octadecylamine—film on AgBr
1749	Acetic acid salt of hexadecylamine
1750	Acetic acid salt of hydrogenated coco amine
1751	Aluminum oleate—KBr disk
1752	Sodium lignosulfonate (3 moles SO_3Na)
1753	Sodium lignosulfonate (14.3% SO_3Na)
1754	$NaSO_3$-naphthalene-formaldehyde condens.
1755	Modified glyceryl phthalate resin

NUMERICAL INDEX TO SPECTRA 1–2130

Spectra Numbers	Compound	Spectra Numbers	Compound
1756	Polyethoxylated stearyl amine (50 moles EtO)	1782	Henkel sodium lauryl sulfate on KBr disk
1757	Polyethoxylated tertiary amine (25 moles EtO)	1783	Glycerol, film on KCl
1758	Polyethoxylated coco amine (2 moles EtO)	1784	FC1802 fluorinated surfactant, cast film
1759	Polyethoxylated coco amine (15 moles EtO)	1785	Polyoxyethylated tridecyl alcohol (9 EtO)
1760	P.O.E. Soya amine (2 moles EtO)—liq. film on AgBr	1786	Sulfated oleic acid (org. SO_3 4.5%), sodium sulfonated red oil
1761	P.O.E. Tallow amine (5 moles EtO)—liq. film on AgBr	1787	Sodium mono- and diamylnaphthalene sulfonates
1762	P.O.E. Soya amine (5 moles EtO)—liq. film on AgBr	1788	Calcium docylbenzene sulfonate (70% in oil)
1763	P.O.E. Soya amine (15 moles EtO)—liq. film on AgBr	1789	Ammonium alkylbenzimidazole sulfonate
1764	P.O.E. Rosin amine (5 moles EtO)—liq. film on AgBr	1790	BASF/PE:PET/CS-2/H_2O
1765	Cetyldimethylethyl ammonium bromide	1791	BASF/PE:PET/CS-2/MeOH/$MeCl_2$
1766	Dilauryldimethyl ammonium bromide	1792	BASF/PE:PET/CS-2/MeOH/hex
1767	Pluronic 10R8prill surfactant—cast film	1793	HC/PE:PET/33514/H_2O
1768	Gemtex SM-33, liquid film on KCl	1794	HC/PE:PET/33514/MeOH/$MeCl_3$
1769	Triton X-100, liquid film on KCl	1795	HC/PE:PET/33514/MeOH/hex
1770	PEG-75 lanolin, liquid film on KCl	1796	HC/PE:PET/DF059/MeOH
1771	Miranol C2MNPLV, cocoamphocarboxy glycine	1797	HC/PE:PET/DF059/MeOH/$MeCl_3$
1772	Sequestrene Na_3, KBr disk	1798	HC/PE:PET/DF059/MeOH/hex
1773	Cetiol 1414-E, myreth-3-myristate	1799	HC/PET:PET/DF059 + DF018/MeOH
1774	Emerest 2316, isopropyl palmitate	1800	HC/PE:PET/DF059 + DF018/MeOH/$MeCl_3$
1775	Emerest 2620, PEG (200) monolaurate	1801	HC/PE:PET/DF059 + DF018/MeOH/hex
1776	Emsorb 2515, (SML) sorbitan monolaurate	1802	HC/PE:PET/DF059 + DF018/MeOH/hex
1777	Lauric acid, ethyl ester	1803	Chisso/PE:PET/hydrophil/H_2O
1778	Lauricidin, fatty acid(s) monoglyceride	1804	Chisso/PE:PET/hydrophil/H_2O
1779	1-Monolauroyl-RAC-glycerol	1805	Chisso/PE:PET/hydrophil/MeOH/$MeCl_3$
1780	Trycol 5966, ethoxylated lauryl alcohol	1806	Chisso/PE:PET/hydrophil/MeOH/hex
1781	Tagat L-2, P.O.E. glycerol fat. acid esters	1807	9509021 Surfynol 504, liquid film
		1808	Trycol 5966, laureth-3, liquid film
		1809	Mackam 151L, lauramino propionic acid
		1810	Deriphat160-C sodium lauriminodipropionic acid

NUMERICAL INDEX TO SPECTRA 1–2130

Spectra Numbers	Compound
1811	Hamposyl L-30, sodium lauroyl sacosinate
1812	Standapol SH124–33, Na_2 laureth(3) sulfate
1813	Sulfochem TLES, TEA laureth sulfate
1814	Mackanate LM-40, Na_2 lauramido MEA sulfonate
1815	Schercozol. L, lauryl hydroxyethyl imidazoline
1816	Trycol 5882, laureth-4, liquid film
1817	Aethoxal B, PPG-5 laureth-5, liquid film
1818	Polyoxyethylated tridecyl alcohol (9 moles EtO)
1819	Laurylisoquinolinium bromide
1820	Ammonium undecylenate (in aqueous alcohol)
1821	Diethanolamine-coconut fatty acid condensate
1822	Potassium monoethylphenylphenol monosulfonate
1823	Ammonium monoethylphenylphenol monosulfonate
1824	n-Oleoyl sarcosine
1825	Cetyldimethylamine oxide
1826	Polyethoxylated tertiary amine
1827	Diisobutylphenoxyethoxyethyl dimethylbenzyl ammonium chloride
1828	Diisobutylcresoxyethoxyethyl Dimethylbenzyl ammonium chloride
1829	Alkyl phosphonamide $RNHP(O-(OR')ONH_3R)$; R is $C_{12}H_{25}$; R' is water solubilizing
1830	Pentaerythritol dioleate
1831	P.O.E. Lauric acid (9 moles EtO)
1832	Polyoxypropylene & 20% EtO (M.W. ~1750)
1833	Polyoxyethylated nonylphenol (10–11 moles EtO)
1834	Polyoxyethylated nonylphenol (8 moles EtO)
1835	Polyoxyethylated oleyl alcohol (20 moles EtO)

Spectra Numbers	Compound
1836	Polyoxyethylated oxypropylated stearic acid
1837	Polyoxyethylated tert-octylphenol (9–10 moles EtO)
1838	Polyoxyethylated tert-octylphenol (3 moles EtO)
1839	Polyoxyethylated tert-octylphenol (30 moles EtO)
1840	Polyoxyethylated tridecyl alochol (12 moles EtO)
1841	Ricinoleate of propylene glycol
1842	Stearate of propylene glycol
1843	Sulfated ethanolamine–lauric acid condensate, Na salt
1844	Tallow acids (distilled)
1845	Ammonium linoleate
1846	Ammonium naphthenate (in aqueous alochol)
1847	BASF/PE:PET/CS-1/H_2O
1848	HC/PE:PET/33514/H_2O
1849	Danaklon/PE:PP/hydrophil/H_2O
1850	HC/PE:PET/33514/MeOH/$MeCl_3$
1851	Chisso/PE:PP/hydrophil/MeOH/$MeCl_3$
1852	Chisso/PE:PET/hydrophil/H_2O
1853	Chisso/PE:PET/hydrophil/MeOH/hex
1854	Chisso/PE:PP/HR_5/MeOH
1855	Chisso/PE:PP/P2/MeOH
1856	HC/PET:PET/DF059 + DF018/MeOH
1857	HC/PE:PET/DF059 + DF018/MeOH/$MeCl_3$
1858	BASF/PE:PET/CS-1/MeOH/$MeCl_3$
1859	HC/PE:PET/DF059/MeOH
1860	HC/PE:PET/DF059/MeOH/hex
1861	Chisso/PE:PP/HR_5/MeOH/hex
1862	Chisso/PE:PET/hydrophil/MeOH/$MeCl_3$
1863	Danaklon/PE:PP/hydrophil/H_2O
1864	Danaklon/PE:PP/hydrophil/H_2O
1865	Danaklon/PE:PP/hydrophil/MeOH
1866	Chisso/PE:PET/hydrophil/H_2O
1867	Danaklon/PE:PP/super 33/MeOH

NUMERICAL INDEX TO SPECTRA 1–2130

Spectra Numbers	Compound
1868	Danaklon/PE:PP/hydrophil/H$_2$O/hex
1869	Chisso/PE:PET/hydrophil/MeOH/MeCl$_3$
1870	Danaklon/PE:PP/hydrophil/MeOH/hex
1871	Chisso/PE:PP/P2/MeOH/MeCl$_3$
1872	Chisso/PE:PP/HR$_5$/MeOH/MeCl$_3$
1873	Danaklon/PE:PP/hydrophil/H$_2$O/MeCl$_3$
1874	BASF/PE:PET/CS-2/MeOH/MeCl$_2$
1875	Danaklon/PE:PP/super 33/MeOH/MeCl$_3$
1876	Chisso/PE:PP/hydrophil/H$_2$O
1877	BASF/PE:PET/CS-2/H$_2$O
1878	HC/PE:PET/33514/MeOH/hex
1879	Danaklon/PE:PP/hydrophil/MeOH/MeCl$_3$
1880	Chisso/PE:PET/hydrophil/MeOH/MeCl$_3$
1881	HC/PE:PET/DF059/MeOH/MeCl$_3$
1882	BASF/PE:PET/CS-1/MeOH/hex
1883	HC/PE:PET/DF059 + DF018/MeOH/hex
1884	Chisso/PE:PET/hydrophil/MeOH/hex
1885	Chisso/PE:PP/hydrophil/MeOH/hex
1886	Chisso/PE:PP/P2/MeOH/hex
1887	Chisso/PE:PP/hydrophil/H$_2$O
1888	Danaklon/PE:PP/super 33/MeOH/hex
1889	Phospholipid CDM
1890	Dodecyl sulfate sodium salt in KBr
1891	Alpha-methylstyrene resin
1892	Polyterpene resin
1893	Modified polyterpene
1894	Synthetic polyterpene
1895	Polyalphamethylstyrene
1896	PVT/alpha-methylstyrene
1897	PVT/alpha-methylstyrene
1898	Modified alpha-methylstyrene
1899	Hydrogenated hydrocarbon—Escorez 5300
1900	Hydrogenated hydrocarbon—Escorez 5320
1901	Hydrogenated hydrocarbon—Escorez 5380
1902	Coumarone-indene resin
1903	Polyethyloxazoline
1904	Styrene 17%/isoprene 83%—SIS block
1905	Styrene 14%/isoprene 86%—SIS block
1906	Styrene 14%/isoprene 86%—SIS block
1907	Styrene 10%/isoprene 90%—SIS block
1908	Styrene 21%/isoprene 79%—SIS block
1909	Styrene 28%/butadiene 72%—SBS block
1910	Hydrogenated rosin ester
1911	Vistalon—ethylene propylene copolymer
1912	Indopol L-50—polybutylene polymer
1913	Plasthall P-670—plasticizer
1914	Paraplex G-30—plasticizer
1915	Terpene phenol resin
1916	Terpene phenol resin, type II
1917	Tetrakis[methylene(3,5-di-*t*-butyl-4-hydroxyhydrocinnamate)] methane
1918	Ethanox 330 antioxidant
1919	Octadecyl 3,5-di-*t*-butyl-4-hydroxycinnamate
1920	Pebax 2533 polyether/polyamide
1921	Pebax 3533 polyether/polyamide
1922	Pebax 4011 polyether/polyamide
1923	Styrene 14%/ethylene:butylene block 86%
1924	Styrene 29%/ethylene:butylene block 71%
1925	Styrene 48%/butadiene 52%
1926	Styrene 28%/ethylene:butylene block 72%
1927	Nirez V-2150—terpene phenol resin

NUMERICAL INDEX TO SPECTRA 1–2130

Spectra Numbers	Compound
1928	Arkon P-125—hydrogenated hydrocarbon
1929	Kristalex 5140—modifed alpha-methylstyrene
1930	Piccotex LC—PVT/alpha-methylstyrene
1931	Known Kraton 1111-0 styrene/isoprene
1932	Known Arkon p70 tackifier resin
1933	Polystyrene:polyisoprene, polyterpene tackifier resin
1934	Beta pinene
1935	Alpha pinene
1936	Polystyrene:polyisoprene with polyterpene tackifier CHCl$_3$ extract
1937	Polystyrene:polybutadiene, hydrocarbon tackifier resin and possibly oil CHCl$_3$
1938	Modified polyacrylamide
1939	Arkon Superester A-100
1940	Styrene/isoprene copolymer
1941	Known Eastobond M500 APP
1942	Pure Zonester 100 melt
1943	Pure Zonarez 7115 melt
1944	Alpha-methylstyrene monomer resin
1945	Hydrogenated rosin ester
1946	Pure Zonarez B-115
1947	Polyterpene resin
1948	Bareco CP-7 PE:PP wax
1949	Kodak Epolene C-16 polyethylene wax
1950	Petroleum hydrocarbon resin, hydrogenated
1951	Hydrogenated hydrocarbon resin
1952	Pure Tuffalo oil
1953	Pure Piccotac B
1954	Polystyrene:polyisoprene, polyterpene tackifier resin
1955	Polystyrene:polyisoprene, polycyclopentadiene resin
1956	Known polyvinyltoluene—mixed isomers
1957	Known Elvax 260 28% VA melt

Spectra Numbers	Compound
1958	Bis(2-ethylhexyl)phthalate, capillary film
1959	Polystyrene:polyisoprene, polycyclopentadiene resin polyterpene tackifier resin
1960	Air prod. airvol PVOH/PVAc—cast film
1961	Pure Eastman H100 resin melted film
1962	Acrylic emulsion
1963	Polystryene:polyisoprene resin
1964	Kraton 1102, 28% polystyrene/72% polybutadiene
1965	Polystyrene:polyisoprene with hydrogenated polycyclopentadiene resin + oil
1966	Polystyrene:polybutadiene and polystyrene:polyisoprene with polyvinyl toluene
1967	Poly(styrene):poly(isoprene)-based with poly(terpene) tackifier resin
1968	Polystryene:polybutadiene rosin, acid ester tackifier resin, polyvinyl toluene
1969	Food-grade paraffin wax
1970	Styrene-butadiene copolymer
1971	Polyvinyl alcohol/polyvinyl acetate
1972	Styrene-butadiene-styrene
1973	Polyisoprene
1974	Styrene-butadiene-styrene
1975	Alpha-methylstyrene
1976	Styrene/isoprene/styrene
1977	Atactic polypropylene/butylene
1978	Synthetic or modified polyterpene
1979	Styrene/isoprene/styrene
1980	Phthalate-based tackifier resin
1981	Styrene/butadiene/styrene
1982	Toluene-based tackifier resin
1983	Polycyclopentadiene resin
1984	C9 and C5 hydrocarbon resins, polystyrene resin
1985	Kraton styrene ethylene butylene styrene, oil, C5 hydrocarbon resin

Spectra Numbers	Compound	Spectra Numbers	Compound
1986	Styrene, ethylene butylene, styrene copolymer, oil, C5/C9 hydrocarbon resin	1993	Styrene butadiene styrene, oil, C5 hydrocarbon resin (C9 and polyterpene)
1987	Polybutylene and polyethylene wax	1994	C5 hydrocarbon resin
1988	Atatic polypropylene	1995	Polystyrene: polyisoprene base polymer
1989	Styrene, ethylene butylene, styrene copolymer	1996	Polystyrene: polyisoprene base polymer with polycyclopentadiene resin
1990	Kraton styrene isoprene styrene		
1991	Styrene, ethylene butylene, styrene copolymer	1997	Hydrogenated rosin ester tackifier resin
1992	Styrene, ethylene butylene, styrene copolymer, oil, C5 hydrocarbon resin	1998	Findley adhesive H 2253 tackifier resin
		1999	Styrene, ethylene butylene, styrene copolymer

SPECTRA NUMBERS 2001–2130

Raman (4000–500 cm^{-1}): Organic Compounds and Polymers

Spectra Numbers	Compound	Spectra Numbers	Compound
2000	Ethylene vinyl acetate	2017	Ethyl alcohol
2001	Methyl paraben, methyl *p*-benzoate	2018	Propylene glycol
2002	Propyl paraben, propyl *p*-benzoate	2019	Chloroform
2003	Butyl paraben, butyl *p*-benzoate	2020	Benzene
2004	Methyl alcohol	2021	Iron oxide
2005	Chloroform, trichloromethane	2022	Ethylene vinyl acetate 18%
2006	Isopropanol, isopropyl alcohol	2023	Rubber modified polypropylene
2007	Acetone, dimethylketone, 2-propanone	2024	Polypropylene
		2025	Calcium carbonate
2008	Perfluoro-1-octanesulfonic acid tetraethylammonium salt	2026	Titanium dioxide
2009	Freon, 1,1,2-trichloro-1,2,2-trifluoroethane	2027	Rosin soap
		2028	Polypropylene and silicates
2010	Miranol	2029	Polyester
2011	Polyester thread	2030	Polypropylene
2012	Styrofoam, white	2031	Cellophane
2013	Styrofoam, red dye	2032	Cellulose
2014	1,2,4-Trichlororbenzene	2033	Sulfonated cellulose
2015	Chlorobenzene	2034	Aqualon
2016	Toluene, methylbenzene, phentlmethane	2035	Sodium sulfate

NUMERICAL INDEX TO SPECTRA 1–2130

Spectra Numbers	Compound
2036	Silicate
2037	Poly(ethylene), high density
2038	Cotton (cellulose)
2039	Polyester
2040	Whole milk
2041	Citric acid, 2-hydroxy-1,2,3-propanetricarboxylic acid
2042	Sodium lauryl sulfate—sulfuric acid monododecyl ester sodium salt
2043	Malic acid, hydroxy-butanedioic acid
2044	Dextrose, glucose
2045	Chitosan
2046	BTC-50 quaternary amine
2047	Acrylic polymer—ethyl acrylate
2048	Carboxylated styrene-acrylonitrile
2049	Styrene/butadiene
2050	Acrylic latex—acrylic ester
2051	Bisphenol-A polyglycidyl ether
2052	Ethyl acetate
2053	Adipic acid
2054	Galatose
2055	Sucrose
2056	Glucose
2057	Fructose
2058	Surfynol 504
2059	Polyacrylanitrile
2060	Titanium dioxide
2061	Jeffamine
2062	Ethylene vinyl acetate 18%
2063	Nylon salt
2064	Jeffamine 149
2065	Polyamide polymer
2066	Pyridylazo naphthol
2067	Berocel 596
2068	o-Me Galactoside-6-acrylate with 1% crosslinker
2069	Sucrose acrylate hydrogel with 3.5% diacrylate crosslinker

Spectra Numbers	Compound
2070	a-Me Glucoside-6-acrylate with 1.5% diacrylate crosslinker
2071	Calcium oxalate
2072	Phthalate extracted from Tygon tubing
2073	Ethylene glycol
2074	Sodium lauryl sulfate 98%
2075	Poly(ethylene glycol) 600
2076	Dehydroacetic acid
2077	Tygon tubing minus phthalate
2078	Phthalate
2079	Teflon
2080	Beeswax (paraffin)
2081	Houghton release agent 564
2082	Dioctyl phthalate
2083	PVC organasol adhesive
2084	PVAc latex
2085	Mineral oil
2086	Nylon fabric
2087	89% Nylon/11% Lycra
2088	92% Nylon/8% Lycra
2089	Uric acid
2090	87% Nylon/13% Lycra
2091	86% Nylon/14% Lycra
2092	100% Polyester
2093	Silicone sealant
2094	Poly(vinyl alcohol)
2095	Poly(acrylic acid)
2096	Poly(hydroxy ethyl methacrylate)
2097	Pyridine, anhydrous
2098	Phospholipid PTC
2099	Polypropylene glycol methacrylate
2100	Polypropylene glycol
2101	Sodium hydroxide pellets
2102	Sodium iodate
2103	Potassium iodate
2104	Sodium periodate
2105	Aluminum ammonium sulfate · 12H$_2$O

NUMERICAL INDEX TO SPECTRA 1–2130

Spectra Numbers	Compound
2106	Aluminum K sulfate, AlK(SO$_4$)$_2$·12H$_2$O
2107	Stearyl alcohol
2108	White ceresin wax
2109	Kadol mineral oil
2110	Air products Surfynol 504
2111	Myreth-3-myristate
2112	Disodium laureth sulfosuccinate
2113	Sodium lauroyl sarcosinate
2114	Triethanolamine laureth sulfate
2115	Disodium lauramido MEA sulfosuccinate
2116	Lauramino propionic acid
2117	Sodium lauriminodipropionic acid
2118	Lauryl hydroxyethyl imidazoline
2119	Silica glass (empty Raman sample vial)
2120	Polyethylene
2121	Polyamide-epichlorohydrin resin
2122	Aluminum sodium sulfate
2123	Aluminum potassium sulfate
2124	Benzoic acid
2125	Sodium bicarbonate
2126	Polycarbonate with TiO$_2$
2127	50/50 Mixture of silicone and menthol
2128	Poly(isoprene) elastic, clay filled
2129	Poly(ether) urethane, clay filled
2130	Poly(propylene) + TiO$_2$

ALPHABETICAL INDEX TO SPECTRA 1–2130

Compound	Spectrum Numbers
Absorbent Polymer Carboxymethyl cellulose (CMC)	1020
Acetic Acid	18–20
Acetic Acid	603–604
Acetic Acid	733–734
Acetic acid salt of dodecylamine	1429
Acetic acid salt of hexadecylamine	1749
Acetic acid salt of hydrogenated cocoamine	1750
Acetic acid salt of hydrogenated tallow amine	1612
Acetic acid salt of octadecylamine	1636
Acetic acid salt of oleylamine	1574
Acetic acid salt of soya amine	1615
Acetic acid salt of tallow amine	1635
Acetic Anhydride	42–44
Acetoin (3-Hydroxy-2-butanone) (ATR)	1155
Acetone (ATR)	1185
Acetone	3–5
Acetone	561–562
Acetone	593–594
Acetone, dimethylketone, 2-propanone	2007
Acrylic emulsion	1962
Acrylic latex—acrylic ester	2050
Acrylic polymer—ethyl acrylate	2047
Acrylonitrile/butadiene/styrene resin	1227
Acrylonitrile/butadiene/styrene resin	204–207
Acrylonitrile/butadiene/styrene resin	811–812
Adipic acid	2053

ALPHABETICAL INDEX TO SPECTRA 1–2130

Compound	Spectrum Numbers
Aethoxal B, PPG-5 laureth-5, liquid film	1817
Air prod. airvol PVOH/PVAc—cast film	1960
Air products Surfynol 504	2110
Alcohol ethoxylate	791–792
ALDOL	1177
Alginic acid, sodium salt	1228
Alginic acid, sodium salt	208–211
Alginic acid, sodium salt	813–814
Alkenyldimethylethyl ammonium bromide	1657
Alkyl phosphonamide RNHP(O–(OR')ONH$_3$R	1650
Alkyl phosphonamide RNHP(O-(OR')ONH$_3$R; R is C$_{12}$H$_{25}$; r' is water solubilizing	1829
Alkyl polyphosphate Na$_5$R$_5$(P$_3$O$_{10}$)$_2$ R—second sample	1630
Alkyl polyphosphate Na$_5$R$_5$(P$_3$O$_{10}$)$_2$ R	1629
Alkylarylpolyether sulfonate—sodium salt	1447
Alkyldimethyl-3,4-dichlorobenzyl ammonium chloride	1403
Alkyl-NH$_4$ dodecylbenzene sulfonate	1488
Aloe vera, capillary film between KCl	1096
Alpha pinene	1935
Alpha-methylstyrene	1975
Alpha-methylstyrene monomer resin	1944
Alpha-methylstyrene resin	1891
Aluminum ammonium sulfate (in KBr)	1097
Aluminum K sulfate, AlK (SO$_4$)$_2$·12H$_2$O	2106
Aluminum ammonium sulfate ·12H$_2$O	2105
Aluminum oleate—KBr disk	1751
Aluminum oleate	797–798
Aluminum palmitate on KBr disk	1736
Aluminum potassium sulfate (in KBr)	1098
Aluminum potassium sulfate	2123
Aluminum sodium sulfate in KBr	1099
Aluminum sodium sulfate	2122
Aluminum stearate in KBr	1515
a-Me Glucoside-6-acrylate with 1.5% diacrylate crosslinker	2070
Ammonium abietate, cast film on AgBr	1710
Ammonium alkylbenzimidazole sulfonate	1789
Ammonium Bicarbonate in KBr	1073
Ammonium caprate, liquid film on AgBr	1716
Ammonium lauryl sulfate—film on AgBr	1746
Ammonium linoleate	1845

ALPHABETICAL INDEX TO SPECTRA 1–2130

Compound   Spectrum Numbers

Ammonium monoethylphenylphenol monosulfonate   1482

Ammonium monoethylphenylphenol monosulfonate   1823

Ammonium myristate, cast film on AgBr   1722

Ammonium naphthenate (in aqueous alcohol)   1846

Ammonium oleate, cast film on AgBr   1720

Ammonium palmitate   1327

Ammonium petroleum sulfonate (M.W. ~445)   1742

Ammonium+ alkylbenzimidazole sulfonate   1450

Ammonium ricinoleate, liquid film on AgBr   1709

Ammonium stearate, cast film on AgBr   1721

Ammonium undecylenate (in aqueous alcohol)   1820

Aniline, liquid   1108

Anisaldehyde (ATR)   1181

Anisaldehyde   725–726

Aqualon   2034

Arkon P-125—hydrogenated hydrocarbon   1928

Arkon Superester A-100   1939

Atactic polypropylene/butylene   1977

Atatic polypropylene   1988

Bareco CP-7 PE:PP wax   1948

Barium naphthenate, liquid on AgBr   1725

Barium petroleum sulfonate (ave. M.W. 1000)   1741

Barium stearate, KBr disk   1737

BASF/PE:PET/CS-1/H_2O 1847

BASF/PE:PET/CS-1/MeOH/hex   1882

BASF/PE:PET/CS-1/MeOH/$MeCl_3$ 1858

BASF/PE:PET/CS-2/H_2O 1790

BASF/PE:PET/CS-2/H_2O 1877

BASF/PE:PET/CS-2/MeOH/hex   1792

BASF/PE:PET/CS-2/MeOH/$MeCl_2$ 1791

BASF/PE:PET/CS-2/MeOH/$MeCl_2$ 1874

Beeswax   2080

Behenic acid   1706

Benzaldehyde (ATR)   1178

Benzaldehyde   177–179

Benzaldehyde   721–722

Benzene   2020

Benzethonium chloride (in KBr)   1131

Benzoic acid   2124

ALPHABETICAL INDEX TO SPECTRA 1–2130

Compound	Spectrum Numbers
Benzoic acid, 99%+ in KBr	1072
Benzoyl peroxide, 97% in KBr	1095
Benzyl alcohol (ATR)	1164
Benzyl alcohol	138–140
Benzyl alcohol	697–698
Berocel 596	2067
Beta pinene	1934
b-Hydroxyethylo"coco"imidazolinium chloride	1655
Bis(2-ethylhexyl) phthalate, capillary film	1030
Bis(2-ethylhexyl) phthalate, capillary film	1958
Bisphenol-A polyglycidyl ether	2051
Boric acid (in KBr)	1124
Bromophenol blue	1116
Brown kraft bag, fiber in KBr	1057
BTC-50 quaternary amine	2046
1,3-Butanediol (ATR)	1157
1,3-Butanediol	114–116
1,3-Butanediol	685–686
Butyl alcohol (ATR)	1137
Butyl alcohol	54–56
Butyl alcohol	651–652
Butyl methacrylate/isobutyl methacrylate copolymer	1229
Butyl methacrylate/isobutyl methacrylate copolymer	212–215
Butyl methacrylate/isobutyl methacrylate copolymer	815–816
Butyl paraben, butyl p-benzoate	2003
Butylparabenzoic acid (in KBr)	1068
Butyric acid	30–32
Butyric acid	607–608
Butyric anhydride	45–47
Butyric anhydride	645–646
C35-C60 hydrocarbon wax (ATR)	1186
C5 hydrocarbon resin	1994
C9 and C5 hydrocarbon resins, polystyrene resin	1984
Ca lignosulfonate (12.2% calcium sulfate groups)	1622
Ca Polymerized alkylbenzene sulfonate	1555
Ca Polymerized alkylbenzene sulfonate	1617
Calcium carbonate in KBr	1034
Calcium carbonate	2025
Calcium docylbenzene sulfonate (70% in oil)	1788

ALPHABETICAL INDEX TO SPECTRA 1–2130

Compound	Spectrum Numbers
Calcium linoleate, liquid on AgBr	1726
Calcium naphthenate, calcium liquid 4%	1531
Calcium oleate	1060
Calcium oleate KBr	1580
Calcium oxalate	2071
Calcium oxide (in KBr)	1092
Calcium petroleum sulfonate	1556
Calcium ricinoleate in KBr	1518
Calcium stearate in KBr	1028
Calcium stearate in KBr	1517
Camphor in KBr	1125
Camphor	773–774
Capric acid	1478
Carboxylated styrene-acrylonitrile	2048
Cellophane	1019
Cellophane	2031
Cellulose (63%) and polypropylene (37%)	765–766
Cellulose (natural cotton)	1120
Cellulose	2032
Cellulose 35 by 70 micron single fiber	1010
Cellulose acetate	1230
Cellulose acetate	216–219
Cellulose acetate	817–818
Cellulose acetate butyrate	1231
Cellulose acetate butyrate	220–223
Cellulose acetate butyrate	819–820
Cellulose propionate	1232
Cellulose propionate	224–227
Cellulose propionate	821–822
Cellulose triacetate	1233
Cellulose trlacetate	228–230
Cellulose triacetate	823–824
Cellulose, 100% soft wood bleached kraft	1013
Ceresin wax	1083
Cetiol 1414-E, myreth-3-myristate	1773
Cetyldimethylamine oxide	1463
Cetyldimethylamine oxide	1825
Cetyldimethylethyl ammonium bromide	1765
Cetylpyridinium bromide in KBr	1660

ALPHABETICAL INDEX TO SPECTRA 1–2130

Compound	Spectrum Numbers

Cetyltrimethyl ammonium chloride in KBr   1658

Chisso/PE:PET/hydrophil/H_2O 1803

Chisso/PE:PET/hydrophil/H_2O 1804

Chisso/PE:PET/hydrophil/H_2O 1852

Chisso/PE:PET/hydrophil/H_2O 1866

Chisso/PE:PET/hydrophil/MeOH/hex   1806

Chisso/PE:PET/hydrophil/MeOH/hex   1853

Chisso/PE:PET/hydrophil/MeOH/hex   1884

Chisso/PE:PET/hydrophil/MeOH/$MeCl_3$ 1805

Chisso/PE:PET/hydrophil/MeOH/$MeCl_3$ 1862

Chisso/PE:PET/hydrophil/MeOH/$MeCl_3$ 1869

Chisso/PE:PET/hydrophil/MeOH/$MeCl_3$ 1880

Chisso/PE:PP/HR_5/MeOH   1854

Chisso/PE:PP/HR_5/MeOH/hex   1861

Chisso/PE:PP/HR_5/MeOH/$MeCl_3$ 1872

Chisso/PE:PP/hydrophil/H_2O 1876

Chisso/PE:PP/hydrophil/H_2O 1887

Chisso/PE:PP/hydrophil/MeOH/hex   1885

Chisso/PE:PP/hydrophil/MeOH/$MeCl_3$ 1851

Chisso/PE:PP/P2/MeOH   1855

Chisso/PE:PP/P2/MeOH/hex   1886

Chisso/PE:PP/P2/MeOH/$MeCl_3$ 1871

Chitosan   2045

Chlorobenzene   2015

Chloroform (ATR)   1187

Chloroform   2019

Chloroform   597–598

Chloroform   6–8

Chloroform, trichloromethane   2005

Citral (ATR)   1171

Citric acid, 2-hydroxy-1,2,3-propanetricarboxylic acid   2041

Citric acid, cast film on AgBr   1069

Citronellal (ATR)   1172

Citronellal   159–161

Citronellal   711–712

Cocoamine (95% primary)   1642

Coconut acids   1490

Coconut f.a. ester-2-hydroxyethane sulfamide   1459

Coffee, instant, on AgBr window   1047

ALPHABETICAL INDEX TO SPECTRA 1–2130

Compound	Spectrum Numbers

Coffee, on AgBr window   1046

Cola, Coke Classic, cast film on AgBr   1037

Cola, generic   1105

Copper naphthenate, liquid on AgBr   1727

Copper oleate in KBr   1550

Corn syrup, dark, cast film on AgBr   1038

Corn syrup, light, cast film on AgBr   1039

Cotton   2038

Cotton seed oil, between KCl windows   1044

Coumarone-indene resin   1902

Crotonaldehyde (ATR)   1173

Crotonaldehyde   162–164

Crotonaldehyde   713–714

Cyclohexane   563–564

D-Sorbitol, 99% (in KBr)   1031

Danaklon/PE:PP/hydrophil/H_2O 1849

Danaklon/PE:PP/hydrophil/H_2O 1863

Danaklon/PE:PP/hydrophil/H_2O 1864

Danaklon/PE:PP/hydrophil/H_2O/hex   1868

Danaklon/PE:PP/hydrophil/H_2O/$MeCl_3$ 1873

Danaklon/PE:PP/hydrophil/MeOH   1865

Danaklon/PE:PP/hydrophil/MeOH/hex   1870

Danaklon/PE:PP/hydrophil/MeOH/$MeCl_3$ 1879

Danaklon/PE:PP/super 33/MeOH   1867

Danaklon/PE:PP/super 33/MeOH/hex   1888

Danaklon/PE:PP/super 33/MeOH/$MeCl_3$ 1875

Decane, liquid   1110

Dehydroacetic acid   2076

Deriphat160-C sodium lauriminodipropionic acid   1810

Dextrose, cast film on AgBr   1041

Dextrose, cast film on AgBr   1104

Dextrose, glucose   2044

Diacetone alcohol (ATR)   1156

Diacetone alcohol   111–113

Diacetone alcohol   683–684

Diatomaceous earth in KBr   1052

2,3-Dichloro-1-propanol (ATR)   1154

1,3-Dichloro-2-propanol (ATR)   1153

Dichloroacetic acid   21–23

Compound	Spectrum Numbers

Dichloroacetic acid 637–638

Dicoco amine—85% secondary 1581

Di"coco" dimethyl ammonium chloride 1405

Diethanolamine lauric acid condensate, 90% 1682

Diethanolamine myristic acid condensate 86% 1679

Diethanolamine oleic acid condensate 1683

Diethanolamine stearic acid condensate 1672

Diethanolamine-coconut fatty acid condensate (90%) 1352

Diethanolamine-coconut fatty acid condensate 1821

Diethanolamine–oleic acid condensate 1329

Diethanolammonium lauryl sulfate 1596

Diethyl heptadecyl imidazolinium ethyl sulfate 1397

Diethylene glycol (ATR) 1162

Diethylene glycol 126–128

Diethylene glycol 691–692

Diethylene glycol monobutyl ether (ATR) 1148

Diethylene glycol monobutyl ether 673–674

Diethylene glycol monobutyl ether 96–98

Diethylene glycol monoethyl ether (ATR) 1149

Diethylene glycol monoethyl ether 675–676

Diethylene glycol monoethyl ether 99–101

Diethylene glycol monostearate 1673

Dihydrogenated tallow amine (85% sec.) 1634

Di"hydrogenated tallow" dimethyl ammonium chloride 1571

Diisobutylphenoxyethoxyethyl dimethylbenzyl ammonium chloride 1827

Diisobutylcresoxyethoxyethyl dimethylbenzyl ammonium chloride 1828

Diisobutylcresoxy-EtO-Et)dimethyl-0 ammonium chloride 1659

Diiso-C4–0-o-Et-o-C2) dimethyl-0 ammonium chloride 1402

Dilaurate of polyethylene glycol 1540 1595

Dilaurate of polyethylene glycol 200 1342

Dilaurate of polyethylene glycol 300 1686

Dilaurate of polyethylene glycol 400 1428

Dilaurate of polyethylene glycol 600 1687

Dilauryldimethyl ammonium bromide 1766

Dimethicone (silicone) 757–758

Dimethicone, 10,000 cs (ATR) 1188

Dimethicone, 350 cs, film cast on KCl 1071

Dimethyl coco amine—liq. film on AgBr 1744

Dimethyl hexadecyl amine (92% tertiary) 1643

Compound	Spectrum Numbers
Dimethyl octadecyl amine (92% tertiary)	1457
Dimethyl soya amine—92% tertiary	1441
3,6-Dimethyl-4-octyne-3,6-diol	1480
Dioctyl phthalate	2082
Dioleate of polyethylene glycol 1000	1705
Dioleate of polyethylene glycol 1540	1334
Dioleate of polyethylene glycol 200	1336
Dioleate of polyethylene glycol 400	1427
Dioleate of polyethylene glycol 4000	1570
Dioleate of polyethylene glycol 600	1335
Dipropylene glycol (ATR)	1163
Dipropylene glycol	129–131
Dipropylene glycol	693–694
Disodium dibutylphenylphenol disulfonate	1626
Disodium lauramido MEA sulfosuccinate	2115
Disodium laureth sulfosuccinate	2112
Disodium n-lauryl-b-iminodipropionate	1389
Disodium n-tallow-b-iminodipropionate—KBr	1662
Distearate of polyethylene glycol 1000	1339
Distearate of polyethylene glycol 300	1340
Distearate of polyethylene glycol 400	1703
Distearate of polyethylene glycol 600	1588
Distearate of polyethylene glycol 6000	1369
2,6-Di-tert-butyl-p-cresol (BHT)	1015
DL-a-Methylbenzyl alcohol (ATR)	1165
DL-a-Methylbenzyl alcohol	141–143
DL-a-Methylbenzyl alcohol	699–700
Dodecyl sulfate sodium salt in KBr	1890
Dodecyltrimethyl ammonium chloride	1408
Emerest 2316, isopropyl palmitate	1774
Emerest 2620, PEG (200) monolaurate	1775
Emsorb 2515, (SML) sorbitan monolaurate	1776
Erucamide	1043
Erucic acid	1493
Ethanol	63–65
Ethanolamine salt—dibutylnaphthalene, sulfonate	1549
Ethanox 330 antioxidant	1918
2-Ethybutyric acid	611–612
Ethyl acetate	2052

ALPHABETICAL INDEX TO SPECTRA 1–2130

Compound	Spectrum Numbers
Ethyl alcohol (ATR)	1139
Ethyl alcohol	2017
Ethyl alcohol	655–656
Ethyl cellulose	1234
Ethyl cellulose	231–233
Ethyl cellulose	825–826
Ethyl cellulose, chloroform cast film	1080
Ethyl cellulose, chloroform extract cast film	1197
2-Ethyl-1,3-hexanediol	687–688
2-Ethyl-1,3-hexanediol	120–122
2-Ethyl-1-butanol	66–68
2-Ethyl-1-butanol (ATR)	1140
2-Ethyl-1-butanol	657–658
Ethylbenzene	565–566
2-Ethylbutyric acid	39–41
2-Ethyl-1,3-hexanediol (ATR)	1158
Ethylene glycol	2073
Ethylene glycol monobutyl ether (ATR)	1150
Ethylene glycol monobutyl ether	102–104
Ethylene glycol monobutyl ether	677–678
Ethylene glycol monostearate	1591
Ethylene vinyl acetate 18%	2022
Ethylene vinyl acetate 18%	2062
Ethylene vinyl acetate	2000
Ethylene vinyl acetate with C9 and C5 hydrocarbon resins	749–750
Ethylene/acrylic acid copolymer	1235
Ethylene/acrylic acid copolymer	234–236
Ethylene/acrylic acid copolymer	827–828
Ethylene/ethyl acrylate, 82/18 copolymer	1236
Ethylene/ethyl acrylate, 82/18 copolymer	237–240
Ethylene/ethyl acrylate, 82/18 copolymer	829–830
Ethylene/propylene, 60/40 copolymer	1237
Ethylene/propylene, 60/40 copolymer	241–244
Ethylene/propylene, 60/40 copolymer	831–832
Ethylene/vinyl acetate, 60/40 copolymer	1243
Ethylene/vinyl acetate, 60/40 copolymer	265–268
Ethylene/vinyl acetate, 60/40 copolymer	841–842
Ethylene/vinyl acetate, 67/33 copolymer	261–264
Ethylene/vinyl acetate, 67/33 copolymer	1242

Compound	Spectrum Numbers
Ethylene/vinyl acetate, 72/28 copolymer	1241
Ethylene/vinyl acetate, 72/28 copolymer	257–260
Ethylene/vinyl acetate, 72/28 copolymer	839–840
Ethylene/vinyl acetate, 75/25 copolymer	1240
Ethylene/vinyl acetate, 75/25 copolymer	253–256
Ethylene/vinyl acetate, 75/25 copolymer	837–838
Ethylene/vinyl acetate, 82/18 copolymer	1239
Ethylene/vinyl acetate, 82/18 copolymer	249–252
Ethylene/vinyl acetate, 82/18 copolymer	835–836
Ethylene/vinyl acetate, 86/14 copolymer	1238
Ethylene/vinyl acetate, 86/14 copolymer	245–248
Ethylene/vinyl acetate, 86/14 copolymer	833–834
Ethylenediamine tetraacetic acid, 2 sodium salt	1386
2-Ethylhexanoic acid	60–62
2-Ethylhexanoic acid	615–616
Ethylparabenzoic acid in KBr	1066
FC1802 fluorinated surfactant, cast film	1784
FD&C Blue no. 1 dye in KBr	1111
Findley adhesive H 2253 tackifier resin	1998
Fluorocarbon surfactant, cast film	1077
Food-grade paraffin wax	1969
Formaldehyde (ATR)	1174
Formaldehyde	165–167
Formaldehyde	715–716
Formic acid	617–618
Formic acid	72–74
Freon, 1,1,2-trichloro-1,2,2-trifluoroethane	2009
Fructose	1103
Fructose	2057
Galatose	2054
Gasoline (high aromatics content)	569–570
Gasoline (high ethanol content)	567–568
Gasoline (low aromatics content)	571–572
Gemtex SM-33, liquid film on KCl	1768
Gluconic acid	24–26
Gluconic acid	639–640
Glucose	1102
Glucose	2056
Glycerol (ATR)	1189

ALPHABETICAL INDEX TO SPECTRA 1–2130

Compound	Spectrum Numbers

Glycerol, cast film on KCl   1198

Glycerol distearate, cast film on KCl   1033

Glycerol, film on KCl   1783

Glycerol monostearate, cast film on KCl   1032

Glycerol oleate, liquid film   1051

Glyceryl dilaurate   1425

Glyceryl distearate   1664

Glyceryl monolaurate   1426

Glyceryl monoricinoleate   1665

Glyceryl monostearate   1663

Glyceryl trioleate   1367

Glyoxal (ATR)   1175

Glyoxal   171–173

Glyoxal   717–718

Glyoxolated cationic polyamide, cast film   1134

Grease, heavy lubricating, thin film   1058

Grease, light lubricating, thin film   1059

Guanidinium monoethylphenylphenol sulfonate   1627

Gyceryl phthalate   795–796

Hamposyl L-30, sodium lauroyl sacosinate   1811

HC/PE:PET/33514/H$_2$O   1793

HC/PE:PET/33514/H$_2$O   1848

HC/PE:PET/33514/MeOH/hex   1795

HC/PE:PET/33514/MeOH/hex   1878

HC/PE:PET/33514/MeOH/MeCl$_3$ 1794

HC/PE:PET/33514/MeOH/MeCl$_3$ 1850

HC/PE:PET/DF059/MeOH   1796

HC/PE:PET/DF059/MeOH   1859

HC/PE:PET/DF059/MeOH/hex   1798

HC/PE:PET/DF059/MeOH/hex   1860

HC/PE:PET/DF059/MeOH/MeCl$_3$ 1797

HC/PE:PET/DF059/MeOH/MeCl$_3$ 1881

HC/PE:PET/DF059 + DF018/MeOH/hex   1801

HC/PE:PET/DF059 + DF018/MeOH/hex   1802

HC/PE:PET/DF059 + DF018/MeOH/hex   1883

HC/PE:PET/DF059 + DF018/MeOH/MeCl$_3$ 1800

HC/PE:PET/DF059 + DF018/MeOH/MeCl$_3$ 1857

HC/PET:PET/DF059 + DF018/MeOH   1799

HC/PET:PET/DF059 + DF018/MeOH   1856

Compound	Spectrum Numbers

Henkel sodium lauryl sulfate on KBr disk   1782

Heptane   1106

Heptanoic acid   93–95

Heptanoic acid   619–620

Hexadecyltrimethyl ammonium chloride   1654

1,2,6-Hexanetriol (ATR)   1159

Hexanoic acid   36–38

Hexanoic acid   613–614

Hexanol   1101

Houghton release agent 564   2081

Hydrogenated hydrocarbon—Escorez 5300   1899

Hydrogenated hydrocarbon—Escorez 5320   1900

Hydrogenated hydrocarbon—Escorez 5380   1901

Hydrogenated hydrocarbon resin   1951

Hydrogenated rosin ester   1910

Hydrogenated rosin ester   1945

Hydrogenated rosin ester tackifier resin   1997

Hydrogenated tallow amine 95% primary   1632

Hydroxybutyl methyl cellulose, 8% hydroxybutyl, 20% methoxyl   1244

Hydroxybutyl methyl cellulose, 8% hydroxybutyl, 20% methoxyl   269–271

Hydroxybutyl methyl cellulose, 8% hydroxybutyl, 20% methoxyl   843–844

Hydroxypropyl cellulose   1245

Hydroxypropyl cellulose   272–274

Hydroxypropyl cellulose   845–846

Hydroxypropyl methyl cellulose, 10% hydroxypropyl, 30% methoxyl   1246

Hydroxypropyl methyl cellulose, 10% hydroxypropyl, 30% methoxyl   275–277

Hydroxypropyl methyl cellulose, 10% hydroxypropyl, 30% methoxyl   847–848

Imidazoline-based debonder, cast film on KCl   1135

Indopol L-50—polybutylene polymer   1912

Iron naphthenate, 6% Fe   1532

Iron oxide   2021

Iron stearate KBr   1572

iso-Butyric acid   33–35

iso-Butyric acid   609–610

Isooctane   755–756

Isopropanol (ATR)   1190

Isopropanol   12–14

Isopropanol   573–574

Isopropanol   601–602

ALPHABETICAL INDEX TO SPECTRA 1–2130

Compound	Spectrum Numbers

Isopropanol, isopropyl alcohol   2006

Isopropylamine salt–sulfonated petroleum   1546

iso-Valeric acid   201–203

Iso-Valeric acid   635–636

Jeffamine 149   2064

Jeffamine   2061

K Monoethyl phenylphenol monosulfonate   1743

K Polymerized alkylnaphthalene sulfonate   1552

Kadol mineral oil   2109

Known abietic acid   1569

Known aerosol OT WS6651   1343

Known Arkon P70 tackifier resin   1932

Known calcium stearate in KBr   1438

Known Eastobond M500 APP   1941

Known Elvax 260 28% VA melt   1957

Known glyceryl monooleate   1510

Known Kraton 1111–0 styrene/isoprene   1931

Known kristalex 3100 resin melted film   1432

Known myristic acid—95% purity   1380

Known oleic acid   1506

Known polyvinyltoluene—mixed isomers   1956

Known ricinoleic acid   1526

Known sodium lauryl sulfate   1465

Known sodium stearate   1421

Known sorbitan monooleate   1444

Known sulfated oleic acid—sodium salt   1603

Known Vinol 165 polyvinyl alcohol   1345

Known zinc stearate in KBr   1418

Kodak Epolene C-16 polyethylene wax   1949

Kraton 1102, 28% polystyrene/72% polybutadiene   1964

Kraton styrene ethylene butylene styrene, oil, C5 hydrocarbon resin   1985

Kraton styrene isoprene styrene   1990

Kristalex 5140—modifed alpha-methylstyrene   1929

Kymene   1007

Lactic Acid   641–642

Lauramino propionic acid   2116

Laurate of diethylene glycol   1347

Laurate of polyethylene glycol 1540   1433

Laurate of polyethylene glycol 200   1346

ALPHABETICAL INDEX TO SPECTRA 1–2130

Compound	Spectrum Numbers

Laurate of polyethylene glycol 300   1684

Laurate of polyethylene glycol 400   1548

Laurate of polyethylene glycol 600   1685

Lauric acid   1614

Lauric acid, ethyl ester   1777

Lauricidin, fatty acid(s) monoglyceride   1778

Lauryl hydroxyethyl imidazoline   2118

Lauryldimethylbenzyl ammonium chloride   1404

Laurylisoquinolinium bromide   1395

Laurylisoquinolinium bromide   1819

Laurylpyridinium chloride   1396

Lead naphthenate, liquid on AgBr   1728

Lead stearate in KBr   1519

Lemon-lime soda, cast film on AgBr   1042

Linoleic acid   117–119

Linoleic acid   1487

Linoleic acid   623–624

Lithium stearate—KBr disk   1495

Lycra, elastic thread   1133

Machine oil, capillary film between KCl windows   1055

Mackam 151L, lauramino propionic acid   1809

Mackanate LM-40, Na_2 lauramido MEA sulfonate   1814

Magnesium lauryl sulfate   1540

Magnesium silicate (talc) in KBr   1035

Magnesium stearate KBr   1566

Malic acid (in KBr)   1076

Malic acid (in KBr)   1130

Malic acid, hydroxy-butanedioic acid   2043

Manganese naphthenate, liquid on AgBr   1729

Menthol   1082

Menthol in KBr   1123

Menthol   761–762

Methanol (ATR)   1191

Methanol   781–782

Methoxacetic acid   27–29

Methoxypolyethylene glycol 500 "cocoate"   1467

3-Methoxy-1-butanol (ATR)   1151

3-Methoxy-1-butanol   105–107

3-Methoxy-1-butanol   679–680

ALPHABETICAL INDEX TO SPECTRA 1–2130

Compound	Spectrum Numbers

1-Methoxy-2-propanol (ATR)   1152

1-Methoxy-2-propanol   108–110

1-Methoxy-2-propanol   681–682

Methoxyacetic acid   643–644

Methyl alcohol   2004

Methyl cellulose   1247

Methyl cellulose   278–280

Methyl cellulose   849–850

Methyl paraben, methyl *p*-benzoate   2001

Methyl vinyl ether/maleic acid, 50/50 copolymer   1248

Methyl vinyl ether/maleic acid, 50/50 copolymer   281–283

Methyl vinyl ether/maleic acid, 50/50 copolymer   851–852

Methyl vinyl ether/maleic anhydride, 50/50 copolymer   1249

Methyl vinyl ether/maleic anhydride, 50/50 copolymer   284–286

2-Methyl-1-butyn-2-ol   69–71

2-Methyl-3-butyn-2-ol (ATR)   1141

2-Methyl-3-butyn-2-ol   659–660

Methylacrylate, neat liquid   1093

Methylal   15–17

Methylal   605–606

3-Methylcyclohexanol (ATR)   1142

4-Methylcyclohexanol (ATR)   1143

3-Methylcyclohexanol   661–662

4-Methylcyclohexanol   663–664

3-Methylcyclohexanol   75–77

4-Methylcyclohexanol   78–80

Methylene (3,5-di-*tert*-butyl-4-hydroxyhydrocinnamate)   1114

Methylparabenzoic acid (in KBr)   1065

Mineral oil   2085

Miranol   2010

Miranol C2M salt of lauryl sulfate   1739

Miranol C2MNPLV, cocoamphocarboxy glycine   1771

Miranol SM salt of lauryl sulfate   1738

50/50 Mixture of silicone and menthol   2127

Modified alpha-methylstyrene   1898

Modified glyceryl phthalate resin   1755

Modified polyacrylamide   1938

Modified polyterpene   1893

Molasses, cast film on AgBr   1040

Compound	Spectrum Numbers

Monoethanolamide lauric acid   1353

Monoisopropanolamide-lauric acid   1348

1-Monolauroyl-RAC-glycerol   1779

Monorincinoleate of ethylene glycol   1424

Morpholine abietate   1508

Morpholine laurate, liquid film on AgBr   1711

Morpholine linoleate, liquid film–AgBr   1717

Morpholine myristate, liquid film–AgBr   1712

Morpholine naphthenate, liquid on AgBr   1715

Morpholine oleate, liquid film on AgBr   1714

Morpholine palmitate   1505

Morpholine ricinoleate   1507

Morpholine stearate, cast film on AgBr   1713

Morpholine undecylenate   1590

Myreth-3-myristate   2111

Myristic acid   1484

Naphthenic acids   1489

NaSO$_3$-naphthalene-formaldehyde condensate   1754

n-b-Hydroxyethyl coco imidazoline   1409

n-b-Hydroxyethyl oleyl imidazoline   1637

n-b-Hydroxyethyl stearyl imidazoline   1410

n-Butyraldehyde   156–158

n-Butyraldehyde (ATR)   1170

n-Butyraldehyde   709–710

n-Coco-propylenediamine—84% diamine   1440

n-Decane   577–578

n-Decane   753–754

n-Heptane   579–580

n-Hexadecylamine—95% primary   1584

Nickel oleate, liquid on AgBr   1730

Nirez V-2150—terpene phenol resin   1927

N,N-Dimethylacetamide   189–191

N,N-Dimethylformamide   192–194

N,N-Dimethylacetamide (ATR)   1183

N,N-Dimethylacetamide   729–730

N,N-Dimethylformamide (ATR)   1184

N,N-Dimethylformamide   731–732

n-Octadecylamine—film on AgBr   1748

n-Oleoyl sarcosine   1461

ALPHABETICAL INDEX TO SPECTRA 1–2130

Compound	Spectrum Numbers
n-Oleoyl sarcosine	1824
n-Oleoylethylenediamine, formate salt	1401
Nonanoic acid	132–134
Nonanoic acid	621–622
n-Soya-propylenediamine (80% diamine)	1638
n-Stearoylethylenediamine, formate salt	1400
n-Stearoyl-palmitoyl sarcosine	1623
n-Tallow-propylenediamine—80% diamine	1437
n-Tetradecylamine	1631
Nujol oil, between KCl windows	1036
n-Vinyl pyrrolidone/vinyl acetate copolymer	1003–1004
n-Vinyl pyrrolidone/vinyl acetate copolymer	1325
n-Vinyl pyrrolidone/vinyl acetate copolymer	555–557
Nylon 11 (polyundecanoamide)	1256
Nylon 11 (polyundecanoamide)	313–316
Nylon 11 (polyundecanoamide)	865–866
Nylon 12 (polylaurylactam)	1257
Nylon 12 (polylaurylactam)	317–320
Nylon 12 (polylaurylactam)	867–868
Nylon 6 (polycaprolactam)	1250
Nylon 6 (polycaprolactam)	287–290
Nylon 6 (polycaprolactam)	853–854
Nylon 6/10 (polyhexamethylene sebacamide)	1253
Nylon 6/10 (polyhexamethylene sebacamide)	302–305
Nylon 6/10 (polyhexamethylene sebacamide)	859–860
Nylon 6/12 (polyhexamethylene dodecanediamide)	1254
Nylon 6/12 (polyhexamethylene dodecanediamide)	306–309
Nylon 6/12 (polyhexamethylene dodecanediamide)	861–862
Nylon 6/6 (polyhexamethylene adipamide)	1251
Nylon 6/6 (polyhexamethylene adipamide)	291–295
Nylon 6/6 (polyhexamethylene adipamide)	855–856
Nylon 6/9 (polyhexamethylene nonanediamide)	1252
Nylon 6/9 (polyhexamethylene nonanediamide)	296–301
Nylon 6/9 (polyhexamethylene nonanediamide)	857–858
Nylon 6/T (polytrimethyl hexamethylene terephthalamide)	1255
Nylon 6/T (polytrimethyl hexamethylene terephthalamide)	310–312
Nylon 6/T (polytrimethyl hexamethylene terephthalamide)	863–864
Nylon fabric	2086
Nylon salt	2063

Compound	Spectrum Numbers
89% Nylon/11% Lycra	2087
87% Nylon/13% Lycra	2090
86% Nylon/14% Lycra	2091
92% Nylon/8% Lycra	2088
Octadecyl 3,5-di-*t*-butyl-4-hydroxycinnamate	1919
Octadecyltrimethyl ammonium chloride	1407
Octadecyltrimethyl ammonium chloride	1647
Octanoic acid	625–626
2-Octanol (ATR)	1144
2-Octanol	665–666
2-Octanol	81–83
Octyl alcohol (ATR)	1145
Octyl alcohol	667–668
Octyl alcohol	84–86
o-Hydroxyacetophenone	153–155
o-Hydroxyacetophenone (ATR)	1169
o-Hydroxyacetophenone	707–708
Oleate 2-hydroxy-ethane sulfonic acid, sodium salt	1560
Oleate of diethylene glycol	1689
Oleate of ethylene glycol	1338
Oleate of polyethyene gylcol 1000	1592
Oleate of polyethylene glycol 200	1337
Oleate of polyethylene glycol 300	1691
Oleate of polyethylene glycol 400	1368
Oleate of polyethylene glycol 600	1692
Oleic acid	168–170
Oleic acid	627–628
Oleic acid monoisopropanolamide	1578
Oleoyl polypeptide—sodium salt	1417
Oleyl amine—95% primary	1583
o-Me Galactoside-6-acrylate with 1% crosslinker	2068
Palmitic acid—KBr disk	1485
Paraffin (bees) wax (on KCl)	1088
Paraffin wax, food grade	1023
Paraplex G-30–plasticizer	1914
Pebax 2533 polyether/polyamide	1920
Pebax 3533 polyether/polyamide	1921
Pebax 4011 polyether/polyamide	1922
PEG-75 lanolin, liquid film on KCl	1770

ALPHABETICAL INDEX TO SPECTRA 1–2130

Compound	Spectrum Numbers
Pentaerythritol dioleate	1830
Pentaerythritol distearate	1330
Pentaerythritol monolaurate	1670
Pentaerythritol tetrastearate	1671
Pentane	581–582
Perflouro surfactant—cationic	1387
Perflouro surfactant—anionic + B76 ionic	1388
Perfluoro-1-octanesulfonic acid tetraethylammonium salt	2008
Petrolatum (ATR)	1192
Petroleum hydrocarbon resin, hydrogenated	1950
Phenoxy resin	1258
Phenoxy resin	321–324
Phenoxy resin	869–870
2-Phenoxyethanol	705–706
2-Phenoxyethanol	1168
2-Phenoxyethanol	150–152
3-Phenyl-1-propanol (ATR)	1167
3-Phenyl-1-propanol	147–149
3-Phenyl-1-propanol	703–704
2-Phenylethyl alcohol	701–702
2-Phenylethyl alcohol (ATR)	1166
2-Phenylethyl alcohol	144–146
Phospholipid	793–794
phospholipid CDM	1889
phospholipid PTC	2098
Phthalate	2078
Phthalate-based tackifier resin	1980
Phthalate extracted from Tygon tubing	2072
Piccotex LC—PVT/alpha-methylstyrene	1930
Plasthall P-670–plasticizer	1913
Pluronic 10R8prill surfactant—cast film	1767
P.O.E. Castor oil (40 moles EtO)	1365
P.O.E. Coco amine (5 moles EtO)	1414
P.O.E. Coco fatty acids (15 moles EtO)	1373
P.O.E. Duomeen t (3 moles EtO)	1411
P.O.E. Hydrogenated tallow amide	1349
P.O.E. Hydrogenated tallow amide (5 moles EtO)	1568
P.O.E. Lauric acid (14 moles EtO)	1377
P.O.E. Lauric acid (9 moles EtO)	1831

ALPHABETICAL INDEX TO SPECTRA 1–2130

Compound	Spectrum Numbers

P.O.E. Lauric amide (5 moles EtO)   1616

P.O.E. Lauryl alcohol (23 moles EtO)   1430

P.O.E. Lauryl alcohol (4 moles EtO)   1460

P.O.E. Oleic acid (6 moles EtO)   1374

P.O.E. Oleyl alcohol (20 moles EtO)   1604

P.O.E. Oleyl amine (5 moles EtO)   1416

P.O.E. Rosin amine (5 moles EtO) liq. film on AgBr   1764

P.O.E. Sorbitan monolaurate   1364

P.O.E. Sorbitan monolaurate HLB 13.3   1586

P.O.E. Sorbitan monolaurate HLB 13.3   1666

P.O.E. Sorbitan monooleate (20 moles EtO)   1344

P.O.E. Sorbitan monooleate HLB 13.9   1576

P.O.E. Sorbitan monooleate HLB 10.0   1651

P.O.E. Sorbitan monopalmitate HLB 15.6   1667

P.O.E. Sorbitan monostearate   1363

P.O.E. Sorbitan monostearate (20 moles EtO)   1379

P.O.E. Sorbitan trioleate (20 moles EtO)   1359

P.O.E. Sorbitan tristearate (20 moles EtO)   1362

P.O.E. Soya amine (10 moles EtO)   1412

P.O.E. Soya amine (15 moles EtO), liq. film on AgBr   1763

P.O.E. Soya amine (2 moles EtO), liq. film on AgBr   1760

P.O.E. Soya amine (5 moles EtO), liq. film on AgBr   1762

P.O.E. Stearic acid (15 moles EtO)   1375

P.O.E. Stearic acid (9 moles EtO)   1376

P.O.E. Stearyl alcohol (20 moles EtO)   1446

P.O.E. Stearyl amine (15 moles EtO)   1423

P.O.E. Stearyl amine (5 moles EtO)   1434

P.O.E. Tall oil (16 moles EtO)   1372

P.O.E. Tallow amine (15 moles EtO)   1413

P.O.E. Tallow amine (5 moles EtO), liq. film on AgBr   1761

P.O.E. Tertiary amine C12–14H25–29NH(C_2H_4O)$_5$H   1462

P.O.E. Tertiary amine C18–24H37–49NH(C_2H_4O)$_{15}$H   1415

P.O.E. Tertiary-dodecyl mercaptan (12 moles EtO)   1476

P.O.E. Tertiary-octylphenol (16 moles EtO)   1695

P.O.E. Tertiary-octylphenol (7–8 moles EtO)   1694

P.O.E. Tridecyl alcohol (12 moles EtO)   1698

P.O.E. Tridecyl alcohol (6 moles EtO)   1690

P.O.E. Tridecyl alcohol (9 moles EtO)   1575

P.O.E. Trimethylnonyl alcohol (8 moles EtO)   1697

ALPHABETICAL INDEX TO SPECTRA 1–2130

Compound	Spectrum Numbers
Poly(1-butene), isotactic	1266
Poly(1-butene), isotactic	348–351
Poly(1-butene), isotactic	885–886
Poly(2,4,6-tribromostyrene)	1297
Poly(2,4,6-tribromostyrene)	456–458
Poly(2,4,6-tribromostyrene)	947–948
Poly(2,6-dimethyl-*p*-phenylene oxide)	1272
Poly(2,6-dimethyl-*p*-phenylene oxide)	369–371
Poly(2,6-dimethyl-*p*-phenylene oxide)	897–898
Poly(2-hydroxyethyl methacrylate)	1284
Poly(2-hydroxyethyl methacrylate)	411–413
Poly(2-hydroxyethyl methacrylate)	921–922
Poly(4,4-dipropoxy-2,2-diphenyl propane fumarate)	1273
Poly(4,4-dipropoxy-2,2-diphenyl propane fumarate)	372–374
Poly(4,4-dipropoxy-2,2-diphenyl propane fumarate)	899–900
Poly(4-methyl-1-pentene)	1288
Poly(4-methyl-1-pentene)	425–428
Poly(4-methyl-1-pentene)	929–930
Poly(acrylic acid)	1220
Poly(acrylic acid)	1263
Poly(acrylic acid)	2095
Poly(acrylic acid)	338–340
Poly(acrylic acid)	805–806
Poly(acrylic acid)	879–880
Poly(alpha-methylstyrene)	1289
Poly(alpha-methylstyrene)	429–432
Poly(alpha-methylstyrene)	931–932
Poly(butyl acrylate)	1206
Poly(diallyl isophthalate)	1270
Poly(diallyl isophthalate)	363–365
Poly(diallyl isophthalate)	893–894
Poly(diallyl phthalate)	1271
Poly(diallyl phthalate)	366–368
Poly(diallyl phthalate)	895–896
Poly(ether) urethane, clay filled	2129
Poly(ethyl methacrylate)	1274
Poly(ethyl methacrylate)	375–378
Poly(ethyl methacrylate)	901–902
Poly(ethylene glycol) 600	2075

Compound	Spectrum Numbers
Poly(ethylene), high density	2037
Poly(ethylene oxide)	1281
Poly(ethylene oxide)	400–403
Poly(ethylene oxide)	915–916
Poly(ethylene terephthalate)	1283
Poly(ethylene terephthalate)	408–410
Poly(ethylene terephthalate)	919–920
Poly(glycolide-co-lactide), approximately 18% oxylactoyl units	1207
Poly(hydroxy ethyl methacrylate)	1209
Poly(hydroxy ethyl methacrylate)	2096
Poly(hydroxy ethyl methacrylate) melt	1208
Poly(isobutyl methacrylate)	1285
Poly(isobutyl methacrylate)	414–417
Poly(isobutyl methacrylate)	923–924
Poly(isoprene) elastic, clay filled	2128
Poly(lactic acid)	1210
Poly(lactic acid), reflectance	1216
Poly(lactic acid-g-AA)	1211
Poly(lactic acid-g-methyl methacrylate) (30%) and polyvinyl alcohol (70%)	1212
Poly(maleic acid)	1213
Poly(methyl methacrylate)	1214
Poly(methyl methacrylate)	1287
Poly(methyl methacrylate)	421–424
Poly(methyl methacrylate)	927–928
Poly(n-butyl methacrylate)	1267
Poly(n-butyl methacrylate)	352–354
Poly(n-butyl methacrylate)	887–888
Poly(phenylene sulfide)	1291
Poly(phenylene sulfide)	436–438
Poly(phenylene sulfide)	935–936
Poly(p-phenylene ether-sulphone)	1290
Poly(p-phenylene ether-sulphone)	433–435
Poly(p-phenylene ether-sulphone)	933–934
Poly(propylene glycol), M.W. approximately 425	1094
Poly(propylene) + TiO$_2$	2130
Poly(styrene):poly(isoprene)-based with poly(terpene)tackifier resin	1967
Poly(tetrafluoroethylene)	1296
Poly(tetrafluoroethylene)	453–455
Poly(tetrafluoroethylene)	945–946

ALPHABETICAL INDEX TO SPECTRA 1–2130

Compound	Spectrum Numbers
Poly(vinyl acetate)	1298
Poly(vinyl acetate)	459–461
Poly(vinyl acetate)	949–950
Poly(vinyl alcohol)	2094
Poly(vinyl alcohol), 100% hydrolyzed	1299
Poly(vinyl alcohol), 100% hydrolyzed	462–465
Poly(vinyl alcohol), 100% hydrolyzed	951–952
Poly(vinyl alcohol), 98% hydrolyzed	1300
Poly(vinyl alcohol), 98% hydrolyzed	466–469
Poly(vinyl alcohol), 98% hydrolyzed	953–954
Poly(vinyl butyral)	1301
Poly(vinyl butyral)	470–473
Poly(vinyl butyral)	955–956
Poly(vinyl chloride)	1302
Poly(vinyl chloride)	474–477
Poly(vinyl chloride)	957–958
Poly(vinyl chloride), carboxylated	1303
Poly(vinyl chloride), carboxylated	478–481
Poly(vinyl chloride), carboxylated	959–960
Poly(vinyl formal)	1304
Poly(vinyl formal)	482–484
Poly(vinyl formal)	961–962
Poly(vinyl stearate)	1306
Poly(vinyl stearate)	488–491
Poly(vinyl stearate)	965–966
Poly(vinylidene fluoride)	1307
Poly(vinylidene fluoride)	492–494
Poly(vinylidene fluoride)	967–968
Poly(vinyloctadecylether-co-maleic anhydride-co-maleic acid)	1215
Polyacetal	1259
Polyacetal	325–328
Polyacetal	871–872
Polyacrylamide	1009
Polyacrylamide	1260
Polyacrylamide	329–331
Polyacrylamide	873–874
Polyacrylamide, carboxyl modified (high content)	1262
Polyacrylamide, carboxyl modified (high content)	335–337
Polyacrylamide, carboxyl modified (high content)	877–878

ALPHABETICAL INDEX TO SPECTRA 1–2130

Compound	Spectrum Numbers
Polyacrylamide, carboxyl modified (Low content)	1261
Polyacrylamide, carboxyl modified (Low content)	332–334
Polyacrylamide, carboxyl modified (Low content)	875–876
Polyacrylamide, modified dried film	1025
Polyacrylanitrile	2059
Polyalphamethylstyrene	1895
Poly-alpha-olefins, amorphous	741–742
Polyamide—epichlorohydrin resin	1022
Polyamide polymer	2065
Polyamide resin	1264
Polyamide resin	341–343
Polyamide resin	881–882
Polyamide—epichlorohydrin resin	1115
Polyamide—epichlorohydrin resin	2121
1,2-Polybutadiene	1265
1,2-Polybutadiene	344–347
1,2-Polybutadiene	883–884
Polybutylene and polyethylene wax	1987
Polycaprolactone	1268
Polycaprolactone	355–358
Polycaprolactone	889–890
Polycarbonate	1109
Polycarbonate resin	1269
Polycarbonate resin	359–362
Polycarbonate resin	891–892
Polycarbonate with TiO_2	2126
Polycyclopentadiene resin	1983
Polyester	2029
Polyester	2039
100% Polyester	2092
Polyester	595–596
Polyester thread	2011
Polyethoxylated coco amine (15 moles EtO)	1759
Polyethoxylated coco amine (2 moles EtO)	1758
Polyethoxylated oleamide (5 moles EtO)	1351
Polyethoxylated stearyl amine (10 moles EtO)	1745
Polyethoxylated stearyl amine (50 moles EtO)	1756
Polyethoxylated tertiary amine (15 moles EtO)	1640
Polyethoxylated tertiary amine (25 moles EtO)	1757

Compound	Spectrum Numbers
Polyethoxylated tertiary amine	1826
Polyethylene (high density)	1201
Polyethylene (high density) and polypropylene	1200
Polyethylene	2120
Polyethylene film	1029
Polyethylene, chlorinated (36% Cl)	1277
Polyethylene, chlorinated (36% Cl)	387–389
Polyethylene, chlorinated (36% Cl)	907–908
Polyethylene, chlorinated (42% Cl)	1278
Polyethylene, chlorinated (42% Cl)	390–393
Polyethylene, chlorinated (42% Cl)	909–910
Polyethylene, chlorinated (48% Cl)	1279
Polyethylene, chlorinated (48% Cl)	394–396
Polyethylene, chlorinated (48% Cl)	911–912
Polyethylene, chlorinated, (25% Cl)	1276
Polyethylene, chlorinated, (25% Cl)	383–386
Polyethylene, chlorinated, (25% Cl)	905–906
Polyethylene, chlorosulfonated	1280
Polyethylene, chlorosulfonated	397–399
Polyethylene, chlorosulfonated	913–914
Polyethylene, high density	1275
Polyethylene, high density	379–382
Polyethylene, high density	903–904
Polyethylene, oxidized	1282
Polyethylene, oxidized	404–407
Polyethylene, oxidized	917–918
Polyethylene/polyvinyl acetate standard	1203
Polyethyleneterphthalate fiber, melted film	1014
Polyethyloxazoline	1903
Polyglycerol ester of oleic acid	1366
Polyisoprene	1973
Polyisoprene, chlorinated	1286
Polyisoprene, chlorinated	418–420
Polyisoprene, chlorinated	925–926
Polylactic acid (30%) and polyvinyl alcohol (70%) blend	1204
Polyoxyethylated castor oil (20 moles EtO)	1609
Polyoxyethylated cetyl alcohol (20 moles EtO)	1470
Polyoxyethylated coco amide (5 moles EtO)	1350
Polyoxyethylated coco amine (10 moles EtO)	1646

ALPHABETICAL INDEX TO SPECTRA 1–2130

Compound	Spectrum Numbers
Polyoxyethylated coco fatty acid (5 moles EtO)	1469
Polyoxyethylated lanolin	1513
polyoxyethylated nonylphenol (10–11 moles EtO)	1833
Polyoxyethylated nonylphenol (1–2 moles EtO)	1523
Polyoxyethylated nonylphenol (6 moles EtO)	1696
polyoxyethylated nonylphenol (8 moles EtO)	1834
Polyoxyethylated nonylphenol (9–10 moles EtO)	1464
Polyoxyethylated nonylphenol	1448
Polyoxyethylated nonylphenol (15 moles EtO)	1573
Polyoxyethylated nonylphenol (20 moles EtO)	1483
Polyoxyethylated nonylphenol (30 moles EtO)	1435
Polyoxyethylated nonylphenol (4 moles EtO)	1587
Polyoxyethylated octyl phosphate	1477
Polyoxyethylated octylphenol (12–13 moles EtO)	1599
Polyoxyethylated octylphenol (3 moles EtO)	1610
Polyoxyethylated octylphenol (30 moles EtO)	1598
Polyoxyethylated oleic acid (9 moles EtO)	1449
Polyoxyethylated oleyl alcohol (20 moles EtO)	1471
Polyoxyethylated oleyl alcohol (20 moles EtO)	1835
Polyoxyethylated oleyl amine (2 moles EtO)	1644
Polyoxyethylated oxypropylated stearic acid	1836
Polyoxyethylated red oil (10 moles EtO)	1466
Polyoxyethylated rosin	1445
Polyoxyethylated stearic acid (10 moles EtO)	1677
Polyoxyethylated stearic acid (40 moles EtO)	1678
Polyoxyethylated stearic acid (5 moles EtO)	1676
Polyoxyethylated *t*.octylphenol (9–10 moles EtO)	1455
Polyoxyethylated tall oil (12 moles EtO)	1611
Polyoxyethylated tallow amine (2 moles EtO)	1639
Polyoxyethylated tertiary amine (5 moles EtO)	1645
Polyoxyethylated tertiaryoctylphenol (5 moles EtO)	1577
Polyoxyethylated *tert*-octylphenol (3 moles EtO)	1838
Polyoxyethylated *tert*-octylphenol (30 moles EtO)	1839
Polyoxyethylated *tert*-octylphenol (9–10 moles EtO)	1837
Polyoxyethylated tetradecyl alcohol (7 moles EtO)	1704
Polyoxyethylated tridecyl alcohol (15 moles EtO)	1579
Polyoxyethylated tridecyl alcohol (3 moles EtO)	1451
Polyoxyethylated tridecyl alcohol (9 moles EtO)	1785
Polyoxyethylated tridecyl alcohol (9 moles EtO)	1818

ALPHABETICAL INDEX TO SPECTRA 1–2130

Compound	Spectrum Numbers
Polyoxyethylated tridecyl alochol (12 moles EtO)	1840
Polyoxypropylene & 10% EtO (M.W. ~1750)	1473
Polyoxypropylene & 20% EtO (M.W. ~1750)	1832
Polyoxypropylene & 40% EtO (M.W. ~1200)	1472
Polyoxypropylene & 40% EtO (M.W. ~1750)	1474
Polyoxypropylene & 50% EtO (M.W. ~2100)	1456
Polyoxypropylene & 80% EtO (M.W. ~1750)	1479
Polypropylene (66%) and polyester (34%)	1084
Polypropylene (66%) and polyester (34%)	763–764
Polypropylene	2024
Polypropylene	2030
Polypropylene and polyethylene	767–768
Polypropylene and silicates	2028
Polypropylene glycol	2100
Polypropylene glycol methacrylate	2099
Polypropylene standard	1199
Polypropylene with trace polyethylene	1225
Polypropylene, atactic	747–748
Polypropylene, crystalline	785–786
Polypropylene, isotactic	1293
Polypropylene, isotactic	442–445
Polypropylene, isotactic	939–940
Polypropylene, isotactic, chlorinated	1292
Polypropylene, isotactic, chlorinated	439–441
Polypropylene, isotactic, chlorinated	937–938
Polypropylene/polyethylene (60%) and polyester (40%)	771–772
Polypropylene/polyethylene (60%), and polyester (40%)	1132
Polypropylene/polyethylene	1087
Polystyrene	1294
Polystyrene	446–449
Polystyrene	941–942
Polystyrene (17%), and Polyisoprene (83%)	1221
Polystyrene (32 cm^{-1} resolution)	807–808
Polystyrene (4 cm^{-1} resolution)	809–810
Polystyrene (Dow)	1224
Polystyrene:polybutadiene and polystyrene:polyisoprene with polyvinyl toluene	1966
Polystyrene:polybutadiene copolymer	1222
Polystyrene:polybutadiene, hydrocarbon tackifier resin	1937
Polystyrene:polybutadiene/polystyrene:polyisoprene	739–740

Compound	Spectrum Numbers

Polystryene:polybutadiene rosin, acid ester tackifier resin, polyvinyl toluene   1968

Polystyrene: polyisoprene base polymer   1995

Polystyrene: polyisoprene base polymer with polycyclopentadiene resin   1996

Polystryene:polyisoprene resin   1963

Polystyrene:polyisoprene with hydrogenated polycyclopentadiene resin + oil   1965

Polystyrene:polyisoprene with polyterpene tackifier CHCl$_3$ extract   1936

Polystyrene:polyisoprene, polycyclopentadiene resin   1955

Polystyrene:polyisoprene, polycyclopentadiene resin   1959

Polystyrene:polyisoprene, polyterpene tackifier resin   1933

Polystyrene:polyisoprene, polyterpene tackifier resin   1954

Polystyrene:polyisoprene/polycyclopentadiene resin   745–746

Polysulfone resin   1295

Polysulfone resin   450–452

Polysulfone resin   943–944

Polyterpene resin   1892

Polyterpene resin   1947

Polyvinyl alcohol   1202

Polyvinyl alcohol, 78,000 M.W.   1064

Polyvinyl alcohol, standard   1008

Polyvinyl alcohol/polyvinyl acetate   1045

Polyvinyl alcohol/polyvinyl acetate   1226

Polyvinyl alcohol/polyvinyl acetate   1971

Polyvinyl methyl ether/isobutyl vinyl ether (12.0)   1217

Polyvinyl methyl ether/isobutyl vinyl ether (4.0)   1218

Polyvinyl methyl ether/isobutyl vinyl ether (8.3)   1219

Polyvinyl pyrrolidone   1205

Polyvinyl pyrrolidone   1305

Polyvinyl pyrrolidone   485–487

Polyvinyl pyrrolidone   963–964

Polyvinyltoluene, mixed isomers   1027

Poneac S red, dye   1117

Potassium abietate   1504

Potassium iodate in KBr   1091

Potassium iodate   2103

Potassium laurate   1585

Potassium linoleate   1601

Potassium monoethylphenylphenol monosulfonate   1822

Potassium myristate   1500

Potassium naphthenate   1503

Compound	Spectrum Numbers
Potassium palmitate	1499
Potassium ricinoleate, cast film on AgBr	1708
Potassium stearate—KBr disk	1496
Potassium undecylenate	1702
Prolpylene glycol	779–780
1,3-Propanediol (ATR)	1160
1,2-Propanediol, capillary film	1012
Propionaldehyde (ATR)	1176
Propionaldehyde	174–176
Propionaldehyde	719–720
Propionic acid	186–188
Propionic acid	629–630
Propionic anhydride	48–50
Propionic anhydride	647–648
Propyl alcohol (ATR)	1146
Propyl alcohol	669–670
Propyl alcohol	87–89
Propyl ester of sulfooleic acid—sodium salt	1458
Propyl paraben, propyl *p*-benzoate	2002
Propylene glycol (ATR)	1161
Propylene glycol	123–125
Propylene glycol	2018
Propylene glycol	689–690
Propylparabenzoic acid (in KBr)	1067
2-Propyn-1-ol (ATR)	1147
2-Propyn-1-ol	671–672
2-Propyn-1-ol	90–92
Protein, human blood, cast film on AgBr	1081
Protein, human hair (IR microspectroscopy)	1024
P-Tolualdehyde (ATR)	1180
Pure Eastman H100 resin melted film	1961
Pure Piccotac B	1953
Pure Triton X-102	1419
Pure Tuffalo oil	1952
Pure Zonarez 7115 melt	1943
Pure Zonarez B-115	1946
Pure Zonester 100 melt	1942
PVAc latex	2084
PVC organosol adhesive	2083

Compound	Spectrum Numbers

PVT/alpha-methylstyrene 1896

PVT/alpha-methylstyrene 1897

p-Xylene 583–584

Pyridine, anhydrous 2097

Pyridylazo naphthol 2066

Quartz cuvet 1–2

Quaternary imidazolinium salt—stearic acid 1398

Rayon 787–788

Rayon and polyester 769–770

Ricinoleate of diethylene glycol 1693

Ricinoleate of polyethylene glycol 600 1442

Ricinoleate of propylene glycol 1841

Ricinoleate of polyethylene glycol 400 1593

Rosin soap 2027

Rubber modified polypropylene 2023

Salicylaldehyde 183–185

Salicylaldehyde 727–728

Salicylaldehyde (ATR) 1182

SC10–008 Cetiol 1414E surfactant 1328

Schercozol. L, lauryl hydroxyethyl imidazoline 1815

Sequestrene Na_3, KBr disk 1772

Sesame seed oil, between KCl windows 1054

Silica glass (empty Raman sample vial) 2119

Silica, flint glass 1119

Silicate 2036

Silicon wax (ATR) 1193

Silicone (FTS-226), between KCl windows 1078

Silicone defoamer—water dispersible 1385

Silicone fluid (Dow 2–1922) 801–802

Silicone fluid (Dow, 1000 cs) 799–800

Silicone fluid (Dow, 350 cs) 775–776

Silicone fluid (SWS, 350 cs) 777–778

Silicone fluid, 1000 cs 1127

Silicone fluid, 350 cs 1128

Silicone fluid, Dow 2–1922 1129

Silicone sealant 2093

Silicone, Y-12226, 1061

Silk 1121

Siloxane, wetting agent 1075

ALPHABETICAL INDEX TO SPECTRA 1–2130

Compound	Spectrum Numbers
Sodium 2-ethylhexyl sulfate	1537
Sodium benzylnaphthalene sulfonate	1613
Sodium bicarbonate (in KBr)	1074
Sodium bicarbonate	2125
Sodium bisulfite	1118
Sodium borate, tetra (sodium borate·10HOH)	1384
Sodium caprate	1699
Sodium carbonate	1017
Sodium carbonate	1383
Sodium cetyl sulfate	1528
Sodium di(2-ethylhexyl) phosphate	1602
Sodium diamyl sulfosuccinate	1618
Sodium dibutylnaphthalene sulfonate	1527
Sodium dihexyl sulfosuccinate	1624
Sodium dihydrogen phosphite (in KBr)	1062
Sodium dihydroxyethyl glycinate	1652
Sodium diisobutyl sulfosuccinate	1454
Sodium diisopropylnaphthalene sulfonate	1551
Sodium dioctyl sulfosuccinate	1453
Sodium dioctyl sulfosuccinate	789–790
Sodium ditridecyl sulfosuccinate	1331
Sodium dodecylbenzene sulfonate	1525
Sodium hydroxide pellets	2101
Sodium iodate	2102
Sodium iodate in KBr	1090
Sodium kerylbenzene sulfonate	1553
Sodium laurate—KBr disk	1497
Sodium lauriminodipropionic acid	2117
Sodium lauroyl sarcosinate	2113
Sodium lauryl sulfate—sulfuric acid monododecyl ester sodium salt	2042
Sodium lauryl sulfate (in KBr)	1079
Sodium lauryl sulfate 98%	2074
Sodium lignosulfonate (14.3% SO$_3$Na)	1753
Sodium lignosulfonate (3 moles SO$_3$Na)	1752
Sodium lignosulfonate (5.4% sodium sulfate)	1621
Sodium lignosulfonate, 2 moles sodium/lignin unit	1332
Sodium linoleate	1498
Sodium lorol sulfoacetate	1561
Sodium mono- and diamylnaphthalene sulfonates	1787

Compound	Spectrum Numbers
Sodium monobutyldiphenyl sulfonate	1554
Sodium monobutylphenylphenol monosulfonate	1439
Sodium myristate	1707
Sodium *N*-alkylsulfoacetamide	1452
Sodium naphthenate (4% sodium)	1420
Sodium *N*-coco-b-aminopropionate	1390
Sodium *N*-cyclohexyl-*N*-palmitoyl taurate	1747
Sodium *N*-lauroyl sarcosinate	1628
Sodium *N*-lauryl-myristyl-b-aminopropionate	1661
Sodium *N*-methyl-*N*-oleyl taurate	1333
Sodium *N*-methyl-*N*-palmitoyl taurate	1619
Sodium *N*-methyl-*N*-tall-oil-acid taurate	1625
Sodium *N*-methyl-*N*-tallow acid taurate	1620
Sodium *n*-octyl sulfate	1536
Sodium oleate	1491
Sodium oleyl-stearyl sulfate	1539
Sodium palmitate	1501
Sodium periodate	2104
Sodium periodate in KBr	1089
Sodium petroleum sulfonate	1564
Sodium petroleum sulfonate (M.W. 340–360)	1545
Sodium petroleum sulfonate (M.W. 513)	1740
Sodium phosphate, tribasic (sodium phosphate·12HOH)	1382
Sodium poly-alkylbenzene sulfonate	1422
Sodium poly-alkylnaphthalene sulfonate	1436
Sodium resinate (abietate)	1701
Sodium ricinoleate	1492
Sodium *sec*-tetradecyl sulfate	1538
Sodium silicate—soluble	1381
Sodium sulfate in KBr	1016
Sodium sulfate in KBr	1502
Sodium sulfate	2035
Sodium sulfonated neatsfoot oil, cap. film	1733
Sodium sulfonated sperm oil, liquid on AgBr	1735
Sodium sulfooleate	1443
Sodium toluene sulfonate	1558
Sodium undecylenate	1700
Sodium xylene sulfonate	1557
Sodium-sulfated ethanolamine-lauric acid con.	1605

ALPHABETICAL INDEX TO SPECTRA 1–2130

Compound	Spectrum Numbers
Solium silicate, soluble	1018
Sorbic acid in KBr	1070
Sorbitan monolaurate	1358
Sorbitan monopalmitate	1378
Sorbitan monostearate	1361
Sorbitan sesquioleate	1680
Sorbitan trioleate	1357
Sorbitan trioleate	1681
Sorbitan tristearate	1668
Soy phosphotides (95%) (lecithin)	1608
Soy sterol (ATR)	1194
Soya amine 95% primary	1633
Soya trimethyl ammonium chloride	1406
Standapol SH124–33, Na_2 laureth(3) sulfate	1812
Starch	1011
Starch	759–760
Starch, cast film on AgBr	1063
Starch, extract cast film on AgBr	1107
Stearamido propyldimethyl-b-hydroxethyl	1399
Stearate of polyethylene glycol 1540	1589
Stearate of polyethylene glycol 200	1341
Stearate of polyethylene glycol 300	1674
Stearate of polyethylene glycol 4000	1370
Stearate of polyethylene glycol 600	1371
Stearate of polyethylene glycol 6000	1675
Stearate of propylene glycol	1842
Stearic acid	1562
Stearyl alcohol	1113
Stearyl alcohol	2107
Stearyldimethylbenzyl ammonium chloride	1648
Styrene/acrylonitrile, 70/30 copolymer	1309
Styrene/acrylonitrile, 70/30 copolymer	499–501
Styrene/acrylonitrile, 70/30 copolymer	971–972
Styrene/acrylonitrile, 75/25 copolymer	1308
Styrene/acrylonitrile, 75/25 copolymer	495–498
Styrene/acrylonitrile, 75/25 copolymer	969–970
Styrene/allyl alcohol copolymer	1310
Styrene/allyl alcohol copolymer	502–505
Styrene/allyl alcohol copolymer	973–974

ALPHABETICAL INDEX TO SPECTRA 1–2130

Compound	Spectrum Numbers
Styrene/butadiene	2049
Styrene/butadiene, ABA block copolymer	1311
Styrene/butadiene, ABA block copolymer	506–509
Styrene/butadiene, ABA block copolymer	975–976
Styrene-butadiene copolymer	1970
Styrene-butadiene-styrene	1972
Styrene-butadiene-styrene	1974
Styrene/butadiene/styrene	1981
Styrene:butadiene:styrene ABA film	735–736
Styrene butadiene styrene, oil, C5 hydrocarbon resin (C9 and polyterpene)	1993
Styrene/butyl methacrylate copolymer	1312
Styrene/butyl methacrylate copolymer	510–513
Styrene/butyl methacrylate copolymer	977–978
Styrene/ethylene/butylene, ABA block copolymer	1313
Styrene/ethylene/butylene, ABA block copolymer	514–517
Styrene/ethylene/butylene, ABA block copolymer	979–980
Styrene:ethylene butylene:styrene copolymer	743–744
Styrene:ethylene butylene:styrene copolymer II	751–752
Styrene:ethylene/butylene (86%), and polystyrene (14%)	1223
Styrene, ethylene butylene, styrene copolymer	1989
Styrene, ethylene butylene, styrene copolymer	1991
Styrene, ethylene butylene, styrene copolymer	1999
Styrene, ethylene butylene, styrene copolymer, oil, C5 hydrocarbon resin	1992
Styrene, ethylene butylene, styrene copolymer, oil, C5/C9 hydrocarbon resin	1986
Styrene/isoprene copolymer	1940
Styrene/isoprene, ABA block copolymer	1314
Styrene/isoprene, ABA block copolymer	518–520
Styrene/isoprene, ABA block copolymer	981–982
Styrene:isoprene:styrene	737–738
Styrene/isoprene/styrene	1976
Styrene/isoprene/styrene	1979
Styrene/maleic anhydride, 50/50 copolymer	1315
Styrene/maleic anhydride, 50/50 copolymer	521–524
Styrene/maleic anhydride, 50/50 copolymer	983–984
Styrene 10%/isoprene 90%—SIS block	1907
Styrene 14%/ethylene:butylene 86% block	1923
Styrene 14%/isoprene 86%—SIS block	1906
Styrene 14%/isoprene 86%—SIS block	1905
Styrene 17%/isoprene 83%—SIS block	1904

ALPHABETICAL INDEX TO SPECTRA 1–2130

Compound	Spectrum Numbers

Styrene 21%/isoprene 79%—SIS block   1908

Styrene 48%/butadiene 52%   1925

Styrene 28%/butadiene 72%—SBS block   1909

Styrene 28%/ethylene:butylene 72% block   1926

Styrene 29%/ethylene:butylene 71% block   1924

Styrofoam, red dye   2013

Styrofoam, white   2012

Substituted imidazolinium salt   1391

Substituted imidazolinium salt, example 2   1392

Substituted oxazoline (Alkaterge A)   1656

Substituted oxazoline (Alkaterge T)   1649

Substituted oxazoline (Alkaterge C)   1394

Substituted oxazoline (Alkaterge E)   1393

Sucrose   1100

Sucrose   2055

Sucrose acrylate hydrogel with 3.5% diacrylate crosslinker   2069

Sucrose dioleate   1355

Sucrose monomyristate   1688

Sucrose monooleate   1669

Sucrose monopalmitate   1360

Sucrose monostearate   1356

Sucrose monotallowate   1354

Sugar, brown, cast film on AgBr   1050

Sulfated 9-phenyl ether-4-ethylene glycol, ammonium salt   1543

Sulfated amyl oleate, sodium salt   1534

Sulfated butyl oleate, sodium salt   1607

Sulfated castor oil fatty acid, sodium salt   1544

Sulfated castor oil, sodium salt (org. SO_3—2%)   1600

Sulfated cod oil, sodium salt   1535

Sulfated ethanolamine-lauric acid condensate, Na salt   1843

Sulfated ethanolamine-myristic acid, sodium salt   1524

Sulfated glyceryl trioleate, sodium salt   1606

Sulfated isopropyl oleate—sodium salt, cap. film   1732

Sulfated laural ether of tetraethylene glycol, sodium salt   1431

Sulfated oleic acid (org. SO_3 4.5%), sodium sulfonated red oil   1786

Sulfated polyoxyethylated nonylphenol, sodium salt   1542

Sulfated polyoxyethylated octylphenol-sodium   1563

Sulfated propyl oleate, sodium salt   1533

Sulfated rice brand oil—sodium salt, cap. film   1734

ALPHABETICAL INDEX TO SPECTRA 1–2130

Compound	Spectrum Numbers

Sulfated soybean oil, sodium salt (org. SO$_3$—4%)   1597

Sulfated tallow, sodium salt   1594

Sulfochem TLES, TEA laureth sulfate   1813

Sulfonated cellulose   2033

Sunflower seed, oil (ATR)   1195

Surfynol 504   2058

950902 Surfynol 504, liquid film   1807

Synthetic or modified polyterpene   1978

Synthetic polyterpene   1894

Tagat L-2, P.O.E. glycerol fat. acid esters   1781

Tall oil fatty acids   1516

Tallow acids (distilled)   1494

Tallow acids (distilled)   1844

Tallow amine—95% primary   1582

Tea, water extract, cast film   1048

Teflon   2079

Terpene phenol resin   1915

Terpene phenol resin, type II   1916

tert-Amyl alcohol   649–650

tert-Amyl alcohol (ATR)   1136

tert-Amyl alcohol   51–53

tert-Butanol   585–586

tert-Butyl alcohol (ATR)   1138

tert-Butyl alcohol   57–59

tert-Butyl alcohol   653–654

tert-Butyl methyl ether   575–576

Tertiary-C11–14 H23–29 amine   1468

Tertiary-C18–24 H37–49 amine   1641

2-Tertiary-dodecylmercaptoethanol   1475

Tetrakis[methylene(3,5-di-*t*-butyl-4-hydroxyhydrocinnamate)] methane   1917

2,4,7,9-Tetramethyl-5-decyne-4,7-diol   1481

Titanium dioxide   2026

Titanium dioxide   2060

Toluene   1112

Toluene   587–588

Toluene   783–784

Toluene-based tackifier resin   1982

Toluene, methylbenzene, phentlmethane   2016

trans-Cinnamaldehyde   180–182

ALPHABETICAL INDEX TO SPECTRA 1–2130

Compound   Spectrum Numbers

trans-Cinnamaldehyde (ATR)   1179
trans-Cinnamaldehyde   723–724
1,2,4-Trichlororbenzene   2014
Triethanolamine   1126
Triethanolamine   803–804
Triethanolamine abietate, liquid on AgBr   1724
Triethanolamine caprate, cast film on AgBr   1718
Triethanolamine laurate   1509
Triethanolamine laureth sulfate   2114
Triethanolamine linoleate   1529
Triethanolamine myristate   1511
Triethanolamine naphthenate   1530
Triethanolamine oleate   1514
Triethanolamine palmitate   1512
Triethanolamine ricinoleate, liquid film   1723
Triethanolamine salt—sulfonated petroleum   1547
Triethanolamine stearate   1567
Triethanolamine undecylenate, film on AgBr   1719
Triethanolammonium lauryl sulfate   1541
Triethanol-ammonium dodecylbenzene sulfonate   1559
Triethylene blycol   135–137
Triethylene glycol   695–696
Trimethyl pentane   589–590
Trisodium nitrilotriacetate   1653
Triton X-100, liquid film on KCl   1769
Triton X-102   1026
Trycol 5882, laureth-4, liquid film   1816
Trycol 5966, ethoxylated lauryl alcohol   1780
Trycol 5966, laureth-3, liquid film   1808
Tygon tubing minus phthalate   2077
10-Undecenoic acid   631–632
10-Undecenoic acid   195–197
Undecylenic acid   1486
Uric acid   2089
Urine, synthetic, cast film on AgBr   1049
Valeric acid   198–200
Valeric acid   633–634
Vinyl alcohol/vinyl butyral copolymer (80% vinyl butyral)   1316
Vinyl alcohol/vinyl butyral copolymer (80% vinyl butyral)   525–528

ALPHABETICAL INDEX TO SPECTRA 1–2130

Compound	Spectrum Numbers

Vinyl alcohol/vinyl butyral copolymer (80% vinyl butyral)　　985–986

Vinyl chloride/vinyl acetate copolymer (81% vinyl chloride)　　1317

Vinyl chloride/vinyl acetate copolymer (81% vinyl chloride)　　529–531

Vinyl chloride/vinyl acetate copolymer (81% vinyl chloride)　　987–988

Vinyl chloride/vinyl acetate copolymer (88% vinyl chloride)　　1318

Vinyl chloride/vinyl acetate copolymer (88% vinyl chloride)　　532–535

Vinyl chloride/vinyl acetate copolymer (88% vinyl chloride)　　989–990

Vinyl chloride/vinyl acetate copolymer (90% vinyl chloride)　　1319

Vinyl chloride/vinyl acetate copolymer (90% vinyl chloride)　　536–538

Vinyl chloride/vinyl acetate copolymer (90% vinyl chloride)　　991–992

Vinyl chloride/vinyl acetate copolymer carboxylated (86% vinyl chloride)　　539–541

Vinyl chloride/vinyl acetate copolymer carboxylated (86% vinyl chloride)　　993–994

Vinyl chloride/vinyl acetate copolymer carboxylated (86% vinyl chloride)　　1320

Vinyl chloride/vinyl acetate/hydroxypropyl acrylate terpolymer (80% vinyl chloride)　　542–545

Vinyl chloride/vinyl acetate/hydroxypropyl acrylate terpolymer (80% vinyl chloride)　　995–996

Vinyl chloride/vinyl acetate/hydroxypropyl acrylate terpolymer (80% vinyl chloride)　　1321

Vinyl chloride/vinyl acetate/vinyl alcohol terpolymer (91% vinyl chloride)　　546–548

Vinyl chloride/vinyl acetate/vinyl alcohol terpolymer (91% vinyl chloride)　　997–998

Vinyl chloride/vinyl acetate/vinyl alcohol terpolymer (91% vinyl chloride)　　1322

Vinylidene chloride/acrylonitrile copolymer (20% acrylonitrile)　　1323

Vinylidene chloride/acrylonitrile copolymer (20% acrylonitrile)　　549–551

Vinylidene chloride/acrylonitrile copolymer (20% acrylonitrile)　　999–1000

Vinylidene chloride/vinyl chloride copolymer (5% vinylidene chloride)　　552–554

Vinylidene chloride/vinyl chloride copolymer (5% vinylidene chloride)　　1001–1002

Vinylidene chloride/vinyl chloride copolymer (5% vinylidene chloride)　　1324

Vistalon—ethylene propylene copolymer　　1911

Water (in KBr)　　1056

Water, deionized (0.2 cm)　　599–600

Water, deionized (ATR)　　1196

Water, deionized　　591–592

Water, deionized　　9–11

White ceresin wax　　2108

Whole milk　　2040

Wood pulp (63%), polyethylene (21%), polypropylene (16%)　　1085

Wood pulp (68%), polypropylene (32%)　　1086

Wood pulp, thermomechanical (in KBr)　　1021

Wool　　1122

Zein, purified　　1005–1006

Zein, purified　　1326

Compound	Spectrum Numbers
Zein, purified	558–560
Zinc linoleate KBr	1565
Zinc oleate, liquid on AgBr	1731
Zinc palmitate in KBr	1520
Zinc resinate in KBr	1522
Zinc stearate in KBr	1521
Zinc sulfate in KBr	1053

ABBREVIATIONS USED IN SURFACTANT AND POLYMER SPECTRA NAMES

AgBr—Silver bromide
Alk—Alkyl
ATR—Attenuated total reflectance (i.e., an internal reflectance technique used for infrared)
BASF—Badische Anilin-und Soda-Fabrik Chemical Company, Germany
C-#—Carbon number
Ca—Calcium
CHISSO—Chisso Corporation, Tokyo, Japan
CS—Sulfides
EtO—Ethylene oxide
Fe—Iron
H-#—Hydrogen number
HATR—Horizontal attenuated total reflectance
Hex or hex—Hexane
HLB—Hydrophile-lipophile balance (i.e., hydrophilic tendency)
K—Potassium
KBr—Potassium bromide
KCl—Potassium chloride
M.W.—Molecular weight in a.m.u. or daltons
Me—Methyl
MEA—Methylethylamine
$MeCl_2$—Dichloromethane
$MeCl_3$ ($CHCl_3$)—Chloroform, Trichloromethane
MeOH—Methanol
P.E.O.—Poly(ethylene oxide)
P.O.E.—Polyoxyethylated (a.k.a., Polyethoxylated)
PAN—Polyacrylonitrile
PAP—Polyoxyethylated alkyl phosphate(s)
PCL—Polycaprolactone
PDMS—Polydimethylsiloxane
PE—Polyethylene
PEG—Polyethylene glycol
PEI—Polyethyleneimine
PET—Poly(ethylene terephthalate)
PMMA—Poly(methyl methacrylate)
POP—Polyoxypropylene

ABBREVIATIONS USED IN SURFACTANT AND POLYMER SPECTRA NAMES

PP—Polypropylene
PS—Polystyrene
PTFE—Polytetrafluoroethylene
PVA—Polyvinyl alcohol
PVAc—Polyvinyl acetate
PVC—Polyvinyl chloride
PVOH—Polyvinyl alcohol
PVT—Polyvinyl terepthalate
SH—Thiols, thiol alcohols, or mercaptans (-SH)
T—Terephthalamide
TiO_2—Titanium dioxide
VA—Vinyl acetate

INDEX

A

Absorbance spectrum, 4
Absorptivity (ε), 3, 65
Acetylenes, IR and Raman, 218
Acyl halides, IR, 233
Acyl halides, NIR, 187
Acyl halides, Raman, 247
Addition reaction, 265
Agricultural applications, NIR, 170
Agricultural sciences, NIR and IR spectroscopic applications, 82
Alcohols, chemical identification tests, 39, 46
Alcohols, IR, 234
Alcohols, NIR, 169–170, 188
Alcohols, Raman, 248
Aldehydes, chemical identification tests, 39, 43, 44
Aldehydes, IR, 232
Aldehydes, NIR, 186
Aldehydes, Raman, 246
Alignment, infrared microspectroscopy, 29
Aliphatic hydrocarbons (n-alkanes), 156–158
Alkenes (olefins), NIR, 159
Amides, IR and Raman, 222, 224
Amides, IR, 233
Amides, NIR, 162, 187
Amides, Raman, 247
Amines (N-H stretch), Raman, 248, 249
Amines, (N-H stretch), IR, 235
Amines, chemical identification tests, 39, 44, 45
Amines, IR and Raman, 224
Amines, NIR, 166, 189
Anharmonic oscillator 149, 153

Anhydrides, IR and Raman, 222
Anhydrides, IR, 233
Anhydrides, NIR, 187
Anhydrides, Raman, 247
Animal sciences, NIR and IR spectroscopic applications, 83
Aperture diameter, 13
Aperture, infrared microspectroscopy, 30
Applications of NIR and IR spectroscopy, 82
Area normalization, 297
Aromatic compounds, NIR, 160–161
Aromatic systems, IR and Raman, 219
Aromaticity, chemical identification tests, 39–41
ATR, objectives, infrared microspectroscopy, 32
Attenuation (i.e., light or energy loss), 16
Autoscaling, 295

B

Bandpass interference filters, 9
Bandpass, spectrometers, 14, 15, 16
Baseline correction, 298
Baseline offset correction, 298
Beam splitters, 10
Beer's law, 3
Benzene, vibrational states, IR and Raman, 220
Beverages, NIR and IR spectroscopic applications, 88
Biotechnology, NIR and IR spectroscopic applications, 83
Black body, 8
Bond moment angles, 37

Bouguer, Lambert, and Beer relationship (see Beer's law), 3
Boxcar smoothing, 296

C

C=C stretch, Raman, 254
Calibration lamps, for UV-Vis-NIR, 52
Calibration transfer, 319
CANDECOMP (canonical decomposition), 342
Canonical decomposition (see CANDECOMP)
Carbohydrates, chemical identification tests, 39, 45, 46
Carbonates, IR, 236
Carbonyl (C=O stretch), IR, 231
Carbonyl (C=O stretch), NIR, 162, 185
Carbonyl (C=O stretch), Raman, 245, 253
Carbonyl compounds, IR and Raman, 221
Carboxylic acids, IR, 233
Carboxylic acids, NIR, 187
Carboxylic acids, Raman, 247
C-C stretch, IR, 231
C-C stretch, NIR, 185
C-C stretch, Raman, 245
Cellulose, NIR, 172
Centering, 347
C-F bend, IR, 236
C-F stretch, IR, 236
C-H bend, alkanes, NIR, 159
C-H bend, IR, 230, 238–242
C-H bend, NIR, 184
C-H bend, olefins, NIR, 159
C-H bend, Raman, 244–245
C-H stretch, alkanes, NIR, 159
C-H stretch, alkynes, NIR, 159
C-H stretch, IR and Raman, 216
C-H stretch, IR, 230, 238–242
C-H stretch, NIR, 184
C-H stretch, olefins, NIR, 159
C-H stretch, Raman, 244, 252
Chemical identification tests, 39–46
Chemical production, NIR and IR spectroscopic applications, 85
Chemometrics, 293
Chemometrics, books, 303
Chemometrics, history, 304
Chemometrics, internet (web) resources, 302
Chemometrics, journals, 304
Chemometrics, official methods, 305
Chemometrics, reviews, 301

Chemometrics, software, 302
Chemometrics, trends, 301
Chemometrics, tutorials, 302
Clear samples, optical properties, 55
Clinical chemistry, NIR and IR spectroscopic applications, 99
Cluster analysis, 315
Cluster/discriminant analysis, 318
Colored samples, optical properties, 55
Compensation ring, infrared microspectroscopy, 30
Compression cell, infrared microspectroscopy, 31
Condensation reaction, 265
Continuum regression, 321
Copolymers, 265
Correlation coefficient, uncorrected, 342
Correlation matching, 309
Cosmetics, NIR and IR spectroscopic applications, 84
Coupling (of vibrations), 154
Coupling, IR and Raman, 213
Creams sample preparation, 24, 25
C-Si stretch, IR, 236
Cumulated double-bonds, IR and Raman, 219
Cut-off filters, 9
Cut-on filters, 9
Cuvet cleaning, for UV-Vis-NIR, 54
Cuvet materials, for UV-Vis-NIR, 52, 53
C-X stretch (halogens), NIR 190

D

D^* (D-star), 8
Dark current, 4
Data preprocessing, 327
Data preprocessing, comparison of methods, 328
Data preprocessing, traditional NIR, 299, 300
Data processing, 293
Data transformations, 329
DATAN algorithm, 345
Degeneracy, hard (true), 342
Degeneracy, soft (pseudo), 342
Denoising, of spectra, 328
Derivatives, of spectra, 298, 329
Detection limit, 4
Detectors (see specific measurement technique)
Detectors, 8, 52

Dichroic measurements and reporting of data, 38
Dichroic measurements in polymer films, 35–36
Dichroic parameters, 36
Dichroism, infrared microspectroscopy, 31
Dielectric constant, 270, 272
Dielectric constants of materials, 275–292
Dielectric loss, 270, 273
Dielectric spectroscopy, 267, 269
Diffraction gratings, 10
Diffuse reflectance correction, 334
Digital filtering, 333
Diode array spectrometers (spectrophotometers), 5, 50
Direct trilinear decomposition (see DTD)
Discrete photometers, 4
Discrete quanta potential energy curves, 152
Discriminant analysis, PCA-based, 309
Discriminant analysis, wavelength-based, 308
Dispersion (diffraction grating), 10
Double monochromator systems, 5, 50
Double-beam spectrometer systems, 6
Double-beam/dual-wavelength spectrometers, 6
Double-centering, 347
DTD, 343
Dynamic range, 18

E

ε (absorptivity), 65
Earth Sciences, NIR and IR spectroscopic applications, 84
Edge filters, 9
Electric modulus, 274
Electronic components (spectrometers), 13
Electronics, NIR and IR spectroscopic applications, 108
Emission sources (see specific measurement technique)
Emission sources, 21
Emission spectrometers, 6
Emissivity, 8
Entrance pupil, 13
Environmental sciences, NIR and IR spectroscopic applications, 85
Enzymatic analysis methods, 57
Esters, IR and Raman, 222
Esters, IR, 232
Esters, NIR, 186

Esters, Raman, 246
Etendue, 16, 17
Ethers, IR and Raman, 222
Ethers, Raman, 247
Exit pupil, 13
Expert calibration systems, IR, 318
Explosives, NIR and IR spectroscopic applications, 101

F

Fabry-Perot cavity, 9
Fats, Oils, Waxes, NIR, 171
Fermi resonance (second-order coupling), 154
Fermi resonance, IR and Raman, 226
Fiber-optic materials, 23
Fiber-optic materials, for UV-Vis-NIR, 53
Fieser-Kuhn rules, 66–67
Filters, 9
Fine chemicals, NIR and IR spectroscopic applications, 85
Fine scattering particulates, optical properties, 55
Fingerprint frequencies, 154
Fingerprint frequencies, IR and Raman, 212
Flat baseline correction, 298
Flat correction, of spectra, 299
Flour analysis using NIR, 146
Fluorescence, multiway methods, 348
Food applications, NIR, 170
Foods, NIR and IR spectroscopic applications, 88
Forage analysis using NIR, 146
Forensic science, NIR and IR spectroscopic applications, 91
Fourier transform spectrometers (spectrophotometers), 6, 7
Fourier-domain smoothing, 296
Fourth overtone C-H stretch, 135–136
Fuels research, NIR and IR spectroscopic applications, 101
FWHM (Full bandwidth at half maximum peak height), 9

G

Gain setting, infrared microspectroscopy, 31
Gas-phase analysis, NIR and IR spectroscopic applications, 92
Generalized rank annihilation factor analysis (see GRAFA)

Generalized rank annihilation method (see GRAM)
GRAFA, 343
Grain analysis using NIR, 146
GRAM, 342
Gratings, 9
Grazing angle, objectives, infrared microspectroscopy, 32
Group frequencies 155, 156
Group frequencies, IR and Raman, 214

H

Half-power penetration depth, dielectric, 273
Halogens (X-H stretch), Raman, 249
Harmonic oscillator, 151
High absorptivity samples, optical properties, 55
Homopolymers, 265
Hydrocarbon analysis using NIR, 146
Hydrogen bonding, NIR, 163

I

Ideal harmonic oscillator, 148, 151, 153
Image analysis, 315
Imaginary [permittivity], 273
Imaginary part of electric modulus, 271
Imaging spectroscopy, 80, 81
Infrared (see IR)
Infrared microspectroscopy, 29–33
Infrared polarizers, 37
Infrared spectroscopy, interpretive, 209
Infrared-active molecule, 148
Instrument physics, NIR and IR spectroscopic applications, 92
Instrument standardization, 320
Intensity-distribution for diffraction gratings, 10
Interference filter photometer, 4, 51
Interferogram, 12, 13
Interferometer assemblies, 11
Interferometer-based spectrometers, 6, 7, 51
Interpretive NIR spectroscopy, 143
IR (infrared) detectors, 22
IR (infrared) emission sources, 21
IR (infrared) measurement region, sample preparation for, 24–25
IR (infrared) measurement region, solvents for, 24
IR (infrared) window materials, 22
IR (see infrared)
IR chemometrics, 316
IR functional groupings and calculated band locations, 229
IR spectral correlation charts, 237
IR spectroscopy, 77

K

Ketones, chemical identification tests, 39, 43, 44
Ketones, IR, 231
Ketones, NIR, 185
Ketones, Raman, 245, 246
Kirchoff's law, 8
Kramers-Kronig correction, 298
Kubelka-Munk transformation, 298

L

Large scattering particulates, optical properties, 55
Light (energy) sources, 7, 52
Light sources, infrared microspectroscopy, 31
Liquids sample preparation, 24, 25
Long wavelength near infrared (see LW-NIR)
Loss tangent, 271, 273
LW-NIR (long wavelength NIR), 198–208
LW-NIR (see long wavelength near infrared)
LW-NIR band locations, 58, 61

M

Mahalanobis distance, 310
Manual subset selection, 307
Mapping spectroscopy, 82
Mean centering, 295, 334
Measurement techniques data, 23
Medicine, NIR and IR spectroscopic applications, 99
Michelson interferometer, 6, 7, 12, 13, 51
Microspectroscopy (see infrared microspectroscopy)
Military, NIR and IR spectroscopic applications, 101
Mineralogy, NIR and IR spectroscopic applications, 84
Moisture measurements, NIR and IR spectroscopic applications, 101
Molar extinction coefficient (see Absorptivity), 3

Monomers, 264
Motorized stage, infrared microspectroscopy, 31
MSC (see multiplicative signal correction)
Multiple cosine, 342
Multiplicative signal correction (MSC), 297
Multiway, books, 341
Multiway, data analysis, 339
Multiway, models, 341
Multiway, reviews, 341
Multiway, software, 340
Multiway, tutorials, 341

N

N.A. (numerical aperture), 16
n-alkanes (see aliphatic hydrocarbons)
Natural gas, NIR and IR spectroscopic applications, 101
Natural products, NIR, 170
Near infrared (see NIR)
Network analyzers, 269
Neural networks, 315
N-H bend, IR, 234
N-H group frequencies, IR and Raman, 224
N-H group frequencies, NIR, 163, 165
N-H stretch, IR, 234
N-H stretch, Raman, 253
Nielsen's rules, 67
NIR (near infrared) detectors, 22
NIR (near infrared) emission sources, 21
NIR (near infrared) measurement region, sample preparation for, 24–25
NIR (near infrared) measurement region, solvents for, 23, 24
NIR (near infrared) window materials, 22
NIR (see near infrared)
NIR chemometrics, 313
NIR filter choices, 59–60
NIR, history of, 133, 144
NIR spectral features, 60–61
NIR spectroscopy, 49, 77
NIR, IR, and Raman spectra comparisons, 77
Nitriles, IR and Raman, 219
Nitrogen containing compounds, unsaturated, IR, 235
Nitrogen containing compounds, unsaturated, NIR, 189
Noise compensation, 331
Noise estimation, 327
Noise measurement, infrared microspectroscopy, 32
Normalization, 297
Numerical aperture (N.A.), 16

O

Objectives, infrared microspectroscopy, 32–33
O-H stretch, IR and Raman, 223
O-H stretch, NIR, 163, 164
O-H stretch, Raman, 248
Oils, NIR, 171
Olefinic C=C stretch, IR and Raman, 216, 217
Olefinic C-H bend, IR and Raman, 218
Olefinic C-H stretch, IR and Raman, 216
Olefins (see alkenes)
Open-path spectrometers, 6, 7
Optical throughput, 3
Organic compound band assignments, NIR, 197–208
Overtones, NIR, 137–138

P

Paper industry, NIR and IR spectroscopic applications, 105
PARAFAC (parallel factor analysis), 342
Parallel factor analysis (see PARAFAC)
Paraxial optical theory, 3
Pastes, sample preparation, 24, 25
Pathlength correction, 297
Pathlength selection data, 23
Pathlength selection, for UV-Vis-NIR, 55
Pathlength, 3
PCA, 342
Penetration depth, dielectric, 273
Permeability, 273
Peroxides, IR and Raman, 223
Petroleum, NIR and IR spectroscopic applications, 101
Pharmaceutical industry, NIR and IR spectroscopic applications, 102
Phenolics, IR, 234
Phenolics, NIR, 188
Phenolics, Raman, 248
Phosphorus compounds, Raman, 250
Photoemissive detectors, 8
Photometric accuracy, 19
Photometric repeatability, 19
Photon flux density, 8
Planck's constant, 8
Planck's law, 8
Planck's radiation formula, 8

Plant sciences, NIR and IR spectroscopic applications, 105
Plastics, 264
Polarizers, 13
Polymer compound band assignments, NIR, 197–208
Polymer science, NIR and IR spectroscopic applications, 105
Polymer stretching, 38
Polymerization, 265
Polymers, alternating, 265
Polymers, atactic, 264
Polymers, block, 265
Polymers, description, 264, 265
Polymers, graft, 265
Polymers, isotactic, 264
Polymers, random, 265
Polymers, syndiotactic, 264
Potential energy curve, 150
Powders, sample preparation, 24, 25
Power dissipation factor (see loss tangent), 273
Preprocessing (see data preprocessing)
Principal component(s) analysis (see PCA)
Prisms, 10
Process spectroscopy, NIR and IR, 79
Propellants, NIR and IR spectroscopic applications, 101
Protein (amino acids), NIR, 173–176
Pulp industry, NIR and IR spectroscopic applications, 105

Q

Qualitative analysis, 306
Qualitative analysis, multiway methods, 348
Quantitative analysis, 305

R

Radiant power, 8
RAFA (rank annihilation factor analysis), 343
Raman detectors, 22
Raman emission sources, 21
Raman functional groupings and calculated band locations, 243
Raman measurement region, sample preparation for, 24–25
Raman measurement region, solvents for, 25
Raman spectral correlation charts, 251
Raman spectroscopy, 77
Raman spectroscopy, interpretive, 209
Raman window materials, 22
Random sample selection, 307
Rank annihilation factor analysis (see RAFA)
Real [permittivity], 273
Redundant apertures, infrared microspectroscopy, 33
Reference band normalization, 297
Reflectance spectrum, 4
Refraction of light, infrared microspectroscopy, 33
Resolution for diffraction gratings, 10
Resolution, spectrometer, 14, 15, 16
Reviews, NIR and IR, 134, 145
RMS (root mean square), 17
Rotation angle for polarizers, 13
Rubber compound band assignments, 197–208

S

Sample optical properties, for UV-Vis-NIR, 55
Sample preparation (see specific measurement technique)
Sample presentation geometry, for UV-Vis-NIR, 54
Sample selection, 306
Savitsky-Golay smoothing, 296
Scatter correction (see MSC), 334
S-CN stretch, IR, 236
Second-order calibration, 312
Second-order coupling (see Fermi resonance)
Selection rule, 152
Selectivity, 4
Semiconductors, NIR and IR spectroscopic applications, 108
Sensitivity, 4
Short wavelength near infrared (see SW-NIR)
Short-wave NIR, 139, 145
Signal-to-noise (S/N) ratio, 17
Single monochromator systems, 5, 50
Single-beam spectrometers, 4
Single-centering, 347
Si-O stretch, IR, 236
Small training sets, 309
Smoothing, 296
Snell's law, 11
Solid state detectors, 8

INDEX

Solvents (see specific measurement technique)
Solvent-soluble materials, sample preparation, 24, 25
Specific detectivity (see D*), 8
Spectral enhancement, 328
Spectral matching (see correlation matching), 311
Spectral radiance, 8
Spectral reconstruction, 312
Spectral searching algorithms, 319
Spectral subtraction, 308
Spurious energy, infrared microspectroscopy, 33
Standardization (see instrument standardization)
Starches, NIR, 171
Stefan Boltzmann law, 8
Stefan's constant, 8
Stratified sample selection, 308
Stray light (i.e., stray radiant energy), 18
Stray light, infrared microspectroscopy, 33
Stray radiant energy (see stray light)
Sugars, NIR, 171
Sulfur compounds, IR, 236
Sulfur compounds, NIR, 190
Sulfur compounds, Raman, 249, 250
Surface analysis, NIR and IR spectroscopic applications, 109
Surfactant groups, 255
Surfactants, acylated amino acids, salts, 262
Surfactants, acylated polyamines, salts 263
Surfactants, acylated polypeptides, salts, 262
Surfactants, alkoxylated amides, 256
Surfactants, alkyl phosphonamides, 263
Surfactants, alkylolamine-fatty acid condensates (oxazolines), 263
Surfactants, amine salts, 262
Surfactants, amines, 262
Surfactants, amphoterics, 264
Surfactants, anionic, 258
Surfactants, carboxylic acids and soaps, 258, 259
Surfactants, cationics, 262
Surfactants, definition, 255
Surfactants, esters of polyhydric alcohols, 255
Surfactants, esters of polyoxyalkylene glycols, 256
Surfactants, ethers of polyoxyalkylene glycols, 257
Surfactants, heterocyclic amines, salts, 263
Surfactants, inorganics, 264
Surfactants, nonionic, 255
Surfactants, perfluoro compounds, 264
Surfactants, phosphates and phosphatides, 262
Surfactants, polyoxyethylated alkyl phosphates, 258
Surfactants, quaternary ammonium salts, 263
Surfactants, sequestrants, 264
Surfactants, silicones, 264
Surfactants, sulfated alcohols, salts, 260
Surfactants, sulfated amides, salts, 260
Surfactants, sulfated carboxylic acids, salts, 260
Surfactants, sulfated esters, salts, 260
Surfactants, sulfated ethers, 260
Surfactants, sulfonated aliphatic hydrocarbons, salts or acids, 261
Surfactants, sulfonated amides, salts, 261
Surfactants, sulfonated amines, salts, 261
Surfactants, sulfonated aromatic hydrocarbons, salts, 261
Surfactants, sulfonated carboxylic acids, salts, 261
Surfactants, sulfonated esters, salts, 261
Surfactants, sulfonated ethers, salts, 261
Surfactants, sulfonated lignins, salts, 261
Surfactants, sulfonated petroleum, salts, 260
Surfactants, sulfonated phenols, salts, 261
Surfactants, tertiary acetylenic glycols, 258
Surfactants, trialkylamine oxides, 262
SW-NIR (see short wavelength near infrared)
SW-NIR spectra, 60–61

T

Terminal acetylene bonds, chemical identification tests, 39, 42
Terpolymers, 265
Textiles, NIR and IR spectroscopic applications, 110
Throughput, 17
TLD, 343
Transmittance (transmission) spectrum, 4
Trilinear decomposition (see TLD)
Triple-centering, 347

TUCKALS2 algorithm, 345
TUCKALS3 algorithm, 345

U

Ultraviolet (see UV-Vis), 21, 22
Ultraviolet detectors, 22
Ultraviolet emission sources, 21
Ultraviolet measurement region, sample preparation for, 24–25
Ultraviolet measurement region, solvents for, 23
Ultraviolet window materials, 22
Unsaturated nitrogen compounds, Raman, 249
Unsaturation, chemical identification tests, 39, 41, 42
UV (see ultraviolet)
UV-Vis chemometrics, 312
UV-Vis chromophores, 56
UV-Vis solvents, 56
UV-Vis spectral evaluation, 65
UV-Vis spectroscopy charts, 63
UV-Vis spectroscopy, 49
UV-Vis spectroscopy, functional groupings and band assignments, 69–70
UV-Vis spectroscopy, relative band intensities, 74
UV-Vis spectroscopy, spectra correlation charts, 71–73
UV-Vis, multiway methods, 347
UV-Vis-NIR measurement modes, 49

V

Vis (visible), 22, 23
Visible (see Vis)
Visible detectors, 22
Visible emission sources, 21
Visible measurement region, sample preparation for, 24–25
Visible measurement region, solvents for, 23
Visible window materials, 22
Visible, objectives, infrared microspectroscopy, 33

W

Water, NIR, 168
Wave propagation angle (prism), 11
Wavelength accuracy, 19
Wavelength repeatability, 19
Waxes, NIR, 171
Window materials (see specific measurement technique)
Window materials, for UV-Vis-NIR, 52, 53
Window materials, infrared microspectroscopy, 33
Woodward-Fieser rules, 66–67

X

X=O stretch, Raman, 254
X-H functional groups, NIR, 163
X-S stretch, Raman, 254

ISBN 0-12-763561-0